Flame Structure and Processes

THE JOHNS HOPKINS UNIVERSITY

Applied Physics Laboratory Series in Science and Engineering

SERIES EDITOR: John R. Apel

William H. Avery and Chih Wu
Renewable Energy from the Ocean: A Guide to OTEC

Bruce I. Blum
Software Engineering: A Holistic View

R. M. Fristrom
Flame Structure and Processes

Richard A. Henle and Boris W. Kuvshinoff
Desktop Computers: In Perspective

Vincent L. Pisacane and Robert C. Moore (eds.)
Fundamentals of Space Systems

Bruce I. Blum
The Foundations of Holistic Design and Fragment-Based Specifications

Flame Structure and Processes

R. M. FRISTROM

The Johns Hopkins University
Applied Physics Laboratory

New York Oxford
OXFORD UNIVERSITY PRESS
1995

Oxford University Press

Oxford New York Toronto
Delhi Bombay Calcutta Madras Karachi
Kuala Lumpur Singapore Hong Kong Tokyo
Nairobi Dar es Salaam Cape Town
Melbourne Auckland Madrid

and associated companies in
Berlin Ibadan

Published by Oxford University Press, Inc.,
200 Madison Avenue, New York, New York 10016

Library of Congress Cataloging-in-Publication Data
Fristrom, R. M.
Flame structure and processes / R.M. Fristrom.
p. cm. (Johns Hopkins University/Applied Physics Laboratory series in science and engineering)
Includes bibliographical references and index.
ISBN 0-19-507151-4
1. Flame
I. Title. II. Series.
QD516.F735 1995 541.3'61--dc20
93-46015

1 3 5 7 9 8 6 4 2

Printed in the United States of America
on acid-free paper

ACKNOWLEDGMENTS

This book began as a revision of the earlier work *Flame Structure* which was coauthored with Dr. A. A. Westenberg. His early retirement to Vermont has made him the envy of many of his colleagues, but without his collaboration the project went more slowly than desired. During the twenty-nine years since the first volume the field expanded exponentially. This made it necessary to change and restrict the coverage. To reflect this the title was changed to *Flame Structure and Processes*. The objectives remain unchanged, however: to offer a guide to the literature, give practical advice on experimental methods, and provide a basic understanding of the physics and chemistry of flames. The results should be of interest to scientists and engineers interested in high temperature spectroscopy, transport thermodynamics, chemical kinetics, and the flow of reacting gases. The book presents flame structure studies systematically, so that a student entering the field can better appreciate how his efforts contribute to the overall understanding of combustion.

The author has been helped by many friends, colleagues, and institutions during the preparation of this book and would like to offer his grateful thanks. To any friend whom he has forgotten to mention he can only plead good intentions and a bad memory. Several colleagues have made direct contributions to the text and deserve special thanks. Prof. O. I. Smith wrote chapter seven on probe sampling. Drs. M. Lapp and P. Witze of Sandia National Laboratory and Prof. R. Lucht of the University of Illinois made a number of corrections and additions to the chapter on laser diagnostics and reviewed the laser aspects of the temperature and composition chapters. These contributions are greatly appreciated and have greatly improved the coverage in these rapidly developing areas. Prof. G. Dixon-Lewis made constructive suggestions on the manuscript. In addition, Drs. E. S. Oran, K. Kailasanath, and J. P. Boris of the Naval Research Laboratory prepared a section on numerical modeling which could not be included because of space restrictions. The author appreciates their effort and regrets the necessity of contracting the book. The coverage is poorer for this loss.

The original work at APL was a part of the Navy Bumblebee propulsion research program developed by Dr. W. H. Avery. He has been an inspiration and constant supporter of our research over the years. The Navy has supported the author's research either directly or indirectly throughout most of his career, and he is grateful for their continuing constructive view of scientific research. The flame studies were made in collaboration with Dr. A. A. Westenberg and a major part of the experimental studies were carried out by Mr. C. Grunfelder whose ingenuity made the ideas a reality. The computational and mathematical aspects have been in the capable hands of

Mr. S. Favin who has been a valued advisor over the years. More recently the author has had the pleasure of working with Mr. H. Hoshall who has made many contributions to this book. Dr. W. G. Berl has been a constant, valued advisor over the years. Many other past and present members of APL have made contributions, notably: P. Breisacher, N. J. Brown, N. deHaas, S. N. Foner, G. A. Fristrom, E. L. Gayhart, B. Hochheimer, R. L. Hudson, L. W. Hunter, L. Monchick, J. T. O'Donovan, H. L. Olsen, R. Prescott, S. D. Raezer, and W. E. Wilson.

The author owes a debt of gratitude to his supervisor Dr. D. Silver whose advice has been cogent and whose support has been unwavering over this prolonged period of production. The friendship and advice of Dr. M. J. Linevsky is appreciated. The author would also like to express thanks for the support of the several directors of the laboratory during this period: Drs. R. E. Gibson, A. Kosiakoff, C. O. Bostrom, and G. L. Smith as well as the chairmen of the Milton S. Eisenhower Research Center, Drs. F. T. McClure, R. W. Hart, T. O. Poehler, and D. J. Williams. They have fostered and maintained the professional scientific atmosphere which has made the author proud to have worked at APL. In addition, the author was the recipient of the Janey Fellowship of APL/JHU during a critical period in the preparation of this book.

Over the years the author has had the advice, help, and counsel from many gifted colleagues outside of APL. Dr. W. D. Weatherford of South West Research Institute, Dr. M. F. R. Mulcahy, and Prof. G. Dixon-Lewis made many constructive suggestions and contributions during the formative years of the program. Over the years the author has also had the benefit of conversations with Drs. J. Biordi, J. Hastie, E. Oran, the late Dr. W. Kaskan, Drs. G. Klein, R. Gann, Prof. F. Weinberg, Prof. J. Warnatz, and many others. The advice of Prof. N. Chigier was most useful during the formative period of the book and is greatly appreciated.

Although a major part of the work was done at APL, the author has also had the benefit of extended visits to the laboratories of a number of colleagues. An early visit to the laboratory of the late Prof. Hirschfelder and Prof. C. F. Curtiss at Wisconsin was influential in shaping the author's approach to flame studies. A year at the Johns Hopkins Homewood campus in the Department of Chemical Engineering allowed a pleasant and productive collaboration with Prof. F. Wehner.

The author was awarded the honor of a Humboldt Foundation Prize and the pleasure of working at Göttingen, Germany, with Profs. H. Gg. Wagner and his colleagues Drs. K-H Hoyermann, H. Schacke, and J. Wolfrum. These valued contacts still continue. The author has had the benefit of a long time collaboration with Prof. P. J. Van Tiggelen and Dr. J. Vandooren of Louvain la Neve, Belgium. This was supported by travel grants from NATO. He had the benefit of stimulating discussions with Prof. C.T. Bowmann and his colleagues during a visiting professorship at the Mechanical Engineering Department of Stanford University. The honor of a Springer Professorship at the Department of Mechanical Engineering of the University of California gave the author the stimulus of interacting with Prof. R. F. Sawyer and his colleagues. The same year he was also a visiting scientist at the Combustion Laboratory of the Sandia National Laboratory, Livermore, California, during a stimulating period shortly after its foundation. The author would like to thank Dr. Dan Hartley and his colleagues for their kind hospitality and continuing friendship. Two semesters spent as visiting professor with Prof. O. I. Smith of the Department of Chemical Engineering at UCLA were both pleasant and productive.

Secretaries make any book a reality. The author thanks them for their patience and suggestions which have made this a more readable book. Over most of the period the coordinating force has been Mrs. Anne Landry whose patience, unfailing good humor, and useful suggestions are appreciated. During the past year this work has been taken over by the competent and enthusiastic Mrs. Debbie Coffroad who has brought the project to fruition. Other secretaries in the Research Center made contributions, particularly Mrs. Jeaneen Jernigan in the preparation of some tables.

The many figures in this book have been produced by the APL Illustrations department which is in the capable hands of Mr. Joe Lew. He and his staff have made many useful and attractive illustrations out of what were often very limited sketches.

Thanks are also due to the many competent support workers at our laboratory. The various members of the Research Center Machine shop made many unreasonable designs a reality.

The author would like to thank Dr. John Apel who is the editor of the APL/JHU series and the Oxford University Press staff, especially Mr. J. Robbins and Ms. D. Oetting who made this book a reality. It has been a pleasure to work with them. He would also like to thank the several publishers and their representatives who have graciously given permission to use a number of figures. They include Ms. S. Terpack of the Combustion Institute, Prof. I. Glassman of *Combustion Science and Technology*, the American Chemical Society, and the American Physical Society.

Finally, I would like to thank my wife, Gerrie, for her cheerful support, critical reviewing of the book, and forbearance over the past decade. My son, Rob, has given me much support and advice on computational aspects of the research and book production. This was greatly appreciated.

Although we are now officially on the System International (SI) for scientific units, the reader will notice that this book contains mixed metric units and that calories and atmospheres are used. Much of the combustion literature is in these older units and the author has not systematically changed figures from their original grams and centimeters to meters and kilograms. It is hoped that this will not inconvenience the reader.

CONTENTS

FREQUENTLY USED SYMBOLS

a	Cross-sectional area of stream tube
A	Stream-tube area ratio Preexponential (frequency) factor in Arrhenius rate constant Orifice throat area Parameter in exponential repulsive potential
C	Heat capacity per mole at constant pressure Orifice discharge coefficient
d	Parameter in inverse-power repulsive potential Diameter
D	Ordinary binary diffusion coefficient
D^{T}	Thermal binary diffusion coefficient
E	Activation energy Thermodynamic internal energy
f	Mass fraction
G	Fractional mass flux
H	Enthalpy per mole
k	Boltzmann gas constant Specific reaction rate constant
K	Net molar reaction rate per unit volume Equilibrium constant defined with partial pressures of "products" written in numerator
L	Flame zone thickness Length Total number of chemical elements in a mixture
$\dot{M}$	Mass flow rate
m	Mass flow rate per unit area Mass of a molecule
M	Molecular weight
n	Number of molecules per unit volume
N	Number of moles per unit volume

p	Partial pressure
P	Static pressure
q	Energy flux
Q	Heat release rate per unit volume (negative number) Volumetric flow rate
r	Radius Intermolecular distance
R	Molar gas constant
s	Total number of chemical species in a mixture
t	Time
T	Static temperature
v	Mass average velocity
v_o	Flame velocity
V	Volume Diffusion velocity
w	Mass of a chemical species per unit volume
W	Work
X	Mole fraction
x, y, z	Coordinate distances
Z	Collision number
α	Parameter in exponential repulsive potential
δ	Parameter in inverse-power repulsive potential
ϵ	Emissivity Parameter in Lennard–Jones potential
η	Shear viscosity
γ	Specific heat ratio
λ	Thermal conductivity
μ	Reduced mass of pair of molecules Refractive index
υ	Number of atoms of an element in a molecule
ω	Angular frequency
Ω	Collision integral
ϕ	Intermolecular potential function
ρ	Density Parameter in exponential repulsive potential
σ	Collision diameter Stefan–Boltzmann constant
θ	Angle

Subscripts

i, j	Designates species in a mixture
e	Designates a chemical element
f	Designates formation from the elements
m	Designates mixture
o	Designates reference point (usually cold boundary)
∞	Designates hot boundary
p	Designates constant pressure (usually omitted)
r	Designates reaction surface
R	Designates radiation
s	Designates constant entropy
-	Designates reverse reaction

Superscripts

*	Designates multicomponent diffusion coefficient
o	Designates standard state Designates monatomic gas

Special Characters

$\wedge$	Written over a symbol designates the quantity *per gram*
-	Written over a symbol designates an average quantity
x	Vectors are written as boldface letters
P	Tensors are written as sans serif letters

The reader will notice that the book contains mixed metric units and that calories and atmospheres are used. Officially we now employ the System International (SI) units in scientific studies. However, much early combustion literature is in the old cgs units and it has not been practical to change figures from their original grams and centimeters to meters and kilograms.

Flame Structure and Processes

I

THE SCIENCE OF COMBUSTION: ITS HISTORY, SCOPE, AND LITERATURE

Combustion has furnished humans with a major source of energy since they emerged as a separate species. Most of modern technology is based on its use both as an energy source and in various industrial processes. A quantitative understanding of this phenomenon is, therefore, a desirable practical goal beyond its intrinsic scientific interest.

The study of fire probably began as a practical Stone Age problem. Combustion studies also played an important role during the scientific revolution in the birth of chemistry. Despite this long history, the quantitative description of flames is a comparatively recent development. This lag has not stemmed from lack of interest or effort but rather from the complexity of combustion processes. Even the simplest flame involves a number of simultaneous chemical reactions as well as aspects of fluid flow, heat conduction, and molecular transport. In recent years the understanding of these individual processes has improved, and experimental and computational techniques have progressed to the point where many combustion problems can be quantitatively understood in terms of fundamental processes (Hirschfelder, Curtiss, and Bird [1954]; Williams [1985]; Buckmaster and Ludford [1985]; Gardiner [1984]). It is the objective of this book to offer a unified treatment of the microstructure of one-dimensional laminar premixed flames. The characteristics of diffusion flames and turbulent flames will be briefly mentioned. The analysis of more complex combustion systems such as engines, fires, and furnaces will not be attempted. Their treatment from first principles still remains an elusive goal although current progress is encouraging.

Combustion science and technology is a specialized branch of scientific culture. Our views of science are often biased by the recent upsurge associated with the Western Industrial Revolution. In the longer-term worldwide perspective, however, science in general, and combustion in particular, has undergone continuous development since the earliest of times and, although many changes in the rate of accumulation of knowledge have occurred, the general trend has always been upward. A feeling for this can be obtained by considering the temporal relation of combustion technology with the total population and the size of the largest aggregates of people. This is illustrated in Figure I-1. To compress this wide perspective, the logarithm of the number of individuals is plotted as a function of the logarithm of elapsed time BP ("before the present" in the terminology of the archaeologist). Below the population curve is a histogram showing the rough time of introduction of various combustion techniques and inventions.

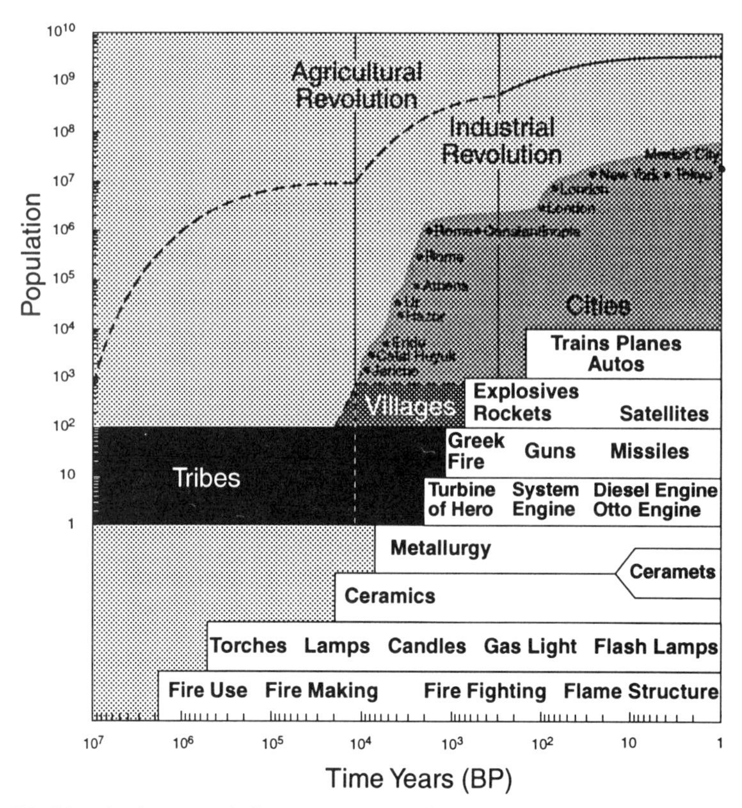

FIG. I-1 World and urban population. Total population and the size of the largest population aggregations are plotted against time BP (before present). The histogram at the bottom shows the accumulation of combustion technologies positioned at the approximate time of initiation.

Several points may be made; the overall population-time behavior is a composite of three Malthusian curves of exponential rises capped by saturation plateaus. This is typical of a species expanding into a new, limited environment and filling it. The two changes that expanded the environment for humans are associated with the Agricultural Revolution, which began some 10,000 years ago, and the Industrial Revolution, which began some 300 years ago. Both of these changes opened new environments and food supplies to man. The dramatic shortening of the time to reach saturation shown by these expansions reflects the changes in the number of people, and in the speeds of communication and transportation. The maximum aggregation is also a series of Malthusian curves controlled by technological and social limitations. Tribes gave way to villages, which in turn gave way to cities. World populations have risen steadily, although during the so called Dark Ages (500–1300 AD) there were significant drops in Europe due to plagues such as the "Black Death" (bubonic plague). This was the result of poor sanitation, the prevalance of which was encouraged by the belief that bathing was harmful to mind and body, a view also promoted by religious extremists who believed in purity of mind rather than body. Cities in China, India, and Byzantium continued to thrive.

Scientific progress has followed similar trends as might be expected. The fraction of the population capable of innovation is small, probably less than one in a million. This probably has not changed significantly with time. The impact of innovations, however, depends on more than gross population. Effective innovation requires collaborators; ideas feed on one another and multiply. Great minds without sympathetic, contemporary collaborators are relatively sterile, as witness the minimal effect of Leonardo da Vinci's mechanical inventions as compared with the influence of the Royal Society group in England at the dawn of the Scientific Revolution. Innovation also depends on communications and public acceptance. Thus the number of effective innovators should show the same increases as the population, modified by such factors as local population density, communication effectiveness, and social climate. It will be fascinating to watch the effects of the Computer Age and the foreshortening of the time scale made possible by communication improvements. Taking the optimistic view that nuclear war and world pollution will be surmounted, man and his innovations have a promising future. The question is where will the new expansion occur—into the ocean, up to the stars, or into some artificial environment?

The recent historical rise of interest in combustion with the Renaissance and Industrial Revolution is illustrated in Figure I–2. This is a histogram of the lifetimes of recent contributors to the science of combustion. The exponential rise is clear and would be even more pronounced if present-day workers were included. Some estimates of population and overall scientific effort are available from historical and archaeological studies, but the number of individuals in combustion studies represents the author's own prejudice.

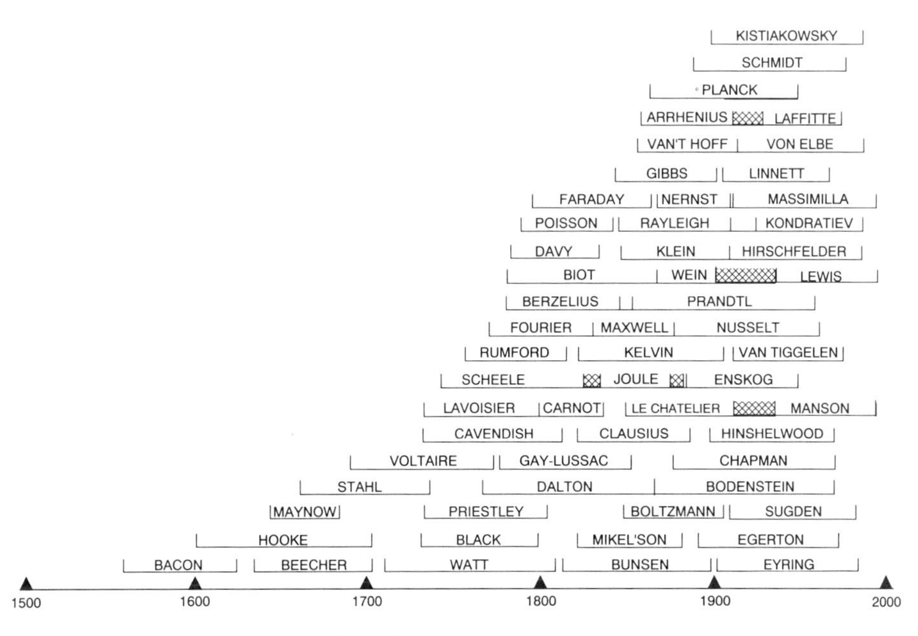

FIG. I-2 Contributors to combustion science and technology during the recent scientific revolution. (Shaded areas represent overlap in life span.)

History of Combustion

Historical expositions covering specialized topics such as combustion are rare. Aside from encyclopedias and general scientific histories (Burstall [1968], Sarton [1953]), the best sources are scattered comments in the treatises of Bone and Townend [1927], Lewis and von Elbe [1986], Jost [1946], Zeldovich [1949], and Gaydon and Wolfhard [1979]. Evans [1952] has provided an excellent critical review on flame theory. Emmons [1980] gives a discussion of fire science; Lienhard [1983] has outlined the development of heat transfer; and Scott [1984] has sketched the history of chemistry from the interesting viewpoint of the early development of ballooning. One of the best short bibliographies on the history and development of science can be found in the classic physics textbook of Millikan, Roller, and Watson [1937]. Partington [1939, 1961] has written a four-volume history of chemistry, which is summarized in a single volume. Lyons [1986] has written a popular discussion on fire.

The discovery and control of fire was one of the first great inventions (Fig. I-1). Its possession and use are some of the criteria used to identify our species. Archaeological evidence places the first use of fire between 600,000 and 1,000,000 BP, with artificial fire coming much later in the Paleolithic era (between 30,000 and 100,000 BP) (Philbean [1984]; Scarre [1988]). The present debate on whether fire-making dates from 30,000 or 100,000 BP hinges on the interpretation of a single fire stick from a Mousterian Neanderthal camp in Yugoslavia, which has been carbon dated at 100,000 BP An illuminating layperson's discussion of these questions as well as those concerning the early use of fire in caves in Chou-kou-tien is given by Purcell [1983]. Speculation on the nature of fire, which probably began in the Stone Age, had by historic times been elaborated into a multitude of myths and religious beliefs. Examples include the Greek legend of Prometheus, the Vestal Virgins who fed the hearth of the Roman state religion, and the Zoroastrian faith of Mithra, in which fire was deified. The first reported example of scientific thinking about fire was by the Greek philosopher Heracleitus (ca. 500 BC), who postulated that fire was the fundamental substance. Empedocles (500–430 BC) chose fire as one of the four elements (air, water, earth, and fire) that made up the universe (Russell [1967]). It is interesting to note the parallelism between his elements and our classification of the states of matter: gas (air), liquid (water), solid (earth), and plasma (fire). The concept of four elements was accepted by Aristotle (384–322 BC), and his great prestige combined with the general acceptance of authority froze these ideas in Western thought until the Renaissance.

Organized fire fighting began in Roman times. Its first serious practitioner was Marcus Licinius Crassus (112–53 BC) (Durant [1944–1963]; Suetonius (Graves) [1979]), from whose name the English word *crass* is derived. He augmented his considerable fortune by the then new technique of organizing a fire brigade of slaves and selling their services to a homeowner during the course of a fire. An alternative technique was to offer to buy endangered houses during the fire, bidding lower and lower amounts as the fire progressed. Once the deal was consummated, he sent in his brigades to quench the fire and added the property to his slumlord holdings. Using this and other manipulations, he amassed a fortune of 170,000,000 sesterces, estimated to exceed the total annual revenues of Rome itself. He also dabbled in politics as an ally of Julius Caesar and maintained his own private army. He was killed at Carrhae while leading a Roman army on a campaign against Parthia. As time went on Rome declined, while in the East the Chinese invented and used gunpowder and rockets, and the Byzantines made extensive engineering studies of Greek fire and kept it as a closely guarded military secret (Durant [1944–1963]).

After the fall of Constantinople in 1453, refugees carried many new ideas to the Western world. This reinforced the Renaissance, where authority was slowly being supplemented by experimentation. This important first step may have been taken in the field of combustion by Francis Bacon (1561–1626), who observed the structure of a candle flame (Sarton [1953]; Burstall [1965]; Bacon [1915]; Partington [1961]) (Fig. I-2). Other investigations were carried out by Boyle (1627–1691), Hooke (1635–1703), and Maynow (1641–1679) (Partington [1961]). The subsequent history of combustion is an integral part of the scientific revolution in which we are still immersed. The field is multidisciplinary, with contributions from chemists, engineers, and physicists. The recent growth of this science can be visualized by reference to Figure I-2, which plots the life spans of some of the more prominent contributors to combustion science. Living contributors are not included, partly because the exponential growth would make the figure ponderous, but primarily because the selection of major contributors would require impossible value judgments.

Early experimental observations in combustion were interpreted in terms of the phlogiston theory originated by Beecher (1635–1682) and championed by Stahl (1660–1734). This was an elaboration of the ancient alchemical idea that flames are due to the flux of an imponderable substance. If heat flux is identified with phlogiston, one can see an analogy with more recent excess enthalpy theories (Lewis and von Elbe [1987]). In 1750 the French Academy of Science took an interest in the problem and sponsored a competition to which Voltaire (1690–1775) and his mistress submitted essays. The collection of essays from this competition presented the contemporary views on combustion (Voltaire [1752]).

Combustion became a major factor in the birth of chemistry as a separate science beginning with the quantitative investigation of gases. A number of elements and gases were discovered by Black (1728–1799), Cavendish (1731–1810), Scheele (1742–1786), and Priestley (1733–1804). These discoveries culminated in the publication by Lavoisier of his classic Reflexions sur le phlogistique (1777), in which he denied the existence of phlogiston and laid the groundwork for modern chemistry (Lavoisier [1970]). By the time of his unfortunate death on the guillotine during the French Revolution, these views were almost universally accepted. In America, during the same period, Ben Franklin (Clark [1983]) improved the candle and invented the stove that bears his name.

For a time combustion was a province of chemistry, and many famous chemists were involved in this field during the Industrial Revolution (1776–1815). Contributors included Volta, Berthelot, Berzelius, Dalton, and Gay-Lussac, who founded the field of fire retardancy while investigating the prevention of theater fires for the French Academy of Science (Lyons [1970]). The next systematic work was made by Benjamin Thompson (Ellis [1876]; Brown [1979]). His career is one of the more interesting stories from this turbulent era. He was born in America, but left under a cloud during the Revolution as a result of favoring Loyalist causes. In Britain he worked for the War Office and did his famous cannon boring experiment, which established the connection between work and heat. In Bavaria, he became "Count Rumford of the Holy Roman Empire" and a practical philanthropist who designed furnaces, soups, and homes for the poor of Munich. His view was that if the poor were warm and fed that perhaps they would become virtuous. He returned to England, founded the Royal Institution, and endowed the Rumford Medal of the Royal Society. Once revolutionary fervor cooled, he was again accepted in America. He endowed a second Rumford Medal in the new country in 1796 through John Adams, then president of the American Academy of Arts and Sciences and later president of the United States. Rumford recommended Humphry Davy [1840] for a position at the Royal Institution, where mine fires were one of his many interests. To mitigate this combustion problem he invented the mine safety lamp. In France Mallard and Le Chatelier [1883] also investigated mine fires and first distinguished flames from detonations. Michael Faraday [1791–1867], Davy's protegé, dabbled in combustion and produced a model for popular scientific exposition in his famous lecture "The Chemical History of a Candle" [1863]. Robert Bunsen [1866], who is best known for the burner named after him which is found in every chemical laboratory, was the first to measure flame temperatures and flame velocities.

During the nineteenth century, combustion turned towards physics, with the application of thermodynamics and fluid mechanics. Newtonian mechanics was extended to continuum flow with the Navier–Stokes equations (Navier [1822]). Only simple problems of perfect fluids, neglecting viscosity, were solved before 1900, but this restriction was removed using the concept of the boundary layer (Schlichting [1968]). The equations are still not solvable for turbulent flow. However, recent progress in distinguishing chaos from random noise appears promising (Gleick [1987]).

The ideas of energy and its transformations were organized into the discipline of thermodynamics. Heat conduction and molecular diffusion were formulated quantitatively by Fourier and Fick. Chemists continued to make contributions with the development in the early 1800s of the atomic theory by Dalton [1975]. This was elaborated by many workers so that the ideas of stoichiometry and atomic weights became established. Toward the end of the nineteenth century, the American genius, J. Willard Gibbs [1928], placed chemical equilibria on a firm thermodynamic basis [1876]. Chemical kinetics was initiated by van't Hoff in a seminal paper in 1884. This was interpreted in molecular terms by his friend Arrhenius, whose name is attached to the common formulation. Arrhenius received the Nobel prize in 1903, but this was for his theory of electrolytic conduction, not his contributions to kinetics (Partington [1961]).

The beginning of the twentieth century found well-established combustion research groups in England, France, Germany, Russia, and the United States. Today there are significant combustion research communities in over fifty countries, as can be seen in the roster of sections of the Combustion Institute.

The ideas of chain reactions and steady state that are critical to the understanding of flames were applied to the hydrogen–bromine system in separate investigations by Christiansen [1919] and Polanyi [1932] and were extended to the more complex hydrogen–oxygen system by Semenov [1935] and Hinshelwood [1941]. They received a Nobel prize for their effort. The interpretation of reaction rates using quantum mechanics was made by Eyring some 30 years after van't Hoff through what he called the theory of rate processes (Glasstone, Eyring, and Laidler [1941]).

A quantitative kinetic theory of molecules was developed by Maxwell, Boltzmann, and Gibbs (Kennard [1938]) just before the beginning of the twentieth century. This was applied to binary systems, first by Enskog [1922] in Sweden and later independently by Chapman in England. Chapman and Cowling systematized the treatment in their seminal book [1939]. The generalization to multicomponent systems was made by Hirschfelder, Curtiss, and Bird in their monumental treatise *Molecular Theory of Gases and Liquids* [1954]. The work was motivated by a desire, originally arising from wartime work, to understand flames and detonations. Thus the basis for the modern understanding of combustion was laid.

In 1928 the First International Symposium on Combustion was held in Swampscott, Maine, and in 1937 the second was held in Rochester, New York. Both were sponsored by the American Chemical Society. The third was held in Madison, Wisconsin in 1948. Following the Fourth Symposium in 1954, The Combustion Institute was formed under the leadership of Bernard Lewis (1909–1992). This pioneer multidisciplinary society has sponsored biannual symposia since that time. During World War II advances were made in fluid mechanics, motivated by needs of the aircraft industry, and significant studies were made of high-temperature thermodynamics, shock, and detonation phenomena. Following the war, a new era began in combustion and other scientific areas. Systematic government support of basic science was pioneered by the Office of Naval Research and extended by the Air Force's Office of Scientific Research, the Army's Research Office, the National Science Foundation, NASA (and its predecessor, the National Advisory Committee on Aeronautics), the Department of Energy, and other government agencies. As a result more work was done in combustion in the 30 years following World War II than in the preceding 300 years, and this increased rate continues to the present day. This pattern is typical of active scientific areas. Progress has been accelerated by increased numbers of workers in the field and through improvements in instrumentation. These include: the mass spectrometer, the gas chromatograph, and, more recently, laser-optical techniques and the ever-increasing capacity of computers. Because of these and other innovations, we produce more and higher-precision studies than our predecessors; however, to bring the situation into perspective, it must be remembered that most of the critical concepts of combustion were formulated by the earlier workers. We are reaping what they have sown.

Flame Structure Studies

There were a number of individual attempts to observe and probe flame structure in the early history of combustion (see Fig. I-3). The first recorded observation was that of Francis Bacon, who probed a candle flame in 1560 with an arrow and observed that the flame was hollow. The next recorded attempt was by Faraday in his famous Christmas lecture in 1863, "The Chemical History of a Candle." Following this a number of research workers attempted to sample flames, but the general conclusion was that flames were so thin and strongly coupled that the gross sampling methods required at that time were considered inadequate because they destroyed the phenomena under observation. These attempts are mentioned in the books by Partington [1961] and Bone and Townend [1927].

Shortly after the end of World War II the U.S. military services as well as those in other countries began a systematic support of science in this and other areas due to their interest in the propulsion of rockets and jets. In the early 1950s a number of

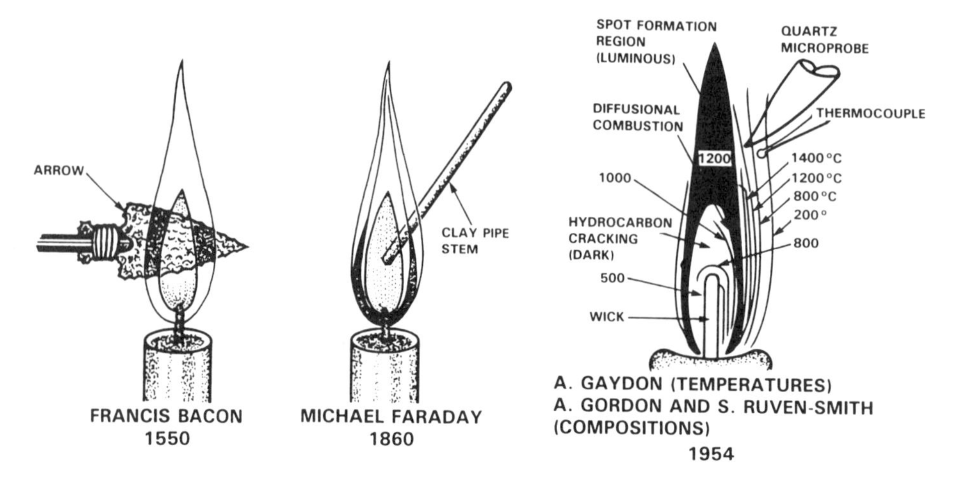

FIG. I-3 Historical Perspective of Flame Structure Studies.

groups became involved in the problem of flame microstructure. The first major meeting where work was reported was the Fourth Symposium (International) on Combustion held at the Massachusetts Institute of Technology in Cambridge, MA. The group reporting on flame structure topics included: Dixon-Lewis, Carpenter, and Smeeton-Leah at Leeds Univ. (England); Friedman at Westinghouse Electric Corp.; Fenimore and Kaskan at The General Electric Corp.; Fristrom, Prescott, Newmann, and Avery at APL/JHU; Gaydon and Weinberg at Imperial College (London); Gilbert at California Institute of Technology; Van Tiggelen at Louvain (Belgium); Wolfhard at RAE Farnborough (England); and H. Gg. Wagner at ICP Göttingen. Hirschfelder's group, called the Naval Research Laboratory of the University of Wisconsin, was particularly influential both in developing flame theory and in training researchers in both theory and experiments. It was supported by the U.S. Naval Bureau of Ordinance through the Applied Physics Laboratory of Johns Hopkins University. Another early influential program in combustion was the Navy's project Squid (for jet propulsion) first under John Fenn at Princeton and later under John Scott at the University of Virginia. To maintain perspective it should be recognized that these efforts are only a small corner of the multitude of scientific efforts undertaken at the time, and the coverage is biased towards those areas where the author is most familiar. Nevertheless, they have influenced the character of present-day government support of the sciences.

The group at the author's laboratory (the Applied Physics Laboratory of the Johns Hopkins University) was headed by Dr. W. H. Avery. Associates included: W. G. Berl, P. Breisacher, N. de Haas, S. Favin, G. Fristrom, R. Fristrom, T. O'Donovan, R. Prescott, C. Grunfelder, S. Raezer, R. Walker, W. D. Weatherford, A. A. Westenberg, and W. E. Wilson.

Modern flame probing began modestly enough with a temperature profile of a low-pressure acetylene flame measured by Klaukens and Wolfhard [1948] using a miniature thermocouple. This pioneer study introduced the concepts of studying reduced pressure flames to widen the flame microstructure together with miniaturized probes.

By 1952 at the Fourth International Symposium on Combustion held at the Massachusetts Institute of Technology, a half-dozen papers on flame microstructure were presented, and a lively discussion on the validity of the techniques ensued. Gerstein [1953] of the NACA Lewis Laboratory gave a survey on laminar flame structure, but a major part of the presentation was concerned with flame shape rather than flame microstructure. The first survey devoted to microstructure was given by Fristrom [1957] at the Sixth Symposium at Yale University.

The techniques were quickly developed and generously shared among the laboratories so that it is not clear in many cases which group pioneered which technique. The author can only speak for his own laboratory. The possible use of quartz microprobe sampling coupled with mass spectrometry was suggested to us by G. B. Kistiakowsky of Harvard University in 1951 and was implemented by R. Prescott using Foner and Hudson's mass spectrometer (Prescott, Hudson, Foner, and Avery [1954]). The particle track technique for velocity measurements (after Andersen and Fein [1949]), miniaturized, coated thermocouples for temperatures (after Klaukens and Wolfhard [1948]), and scavenger microprobes for measuring radical concentrations (Fristrom [1963]) all were developed by Fristrom and his colleagues. Westenberg developed the analytical methods for analyzing the data after the treatment of Hirschfelder and his group at Wisconsin. S. Favin developed data reduction methods. The APL group produced a definitive set of papers on propane–air (1957) and methane–oxygen (1960) (Fristrom et al. [1957–1963], Westenberg et al. [1960–1961]). This represented one of the first quantitatively successful measurements and analyses of flame structure. Much of the improvement of the methane studies relative to an earlier study on the propane system was due to the availability of reliable experimental high-temperature diffusion coefficients by Walker and Westenberg [1960] and thermal conductivities by Westenberg and deHaas [1962].

In the late 1960s a new sampling technique emerged: the molecular beam inlet system combined with mass spectrometric detection. This offered the possibility of direct detection of free radicals. The method is discussed in Chapter VII. The technique had been pioneered in the 1950s by Foner and Hudson [1953] at APL in the era before commercial availability of mass spectrometers, but it only became a common technique after an understanding of supersonic molecular beams emerged. The Göttingen group under Wagner were pioneers in this area. Among many studies, they made an early measurement on a methane flame including its radical profiles (Bonne, Grewer, and Wagner [1960]). Later the group directed their attention to soot formation. Van Tiggelen's group at Louvain, initiated by A. Van Tiggelen and continued by his nephew P. Van Tiggelen, also pioneered molecular beam sampling. At the Midwest Research Institute Milne and Green [1965, 1969] studied methane flames in developing their molecular beam sampling technique. At the National Institute of Standards and Technology (NIST) Hastie [1973b] studied methane and the effects of volatile salts in flames using molecular beam inlet mass spectrometry. His extensive work is summarized in his book on high-temperature vapors (Hastie [1975]). At UCLA Knuth pioneered the development of supersonic molecular beam inlets and applied them to the methane system (Gay, Young, and Knuth [1975]). The work is summarized in a review article (Knuth [1973]). The first completely successful application of this new method to the methane system was reported in two papers at the Fourteenth

Symposium on Combustion by Peeters and Mahnen of Louvain [1973] and Biordi, Lazzara, and Papp of the U.S. Bureau of Mines [1973]. These landmark papers provided complete profiles of several low-pressure methane flames similar to the earlier microprobe studies.

Optical methods in combustion had been pioneered by Weinberg at Imperial College, London. This work is summarized in his book *Optics of Flames* [1963]. In the 1980s the potential of laser techniques began to be applied to flame studies. A number of laboratories began to develop techniques, notably Stanford, University of California, Berkeley, and the General Electric Company. One significant development was the establishment in 1980 of a Combustion Center at Sandia National Laboratory, Livermore, CA. This was sponsored by the Department of Energy for the development of laser methods for combustion. The concept of this laboratory, according to D. Harley, its first director, is to develop lasers and laser methods for combustion research and offer collaborative help to other laboratories that may lack the specialized equipment necessary.

Much of this work was aimed at engine problems and turbulent flames, where pulsed laser techniques offered a powerful new tool. Nevertheless, many significant studies of flame structure have been made. Laser diagnostics are particularly valuable because they do not disturb the flame. Therefore, they can be used to validate other probing techniques. They do not offer a complete panacea because each species requires a separate experiment. Temperature and major species are examined best by Raman scattering, radicals by laser-induced fluorescence (LIF), and trace species by laser absorption spectroscopy. For absorption spectroscopy and LIF of each species it is necessary to identify a suitable transition, making a complete analysis using laser diagnostics cumbersome.

Laser methods have been reviewed by Lapp and Penney [1974], Eckbreth [1981], and Penner, Wang, and Bahadori [1984]. More detailed information can be found in Chapter VIII.

Modeling

The first quantitative treatments of flames comprised the nineteenth century thermal theories developed independently by Mallard and Le Chatelier in France, Haber in Germany, and Mikel'son in Russia (Evans [1952]). The thermodynamic analyses of Chapman [1899] and Jouget [1905] established the distinction between detonations and ordinary deflagrations (flames). The period that followed was dominated by a debate between those who believed that combustion was controlled by thermal conduction and those who believed that it was controlled by diffusion. The Russian school of Frank-Kamenetzki [1955], Semenov [1935], and Zeldovich [1949] pointed out that for species of equal molecular weight and diameter the transport contributions were equal and opposite, and cancelled. This approximation was used by Kelin [1957] to develop an integral, iterative flame theory. The first mathematically rigorous formulations were by Eckart [1940] and Boys and Corner [1949]. These were generalized using multicomponent kinetic theory by Hirschfelder and his collaborators [1954]. This treatment is the basis for most quantitative interpretations of flames, including those in this book. Several approximate solutions, usually involving simplified kinetics with high activation energies, have been suggested by von Karman and Penner, Wilde, and others. The status was reviewed critically by Evans [1952].

Spalding [1956] developed a nonstationary technique called the *marching method,* which was well adapted to computer simulation. Spalding and Stephenson [1971] applied it to the hydrogen bromine flame. In the late 1960s Dixon-Lewis began his seminal studies of the hydrogen system, extending it to carbon monoxide and methane flames. These studies [Dixon-Lewis et al. 1967–1989)] laid the chemical kinetic groundwork for the understanding of the general CHO flame system of oxygen.

A full synthetic model of the methane flame was made by the University of Utah group under Smoot (Smoot, Hecker, and Williams [1976]). They developed a full computer simulation and tested a number of kinetic mechanisms against burning velocity, flame thickness, and flame structure measurements. Their technique was based on the method of Spalding [1956]. They employed a 28 reaction mechanism, five of which they felt could be neglected. The model also was limited by the assumption of unity Lewis number.

Tsatsaronis [1978] spent a period with Dixon-Lewis at Leeds and then returned to Aachen to finish his thesis on flame modeling. He chose the methane flame using a 29 reaction model based on a literature survey with some modifications of rates (within limits of uncertainty) to improve fits to experiments. The predictions were tested against: (1) burning velocity dependence on composition, pressure, and initial temperature; (2) flame structure studies of Fristrom et al. [1960], Peeters and Mahnen [1973], and Dixon-Lewis [1981]. Agreement of modeling with experiments was very satisfactory, although, as the author comments, the kinetic constants may not represent a unique set. Dixon-Lewis extended his flame model for hydrogen and carbon monoxide to include the methane system [1981] using his rigorous transport formulation including thermal diffusion. Warnatz [1981] brought the field to full blossom with his comprehensive modeling study of the lower hydrocarbons. Westbrook and his collaborators at Lawrence Livermore National Laboratories developed a comprehensive combustion modeling program.

Westbrook and Dryer [1981] reviewed the status of modeling of combustion systems in 1981. Other reviews include that of Oran and Boris [1981], Warnatz [1981b], and the book edited by Gardiner [1984]. The most recent review on flame structure, that of Dixon-Lewis [1991], is to be recommended as a well-balanced survey.

The field of flame microstructure underwent three growth periods, each associated with the introduction of a new technique. The initial spurt was associated with microprobing (1950–), the second associated with molecular beam inlet sampling (1970–), and the third with laser diagnostics and computer simulations (1980–). With the increasing power of computer technology in the 1980s there has been an increase of computational simulations at the expense of experimental studies. The author views this trend as a mixed blessing because the flame equations are strictly speaking approximations and many key transport and kinetic parameters that are unavailable are now being used as fitting parameters. This can lead to dangerous extrapolations if used indiscriminately outside the range of experience. To quote the perceptive bard of Avon, "There are more things in heaven and earth, Horatio, than are dreamed of in our philosophy" (Shakespeare [1603]).

Scope of Combustion

It is appropriate at this point to define terms with some care. Let us consider the term *combustion* broadly to include any relatively fast[1] exothermic gas-phase chemical reaction. Note that no reference has been made to oxygen as one of the necessary reactants, even though the great majority of combustion systems involve it. The outstanding case not falling within this definition is the combustion of solid carbon, which is probably heterogeneous because its low vapor pressure precludes prevolatilization.

To answer the question: "What is a flame, and how does it differ from other reacting systems?" a flame may be defined as an exothermic chain reaction that can propagate subsonically through space. It is the property of spatial propagation that distinguishes flames from other combustion reactions. Consider a static system containing a homogeneous mixture capable of chemical reaction. If an ordinary reaction is caused to start in some portion of the system, it will go to completion in this original volume and then stop. A flame, on the other hand, will continue to propagate until the system has been consumed. A nonflame reaction would require ignition of the whole system to accomplish the same thing. Flames are usually accompanied by the emission of visible radiation, but this feature is not essential to the definition.

The propagation of flames results from the strong coupling of the chemical reaction with the transport processes of molecular diffusion and heat conduction and fluid flow. Heat, active species, and radiation can each accelerate chemical reactions, although the latter is usually of importance only in flames containing solids. Reaction raises the temperature and induces gradients in temperature and concentration, which are converted by the transport processes into fluxes of heat and reactive species. These fluxes into the unburned gases speed their reaction. The faster the reaction, the steeper the gradients, and the steeper the gradients, the higher are the fluxes. This positive feedback loop is opposed and limited by the flame propagation. If the feedback exceeds some critical factor, the system will be self-sustaining. If the feedback is not sufficient to bring the volume of gas to the condition from which the transport occurred, then the flame slows up until a balanced steady-state flame is attained. If there is no velocity for which this balance is possible, the flame goes out and the system can be considered to be a reaction perturbed by transport. If the acceleration by feedback exceeds the

1. It is not necessary to be specific about what is meant by "relatively fast" except that the times involved are typically measured in seconds or fractions thereof. This occurs because the time constant for flame reactions must be compatible with the residence time, which in turn is controlled by the propagation velocity, which in turn must balance the velocity of diffusivity. The diffusivity of air at STP is about 0.2 $cm^2\ s^{-1}$, and the maximum thickness for laboratory flames is limited to a few centimeters. Using the approximate relation between diffusivity, time constant, and characteristic distance (Eq. III-1) gives a maximum time of the order of 10 seconds.

$$v \approx 2\frac{D}{zs}; \qquad t \approx \frac{zs}{v}$$

In this equation v is velocity, *zs* is flame thickness, *D* is diffusivity, either average thermal or molecular. If D and *zs* employ the same length units, the time will be in seconds. This is far from a rigorous calculation, but it provides the correct relation between the important factors and provides an order of magnitude estimate primarily because of the partial cancellation of the temperature dependence between velocity and diffusivity.

critical amount, the flame speeds up and in the limit a transition to detonation can occur. Only when an exact balance is attained is the system a steady-state flame.

The system is constrained by conservation of energy and mass of each component element as well as the differential equations of molecular transport and chemical reaction. This produces an over-determined system so that each flame possesses a unique propagation velocity that depends only on its inlet conditions of composition, temperature, and ambient pressure.

If minor differences in stream tube geometry are neglected, the difference between the microstructure of a propagating flame and a Bunsen burner flame becomes only a question of the choice of coordinate system for viewing the processes. This cannot influence the fundamental physics and chemistry of the system. In a steady-state system there must be a single direction of propagation, otherwise the system would dissipate. The ability to define the movement implies that the reaction is confined to a definable zone that is small compared with the dimensions of the apparatus. These constraints suggest that a one-dimensional model might provide a useful approximation for many systems. Ignition, extinction, and diffusion flames require the additional consideration of more than one dimension.

The quantitative description of flames draws on elements of chemistry, fluid mechanics, and molecular physics, combining conservation of energy and elemental mass flux with the differential equations of chemical kinetics, thermal conduction, and molecular diffusion.

Literature of Combustion

A number of scientific disciplines are employed in combustion and, as a result, its literature tends to be scattered through the journals of physics, chemistry, fluid dynamics, and engineering. Some journals specialize in combustion topics, while others carry only an occasional article. A list of common source journals and their abbreviations is given in the bibliography at the end of the book. Combustion topics are abstracted by the usual scientific and technological services. The most generally useful is probably *Chemical Abstracts*. The combustion literature is unusual in the sense that a significant portion has been published as symposia. The oldest of these sources are the volumes of the Symposia (International) on Combustion, usually called the Combustion Symposia. They began in 1928 under the sponsorship of the American Chemical Society. Since 1954 they have been held biannually under the sponsorship of The Combustion Institute, membership in which is open to interested research workers. Papers are reviewed and published the following year. Other regular symposia include those sponsored by AGARD, which is the aeronautical research committee of NATO, and the International Flame Research Foundation. Related fields hold colloquia of interest. Examples include the Colloquia on Dynamics of Explosions, published by the AIAA (American Institute for Aeronautics and Astronautics), and Fire Research Symposia sponsored by the Center for Fire Research of the U.S. National Institute of Standards and Technology (formerly the National Bureau of Standards). In addition, both the Russian and Polish Academies of Science sponsor combustion symposia on a regular basis. One of the most interesting of these is the International Seminars on Flame Structure. Three have been held by the Russian Academy of Sciences in collaboration with IUPAC (the International Union of Pure and Applied Chemistry). Irregularly, meetings of interest are also sponsored by the American Chemical Society, the American Physical Society, the Gordon Conferences, and others. The field seems reasonably well served by the scientific media.

The number of review articles on combustion and related topics has proliferated over the years. This makes it difficult to select a group of reviews that are authoritative, readable, and timely. The compilation begun by the author several decades ago has become too unwieldy, dated, and biased to be generally useful. Even a critical current list is likely to become dated so soon after publication that the reader is better off selecting his own sources. A variety of surveys are now easily accessible through most technical libraries. Books on combustion known to the author are presented in the bibliography at the end of the book. These volumes can provide excellent starting points in special areas. Two good sources of overviews are the reviews that have appeared from time to time in *Science* and *Scientific American.* The journal *Progress in Energy and Combustion Science* carries selected and very useful reviews on a quarterly basis. In addition, the journals *Combustion and Flame* and *Combustion Science and Technology* encourage reviews by competent authorities in the field. The *Symposia (International) on Combustion* offer a reliable source of reviews on topics of current interest, usually together with a collection of related articles. In addition to these combustion-oriented sources, the journals of fluid dynamics, heat transfer, chemical kinetics, and other related fields often offer views of a topic that may broaden the reader's vision of the area as it applies to his or her own interests.

Combustion topics are abstracted by *Chemical Abstracts, Fuel Abstracts,* and *Physics Abstracts.* Special topics are abstracted by *Applied Mechanics Reviews* and elsewhere. The most up-to-date and reliable of these services is *Chemical Abstracts.* Between 1965 and 1978, fire topics were abstracted by *Fire Research Abstracts and Reviews.* Copies can be obtained through the National Technical Information Service (NTIS). *Fire Technology Abstracts,* which was published between 1972 and 1979 and resumed publication in 1983, is also available through NTIS.

There are over a hundred books on combustion topics, and a list of those known to the author is collected in the bibliography at the end of the book.

Summary

This book records the work in this relatively narrow scientific area, and provides some interpretation. It is hoped that value judgments have been restricted to the scientific plane, leaving critical chronicles to scientific historians.

II

MACROSCOPIC FLAME BEHAVIOR

It is not surprising that a subject as complex and ancient as combustion should have acquired a considerable body of phenomenology and qualitative descriptive material. Some of this is only of historic interest, but many aspects remain useful. The main body of this book is devoted to a quantitative discussion of the *microstructure* of flames, that is, the processes taking place within the flame front itself. In other contexts, however, flame structure is used to describe the external geometric shape. This aspect is considered in this chapter. Short discussions are also given of macroscopic properties of flames and related combustion systems. Since this covers most of classical combustion studies, the treatment must be brief and descriptive. The underlying principles will be suggested, but coverage is limited by the complexity of the phenomena. Books on combustion are listed in the bibliography at the end of the book, and these can provide a starting point for finding supplementary information.

What Burns?

In Chapter I combustion was defined as a rapid exothermic reaction and flames as combustion reactions with the ability to propagate through a suitable medium. This propagation results from the strong coupling of the reaction with the molecular transport processes. The rapid reactions produce gradients that the transport processes convert into heat and species fluxes that speed the reactions. The result of this positive feedback is the propagation, which can be either free or balanced by opposing flow. It is logical at this point to ask, "What can burn?". The answer is, "under suitable conditions, almost everything that can undergo a rapid exothermic reaction."

Although there are decomposition flames that involve a single exothermic species, more commonly combustion involves a pair of reactants. More complex mixtures also burn, but reactants can usually be classified as either fuels or oxidizers. Combustion chemistry will be considered in more detail in Chapters X through XIII.

Decomposition Flames

Flames produced from a single initial reactant are called *decomposition flames.* The reaction must be exothermic. Some reactants combine oxidizing and reducing components, such as trinitrotoluene (TNT). Other such flames result from the decomposition of high-energy molecules that may play other roles in other flame systems. For example, ozone can function as a strong oxidizer with suitable fuels, and acetylene is a fuel toward most oxidizers. Decomposition flames are discussed in Chapter XI.

Dual Reactant Flames

The traditional flame has two reactants, a fuel and an oxidizer. A majority of the elements and their compounds are considered fuels relative to oxygen. However, it should be remembered that the designation of fuel and oxidizer is a question of the relative electronegativity of the reactants. For example, although molecular chlorine is normally considered an oxidizer, in the chlorine–fluorine flame it plays the role of a fuel. The intermediate position of nitrogen compounds is noteworthy. Nitrogen forms combustible hydrides such as ammonia (NH_3) and hydrazine (N_2H_4), but also forms strong oxidizers through oxygen and halogen substitution (NO_2, N_2F_4, and NOCl). Hydrazine supports a decomposition flame and can "burn" diborane (Berl and Wilson [1961]). Molecular nitrogen is inert in most flame systems, but dusts of lithium and titanium will burn in it.

Fuels

Volatility is the key to premixed combustion. Sulfur and phosphorus are sufficiently volatile to burn in normal air, while other elements as well as some metals can be burned as low-pressure flames (ca. 10^{-3} atm) with oxygen and other oxidizers. The less volatile elements can be burned as dusts or often in the bulk using oxygen. The alkali metals and alkaline earth metals burn vigorously in the bulk, once ignited, and the more reactive of these (K, Rb, Cs) ignite spontaneously, though often the moisture reaction is the true ignition source. Hydrogen is generated by reaction with the moisture in humid air, producing sufficient heat to ignite a hydrogen flame, which in turn ignites the metal.

The most familiar flame systems are fueled by volatile compounds from Group 3, particularly hydrides. The archetype of these are the hydrocarbons, which form an enormous class. The bulk of this book is devoted to the study of C–H–O–X flames, but it should remembered that this is only a fraction of flame chemistry. Boron, silicon, and phosphorous also form less extensive families of combustible hydrides. Mixed hydrides can be formed and are combustible. In addition to hydrides, several elements form volatile combustible compounds with carbon.

There are many volatile metal organic compounds (Sidgwick [1950]), most of which burn and many of which, such as aluminum trimethyl, ignite in air spontaneously. They are called *hypergolic* fuels.

Low-pressure discharges create atoms and radicals that react to form colorful so-called "atomic flames" (Ferguson and Broida [1956]).

A list of the properties of some fuels considered for air breathing propulsion is given in Table A-4 in the appendix. A systematic discussion of flame chemistries is given in Chapters X through XIII.

Oxidizers

The principal oxidizers are: molecular oxygen, the halogens, and some of their compounds. In addition, there are a number of other strong oxidizers that can support combustion including peroxides, interhalogens, halogen oxides, oxides of nitrogen, nitrogen oxyhalides, and oxyacids such as perchloric and nitric. These systems are discussed in Chapter XIII. High-energy oxidizers were studied extensively during the 1960s as potential rocket propellants. A number of new compounds were prepared and characterized. The field has been reviewed by Lawless and Smith [1975].

The most vigorous oxidizer is elemental fluorine. It will attack most elements and compounds, even other halogens. Fluorine attacks krypton and xenon, but the products are too thermally unstable to form a combustion system. The halogens decline in reactivity with increasing atomic number. Thus, chlorine is less reactive than fluorine. It forms flames with hydrogen and some hydrocarbons, but is unreactive toward many fuels such as carbon monoxide. Bromine is even less reactive. Iodine only forms flames with such electropositive reactants as the alkali metals.

The fuel/oxidizer relation is controlled by relative electronegativity (see Chapter XIII). As a consequence, several elements which are fuels relative to oxygen or fluorine can also function as oxidizers either in their own right or through their oxides or halides. They include boron, sulfur, and phosphorous. Volatile electronegative metals such as the alkali metals can be burned as low pressure flames by such unlikely oxidizers as carbon dioxide and halocarbons. Less volatile reactive metals can be burned as dusts with volatile halogens or form solid state flames with oxides of less negative metals. A common example is the thermite reaction, $Al+Fe^{2}O^{3} => Fe+Al^{2}O^{3}$.

The most common flame oxidizer is molecular oxygen, either neat or diluted in the form of air. It burns most elements and many compounds. The situation can be visualized using the chemists' periodic arrangement of the elements (Fig. II-1). The table can be divided into five sections, each showing a characteristic combustion behavior. The sections are: (1) the inert gases; (2) the oxidizers; (3) elements that form volatile hydrides; (4) metals; and (5) metals with thermally unstable oxides. Oxygen does not form combustion systems with groups 1, 2, and (5). Group 3 elements form the volatile fuel compounds commonly associated with flames. These elements can also be burned as dusts, and many will even sustain vigorous air combustion in the bulk form. Group 4 burn as dusts and often present industrial fire hazards (Palmer et al. [1973]). Many of these elements can be burned in bulk form with oxygen. This is applied in cutting steel plates using an oxygen torch.

Combustion Behavior and Classification

Combustion is a widespread phenomenon. The behavior of the system depends on the relative volatility of reactants, their reactivity, and the flow conditions. There are two major divisions of combustion systems: flames where molecular transport processes are strongly coupled with the reaction, and reactors where flames may exist in the microstructure of the system but are masked by rapid mixing, wall effects, or ignition. Flames are the topic of this book. However, there are many interesting and useful combustion reactors, a selected few of which will be discussed along with some more complex combustion systems such as furnaces and engines.

Common flames involve two reactants. If they are gaseous, premixed on the molecular level, and the flow is laminar, a thin flame reaction sheet of the Bunsen type

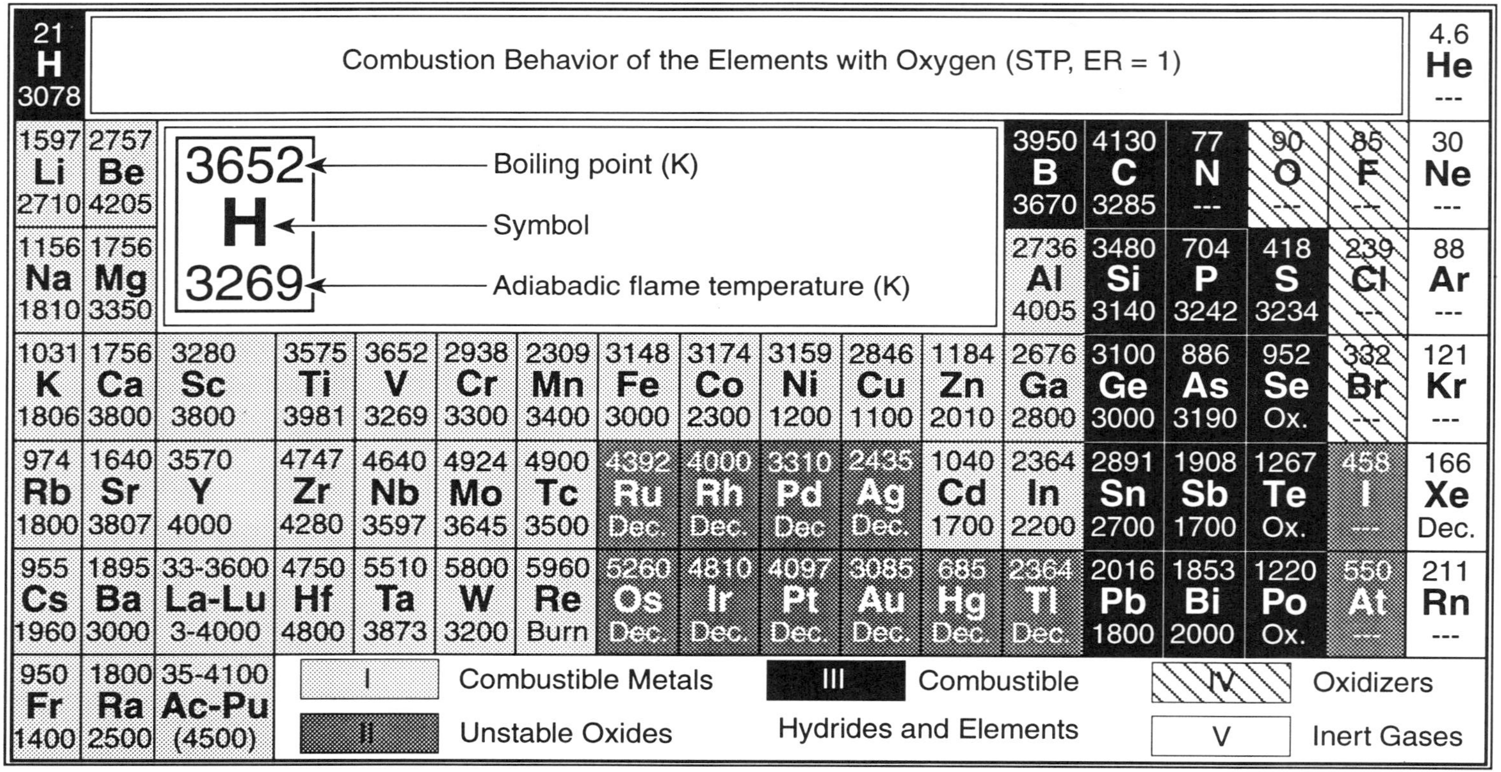

Combustion Behavior of the Elements with Oxygen (STP, ER = 1)

21 H 3078																	4.6 He ---
1597 Li 2710	2757 Be 4205											3950 B 3670	4130 C 3285	77 N ---	90 O ---	85 F ---	30 Ne ---
1156 Na 1810	1756 Mg 3350											2736 Al 4005	3480 Si 3140	704 P 3242	418 S 3234	239 Cl ---	88 Ar ---
1031 K 1806	1756 Ca 3800	3280 Sc 3800	3575 Ti 3981	3652 V 3269	2938 Cr 3300	2309 Mn 3400	3148 Fe 3000	3174 Co 2300	3159 Ni 1200	2846 Cu 1100	1184 Zn 2010	2676 Ga 2800	3100 Ge 3000	886 As 3190	952 Se Ox.	332 Br ---	121 Kr ---
974 Rb 1800	1640 Sr 3807	3570 Y 4000	4747 Zr 4280	4640 Nb 3597	4924 Mo 3645	4900 Tc 3500	4392 Ru Dec.	4000 Rh Dec.	3310 Pd Dec	2435 Ag Dec.	1040 Cd 1700	2364 In 2200	2891 Sn 2700	1908 Sb 1700	1267 Te Ox.	458 I ---	166 Xe Dec.
955 Cs 1960	1895 Ba 3000	33-3600 La-Lu 3-4000	4750 Hf 4800	5510 Ta 3873	5800 W 3200	5960 Re Burn	5260 Os Dec.	4810 Ir Dec.	4097 Pt Dec.	3085 Au Dec.	685 Hg Dec.	2364 Tl Dec.	2016 Pb 1800	1853 Bi 2000	1220 Po Ox.	550 At ---	211 Rn ---
950 Fr 1400	1800 Ra 2500	35-4100 Ac-Pu (4500)															

Key (H): 3652 — Boiling point (K); H — Symbol; 3269 — Adiabadic flame temperature (K)

I — Combustible Metals
II — Unstable Oxides
III — Combustible Hydrides and Elements
IV — Oxidizers
V — Inert Gases

FIG. II-1 Periodic table showing that elements with similar combustion behavior cluster in groups. A similar table could be made for any oxidizer. The identification of a molecule as an oxidizer depends on its electronegativity relative to the fuel considered (see Chapter XIII, Fig. XIII-1). As a consequence a number of elements such as nitrogen, phosphorous, sulfur and chlorine function both as fuels or oxidizers depending on the reactant.

results. Reaction and molecular transport balance one another. The microstructure of this type of flame is the major topic of this book. As will be seen, this is the basic flame reaction region. The microstructure of the other flame types are similar; what varies is the controlling path of mixing (Figs. II-2 and II-3). Diffusion, phase changes, or turbulent flow may dominate, or all may contribute. These factors affect the path of final mixing, but the basic reactions still occur in a restricted zone similar to the premixed

A- DISPERSION SMALLER THAN FLAME THICKNESS

FUEL→ ↓OXIDIZER	GAS	LIQUID	SOLID
GAS	BUNSEN FLAMES	MIST FLAMES	DUST FLAMES
LIQUID	MIST FLAMES	EXPLOSIVES	EXPLOSIVES
SOLID	DUST FLAMES	EXPLOSIVES	GUNPOWDER

B- DISPERSION COMPARABLE TO FLAME THICKNESS

FUEL→ ↓OXIDIZER	GAS	LIQUID	SOLID
GAS	TURBULENT FLAMES	DROPLET FLAMES	COAL FIRES
LIQUID	DROPLET FLAMES	LIQUID ROCKETS	SLURREY EXPLOSIVES
SOLID	H_2-NH_4ClO_4 DUST FIRE	SLURRY EXPLOSIVES	SOLID PROPELLANTS

C- DISPERSION LARGER THAN FLAME THICKNESS

FUEL→ ↓OXIDIZER	GAS	LIQUID	SOLID
GAS	DIFFUSION FLAMES	POOL FIRES	FURNACE FIRES
LIQUID	H_2-HNO_3 POOL FIRE	LIQUID ROCKETS	LIQUID/SOLID ROCKETS
SOLID	H_2-NH_4ClO_4 FIRE	LIQUID/SOLID ROCKETS	THERMITE

FIG. II-2 Combustion behavior classified according to initial phase, dispersion, and flow character.

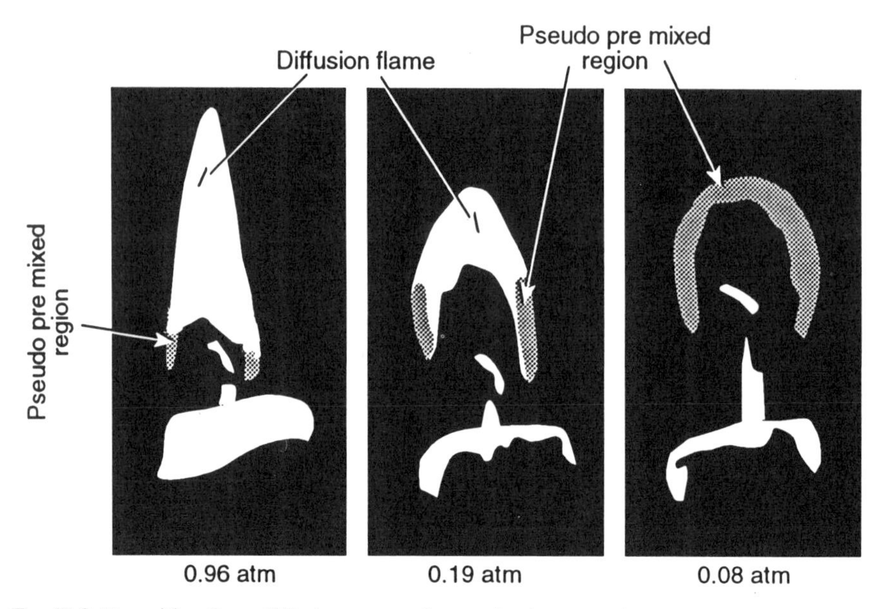

FIG. II-3 Transition from diffusion to pseudopremixed combustion in a candle flame at several pressures. Note the systematic change in the importance of the blue luminosity, which marks pseudopremixed combustion with respect to the yellow soot luminosity, which marks the region of fuel-rich diffusive combustion. The flame standoff and length of uncharred wick provide a measures of the thickness of the preheat region.

flames. Since final mixing must be on the molecular level, diffusion always plays the ultimate role even in turbulent flames. Reactors usually employ rapid mixing and/or homogeneous ignition. Examples of such systems are the low-pressure and turbulent flow reactors. Most practical combustion devices such as furnaces and engines fall into this category. Some systems can be treated as approximately homogeneous. This allows the application of a structureless "black box" characterization using average residence time and reaction conditions (Vulis [1961]). In fast-flow reactors, where transport processes are unimportant, distance replaces time in the analysis. This requires knowledge of velocity, and assumes negligible molecular transport. Simple considerations of conservation and flow continuity often yield detailed descriptions of their behavior, despite internal complexities that may involve interactions among adjacent flames, walls, and radiative transfer. Engine and furnace design is in the transition between macroscopic analyses using overall global parameters and the fundamental approach describing the systems in terms of the basic molecular processes. We are far from this basic understanding at the present time, but the rising tide of fundamental information and computer capabilities makes this approach increasingly practical.

Flames can be classified according to three characteristics: (1) initial mixedness, (2) initial state, and (3) flow character (Fig. II-2). These are the factors that control or limit the mixing of reactants. This book is devoted to microstructure of premixed laminar flames. The other flame classes differ according to which processes dominate the mixing of reactants. Once mixing has occurred, flame reaction zones similar to those in premixed flames are established. Diffusion flames require the additional consideration of lateral transport of energy and species, and ignition and turbulent flames require the introduction of time dependence, but the basic equations are the same in these more complex situations. They can often be treated using a basic one-dimensional model with perturbations.

Mixing controls whether the reactants are homogeneous before entering the reaction zone. If they are, it is called a *premixed flame*. If they are not, it is called a *diffusion flame*, since the mixing of fuel and oxidizer must be accomplished by diffusion. To illustrate the characteristic of "mixedness," consider the familiar laboratory Bunsen burner. In normal operation, the fuel gas entering the bottom entrains air through the side ports, and the two reactants mix during passage up the burner barrel. By the time they reach the burner exit, they are homogeneously mixed (i.e., on a molecular scale), and the resulting flame is of the premixed type. If, however, the air inlet ports are closed off, only the fuel passes out of the exit and must mix with the surrounding air by diffusion in order to burn. The burner thus forms a diffusion flame. The transition is dependent on the ratio of burner dimension to diffusion-limited mixing length (Figs. II-2 and II-3). *The physical state* of the initial reactants—solid, liquid, or gas—is another major factor controlling flame behavior.

The flow character is a third means of classification. This concerns the nature of the flow, that is, whether it is *laminar* or *turbulent*. If flow is laminar, or streamlined, transport is dominated by molecular processes of diffusion and thermal conduction. If flow is turbulent, transport is perturbed and sometimes dominated by macroscopic eddy motion. The addition of turbulence to a flame complicates the situation, and, despite considerable effort, turbulent flames have resisted a definitive quantitative description.

Within these broad classes many permutations (Fig. II-2) can be found, and other factors also enter. However, these three divisions provide the major categories.

Premixed Laminar Flames

Premixed laminar flames are most easily visualized as a thin reaction sheet separating unburned from burned gases and propagating normal to themselves at a characteristic velocity. This surface can be approximately identified with the thin luminous zone characteristic of common flames (see Figs. II-4 and II-5). This simplistic view neglects the detailed microstructure that is the subject of this book, but it provides a good first approximation. The thickness of the microstructure is usually small compared with the other flame dimensions. This suggests that they might be quantitatively modeled using one-dimensional equations, and this indeed is the case. The equations that describe such systems are discussed in Chapter IX. They are boundary-value problems whose solutions are pseudoeigenvalues identified with the experimental rate of propagation. This is commonly called the *burning velocity*. The concept is illustrated for conical flame geometry in Figure II-5. It has been found to depend only on the inlet conditions (pressure, initial temperature, and composition). Burning velocity is, therefore, used to characterize flame systems. This topic is discussed in the section concerned with characterization.

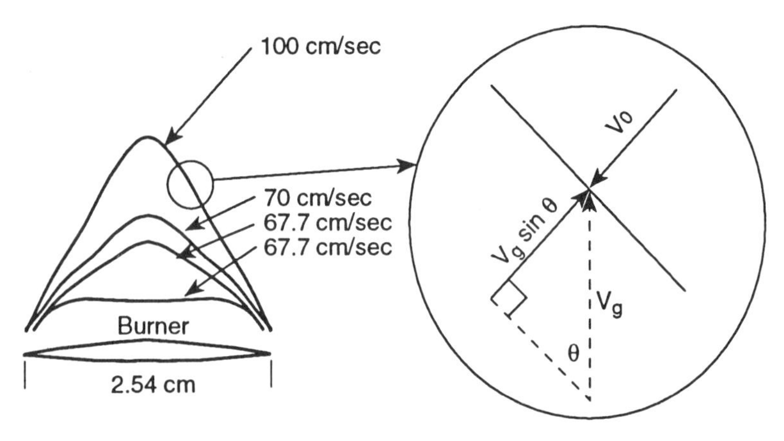

FIG. II-4 Dependence of premixed flame geometry on inlet velocity with uniform approach flow . Stoichiometric half-atmosphere stoichiometric propane–air flame showing the definition of burning velocity.

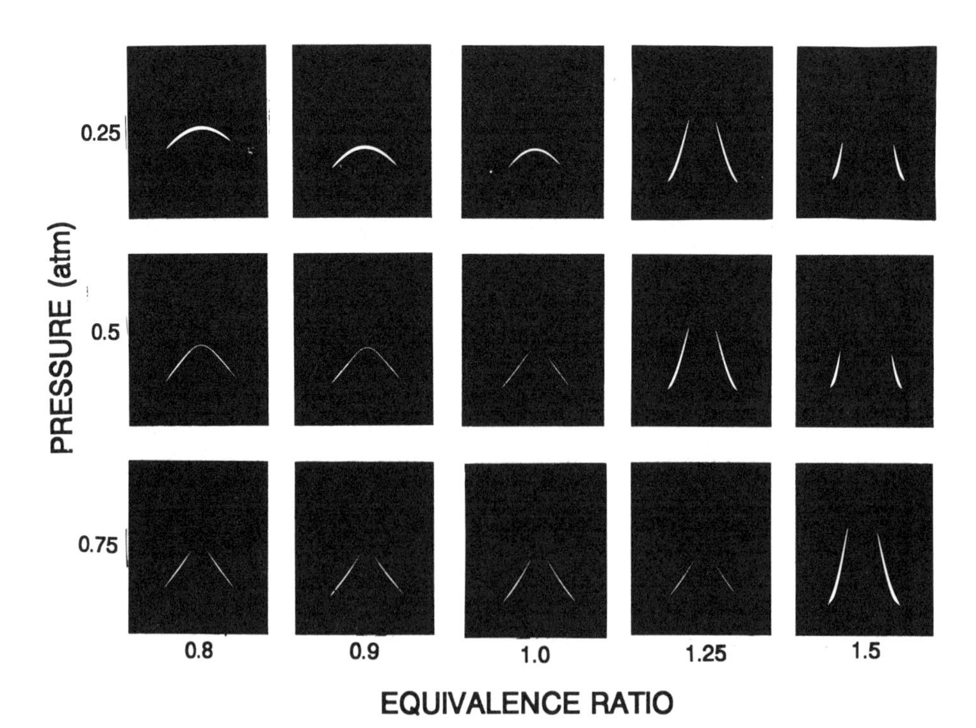

FIG. II-5 Dependence of premixed flame geometry on stoichiometry and pressure for fixed velocity (100 cm/sec). Propane–Air flames; Φ is the equivalence ratio; pressure is in atmospheres.

Low Pressure and "Atomic Flames"

Volatility is a relative term. At pressures below 1 torr many of the reactive metals form flame systems with most of the strong oxidizers and several less likely compounds such as halocarbons and carbon dioxide. In addition, flames can be sustained with the atoms and radicals formed in low-pressure discharges. These so-called atomic flames (Ferguson and Broida [1956]) are very colorful, and most volatile compounds react, since extra energy is available from the radicals. Flames with atomic oxygen, hydrogen, nitrogen, and chlorine as well as hydroxyl radicals are possible. Most laboratory flames operate at low pressure (<1–2 torr) to sustain the discharge, but plasma torches can operate at atmospheric pressure, and the atomic hydrogen torch (Langmuir [1927]) is used commercially for welding, although for safety reasons it is now being displaced by inert gas plasma jets, as mentioned in Chapter III.

Cool Flames

Luminous propagating reactions occur in many fuel-rich C–H–O fuels such as hydrocarbons, ether, and aldehydes. Propagation is usually slower than in normal hot flames, and the final temperatures are much lower, typically 700 K. This is below the threshold of the $H + O_2 \Rightarrow OH + O$ reaction. This reaction is replaced by direct addition of oxygen to olefinic hydrocarbons with the formation of peroxy and hyperoxy radicals RO_2H and RO_2 (Minkoff and Tipper [1962]). Their structure has been studied (Agnew and Agnew [1965]; Carhart, Williams, and Johnson [1959]), but the complexity of the products and intermediates has clouded the identification of the detailed mechanisms. The system is multistaged and shows a multiplicity of products and intermediates, some of which have rates that show negative temperature dependence. This can induce interesting oscillatory behavior (Agnew and Agnew [1965]; Heiss, Dumas, and ben Aim [1976]). These systems have received extensive study because there is a belief that this chemistry may contribute to the understanding of the knock problem in engines (Lewis and von Elbe [1987]; Jost [1946]).

The process is analogous to the *smolder* in solids, which also occurs at low temperatures with complex chemistry, and an apparent cool flame glow has been detected in the study of the oxidation of polyethylene. Phosphorous reacts to form low-temperature flames with oxygen (Lewis and von Elbe [1987], p. 284).

These systems are of interest in their own right, but their chemistries bear little relation to the high-speed combustion of common flames where the residence time is too short for these reactions. We therefore refer the interested reader to the cited references.

Cellular and Separated Flames

An initially premixed laminar system can become separated in either the coordinate parallel to the flame front (cellular or polyhedral flames) or along the propagation direction (sequential flames).

If a substantial difference exists between the diffusivity of fuel and oxidizer, cells are formed. The mechanism is the preferential diffusion induced by flame curvature. The more diffusive reactant migrates, depleting that component. If it is the deficient reactant, the burning velocity is reduced, and in the extreme local extinction can result. This occurs in rich, heavy hydrocarbon and lean hydrogen flames with oxygen,[1] It is observed, in flame tip flow through and as scallops in the sides of conical Bunsen flames, cells on flat flames and scallops on slot burners that are narrower than the cell size (see Fig. III-5). The aerodynamic theory of Markstein [1964] suggests that any flame system will show this instability if the burner is large enough. The extreme is found in fuel-lean hydrogen flames, where the apparent lean limit is significantly extended by local concentration of the fuel by preferential diffusion. The effect is discussed by Lewis and von Elbe [1987].

The driving force for axial separation is the differing rate between sequential reactions. This is an extreme case of zones in flames. An easily understood case occurs when two fuels that react at differing rates are mixed, as occurs in mixtures of diborane and propane with air (Berl and Dembrow [1952]). Diborane reacts with air an order of magnitude more rapidly than does propane. Low concentrations of diborane burn with final temperatures well below the ignition limit of the resulting preheated, but vitiated, hydrocarbon air mixture. Three cases can be distinguished. (1) The final temperature and radical concentration resulting from the diborane component of the flame are so low that thermal and diffusion losses prevent the propane reaction from occurring; that is, its "ignition delay" is too long. (2) The ignition delay is sufficiently short that a secondary flame appears. Such flames will usually go to completion. If the delay is sufficiently long that the two flames are moderately separated, it is possible to "blow off" the secondary flame without materially affecting the primary flame, thus achieving case (1). (3) The ignition delay is so short that the two reaction regions are strongly coupled and either appear as a single reaction region or as twin luminous regions that cannot be separated.

The effect is also seen in hydrocarbon flames where the hydrocarbon reaction zone (sharp luminous region) and the carbon monoxide reaction zone (wide luminous mantle) differ significantly in thickness. It is possible to separate the regions in rich hydrocarbon flames using the Teclu–Smithels-Ingles burner illustrated in Figure III-15.

Diffusion Flames

If the fuel and oxidizer are introduced separately and the flow is laminar, the combustion is usually limited by diffusional mixing of the reactants. This concept was first quantified by Burke and Schumann [1928], who solved the axially symmetric flow-diffusion equation for the limiting case of infinite reaction rate. Despite complications from the temperature dependence of density and diffusion, results predict diffusion flame shape with reasonable fidelity (Fig. II-6). Jost [1946] pointed out that the full equations need not be solved to predict flame height, which was often used to characterize diffusion flames. His approach was to assume that the flame height was the point at which interdiffusion of the fuel and oxidizer first reached the axis. This led to the following simplified equation.

$$h = \frac{\upsilon_g d^2}{2D} \qquad (2\text{–}1)$$

1. In addition to oxygen and air flames, the effect is also observed in hydrogen flames of chlorine and bromine.

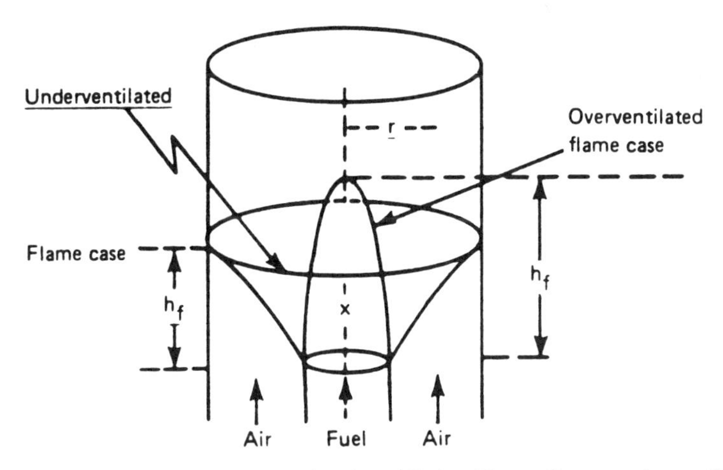

FIG. II-6 Burke–Schumann [1928] Solution for Diffusion Flame Geometry in a uniform flow field with radial symmetry.

In this equation h is the flame height, v_g is the inlet velocity, d is the burner diameter, and D is the fuel–air diffusion coefficient.Condensed-phase fuel sources produce diffusion flames. The complications connected with fuel geometry and volatilization are considered later.

Other Geometries

In the course of research, a number of diffusion flame burners have been developed to aid one or another aspect of their study. Four common types are: (1) parallel fuel jets; (2) the Wolfhard–Parker flat diffusion flame, consisting of parallel streams of fuel and oxidizer separated by a thin septum; (3) the opposed jet flat flame, in which balanced flows are opposed, producing a flat diffusion flame; and (4) the half-sphere or cylinder diffusion flame, in which the fuel (or oxidizer) is introduced from a porous sphere or cylinder in a uniform stream of oxidizer (fuel). These burners and their associated geometries are illustrated in Figure III-6.

Turbulent Flames

Turbulent flames dominate most commercial combustion processes (Spalding [1955]; Jost [1946]; Lewis and von Elbe [1987]). The most obvious effect is the increase in effective flame speed or volumetric heat release rate. Much of this may be geometric in origin rather than fundamental mechanism changes (Figs. II-7 and II-8). The maximum reaction rate obtainable with extreme turbulence occurring in the highly stirred reactor only approaches that observed in premixed laminar flames when proper account is taken of the thickness of the reaction zone (Avery [1955]). This reactor is illustrated in Chapter III. High-intensity turbulence can even extinguish flames. The field has been extensively reviewed by Lewis and von Elbe [1987)]; Jost [1946]; Karlovitz [1954]; Scurlock and Grover [1954]; Williams [1985]; Andrews et al. [1975]; and Peters [1988]. The books of Kuznetov and Sabelnikov [1986] and Libby

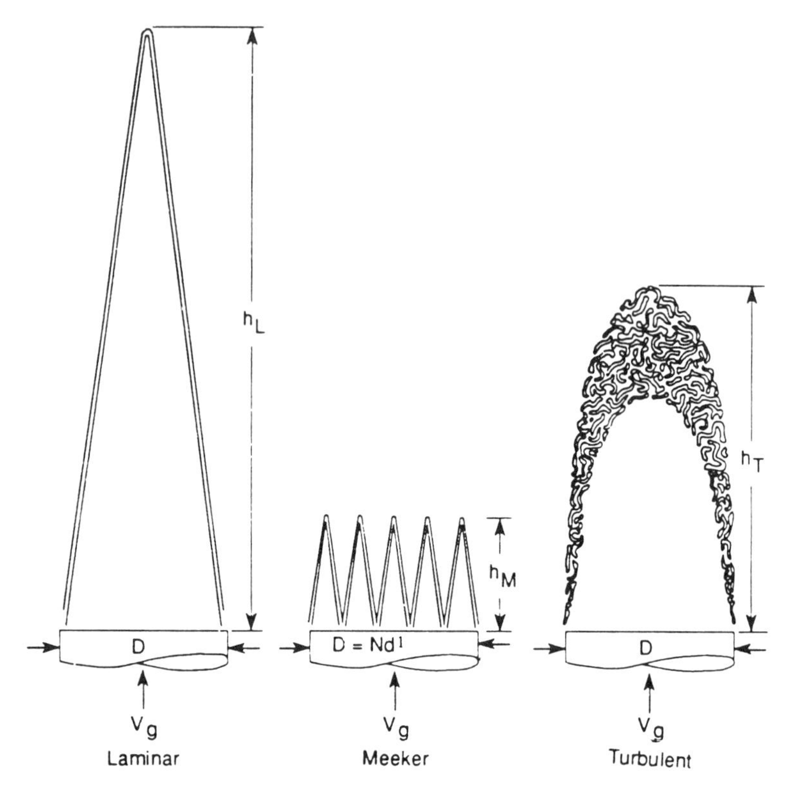

FIG. II-7 Illustration of the difficulties associated with using flame height to estimate turbulent burning velocity. Three burners with equal flow.

and Williams [1980] provide good overviews. The subject is still in flux, and the changes can be followed through the various symposia volumes. The problem is that there is no completely acceptable theory. Progress has been hampered by the lack of fundamental understanding of turbulence. Recently, however, a relationship between mathematical "strange attractors" in dynamic equations and chaotic behavior appears promising (Gleick [1987]), and the ever-increasing capacity of computational facilities should allow exploitation of these concepts. Williams [1985] observes that there are at least seven general aproaches ranging from the purely empirical to the detailed solution of the Boltzmann equations.

In the past experimental studies have been hampered by limitations of diagnostic probes in making time-resolved temperature, velocity, and concentration measurements, but this has been improved by the flood of new laser experimental methods (Chapter VIII). These diagnostics have concentrated on the statistical approach simultaneously measuring properties with limited local correlation and describing them in terms of probability density functions. They allow submicrosecond time resolution with spatial resolution as high as a fraction of a cubic millimeter. Even these capabilities are inadequate for the full turbulent flame problem, since not only is the high resolution necessary, but also cross correlation between these measurements over an extended region of space.

Historically the effect of turbulence on flames was noted by Mallard and Le Chatelier [1883], but the first systematic studies were made by Damkohler in his doctoral thesis [1947], in which he correlated flame behavior with Reynolds number (Fig.II-9). He defined two limiting conditions: The first is at low Reynolds number, where the length scales of the turbulence are large compared with the laminar flame front thickness. In this case turbulence distorts the flame, which has given rise to the common name "wrinkled flame front model." This is primarily a geometric effect (Fristrom [1965]). The second regime occurs for Reynolds numbers where length scales are small compared with flame front thickness. Here it is presumed that turbulence increases heat and mass transport and that the effect of turbulence can be modeled using turbulent eddy diffusivities in place of molecular diffusivities. These and other models using the ideas of Damkohler have had limited success. Karolovitz [1954] obtained data that suggested that increased burning velocity could not be accounted for by these flame theories and suggested that the flame generated turbulence of its own. Westenberg and Rice [1959] and Prudnikof [1959] attempted to measure flame-generated turbulence and concluded that in the systems they studied the effect is minor.

Several other ideas have been suggested in connection with turbulent flames. The "flame stretch" of Karlovitz [1954] is a change in propagation induced by extending or contracting the reaction zone through flow gradients. Such effects are observed in blowoff and in flame tip flow through (Law, Ishizuka, and Cho [1982]). A second model is the extended reaction zone concept of Sommerfield [1956]. A third is the scrolling envelopment picture of Spalding's "Eskimo" model (Ma et al. [1982]). A fourth is the flamelet model of Peters [1988]. Other theories have been developed to use the statistical information provided by the new laser diagnostic techniques using the so-called Favre method of averaging (functions are weighted according to local density).

It should be observed that in any real life situation there is a distribution of eddy sizes and intensities and that intensity as well as scale is likely to be a factor. Damkohler [1939] and Kovasznay [1956] proposed dimensionless parameters. Damkohler proposed considering the ratio of chemical reaction time to turublent fluxctuation time. Kovasznay's parameter, Γ, is the ratio of flow velocity gradient to flame front velocity gradient. It can also be interpreted in the same manner as that of Damkohler as a dimensionless ratio of flow fluctuation time to flame reaction time.

$$\Gamma = \left(\frac{\Delta \upsilon}{\upsilon_0}\right)\left(\frac{t}{L}\right) \qquad (2\text{–}2)$$

where $\Delta\upsilon$ is the flow velocity fluctuation (cm/sec), υ_0 is laminar burning velocity (cm/sec), t is the flame thickness, and L is the scale of velocity fluctuations (cm).

The author's own prejudices on the subject are that quantitative modeling may be possible (for premixed turbulent flames) using the simple "wrinkled flame front" model providing the flow can be characterized. This neglects the complexities where eddy structure falls below flame front thickness as being quantitatively unimportant. This based on the observation that the lifetime of small eddies is limited by momentum diffusivity. Eddies smaller than the flame front thickness have lifetimes that are short compared with their residence time passing through the front. This decay should limit the effect of weak small scale eddies. On the other side small, strong eddies should

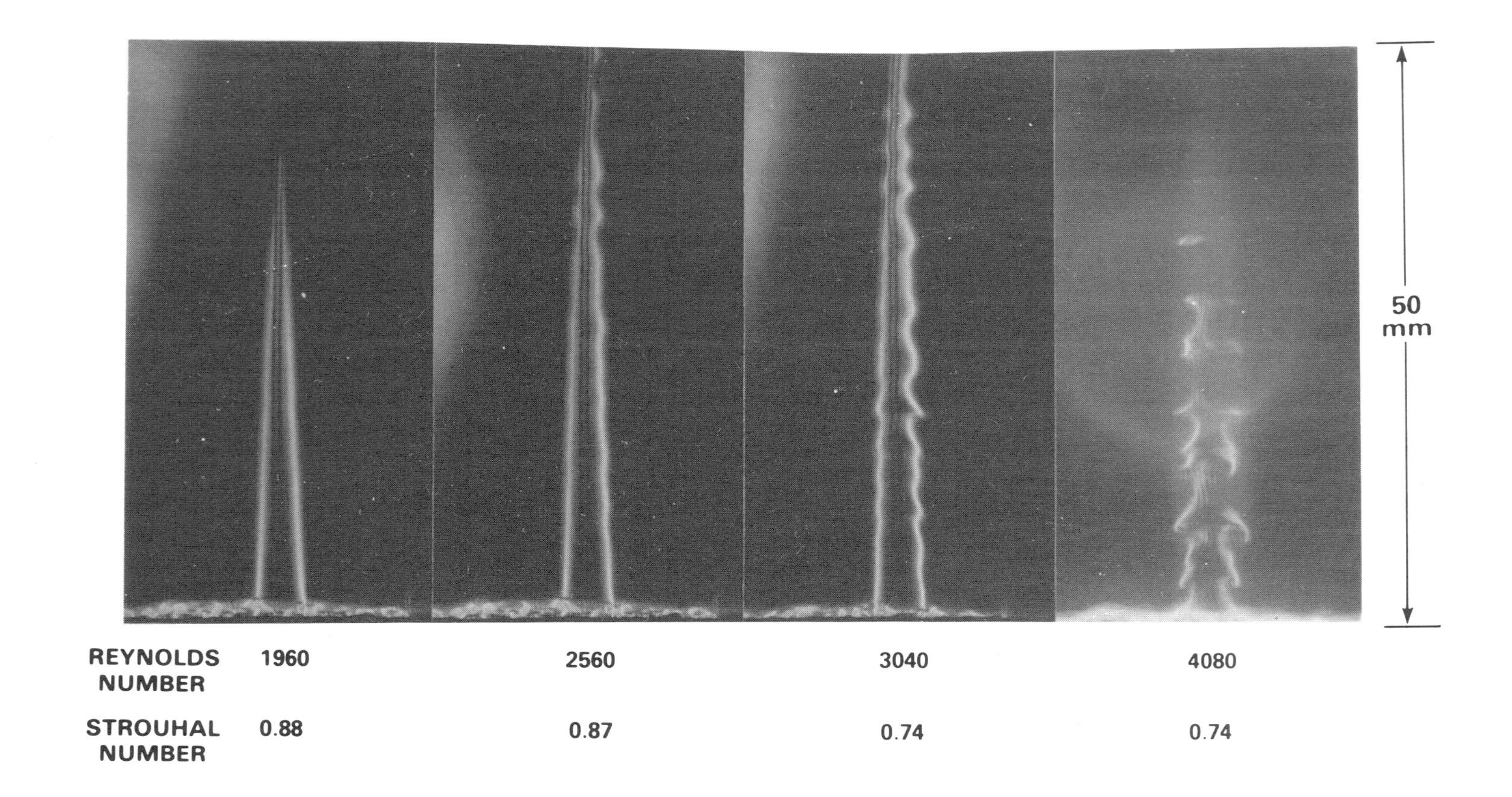
50 mm
REYNOLDS NUMBER
1960
2560
3040
4080
STROUHAL NUMBER
0.88
0.87
0.74
0.74

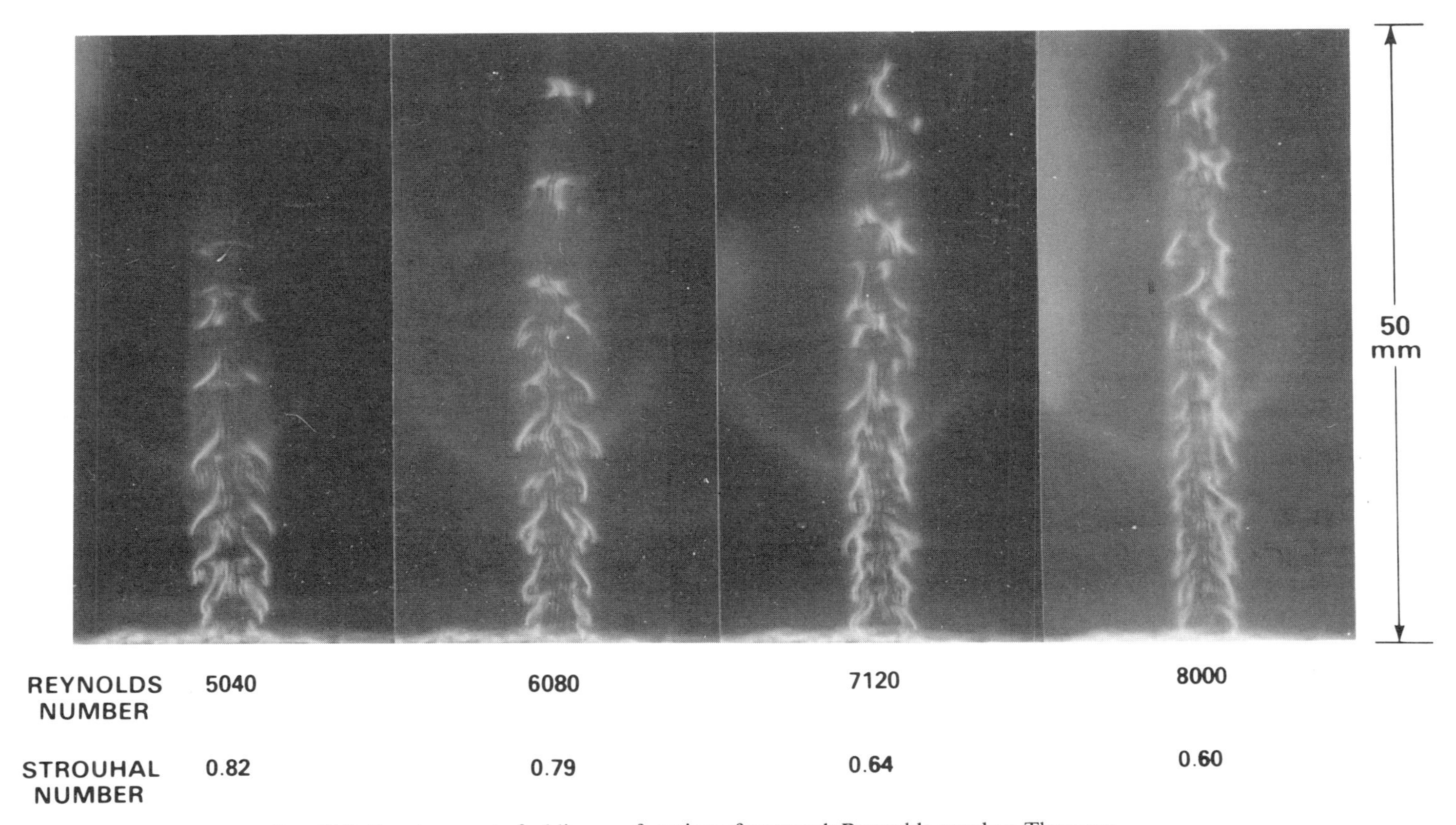

FIG. II-8 Development of eddies as a function of approach Reynolds number. These are instantaneous flash Schlieren pictures. To the eye or with a time exposure all of these flames appear as normal randomly turbulent flames. (After Fristrom, Linevsky, and Hoshall [unpublished work].)

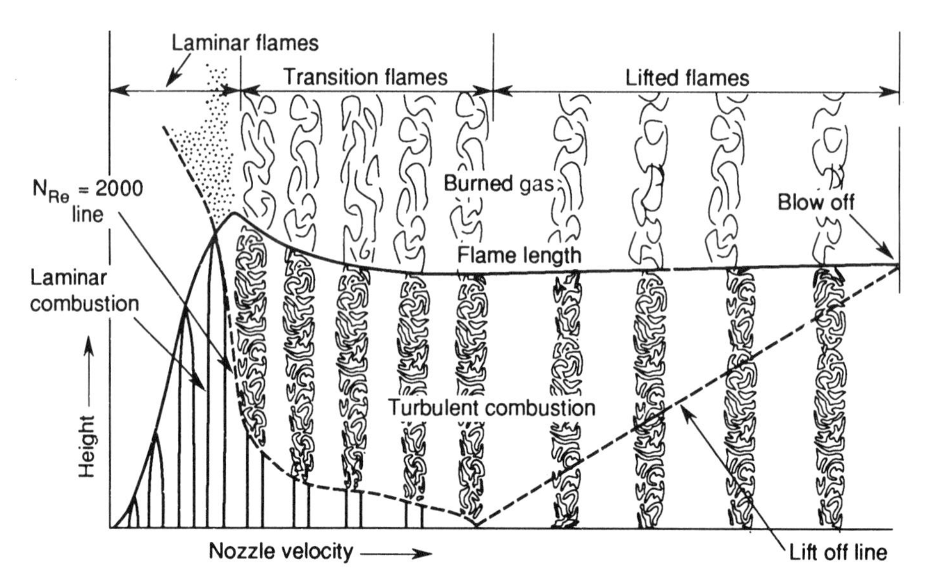

Fig. II-9 Transition from laminar to turbulent flames in open fuel jets (after Hottel and Hawthorne [1949], with modifications).

grow rapidly in scale to become weaker large-scale disturbances due to conservation of angular momentum. Some support of this is given by laboratory studies of acoustic interactions with flame fronts by Linevsky, Fristrom, and Smith [1987], which showed in the system studied that local burning velocity was constant in flame fronts down to curvatures comparable to the thickness of the flame front. Free burning acoustical disturbed flames could be quantitatively modeled using the normal laminar burning velocity and measured flame surface area. These flames differ from ordinary turbulent flames in being quasiperiodic rather than random. These measurements provide some support for simple geometric models such as those of Thomas [1986].

More study is required, but there is a substantial effort underway combining massive computational capacities and modern high-speed experimental diagnostics. One can, therefore, be optimistic about the solution of this long-standing problem.

External Forces

The interaction of flames with gravitational or with electric fields produces aerodynamic forces that affect the gross geometry with little effect on the microstructure.

The force of gravity on flames results from the difference in density of the hot and cold gases (Fig. II-10). This tends to make a freely propagating flame rise. In a stationary flame with flow vertically upward, gravity causes a decrease in the pressure drop across the flame front. This is unimportant except for low-velocity flames. There has been a renewed interest in microgravity combustion (Sachsteder [1991]; Kawakami et al. [1989]), and some shuttle experiments have been projected.

Electric fields accelerate ions in a flame, and since these are usually too low in concentration to disturb the main flame reactions, the principal effect is mechanical, the so-called electrostatic wind that distorts the shape (Fig. II-11). Magnetic fields appear

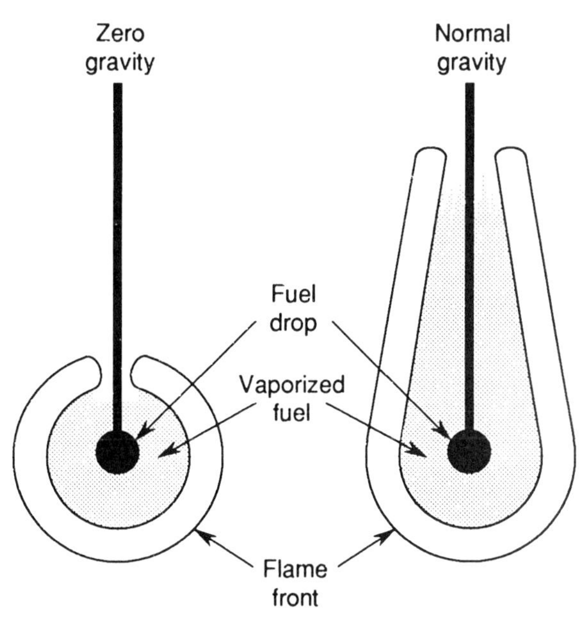

FIG. II-10 Schematic of buoyant effect of gravitational field on shape of droplet diffusion flames (hexane in air). (Sketched from Schlieren photographs by Kumagai and Isoda [1959].)

to have little direct effect on flames, but a field gradient exerts a force on any species that has a magnetic dipole, such as free radicals. Since they are important in flame propagation, it is quite possible that changes in flame microstructure and propagation rate could be brought about by sufficiently strong magnetic gradients. Magnetohydrodynamic generators employ combustion media to drive the flow and produce ions, but the combustion process itself is principally affected by the energy extracted by the generator.

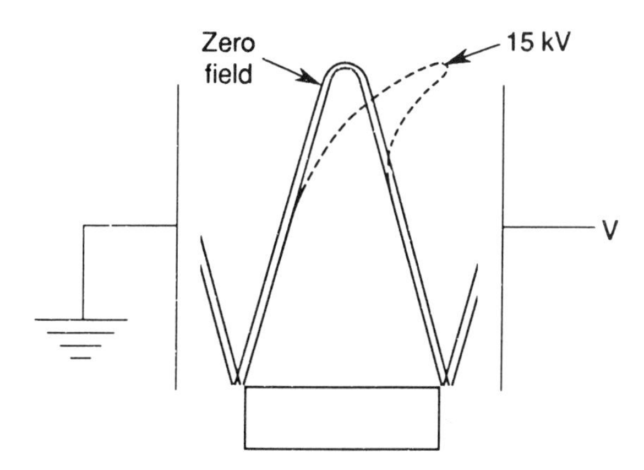

FIG. II-11 Effect of electric field on the shape of a propane–air flame seeded with sodium chloride. (Sketched from photographs by Calcote [1949].)

Heterogeneous Flames

The character of combustion is often controlled by the initial state or dispersion of one of the reactants. Six cases can be distinguished (Fig. II-3): gas–gas, gas–liquid, gas–solid, liquid–liquid, liquid–solid, and solid–solid. The most common flames involve gaseous oxygen since we live in an ocean of air, but it is important to realize that inverse systems exist. A pool of chlorine will burn in an atmosphere of hydrogen or methane, and nitric acid droplets burn in hydrocarbon vapors. Due to differing physical and thermal properties, each of these combustion systems has its own peculiarities, and the handling of these diverse systems is the art and craft of combustion engineering.

Gas–Gas Flames

Gases are completely miscible, and diffusion is moderately rapid at atmospheric pressure. Whether a system burns as a premixed or diffusion flame depends on whether the flame is initiated at a time short or long compared with diffusional mixing time. Gaseous diffusion flames were previously discussed at some length because combustion normally occurs in the gas phase no matter what the initial phases are. The only common exception is the combustion of carbon, where heterogeneous reaction must occur because of the low volatility of graphite. Coal combustion is more complex because adsorbed volatiles can occur in the solid graphitic phase.

Gas–Liquid Flames

Combustible liquids burn readily in reactive gaseous atmospheres. Reaction normally occurs in the gas phase. They form diffusion flames in air (excepting the limiting case of colloidal droplets). A distinction should be made between the *burning or regression rate* (Fig. II-12) and *spread rate* (Fig II-13). *Burning rate* is the rate of consumption of liquid per unit area burning. It is controlled by the available ignited surface, mode of heat feedback, and, occasionally, diffusional mixing. *Spread rate* is the rate at which the combustion region propagates across the liquid surface into nonburning regions. This is controlled by heat transfer from the burning region to vaporize the unburned fuel, which then mixes with air and is ignited by the adjacent fire (Williams [1976]). This complex series of events can be controlled by many factors: radiation, external wind, heat transfer in either gas or liquid, and, interestingly enough, in the case of quiescent liquid fuels, by the Marangoni effect, which is the flow induced by the change of surface tension with temperature (Glassman, Hansel, and Eklund [1969]). If the liquid is dispersed colloidally and evaporates in the preheat zone of the flame, the resulting flame will simulate premixed combustion. With hydrocarbon flames at atmospheric pressure, this behavior occurs with droplet sizes below a few tens of micrometers.

Liquid drops that lie between 3×10^{-3} and 0.1 cm burn as individual spherical diffusion flames with characteristic structure and combustion rate. The burning of fuel droplets can be modeled simply, assuming that the limiting process is liquid evaporation. A straightforward analysis (Spalding [1956]; Williams [1985]) indicates that the drop radius should decrease at a constant rate, with the constant being the ratio of

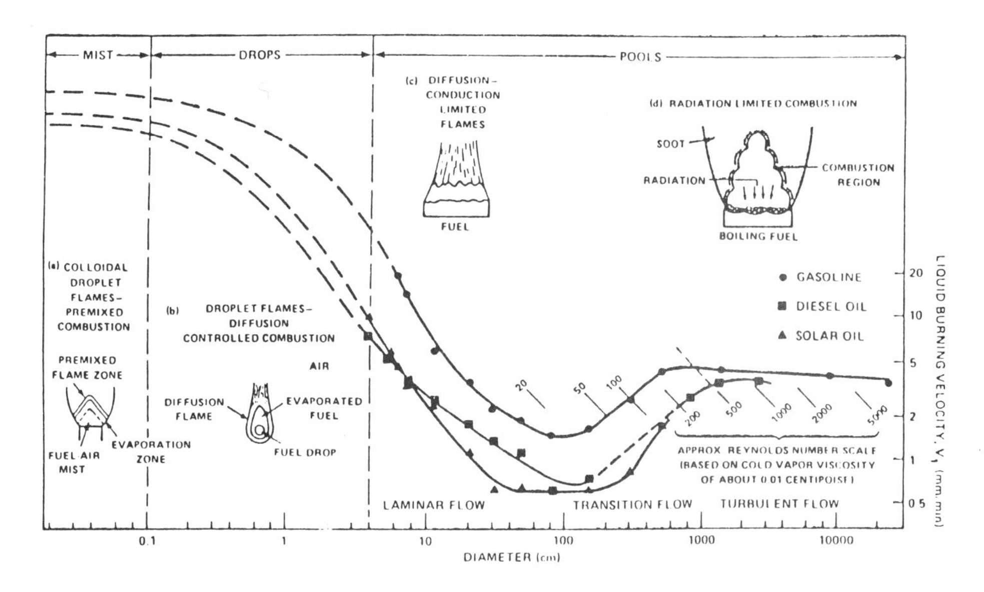

Fig. II-12 Effect of dispersion on regression rates of liquid fuels. (After Hottel [1960], with additions.)

heat liberated by combustion to the heat required to vaporize. This *B* number first introduced by Spalding is a good measure of the combustibility of a material. Its original use with burning liquids has been extended to include solids that vaporize, especially polymers (Table A-5).

Drop size is controlled by the competition between the forces of surface tension, gravity, and shear. They are usually less than a few millimeters in diameter, but this behavior can be extended by reducing gravity or use of a wick, which supports the

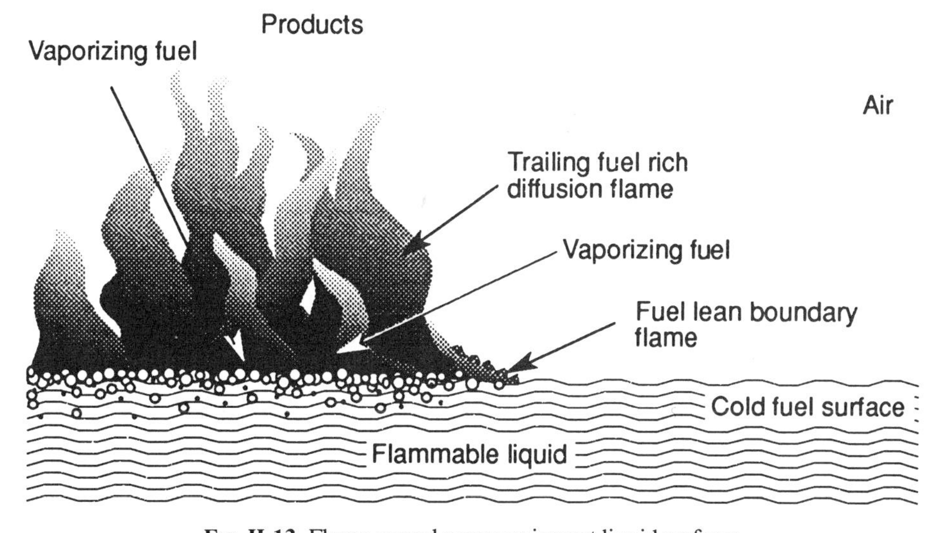

Fig. II-13 Flame spread across quiescent liquid surfaces.

liquid and favors air access. This is the oldest burner used by man and is exemplified today by the kerosene lamp and the candle. The latter is, of course, initiated from the solid phase, but the melting that occurs prior to combustion is normally not the limiting process. Larger aggregates form "pools" rather than drops. Geometry and size restrict air access so that hydrocarbon flames usually produce soot. Soot radiation then becomes the factor that controls the rate of evaporation and ultimate combustion of the fuel. Large pools in which the liquid is confined by a pan while the surface burns are called *pan fires*. This is one of the major classes of accidental fire, and a blazing oil tank fire is an awesome sight. With small pans, sufficient air diffuses in so that combustion occurs close to the surface of the liquid. Neglecting pan losses, such flames obey the same law as droplets. With larger pools, the rising column of hot gas prevents air from reaching the fuel until far above the liquid surface, and the temperature gradient is negligible at the gas–liquid surface. The dominant mechanism of evaporation is radiation. Sooty flames are good radiators; so the mass rate of burning per unit area will approach the upper limit set by radiation.

Very large pans produce turbulent flames that may be so badly underventilated that the effective flame temperature is well below the stoichiometric flame temperature. Large fires of this type are oxygen deficient and act like forest fires in being driven by winds and meteorological factors. In the extreme case where the fire is so large that it produces its own local weather and wind, it is called a *mass fire*.

Gas–Solid Flames

Mankind's first fire involved solid wood burning in air (Chapter I). As was the case with liquid fires, the character of combustion depends on the scale of initial dispersion and the mode of volatilization and final mixing (Williams [1976]) (Figs. II-14, II-15, II-16). The major exceptions are graphitic fuels, where surface reaction must play an important role since carbon is involatile at flame temperatures. The factors previously considered with liquid-fueled flames apply to solids if heat of vaporization is considered in the general sense.

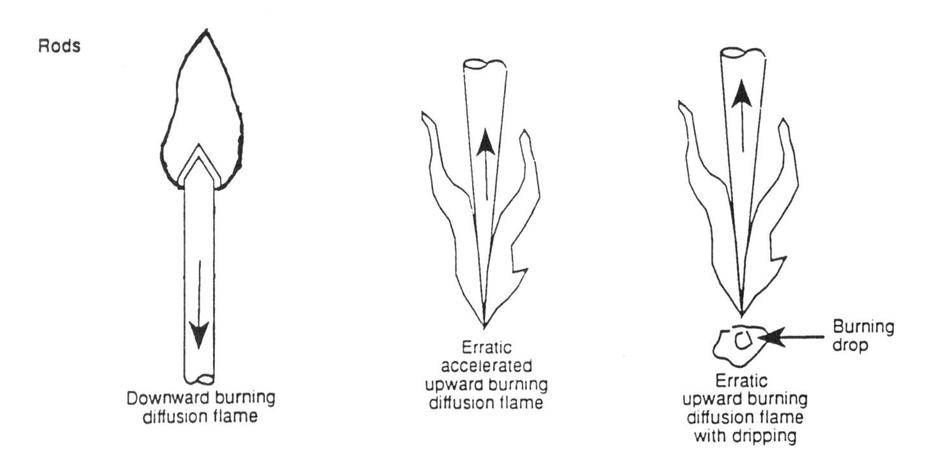

FIG. II-14 Flame spread on vertical rods.

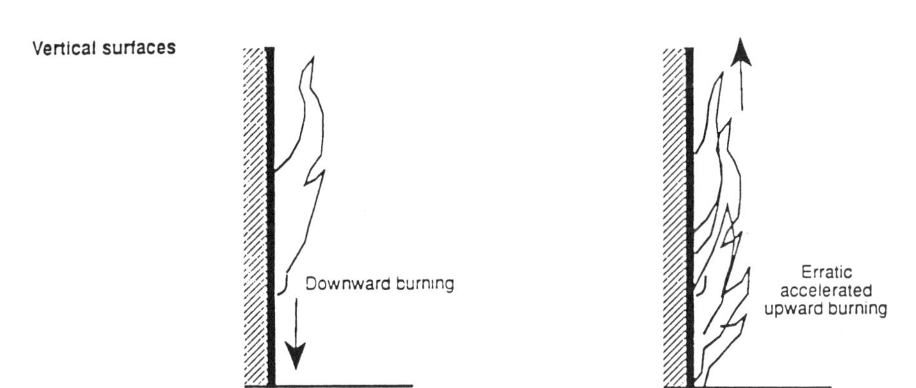

FIG. II-15 Flame spread on vertical surfaces.

Solids offer additional complications. They do not flow, and the initial dispersion size and shape are not changed by gaseous shear forces or by surface tension. Solids that melt before combustion can be considered liquid-fueled flames.

In liquids, dispersion is the controlling factor, and this is also the case with solids. Dusts burn differently from fibers, which burn differently from rods (Fig. II-14), which burn differently from sheets. Gravity forces liquids to burn as horizontal surfaces (excluding droplets, reduced gravity, and wicked flames). By contrast, solids can burn with surfaces at all angles. Due to buoyancy effects, there is a significant difference between upward and downward burning, and the combustion occurring on inverted surfaces such as ceilings is quite unstable (Fig. II-15). Other important factors include the presence or absence of charring and ash formation, the presence or absence of melting and dripping, roughness, edges and porosity, and presence or absence of moisture. Charring and ash formation reduce radiation losses from the solid surface and promote glowing combustion. Materials that melt can be self-extinguishing if they drip away from the flame. On the other hand, burning droplets can also spread fire. Wood burning is particularly complex, since it involves volatile components such as terpenes, pyrolizable components such as cellulose, char-forming components such as lignin, and ash-forming components such as salts. In addition, it has a porous structure, which allows moisture to transfer heat by evaporation and condensation through the pore system, acting as a biological "heat pipe." Additionally, cracking plays a

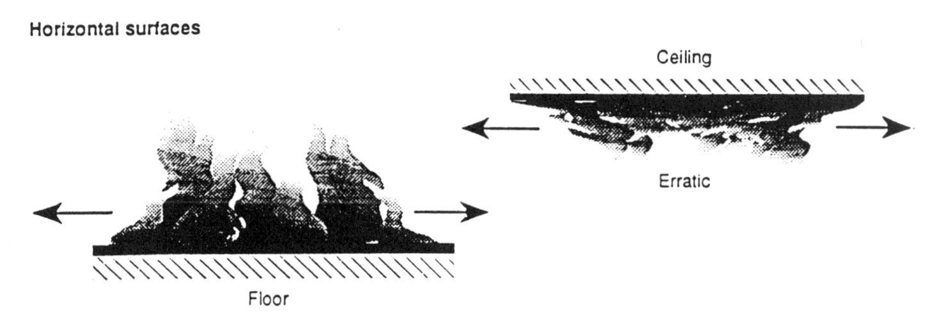

FIG. II-16 Flame spread on horizontal surfaces.

major role in exposing new surfaces, and ash formation acts as an insulator and radiation shield. A log of wood burns much more readily than a nonporous equivalent lump of plastic even though the heat of combustion of the polymer is substantially higher than wood.

Solid fires show critical size properties that are geometrically dependent. They are often limited by the diffusion of oxygen to the solid surface. Because of the importance of radiation losses, most burning solids will not propagate unless two or more reacting surfaces face one another so that radiation losses are reduced. This explains the difficulty of burning a single log or lump of coal, although this is made easier if a surface is coated with ash. In this case the emissivity is lowered while still allowing oxygen to diffuse to the surface. In some metal fires such as aluminum the dust will not ignite in an oxidizing atmosphere unless the melting point of the oxide is exceeded (Friedman and Macek [1963]).

Solid–gas fires are often oxygen deficient so that combustion depends on the rate of air introduction. This explains why such fires are favored by forced draft. The limit for this is reached when air movement is sufficient to maintain air immediately outside the laminar boundary layer. Further heat transfer then cools the surface and reduces the rate of reaction, possibly to extinguishment.

Many fuels that are not volatile enough for normal combustion burn as dust clouds. This represents a major industrial hazard. Their combustion is often augmented by radiative transfer with explosive results. Palmer [1973] has collected an extensive list of the explosive properties of approximately 400 industrial dusts ranging from acetamide to zirconium hydride. His book outlines the mechanisms and properties of dust combustion, safety hazard tests, and methods of suppression.

Many finely divided metals burn with intensely luminous flames. This results from the high temperatures and high emissivity of the solid products. Such flames find applications as flash lamps. Temperatures of metal fires are commonly limited by the oxide volatility. For this reason Zr is preferable to Mg in flash lamps because the boiling point of ZrO_2 produces a higher flame temperature and hence emission. The oxidizer is usually oxygen. Carbon and titanium dioxide powders are produced by combustion processes, and there has been a renewed interest in producing other high-purity powders (Ulrich [1984]).

Smoldering

Smoldering is a nonflaming form of combustion. Its chemistry and mechanisms differ from the normal mode. Surface reactions are significant, and beds or porous material are involved so that diffusion plays an important role in bringing in oxygen and removing products. Air flow can also play a role as, for example, in cigarettes. Usually radiation losses are minimized by ash or intervening porous fuel. Flame temperatures are usually low; only a small part of the fuel is converted by partial oxidation to complex products typical of pyrolysis. Propagation is slow and behavior erratic. It may slow and extinguish, continue at a constant rate or, under some conditions, accelerate into normal flaming combustion. It can be initiated by low-temperature sources and even by the bacterial processes that occur in moist hay and moss (Walker [1967]). A common example of spontaneous combustion is the direct oxygen reaction with unsaturated molecules in oily rags. Smoldering is favored by porous bed configurations where air can diffuse but heat is retained.

Condensed-Phase Combustion

Condensed-phase combustion systems, that is, liquid–liquid, liquid–solid, and solid–solid mixtures, are explosive if the products are gaseous. Examples are gunpowder and nitroglycerine. To be combustible, the system must either be initially in solution or fluidize easily. In the case of solid mixtures at least one of the reactants must have a low melting, boiling, or decomposition point. For example, gunpowder requires the melting of the sulfur and the decomposition of the KNO_3.

There are some combustion systems where both reactants and products are in the condensed phase. The most common example is the thermite reaction between compressed aluminum and iron oxide powders. There has been renewed interest in such systems in the Soviet Union, where they find applications in ceramic production (Merzhanov [1991]).

Characterization of Flames

Many measuring techniques have been used to characterize flame systems. These do not afford the details provided by microstructure studies, but they form a major segment of the combustion literature. Such studies are particularly useful because they offer easy and convenient single-parameter descriptions. Some of the more common techniques are summarized in the following. Data for some common fuel–air systems are collected in Table A-4 in Appendix A.

Thermodynamic Characterization-Heat Release and Adiabatic Flame Temperature

Often one is interested only in the heat that can be extracted from a combustion system or the maximum attainable temperature. These can be calculated with reasonable accuracy from thermodynamic data. The heat release can be calculated directly from the heats of formation of the fuel if the stoichiometry is known. For many purposes C–H–O compounds can be assumed to burn to completion in an excess of air so the maximum heat release per mole of fuel is given by the heats of combustion of water and carbon dioxide corrected for the heats of formation of the fuel. Table A-4 in Appendix A gives some values for common fuels with air. If the mixture is rich and the extraction temperature high, the stoichiometry cannot be written a priori but must be calculated from thermodynamic considerations. This is a straightforward but involved task since composition cannot be determined without knowing the temperature, and vice versa. Fortunately data are available (Stull and Prophet [1986]) and machine programs (Svehela and McBride [1973]) that allow full characterization of combustion systems that are either adiabatic or where the external losses can be specified.

Ignition and Extinction

The initiation and quenching of flames is important to the understanding of combustion. These are nonsteady combustion processes. In ignition the production of heat and active species production exceeds the steady-state value; so the process accelerates. In extinction the heat and radical production fall short of the steady-state values; so the process slows. Ignition proceeds toward steady-state combustion; the flame slows to extinction. Three unsteady states are associated with flow: the accelerating flames associated with the compressibility regime ($\upsilon_0 >$ Mach 0.3), the buoyancy instability of very low-velocity flames ($\upsilon_0 < 3$ cm/sec), and the onset of turbulence, which has been previously discussed.

Minimum Ignition Energy and Quenching Distance

From the standpoint of fire prevention, two important parameters are the quenching distance and the minimum ignition energy of a combustible mixture. The former is the minimal-sized channel that will allow a flame to propagate, while the latter is the minimum amount of energy that will ignite a mixture. These two parameters have been correlated (Lewis and von Elbe [1987]; Barnett and Hibbard [1959]; Brokaw and Gerstein [1957]).

Intuitively one might expect the quenching distance to be related to the thickness of a normal flame front. Flame thickness varies inversely with the burning velocity and pressure. An approximate relation was derived by Lewis and von Elbe [1987] correlating the minimum ignition energy with the quenching distance. This, in turn, is related to minimal flame size and flame thickness. If the logarithm of ignition energy is plotted against the logarithm of the quenching distance, a linear plot is obtained with a slope of -2.2. This is in rough agreement with their ideas. In older work sparks between needle points were used. More recently laser breakdowns have been used.

Ignition Temperature

Early flame theories made use of an ignition temperature defined as the temperature at which the initial reaction occurs. The concept fell into disfavor when it was found to correlate poorly with ignition temperatures measured in homogeneous systems. This point is irrelevant because ignition chemistry differs from flame chemistry (Warnatz [1980]). It occurs because ignition occurs under radical-poor conditions, while in flames radical concentrations are maintained at relatively high values by diffusion. It is a good approximation to say that there is a temperature in a flame below which the reaction rate is negligible. Measured ignition temperatures have a practical application in fire safety, and values for a number of common compounds are listed in Table A-4 in Appendix A. Several books discuss the problem in more detail and provide extensive tables (Palmer [1973]; Mullins [1957].

Ignition Delay

The time delay associated with rapidly mixed thermal ignition is related to the chemistry and is often used to characterize fuels (Mullins [1957]).

Premixed Laminar Flames

The parameter most commonly used to characterize flame systems is the burning velocity υ_0 (Fig. II-4). It may be defined as the velocity component of the *cold* gas *normal* to the one-dimensional flame front (Fristrom [1965]; Dixon-Lewis and Islam [1982]). In an overall flame it is the inlet volumetric flow divided by the flame area. This idea was pioneered by Gouy [1879] and extended by Michelsohn and Mache (Jost [1946], p. 381; Lewis and von Elbe [1987, p. 70]). In the case of a combustible mixture at rest, this would be the velocity with which a plane flame front propagates through the mixture. Mathematically, it is the eigenvalue of the one-dimensional flame equations and is a necessary parameter in characterizing flame structure. Methods for measuring υ_0 are discussed in Chapter IV.

A major consequence of constant burning velocity is that the geometry of premixed flames is determined by the inlet flow. Flames adjust their shape to accommodate local flow. The condition for steady state is that the propagation velocity must everywhere be balanced by the component of gas velocity normal to the local surface (Fig II-4). Axially symmetric uniform flow produces the common conical Bunsen flame (Fig. II-5), but many other geometries are stable. The wealth of variations can be appreciated by consulting Figure III-5, where burner systems are discussed. Flames are usually stabilized by attachment to some surface. This can be at the rim of a tube, the wake behind a rod, or baffle, where the flame interacts and transfers heat to the surface. Stabilization by stream tube expansion is also possible. The conditions for stabilization are discussed in Chapter III (see Fig. III-5). A more detailed discussion can be found in Lewis and von Elbe [1987]. There are, as might be expected, differences between this idealized model due to the finite thickness of the flame microstructure and finite diameter of the flow field or burner. Flame thickness is clearly imaged in the rounded tip found in the common Bunsen flame. The radius of curvature of the flame tip is related to the flame thickness. Even in this region the mass velocity, the product of density, area ratio, and velocity has the same value as in the rest of the flame. This occurs because the change in approach velocity is due to ducting in a region prior to reaction and thus has only second-order effects on the flame eigenvalue in a one-dimensional system. This is the case where the Lewis numbers of the fuel and oxidizer are nearly the same. If the diffusivity of fuel and oxidizer differ, lateral diffusion can change local stoichiometry in the region and through this the local burning velocity at the tip. If diffusion lowers the burning velocity, the tip region can be blown away. This occurs with fuel-lean light fuels such as hydrogen–oxygen or fuel-rich heavy fuels such as, for example, heavy-hydrocarbon–air flames. In the extreme this can produce very complex flame fronts called *cellular* or *polyhedral* flames. There is diffusional and heat exchange with the atmosphere at the boundaries of the burner or flow field. If the atmosphere is air and the flame is fuel rich, this can result in the formation of a secondary diffusion flame (see following section) surrounding the primary premixed flame. This is often observed in laboratory Bunsen flames because it gives the most stable mode of operation. If velocity gradients between the flame and its surroundings are high, chaotic flow often results, and turbulent flames are formed (see following section). Even in this case the concept of a constant local burning velocity on the microscopic level provides a useful model.

Effects of Heat Extraction

Premixed flames are often stabilized on a porous metal plug with inlet flow below the normal burning velocity. Heat extraction by the plug reduces the burning velocity until there is a match with local inlet flow. The resulting flame conforms to the porous surface, which can be of any shape. Burning velocity can be derived by determining the flow where the flame first separates from the plug.

If heat extraction exceeds a critical value, the flame is extinguished. This is the principle of the Davy [1840] safety lamp and explosion suppressors. One can consider these flames as equivalent to precooled flames with the same enthalpy. Burners of this type have been used to measure burning velocity (Botha and Spalding [1954]). Weinberg and Hardesty [1974] discuss the effects of heat feedback on extinction limits.

Diffusion Flames

Diffusion flame geometry is controlled by diffusion, as the name suggests. For this reason gross measurements do not yield direct information on flame chemistry. Microstructural measurements yield information similar to that obtained from premixed flames, although interpretation requires a two-dimensional treatment to include effects of lateral transport (Dixon-Lewis and Missaghi [1989]). Diffusion flames are often used in inhibition and extinction studies since they more nearly simulate some natural fire environments.

The most common method for the characterization of diffusion flames is the "flame strength" method developed by Potter and Butler [1961] (Fig. II-17). The system consists of two opposed jets, which, when flows are balanced, yield a relatively flat flame. If the flows are increased, it is found that beyond a critical flow, a hole appears in the center of the flame sheet. This is interpreted as being the point where residence time falls below the minimum combustion time. Results appear to be independent of flow character. The turbulent version of this apparatus appears related to the Longwell–Weiss stirred reactor [1955] and the Karlovitz "flame stretch concept" [1954]. The design of the burner is discussed in Chapter III.

Turbulent Flames

The most prominent feature of turbulent flames is the apparent increase in burning rate. This is evidenced in a higher volume efficiency, a shorter burnout time, or a shorter burnout length (Fig.II-7), depending on the geometry and flow conditions in the system under study. In the laminar flow regime, the flame height increases proportionally to the inlet velocity and to a crude approximation can be inferred from flame height measurements. Around a Reynolds number of 2300 the jet begins to be turbulent. The region of initiation can be a lateral shear region, as occurs in turbulent fuel jets (Fig. II-9), or at an edge, as occurs in an orifice flame (Fig. II-8). The critical dimension differs in the two cases being jet length in the first and diameter in the second. The velocity for critical Reynolds number differs in the two cases, and the geometry and scale of the turbulence reflects these differences. Turbulent flames show a characteristic brushlike appearance and are shortened significantly. This shortening

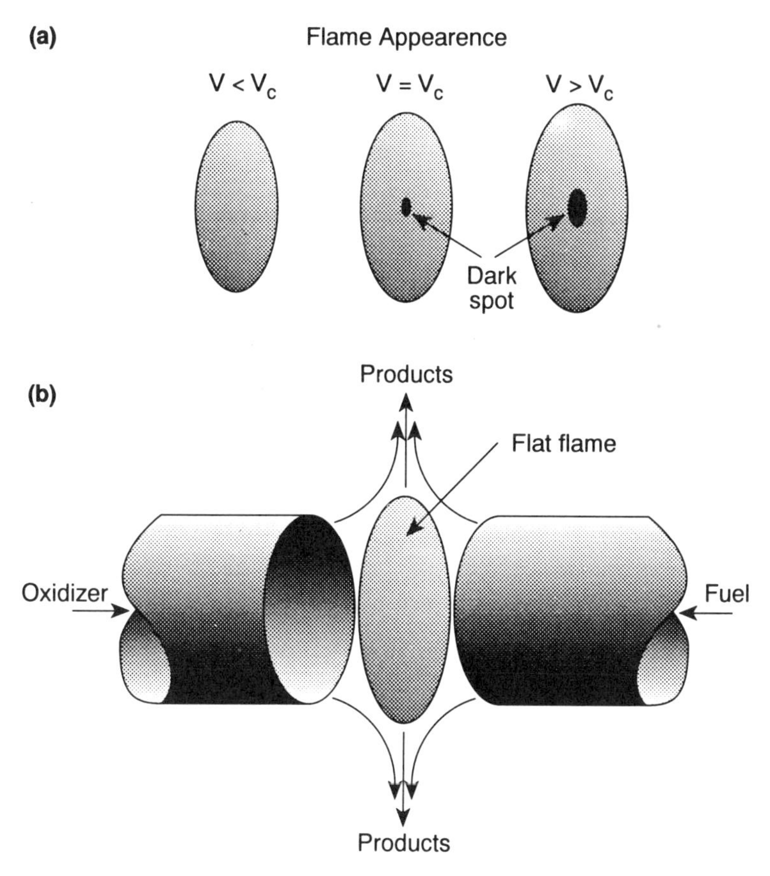

FIG. II-17 Flame strength of opposed jet flames. (a) Flame appearance with velocity near critical point. (b) schematic of apparatus. (After Potter and Butler [1963], with modifications.)

can be interpreted in terms of an effective "turbulent flame velocity" by defining the turbulent burning velocity in the same manner as laminar burning velocity:

$$\begin{aligned} \upsilon_T &= \upsilon_g \sin \theta \\ \upsilon_T &= \tfrac{1}{2}\, \upsilon_g D/H. \end{aligned} \tag{2–3}$$

Turbulent burning velocity inferred in this or some similar manner is usually correlated with local Reynolds number. This naive approach, which is necessary in the absence of information on flame front geometry and area, can lead to biased interpretations, since the local propagation may still be the normal laminar value with a complex geometry (Fig. II-7). It is clear that wherever possible turbulent flame velocity measurements should be made on the basis of flow per unit area of flame front. This approach is not always possible, but, in simple flames such as the orifice burner, where the eddies are uniform and shed periodically, this can be visualized (Linevsky, Fristrom, and Smith [1985]) (Chapt. V).

Turbulent flames can be described in terms of probability density functions measured using randomized pulse diagnostics, but this approaches a detailed structural description that is inappropriate for this chapter.

A recent theory by Thomas characterizes the flames using a simple "wrinkled flame" model [1986]. This has received some experimental support (Linevsky, Fristrom, and Smith [1985]), however, and one can visualize cases where the flow might distort the microstructure.

Limits of Flame Propagation

A few words are in order as to the range of systems that undergo flame reactions. These include the chemical limits, which have been discussed, and the limits of stoichiometry, temperature, velocity, and pressure. Typical stoichiometric limits for hydrocarbon–air systems are shown in Table A-4 in Appendix A. The classic work of Jones, Baker, and Miller [1937] is discussed in some detail by Lewis and von Elbe [1987]. Recently there has been a renewed interest (Blint [1989]) in this practical safety field. Questions of chemical limits have been addressed by several authors (Fristrom and Van Tiggelen [1979]; Law and Egolfopoulos [1991]; Gann [1975a]). Yamaoka and Tsuji [1989] have developed an elegant stagnation flow technique for measuring extinction.

Thermal Limits

The temperature range of flames is wide. Gas-phase flames can be initiated as low as 90 K or as high as 1000 K. Final temperatures can be as low as 500 K, or as high as 5000 K. Normally, however, the incoming gas is around room temperature, with temperature rises of 1000–2000 K. The low-temperature limit for premixed flames is usually dictated by volatility of reactants.[2] The upper limit is usually controlled by the dissociation of the products. This is confirmed by the observation that the hottest flames are not produced by the systems with the greatest heat release, but by those whose final products are the most difficult to dissociate. For example, the (room temperature and atmospheric pressure) hydrogen–fluorine flame with a heat of reaction of 65 kcal per mole of H_2 produces a maximum temperature of 3000 K, limited by the dissociation of HF. On the other hand, the cyanogen–ozone flame with a heat of reaction of only 45 kcal per mole of C_2N_2 has a maximum temperature of 5000 K, because the products, CO and N_2, are more stable (see Table A-10).

Pressure Limits

As far as pressure limits are concerned, flames have been burned over almost a millionfold range in pressure. The lowest reported value is below 10^{-3} atm, while the highest studied is over 100 atm (Diederichsen and Wolfhard [1956]). The practical upper limit appears to be set by the courage (or foolhardiness) of the investigator, while the lower limit is set by the size of the reaction vessel and available pumping speed (see Chapter III).

2. There is also a theoretical low-temperature limit for bimolecular reaction-driven flames [Hirschfelder, Curtiss, and Bird (1964)]. This is not usually a factor.

A flame consisting of bimolecular reactions should have a structure scaling directly with molecular mean free path. That is, flame thickness should vary inversely as the pressure. The time between collisions changes with pressure, but the flame requires a constant number of collisions to go to completion. Two factors modify this simple picture. The first is that three-body reactions are required for the attainment of equilibrium in any flame that involves radicals, as most flames do. At atmospheric pressure, bimolecular collisions are a thousand times more frequent than three-body collisions, and, therefore, the most rapid flame reactions are bimolecular. Flames usually separate into an initial narrow fast reaction where bimolecular reactions dominate, and a wide, slow, secondary reaction zone where termolecular reactions reestablish equilibrium in radical concentrations. Below a tenth of an atmosphere, the three-body recombination reactions are so slow that it is more realistic to consider flames with the bimolecular reactions going to completion rather than to true thermal equilibrium. Most low-pressure flame studies have been carried out in vessels that were too small to allow full thermal equilibrium. This does not prevent the flame from propagating, but should be considered when considering the pressure dependence of flame velocity. At high pressures (thousands of atmospheres), termolecular (and higher) collisions become dominant. This changes the character of combustion reactions.

A second complicating factor is energy loss by radiation. This is controlled by the ratio of excited species lifetimes to the time between molecular collisions. The former is constant, while the latter is inversely proportional to pressure. Therefore, the probability that a given excited species in a flame will radiate before a collision induces a radiationless transition is inversely proportional to pressure, and radiation loss in a flame increases with decreasing pressure. The effect is usually minor because only a small fraction of the molecules are usually excited. If solids are produced, this may become a factor, as suggested in the acetylene decomposition flame (see Chapter XI).

Velocity Limits

Velocity limits for flames occur due to compressibility, buoyancy, and Reynolds number effects (see Chapter III). Laminar flames tend to break spontaneously into turbulent flow near the critical Reynolds number of 2000. This makes it difficult to burn high-pressure laminar flames because of the small burner size required to avoid turbulent flow. However, if minimal burner diameter is used (i.e., the order of ten flame front thicknesses), laminar propagation should be feasible at any pressure where bimolecular reactions control the thickness. If the flow is laminar, flames of any velocity would be possible up to the speed of sound, where the system becomes a detonation. In practice, flames with burning velocities exceeding 10 m/sec appear to be unstable and develop into detonations. The reason for this behavior is that this is the velocity regime where compressibility effects become appreciable. Since compression speeds reaction relative to the initial flow by a factor proportional to the density, any flame that has a greater than linear dependence of propagation rate with respect to pressure will be accelerated by the compressible flow, ultimately reaching the detonation velocity.

On the low-velocity side, the gravitational effects of buoyancy will destroy a flame front if the buoyancy acceleration exceeds the acceleration due to combustion. This is described by the Grashoff number. These forces are responsible for differing limits of propagation in upward and downward directions.

Characterization of Condensed Systems

The combustion of condensed systems in air is important to fire safety, and a multiplicity of methods have been devised to characterize the flammability of solids and fluids. There are tests, official and semiofficial, for many materials, but as Emmons [1971] pointed out, the tests are empirical, and correlation is often poor. The subject is too complex for inclusion here. The reader is referred to specialized texts (Palmer [1973]; Mullins [1957]; Products Research Committee [1980]). One reasonably understood method is the oxygen index (see Table A-5), which is an extinction measurement. Other tests include ignition, spread rates, penetration and endurance heat release, smoke production, and flashover.

Practical Combustion Systems

There are a myriad of combustion devices. Their design depends on application. Some of them are research tools; others are practical. They can be classified according to use. If heat is the desired output, the system is called a *furnace.* If light is desired, it is called a *lamp.* Gas lights are now primarily elegant ornaments, but candles are still in common use, and the Weisbach mantle still serves campers. If mechanical motion or power is the desired output, the system is called an *engine.* These are classified according to the cycle used to extract the energy.

Combustion processes are used to manufacture some chemicals. This is not common because the final combustion products are often uninteresting and the intermediates difficult to extract. Carbon black and titanium dioxide powder for paints are prepared commercially by combustion processes. Proposals have been made to prepare acetylene by combustion, but the method apparently was uneconomic. In the past decade flames have been used for the production of high-purity powders for catalysis (Ulrich [1984]); so the field is expanding.

Research Reactors

A number of reactors are used to study combustion reactions. We will mention a few of these. They use techniques for rapid mixing, rapid homogeneous ignition, or some combination of techniques.

Stirred Reactors and Blowoff Limits

The important parameters of a combustion system can often be derived by an overall treatment, ignoring structural considerations. This approach is not employed in this book, since we are interested in details of mechanism.

In a highly stirred reactor (Fig. II-18a) the incoming air and fuel are strongly mixed while burning and the reacted gas is ejected. The system is characterized by a residence time and average properties, as was the opposed jet flame of (Fig. II-17). A relation was

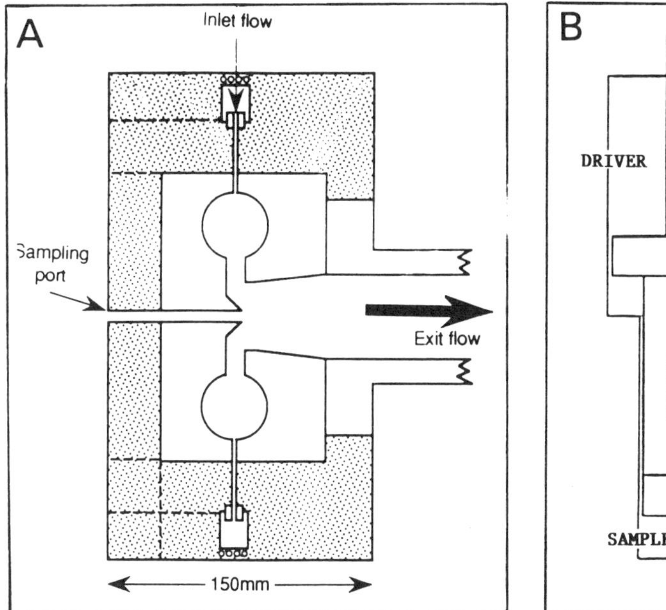

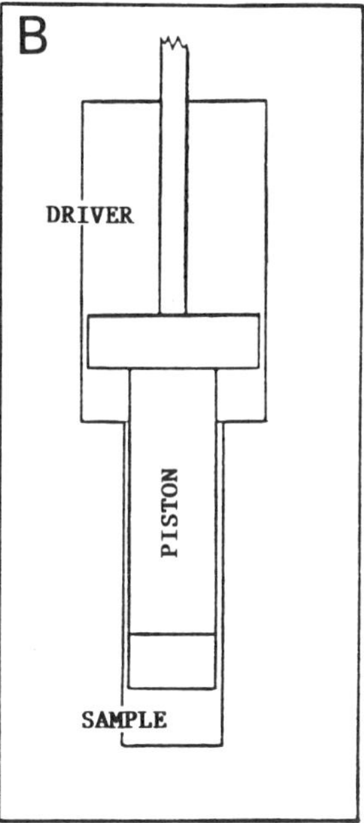

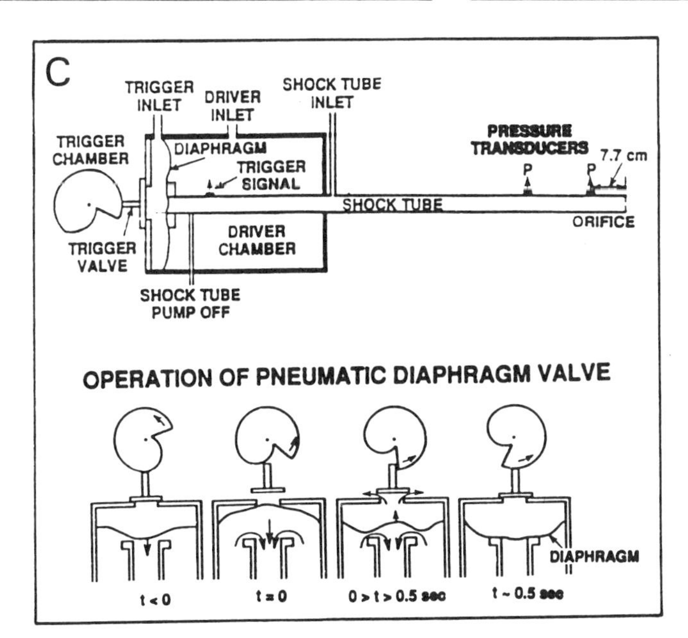

FIG. II-18 Research reactors. (A) Schematic of torroidal well stirred reactor. (After Nenninger, Kridiotis, Chomiak, Longwell, and Sarofim [1985] with modifications.) (B) Adiabatic compression. Rapid initiation of reaction is obtained by compression with the piston. Reaction is followed as a function of time under constant-volume conditions. (C) Repetitive shock tube. Rapid initiation is obtained through the energy carried by the shock wave formed upon the opening of the diaphragm separating the high-pressure driver section from the lower-pressure study section. The insert shows stages in the opening and closing of the diaphragm valve. (After Fristrom, Hoshall, Linevsky, and Vandooren [1993].)

derived by Avery [1955] connecting fuel injection rate with average temperature, concentration, and the overall reaction kinetics. There is an optimum combustor temperature at which maximum throughput occurs. Beyond this point, blowout occurs. The method has been applied to hydrocarbon and air combustion by Longwell and Weiss [1955] to obtain overall rate data for ram jet design. A recently improved toroidal version of this device will be discussed in Chapter V.

Adiabatic Turbulent Flow Reactors

A rapid flow turbulent mixing isothermal reactor has been used by Glassman et al. [1969] and others for following hydrocarbon combustion. High-speed, turbulent flow reduces the complicating effects of diffusion to a minor factor by stretching the flame over many centimeters. Combustion is stabilized by using a fixed injection point in a furnace maintained above the ignition temperature. The method provides mechanistic information for the understanding of complex flame systems.

Adiabatic Compression Machines

One method of obtaining rapid initiation of a reaction is an old device that depends on adiabatic compression by a piston. High transient temperatures and pressures are possible. It evolved from the internal combustion engine and was first used scientifically by Tizard and Pye [1922]. The device was improved by Jost and others. It has been used to study hydrocarbon–air chemistry and ignition and knock problems in engines. The apparatus consists of a cylinder of reactant gases and a compressing piston (Fig. II-18b). Compression can be accomplished by compressed air or a falling weight. The method and problems are discussed by Jost [1949].

Shock Tubes

Another popular device for obtaining transient high temperatures is the shock tube, which was developed in 1899 by Vielle. It consists of a long straight tube divided into two sections by a frangible diaphragm. The two chambers are filled with gas. When the diaphragm is ruptured, a shock wave passes into the low-pressure working section. This produces temperatures up to 20,000 K with optimal geometry, and magnetically driven temperatures up to a million Kelvin are possible. This goes well beyond the range of chemical interest. Ideal compression conditions are calculable, and exact conditions can be inferred from velocity measurements. The design and use of shock tubes has been discussed in detail in several texts (Green and Toennies [1964]; Gaydon and Hurle [1963]; Bradley [1962]). These devices have proved useful in the study of hypersonic aerodynamics and high-temperature chemistry. As mentioned, the difference between adiabatic compression and shock tube heating lies in the flow, which is subsonic for adiabatic compression and supersonic for shock tubes. Due to the mass flow the density changes in adiabatic compression devices. It should be pointed out that at the end of the shock tube the wave is reflected. This compression wave raises the end gas temperature even higher. This effect is absent in adiabatic compression devices. Recently Fristrom, Hoshall, Linevsky, and Vandooren [1993] have introduced a repetitive shock tube that greatly improves the duty cycle (Fig. II-18c).

Flash Photolysis and Low-Pressure Flow Tubes

Chemical kineticists have developed a number of techniques for studying rapid radical reactions (Kaufman [1982]). Flash photolysis is a technique in which the reaction is initiated by radicals produced by flash photolysis. Norrish [1965] received a Nobel Prize with Porter for this technique. It has been used to study a number of combustion reactions. More recently laser shock tube with detection using laser fluorescence has become a standard technique for studying atom and radical reactions (Davidson, Chang, and Hanson [1989]). Morely [1989] introduced a perturbation method for studying combustion reactions.

Low-pressure flow tubes using discharge tubes as atom and radical sources and taking advantage of the rapid diffusion mixing possible at low pressure have been a major tool in studying radical and atom reactions, particularly in the low to more moderate temperature range. Kaufmann [1982] offers two perceptive reviews of the field.

Furnaces

A combustion device operated principally for its heat generation is called a *furnace*, which was among the earliest combustion devices. The Minoans and the Romans had well-developed furnaces. Modern furnaces come in a multiplicity of forms, and new systems are still being developed, pulsed combustion gas furnaces being a recent example. The subject is too involved for the present text; so the interested reader is referred to the book of Thring [1962] and the more recent literature.

Engines

Engines are combustion devices operated to generate power. Combustion engines provide an important source of power for this society, and if heat engines such as steam turbines are included, combustion is dominant.These systems lie beyond the scope of the present book, but surveys can be found in engineering texts.

Jet Engines—The Gas Turbine and Ram Jet Cycles

Jet engines operate using atmospheric air as the oxidizer, and the combustion gases are the engine's working fluid. To obtain efficient operation it is necessary to compress the incoming air so that combustion occurs at higher pressure. This reduces the required combustion chamber size and increases efficiency. The two major methods used are: a turbine input driven by the exhaust (turbo jets), and the compression provided by the shock wave associated with supersonic flight (ram jets) (Fig II-19). Ram jets can operate subsonically as low as Mach 0.6, but they are only efficient supersonically. Typical operation for ram jet missiles is at Mach 2.5.

Rockets

In rockets the fuel and oxidizer are self-contained so an external atmosphere is not required. This allows their use in space flight. Because oxidizer as well as fuel is carried, their weight efficiency is poorer than air-breathing engines. Two general types exist—liquid fueled and solid fueled. The former pumps fuel and oxidizer into a combustion chamber, exhausting products through a nozzle (Fig. II-20); the system is clearly

(A)

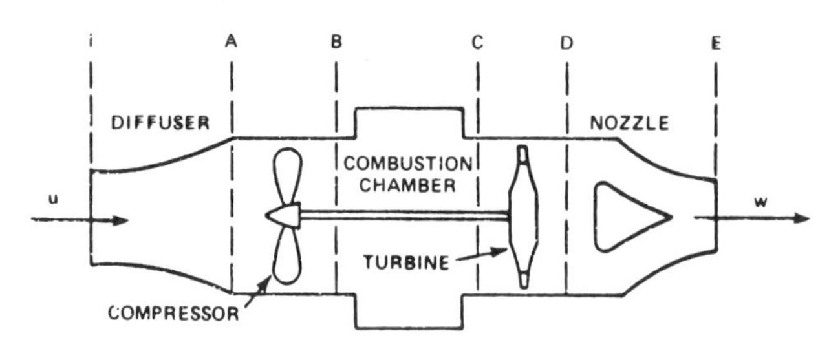

SCHEME OF TURBO-JET ENGINE

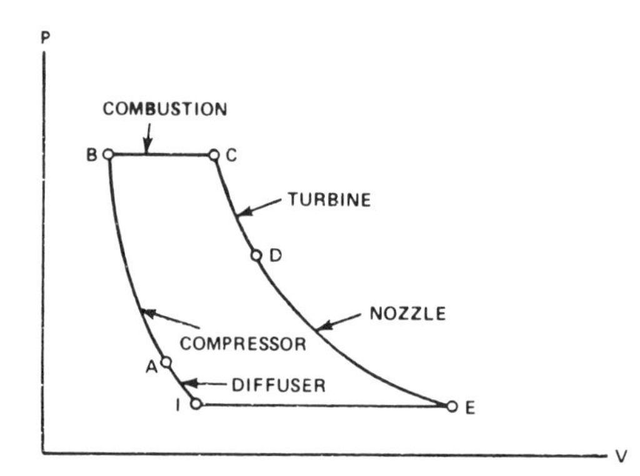

PRESSURE AND SPECIFIC VOLUME IN IDEAL TURBO-JET CYCLE

(B)

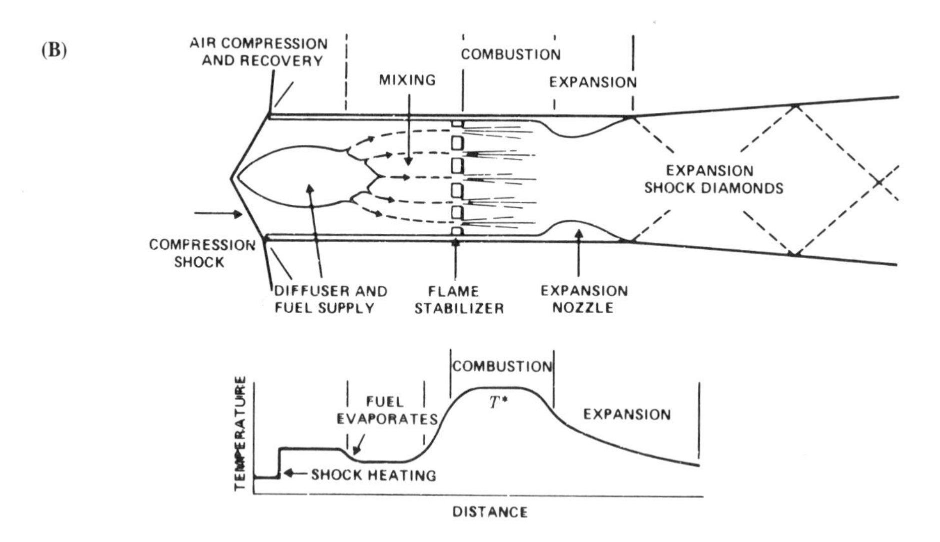

SUPERSONIC RAMJET

Fig. II-19 Jet engines. (A) The turbojet with compression by a turbine. (B) The ram jet with compression from the bow shock. The ram jet is only efficient in supersonic flow, typicaly around Mach 2.

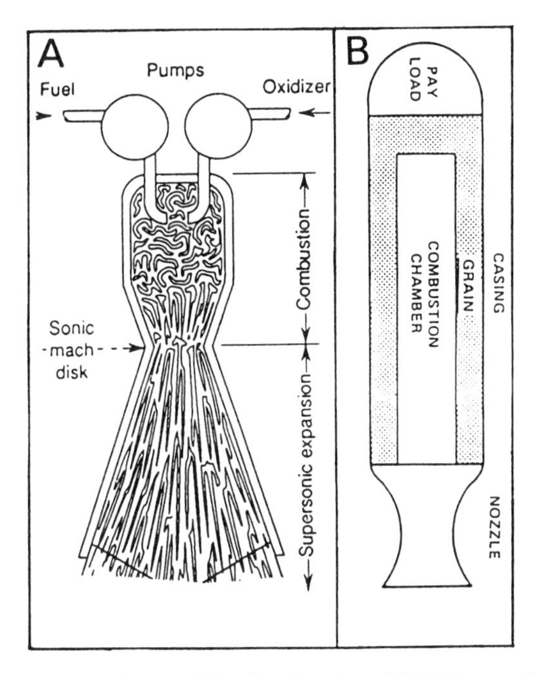

FIG. II-20 Rocket engines: (A) The liquid rocket; (B) Solid propellant rockets.

related to the stirred reactor. Bragg has surveyed liquid rockets [1962]. Solid-fueled rockets consist of a charge that lines the combustion chamber and a nozzle that can be as simple as a hole. Obtaining stable combustion is an art, and many special mixtures and configurations are employed to provide the reliability required for military operation.

Internal Combustion Engines

The common internal combustion or Otto Engine, named after its inventor, is often called a gas engine because the original fuel in English mines was town gas. It commonly employs four cycles (Fig II-21), but two-cycle engines also function. A typical gasoline is a mixture of 25 percent benzene, 25 percent octane, and 50 percent dodecane by weight. Modern engines operate at high compression ratios.

A simple simulation of the thermodynamics (Starkman, Patterson, and Taylor [1961]) of this cycle assumes:

1. Pressure remains above ambient pressure during the intake stroke;
2. The fuel is fully vaporized and that there is burned gas from the previous cycle;
3. The combustion products remain in chemical equilibrium.

Figure II-21 shows specific volume, pressure, and temperature in the engine at five points in the cycle: (1) the start of the compression stroke; (2) the end of the compression stroke; (3) the start of the expansion stroke; (4) the end of the expansion stroke;

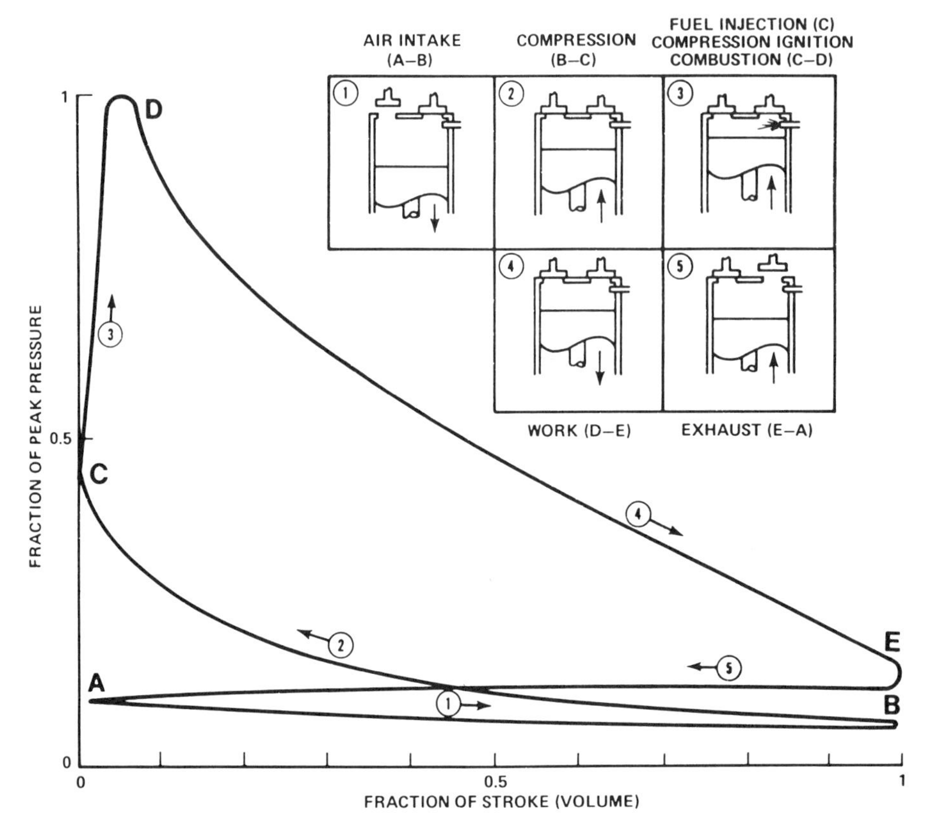

Fig. II-21 Internal combustion engines. Comparison of the Otto and Deisel cycles.

and (5) the end of isentropic blowdown to ambient pressure. This represents an idealization. More realistic calculations allow the fuel to vaporize partially in the intake manifold and permit residual exhaust gases to remain in the engine from cycle to cycle. The diesel cycle is similar except that fuel is injected by a pump during the compression cycle and ignition is by adiabatic compression rather than by spark. Literature is extensive and elementary treatments can be found in engineering texts.

Other Cycles

A number of other engine cycles have been demonstrated and have found limited use. Among these are the Stirling cycle, where the working fluid is confined and heat is applied externally. Another is the Wankel Engine, which is a rotary internal combustion engine. The interested reader is referred to mechanical engineering texts.

Unwanted Fires

Unwanted fires, their avoidance and extinction, are an important societal problem, the major areas being ignition, spread, and extinction. An extensive literature exists

(Hilado [1973]; Madorgsky [1964]; Lyons [1986]; Wall [1972]; Products Research Committee [1980]).

A sustained fire requires three elements: fuel, air, and a mechanism for flame propagation or continuing ignition, which firemen identify with heat. As they are taught, the removal or separation of any of these elements will extinguish a fire. In the early stages of a fire, direct extinguishment of the flame or ignition source is easy, but fires build up exponentially, and direct extinction becomes impractical above some critical size. Beyond this time separation becomes the preferred method. Some common extinguishment methods are: (1) water, which removes heat and interferes with air access, (2) carbon dioxide, whose extinction mechanism combines heat extraction with a blanket of heavy inert gas, which separates air and fuel, and (3) chemical extinguishers such as Freons or powdered salts, which interfere with flame chemistry. We will consider the chemical aspects of flames in some detail in this book. Some powders are very effective on oil fires, but because of the ease of reignition a secondary suppressant is required. One such "dual agent" used by the U.S. Navy is a combination of "purple K" (potassium bicarbonate), which suppresses the flame, with a silicone surfactant, which covers the fuel surface and suppresses vaporization.

A fully developed fire requires the extraction of too much energy to make direct extinction practical. Massive fires are fought by isolation or separation of fuel and oxidizer. Firemen try to isolate a fire and let it go to completion while saving surrounding rooms or in extreme cases saving the surrounding buildings. With forest fires the formation of "fire breaks" using bulldozers, back firing, or simply manual removal are common methods. In ships where airtight metal bulkheads exist, fires are extinguished by sealing off the area from air and waiting for the fire to cool down. This method should not be used blindly, because if the cargo contains oxidizer as well as fuel, it does not make an effective separation. This hazard exists with some otherwise innocuous materials such as ammonium nitrate fertilizer. This traditional method of fire fighting was used with disastrous results on a ship fire in Texas City some years ago, and the resulting explosion wiped out most of the harbor.

A major complication with building fires is that they occur in enclosed spaces. This introduces many problems and makes air access a primary variable. It is instructive to observe that a carbonaceous fuel requires twenty times its own weight of air for complete combustion. The average room containing 2000 cubic feet has only enough oxygen to burn eight pounds of wood, which is half the weight of a kitchen chair. The sealing-off method is not commonly used for house fires because rooms cannot be easily hermetically sealed and floors and walls are often combustible. In addition, there is always the possibility that a room would be sealed off with someone inside. Life saving is the first charge of firemen, with property salvage only a secondary consideration. A typical situation is illustrated Figure II-22(a), which illustrates stages in fire buildup in a room with an open door. In this situation an object is ignited in a room with other combustible objects but with noncombustible walls and ceilings. In the initial stages the fire burns freely, but the rate of involvement is slowed as the soffit (the ledge above most doors) traps the hot, rising, expanding burned gases at the ceiling, forcing air out of the door. This limits the available air supply until the hot gases reach the soffit level. At this time they begin to cascade out in an inverted "waterfall." This induces an inward flow of fresh air through the door, which speeds the combustion. During the entire period radiant heat has been trapped in the room, and wall temperatures

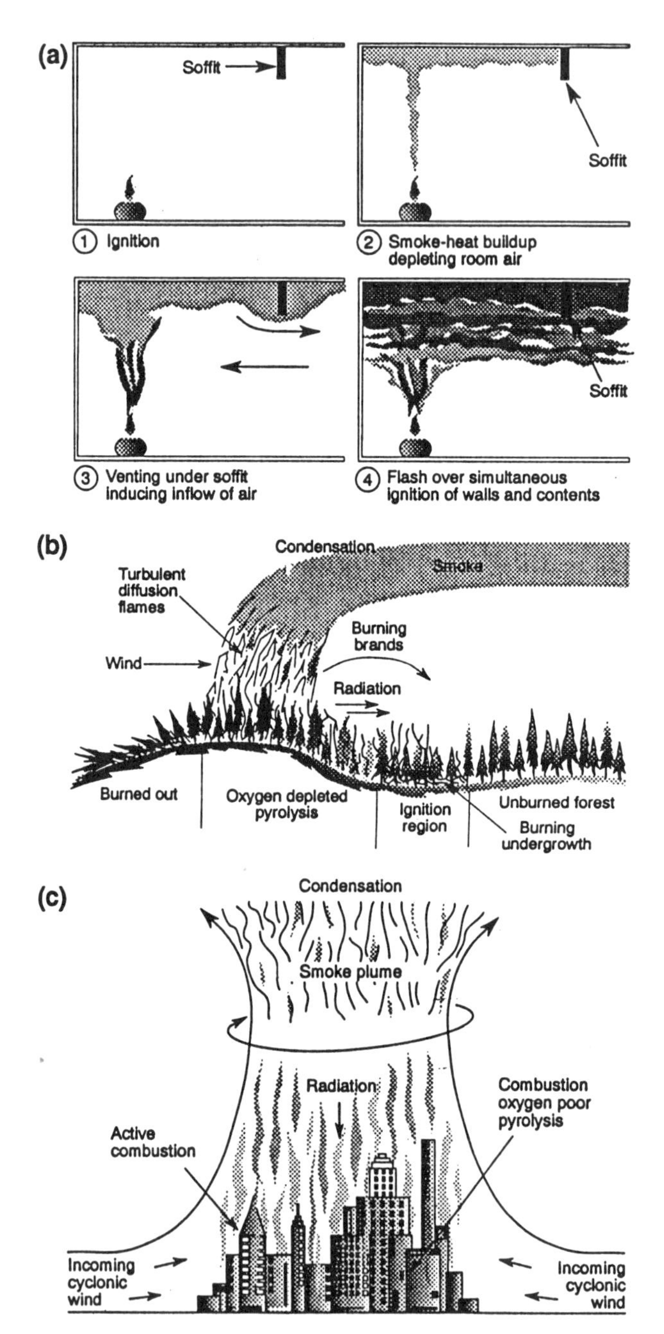

FIG. II-22 Unwanted fires. (a) Schematic of development of a room fire. (b) Schematic of a Forest fire, a large propagating conflagration. (c) Schematic of a Fire Storm, a stationary inwardly burning conflagration that generates its own winds.

increases. The rate of rise increases as the fire grows, and at a critical point when the room temperature reaches the ignition temperature of the combustible objects in the room, they take fire almost simultaneously. This explosive ignition point, called *flashover* is the most dangerous time in a fire. Anyone trapped at this point will almost surely die. The buildup of room fires is deceptively slow, and the difference between an apparently small fire and flashover can be as short as a few seconds. One should never try to fight an established fire in an enclosed space after the first few minutes. Another scenario occurs in a closed room. Fire buildup is slowed by the depletion of air, and soon the poisonous carbon monoxide becomes an important combustion product. In such cases flashover is delayed by lack of oxygen, but it can occur explosively upon opening of the door. It is dangerous to open the door of a burning room.

Open air fires such as forest fires (Fig. II-22b), oil spills, and conflagrations are wind and weather dependent. They are driven by winds and accelerate up slopes driven by buoyancy-induced flow. Forest and grass fires are inhibited by humidity, which raises the ignition temperature of wood debris by increasing its water content. One method used by rangers to evaluate forest fire danger is by measuring the moisture content of fallen wood. This is done by weighing a standard bundle of sticks. The lighter they are, the greater the danger of accidental fire. Most large fires have more or less well-defined fronts and propagate by contact, through flying brands, and by radiation. They might be considered as macroscopic flame fronts. Large propagating fires are called *conflagrations*. Examples include prairie grass fires, forest fires, and the great London fire of 1666.

A large-scale fire involving a whole city simultaneously is called a "fire storm" (Fig. II-22c). These intense fires are localized and intensified by their own rotating hurricanelike wind driven by the rising combustion plume. These should be distinguished from the more frequent propagating conflagrations such as the previously mentioned great London fire. Fire storms require very special initial conditions: low initial wind velocity combined with unstable local air and very large-scale (more than one square kilometer) simultaneous ignitions. Such conditions require human intervention. This has only been realized in a few cities during World War II and, one hopes will never again be seen.

Summary

Combustion is an extraordinarily complex phenomenon that has only been outlined to introduce the reader to the general subject and interest him or her in it. The remainder of the book will be devoted to the measurement and analysis of flame microstructure.

III

BURNER SYSTEMS

The study of flame structure requires burners that produce stable flames with thickness sufficient to allow measurements of the required resolution. Flame thickness varies inversely with pressure. Components for flame studies include burners and devices for pressure measurement, flow measurement, and regulation. Studies at reduced pressure require a housing and a pumping system. Much of this is standard laboratory practice, but some of the techniques are unusual enough to warrant discussion.

Requirements

To be suitable for analysis using the one-dimensional flame equations (Chapter IX), a flame must be one dimensional in the fluid dynamic sense. This requires that the burner be large compared with the thickness of the reaction zone.

The definition of thickness is somewhat arbitrary and usually refers to the primary reaction zone, which is controlled by transport and bimolecular reactions. They are binary collision phenomena. If, on the other hand, the termolecular recombination region is of interest, thickness varies inversely with the second power of the pressure. The definitions refer to the moderate pressure (0.01–1 atm). At high pressure the two regions overlap, while at low pressures the flames are dominantly bimolecular because the termolecular region is large compared with the burner housing. These questions are discussed in Chapter X.

A minimum criterion is that burner radius be three times the length of the region to be studied. An approximate relation for bimolecular flame thickness is given by the following equation, where it is defined as three times the e folding distance at the point of maximum temperature gradient. This is a function of the burning velocity and the thermal diffusivity, which varies inversely with pressure and burning velocity.

$$ZS \approx \frac{C}{\upsilon_0 P} \tag{3–1}$$

In this equation C is a constant that is proportional to the average diffusivity ($l^2\,s^{-1}$) of the flame. For air and oxygen diluted flames $C \approx 3$, P is pressure (atm), υ_0 is the burning velocity ($l\ s^{-1}$) and ZS is the reaction zone thickness (l). The length parameter, l, may be in any consistent set of units.

If a temperature profile is available, thickness can be defined geometrically using the two inflection points in the temperature–distance curve. If composition profiles are available, it can be defined as the distance between the point of initial reaction and the point where radical concentrations pass through a maximum.

Coordinate Systems

To relate measurements it is necessary to establish a coordinate system and obtain congruence between the various measured profiles. The first problem is straightforward, but the second is not.

An ideal reference point should be accurately observable and bear a fixed spatial relationship to true flame coordinates. Several possibilities suggest themselves and have been used: the luminous zone of the flame, the burner screen or edge, a fixed thermocouple in the incoming gas stream, or a fixed gradient or density marked by an optical beam.

The luminous zone provides a visible internal coordinate for flame front structure. This would be ideal, but unfortunately the edges are usually not well enough defined for quantitative work. As an example, the inner edge of the luminous region of the 0.1 atm methane–oxygen flame can only be located within about 100 μm, while the reproducibility of the other structure measurements is about 10 to 20 μm. Since the thickness of the luminous region (and hence the sharpness of the reference boundary) possesses the same dependence on pressure and flame velocity as the main flame structure, it appears that there are few conditions, at least for hydrocarbon flames, where it would be a completely satisfactory reference for quantitative work.

If the flame remains fixed in space, as it must for meaningful measurements, the burner could serve as the reference point or surface for the coordinate origin. This is satisfactory during a single run, but comparing runs can be a problem. Conditions of flow, composition, pressure, initial temperature, and burner cooling must be reproduced to better than one part in a thousand. This is possible, but it requires careful control.

A more satisfactory coordinate reference is provided by a small($< 25\ \mu$m) thermocouple placed in the low-temperature region of the flame ahead of the reaction zone. It can be made of materials with high thermoelectric power such as chromel–alumel, and need not be coated. In the region of initial temperature rise (the first few hundred degrees) the temperature gradient is high, and small flame movements result in substantial temperature changes. For example, in the 0.1 atm, 0.078 methane–oxygen flame (Chapter X), the gradient is such that a movement of 2 μm corresponds to a 1 K temperature change. A fixed thermocouple serves two purposes—it provides a position reference in the flame and also is a sensitive detector of flame movement or oscillation. Measurements can be no better than the spatial stability of the flame. A thermocouple reference allows the small corrections for daily variation of flame position to be made.

An optical reference surface can be provided by a Ronchi-type quantitative Schlieren system (Chapter IV). This allows location of a reproducible gradient point in the approach gases, and physical probes can easily be referenced by the transmitted beam.

Stability and Reproducibility

The reproducibility of position measurements in flames is limited by the flame stability, which, in turn, is controlled by the precision of flow regulation. As was mentioned, stability can be conveniently monitored using a small thermocouple to detect drifts and oscillations of the flame with amplitudes. *The importance of stability to precision measurements cannot be overstated; it is an absolute necessity. Positional reproducibility and spatial resolution are separate concepts. The first implies that if a probe is moved, measurements will change reproducibility. Resolution, on the other hand, implies the ability to measure a true second derivative.* Reproducibility is controlled by the precision of the drive. Resolution depends on position and the dimensions of the sampling region. Spatial reproducibility always exceeds the resolution—usually by an order of magnitude. The reason for the difference is that a probe sample represents an average of the property being measured over the sampling region. The average will change in a reproducible manner with movements that are small compared with the dimensions of the sampling region, and if the second (spatial) derivative is small, a valid determination of the property and its first derivative can be obtained. If the second derivative is large, however, errors will occur. For example, the fixed, monitoring thermocouple is usually 25 μm in diameter. However, it can detect flame movements of a few microns even though the resolution of such a thermocouple is probably no better than 50–100 μm. The effects of probe size on resolution is illustrated in Figure III-1.

Rate measurements in flow systems with transport require second derivatives of concentration. Therefore, the resolution of the measuring technique must be small compared with flame thickness. The minimal requirement for even a single meaningful second derivative is that the resolution be one-third of the flame thickness. A more usable criterion is one-tenth of the flame thickness. This sets another constraint on minimum burner diameter.

$$D_b \gg 50 r_{min} \tag{3–2}$$

As will be seen in succeeding chapters, the resolution of experimental techniques vary between 1 and 0.03 mm. This usually allows burners of reasonable size.

Spatial resolution can be no better than the flame stability and the ability to locate the origin of the coordinate system. This requires that flame movement be less than the required resolution during the period of measurement. Measurement time varies from a fraction of a microsecond for laser pulses to many minutes for some types of flame sampling. Flames stabilized by a heat extraction surface are as space fixed as the flow regulation. There are drifts during warmup periods, but this normally lasts only a few minutes. The coordinate origin problem is complicated by probe distortions, but stability and relative flame position can usually be monitored to a few micrometers.

Congruence

Data interpretation involves relations among several variables and their derivatives. If temperature composition and velocity are all derived from a single measurement tech-

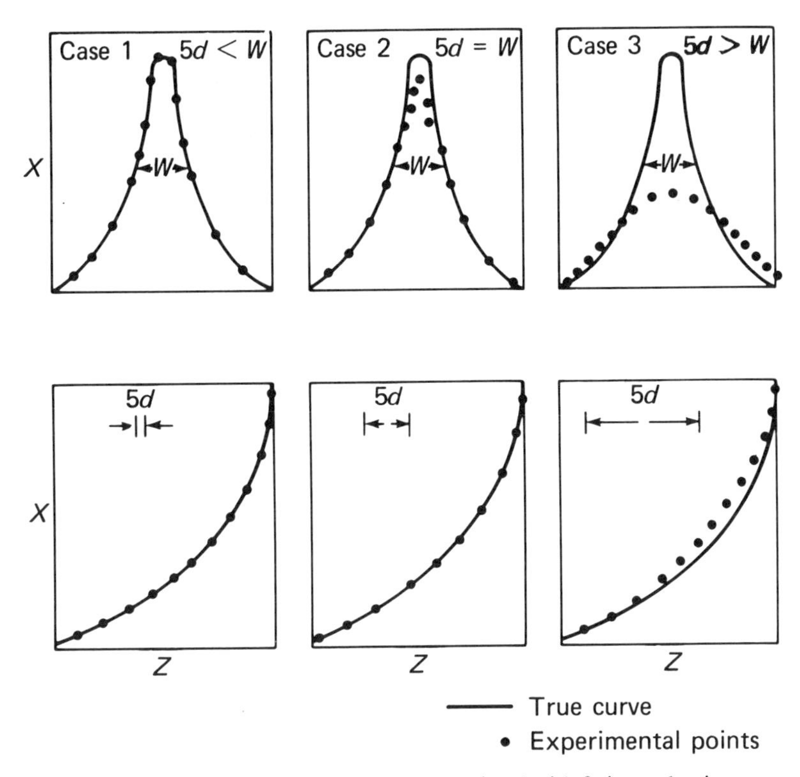

FIG. III-1 Spatial resolution errors associated with finite probe size.

nique, no problem exists. However, more often several techniques are required, and congruence between these profiles becomes a problem. This is not straightforward because the effective position of measurement does not correspond exactly to the physical position of the probe (Fig. III-2). A sampling probe withdraws gas a few diameters upstream of its physical position, while a thermocouple distorts the flame velocity profile locally so that it tends to sample a temperature downstream of its apparent position (Friedman [1953]; Walker [1959]). Particles tend to lag the gas flow so that their position lies upstream of the ideal position. Thermocouples are discussed in Chapter V and sampling probes are discussed in Chapters VI and VII. The displacement of a particle velocity profile due to acceleration lag has been estimated at about 50 μm for particles in a propane–air flame front 3 mm thick (Chapter IV) (Fristrom, Avery, Prescott, and Mattuck [1954]). Optical probes show displacements due to Schlieren effects, although they are small for low-pressure flames. For example, at 0.1 atm the displacement is less than 20 μm across a methane–oxygen flame 5 inches in diameter.

Such estimates indicate that displacements between the several types of profiles can be as much as several tenths of a millimeter. This could introduce serious uncertainties in interpretation. The effects of this on the analysis are discussed in Chapter X.

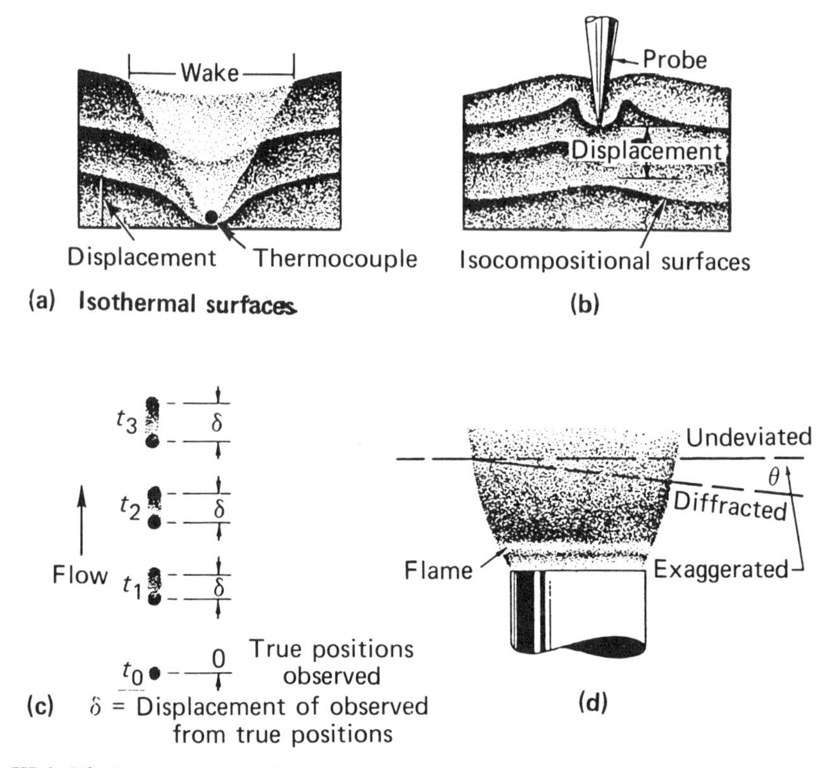

FIG. III-2 Displacement errors: (a) Thermocouples; (b) Probes; (c) Particle tracers; (d) Line of sight optical methods.

Burners

The burner is the basic unit for flame study. Every research worker has his or her own special design, and the literature on special burners is almost as extensive as that of combustion; however, few articles are devoted directly to this single topic.

There are three general flame types: (1) premixed flames; (2) diffusion flames; and (3) turbulent flames. Turbulent flame burners differ principally in the flow velocities used. Therefore, we will concentrate on the first two types.

Premixed combustion can be studied using a flame propagating through a stationary atmosphere or in steady state using one of the many variants of the Bunsen burner. Most structural studies have been made using steady-state burners because temperature and composition measurements can require seconds to minutes. By contrast, freezing a moving flame front can require time resolution of the order of microseconds depending on the burning velocity. Propagating flames, however, are used to measure burning velocities (Chapter IV), and modern laser diagnostics (Chapter VIII) are rapid enough to freeze flame movement. Lasers allow point, linear, planar and even three-dimensional holographic measurements to be made (Chapter VIII). The microstructure of steady-state and propagating flames are almost identical.[1] In one case the flame front lies on a moving coordinate system, while in the other it is fixed in laboratory coordinates.

Propagating Flames

Propagating flames are studied in three environments: (1) constant pressure, (2) constant volume, and (3) through tubes either at constant volume or constant pressure. This controlled environment plays the same role as the burner in the study of steady-state flames.

Constant-pressure operations require an envelope or a slow flowing stream to separate the combustible gases from disturbances of the outside atmosphere. Since the combustion process heats and expands gases, provision must be made for expansion. This can be accomplished using a slack plastic bag or soap bubble as a container or by providing an escape for the gases through a vent. Provision must also be made for providing a known mixture and removing the burned gases after the run. Flame propagation is usually followed by Schlieren or shadow techniques described in Chapter IV.

Constant-volume systems must be sturdy because the pressure rise during combustion can be substantial. These are not idly called combustion bombs. Flame structure and propagation rate both depend on pressure. This changes continuously during constant-volume combustion. Therefore, either a large volume must be used with studies confined to early times when the pressure change can be neglected or corrections must be made for the changes in pressure and temperature induced by adiabatic compression. The second approach provides additional information on the effects of pressure (and preheat), but deconvolution is involved. A short discussion of these problems is given in Chapter IV. Figure III-3 gives the schematic of a constant-volume combustion vessel with a sample pressure trace.

Flame propagation through tubes was used by Mallard and le Chatelier [1883] in their pioneer studies of flames and detonations. Tube flow can be complicated by the effects of boundary layer buildup with resulting complex flame geometries (Jost [1946]; Lewis and von Elbe [1987]), which must be known to derive quantitative burning velocities. However, satisfactory propagating flat flames have been obtained by Fuller, Parks, and Fletcher [1969] using twin opposed spark ignition in a vertical tube (Fig. III-4).

Premixed Steady-State Flames

The classic method for studying steady-state flames is the use of a burner of the type introduced by Bunsen [1866]. Its advantage is that it produces a stable flame, which allows time for detailed measurements. There are many variants of Bunsen's original open tube. They differ principally in the manner in which the flame is stabilized. This is generally by heat extraction either at an edge or boundary, in the incoming gas stream or in the outgoing gas stream. Spatial reference is provided by the point where

1. Gravitational forces introduce minor differences which are usually only significant near the extinction limit.

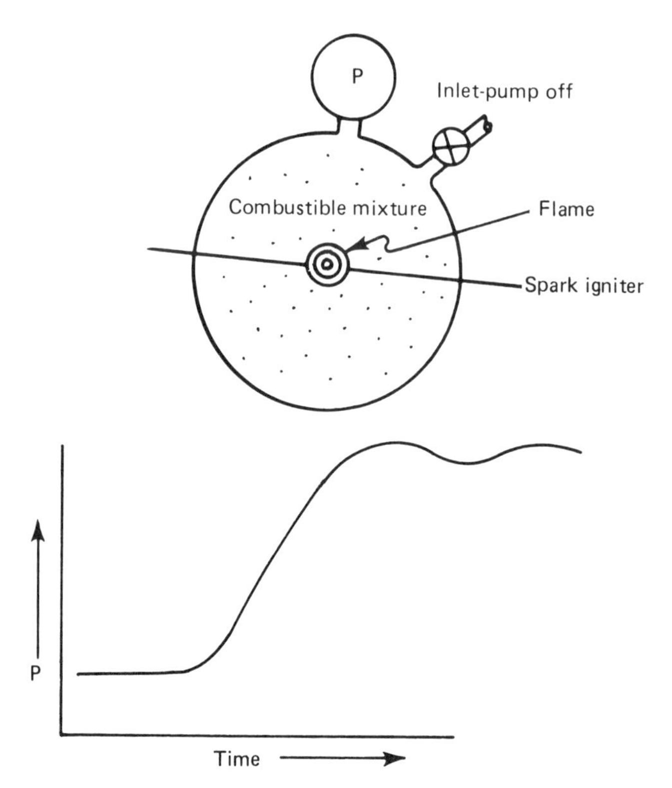

FIG. III-3 Schematic of Constant-Volume Apparatus for Studying Flame Propagation. Pressure Time Trace from a Typical Run.

the heat extraction occurs, due to the balance between flow and propagation with local heat extraction (Lewis and von Elbe, p. 761 [1987]; Jost, p. 220 [1946]). Lifted flames can also be stabilized using divergent flow. In this case heat is lost at the edges through lateral conduction and diffusion.

Tubes and Nozzles

The basic burner is an open tube. Here the flame is stabilized in the boundary layer, where it intersects the wall (Fig. III-5). The detailed mechanisms of stabilization have been discussed by Lewis and von Elbe [1943, 1987] and Wohl [1953]. This can be accomplished in a variety of ways, yielding a range of flame shapes (Fig. III-6).

Tube flames are susceptible to flow disturbances and often flicker and breathe because they are only attached at boundaries or points. However, with proper chimneys and flow regulation flame excursions can be reduced below 100 μm. In Poiseuille flow a parabolic profile is approached after about 100 diameters. Long open burners tend to develop organ pipe oscillations so that shorter tubes are often used. The velocity profile at the throat of such burners lies between a uniform flat and fully developed parabolic velocity distribution so that, in the absence of detailed velocity measurements, the precise flame shape is not predictable. To remedy this, nozzles that provide uniform flat velocity profiles are often used. They are called *Mach–Hebra nozzles*

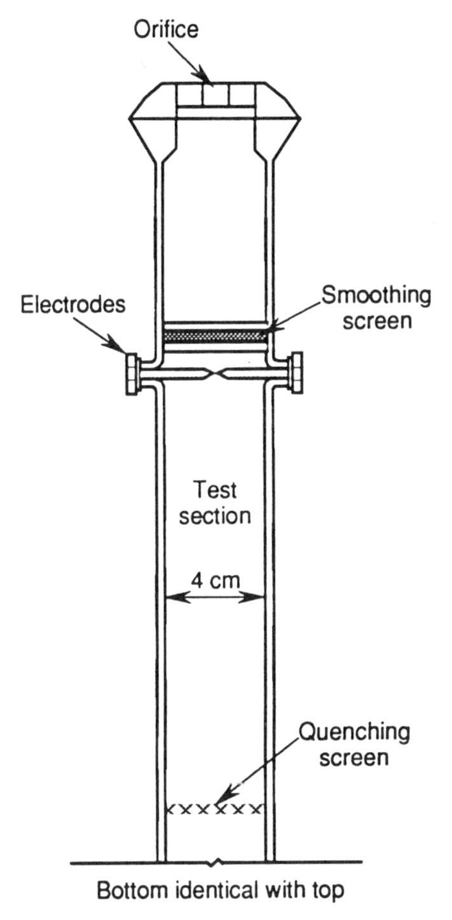

Fig. III-4 Flame Deflagration Tube (Fuller, Parks, and Fletcher [1969]) for studying opposed spark ignited flames. Note that only the upper half of the apparatus is pictured.

(Anderson [1949]) and can be electroformed or spun on a mandril from thin walled stock. Nozzles compress the boundary layer, and this decreases stability. This can be compensated by using a shroud gas, which increases the burning velocity upon diffusing into the boundary, thus anchoring the flame on the rim. Oxygen is useful for this purpose. Flames can also be anchored to a surrounding flat flame. Such a burner was used for studying eddy combustion (see Fig. III-12).

Burners with Heat Extraction

Flames can be stabilized using grids or screens (Fig. III-7). If the grids are thermally isolated, the heat loss can be small, since it is principally through radiation. Where the loss is low enough, the flames can be treated as approximately adiabatic. This loss is, nevertheless, sufficient for stabilization so that very steady flat flames result. This is in accord with theory (Hirschfelder, Curtiss, and Bird [1964]). Flow must be adjusted just below the free-space propagation velocity. This requires accurate flow regulation.

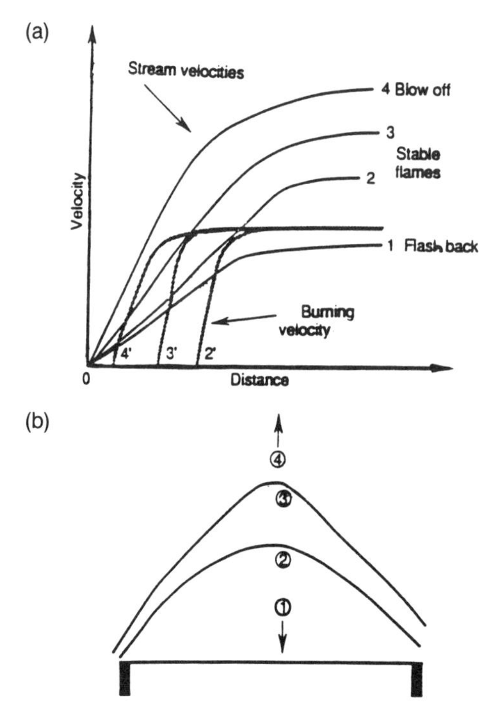

FIG. III-5 Conditions for flashback, blowoff, and stabilization: (a) Local gas velocity and burning velocity in the vicinity of a burner rim; (b) Associated flame shapes.

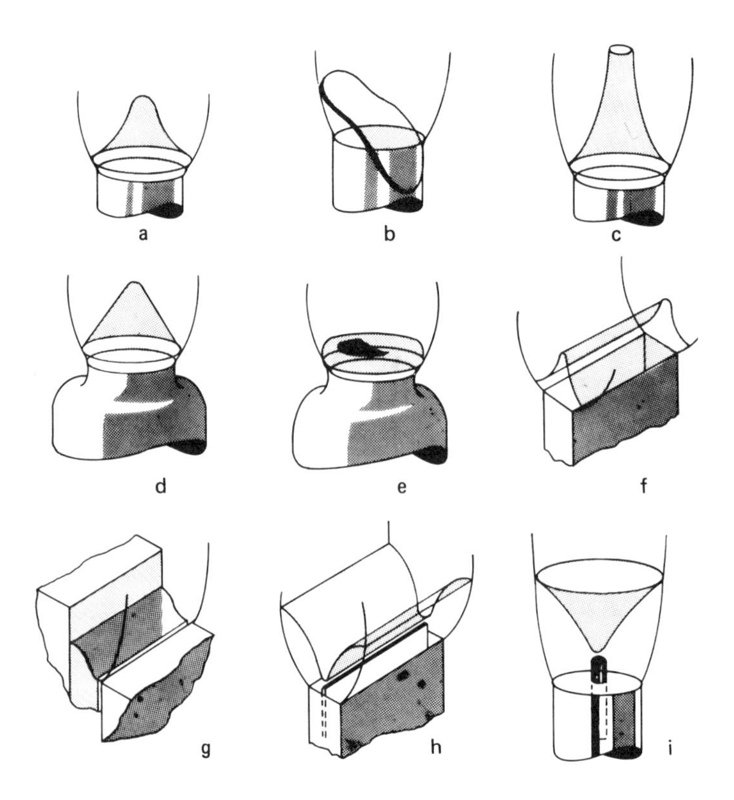

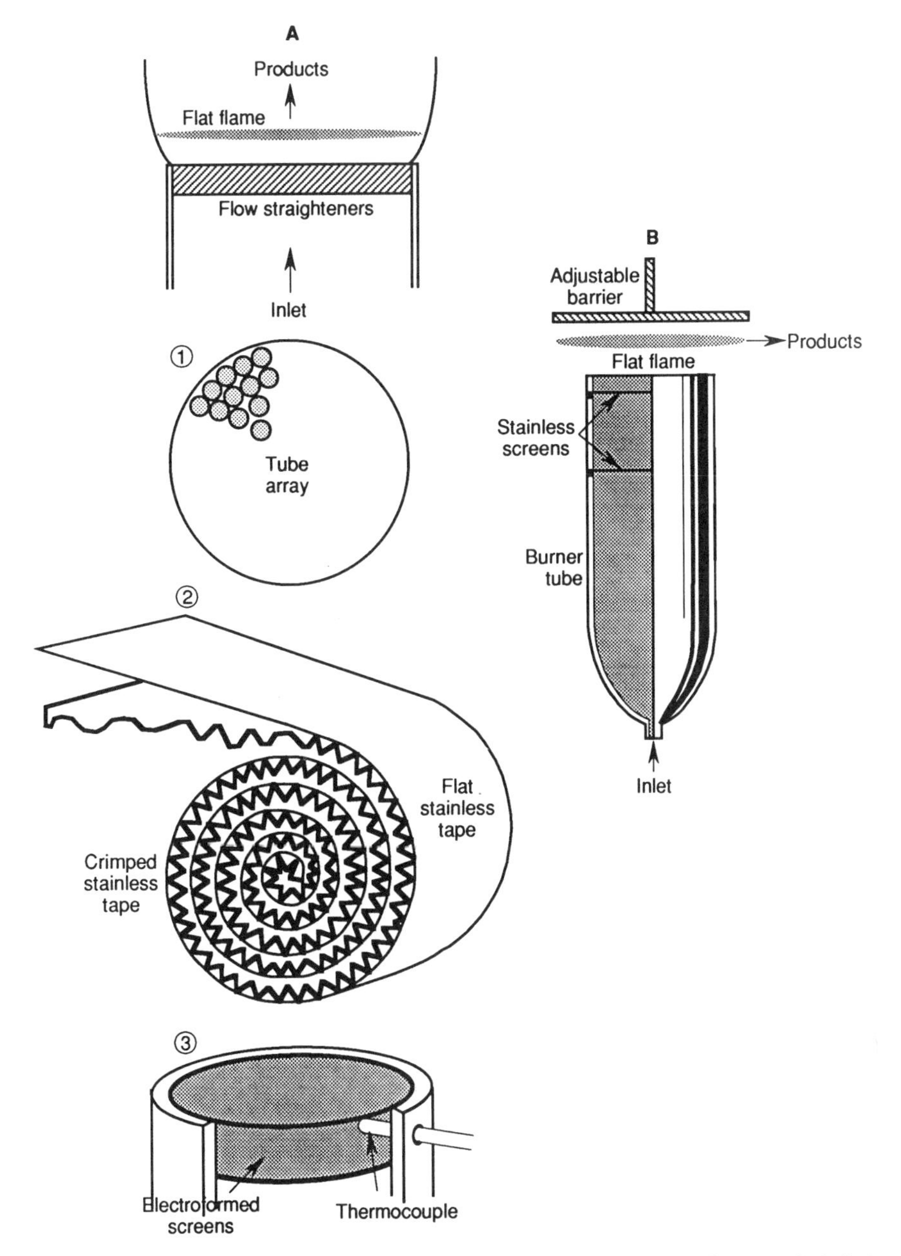

Fig. III-7 Flat Flame Burners with Minimal Heat Extraction: (1) Arrays of Tubes; (2) Spirals of Crimped Metal; (3) Electroformed Screens; (4) Baffle in exit flow.

←

Fig. III-6 Flame Geometries Obtained with Open Tube Burners: (a) Parabolic flow; (b) Low flow with flame partially entering the inlet tube; (c) Tip blowthrough; (d) Constant-velocity nozzle flame; (e) low-velocity nozzle flame of "button" shape; (f) rectangular burner with "tent" flame; (g) Slot burner with single sheet flame; (h) "Inverse tent" flame stabilized above a longitudinal rod; (i) "Inverted conical" flame stabilized above a coaxial rod.

Screen burners are not satisfactory for fast atmospheric pressure flames because thermal gradients are so steep that small excursions overheat the screens irregularly, which then results in warpage. With low-velocity or reduced-pressure flames the temperature gradients are lower, and satisfactory flat flames can be produced with screens. The stabilizing surface can be a grid formed by rolling a corrugated stainless steel ribbon (Powling [1961]) or an electroformed screen (Fristrom [1963]). We prefer the latter because of ease of fabrication.

The stabilizing surface need not be placed in the incoming gas stream. Stable flat flames have been produced using a baffle in the exit stream of a flame (Fig. III-7) (Hoelscher and Biedler [1957]). Here the baffle extracts some heat from the postflame gases, providing a stabilizing reference at the hot boundary. Flames can be extinguished by extracting heat from this boundary, and radiation losses from the hot boundary have been suggested as an extinction mechanism.

Burners with porous metal plates containing cooling coils allow substantial amounts of heat to be extracted (Fig. III-8). This results in stable flat flames, and a majority of flame structure studies have been made using such burners. These flames are well understood (Botha and Spalding [1954]). They can be considered a foreshortened equivalent of precooled flames, with the same enthalpy. Interestingly enough, flames propagate with the enthalpy corresponding to precooling to negative initial temperature. These are not adiabatic flames, and it is necessary to characterize their heat loss. This only requires measurement of the cooling water flow rate and the temperature rise and should always be done.

The geometry of the sintered surface need not be planar and the material need not be metal. Satisfactory spherical (Fristrom [1958]) as well as cylindrical flames (Tsuji [1967]) have been produced. The advantage of these geometries is that the symmetry allows velocity to be derived from temperature (and molecular weight) measurements, making direct aerodynamic measurements unnecessary. Experimental flames show slight asymmetries because the gases must be introduced through a stem and buoyancy effects become significant above a third of the atmosphere.

Diffusion Flames

The structures of diffusion flames are also of interest, and in simple cases a one-dimensional analysis is possible (Chapter IX). In the general case lateral velocity and diffusion must be considered and the analysis becomes a two-dimensional problem. Diffusion flames have two incoming gas streams, the fuel flow and the oxidizer flow. Burners differ in the methods for joining these flows. Flows can be parallel or opposed, and the symmetry can be planar or radial (Fig. III-9). Each arrangement provides special advantages.

Parallel Flow with Radial Symmetry

The traditional diffusion burner is simply an open tube allowing the fuel to flow into an oxidizing atmosphere. The role of fuel and oxidizer can be interchanged. The flow pattern can be simplified by matching the inlet velocity of fuel and oxidizer, but the large density change associated with combustion and the strong coupling of subsonic flow makes even the simplest diffusion flame flow pattern complex. Programs have

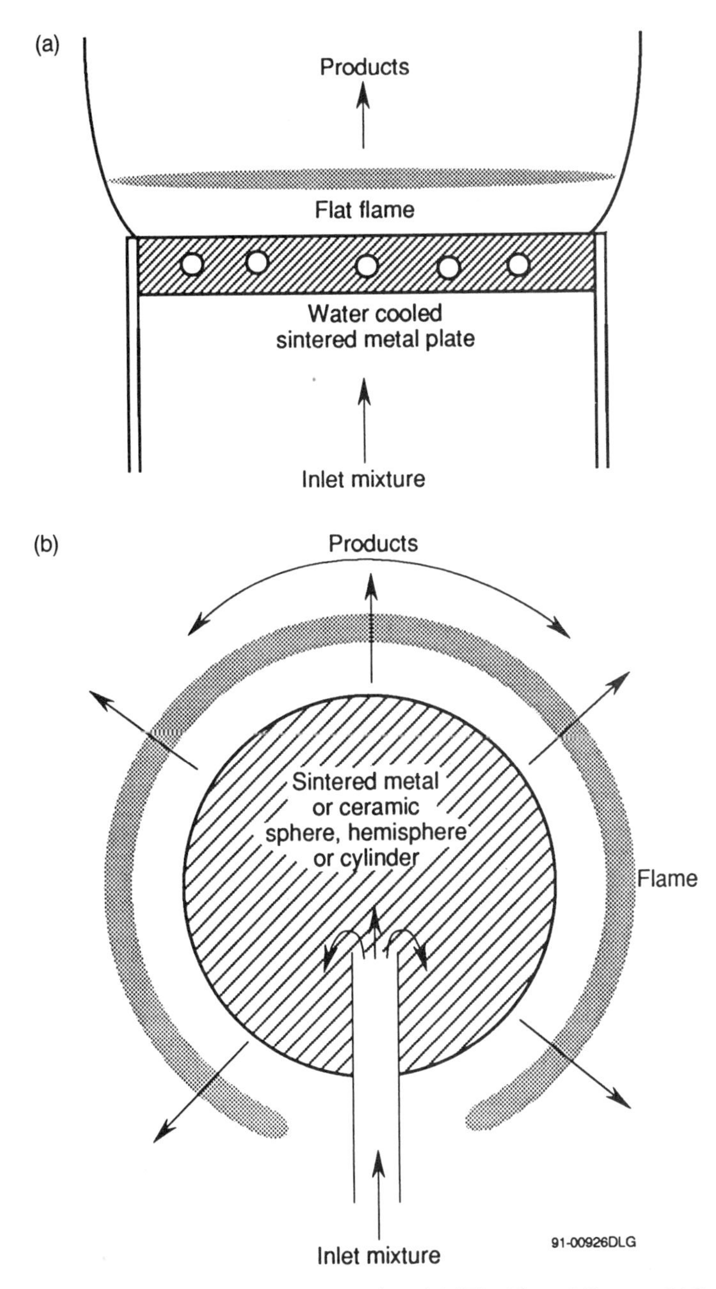

FIG. III-8 Burners with Significant Heat Extraction: (a) "Flat Flame" Burner; (b) Spherical Burner.

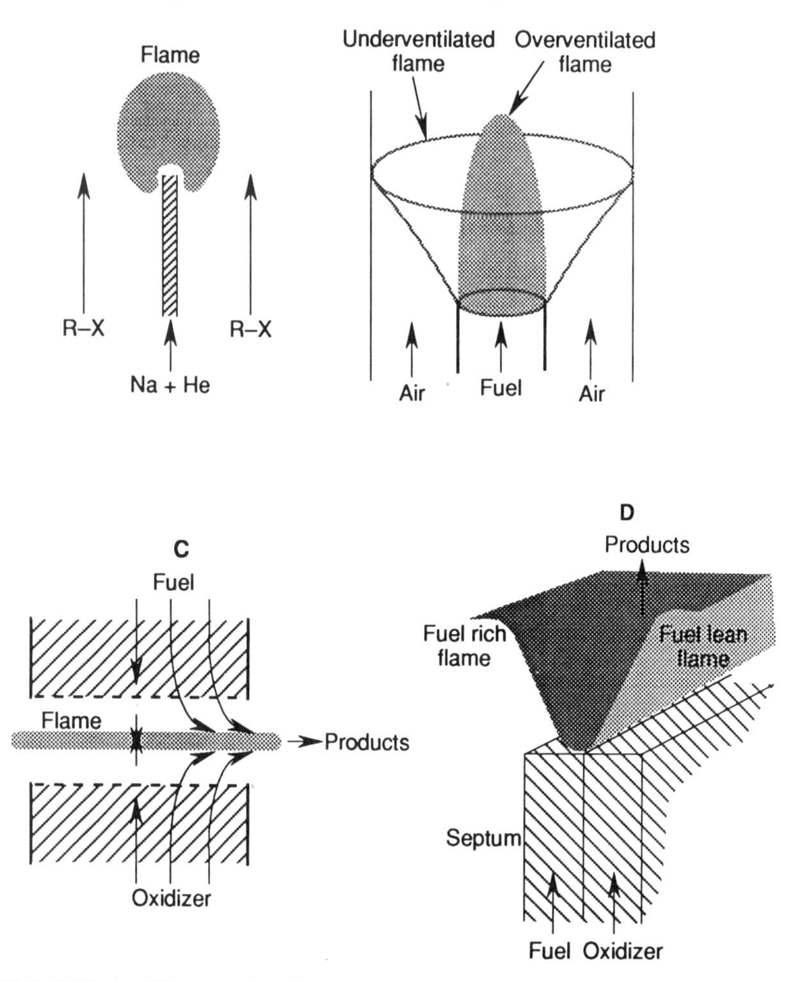

FIG. III-9 Diffusion Flames: (A) Point Source" axial flow (low-pressure sodium–halogen flame); (B) Geometry of a coaxial Fuel Jet Flame; (C) Opposed Jet Flame; (D) Wolfhard–Parker Burner for Optical Studies.

been written to simulate such flows (Mitchell [1953]), and the idealized analysis of Burke and Schumann [1928] provides a reasonable approximation of diffusion flame geometry. This is the limit where reaction is rapid compared with diffusion. Figure III-9(B) shows the axial jet diffusion flame geometry.

An interesting variant of this is the low-pressure diffusion flame developed by Polanyi [1932] to study low-pressure alkali metal–halogen diffusion flames. Here diffusion velocities dominate mass velocities and reaction rate can be deduced from measurements of flow and flame diameter. Spectroscopic–structural studies have been made on flames of this type. Figure III-9(A) shows a schematic of Polanyi's original system.

Parallel Flow with Planar Geometry

Absorption spectroscopic studies require planar geometry to allow probing of homogeneous region unless one is willing to accept the complications of deconvolution, which are characteristic of tomographic studies. Diffusion flames can be studied using parallel flows separated by a thin, flat septum, as in the burner of Wolfhard and Parker [1949] [Fig. III-9D]).

Opposed Jet Burners

Potter and Butler [1961] proposed a burner using opposing flows and matching momentum between fuel and oxidizer flows. He used the burner to measure "flame strength" (see also Chapter II). A similar burner was used by Vinckier and Van Tiggelen [1968] to study the structure of turbulent flames. This device has been elaborated using screens and sintered metal discs to provide uniform flow. Such flat flames are well adapted to optical studies (Pandya and Weinberg [1963], Fig. III-9(C)). Holve and Sawyer [1975] developed an opposed jet burner for solids.

Cylindrical Diffusion Flames

Burners in the form of sintered metal cylinders have been used in an opposed stream to form flames. Their structure can be measured and analyzed along the stagnation axis (Tsuji [1967]). They have been also used to study burning velocity and extinction (Tsuji [1982]).

Other Burners

Many special-purpose burners have been devised. Some illustrations are given below.

Separated Flame Burners

Markstein [1958; 1964] devised a slot burner (Figure III-10) that produces a linear array of cellular flames for the study of flame separation phenomena. Some structural studies have been made. The slot width should be between one and two times the cell wavelength.

Carhart, Williams, and Johnson [1959] developed a "vertical tube reactor" for studying multiple stage cool flames.

Multiple Diffusion Burners

Some combustible mixtures are so reactive that they are difficult to premix without ignition. These are called *hypergolic systems* (Chapter II). Berl and Wilson [1961] used a multiple-tube burner for the study of borane flames [see Fig. III-11(a)]. Maclean and Tregay [1971] used a parallel-plate analog with stainless steel shims as septa to study low-pressure hydrogen fluorine flames. Linevsky and Carabetta [1973] developed a ceramic burner for their carbon disulfide flame laser.

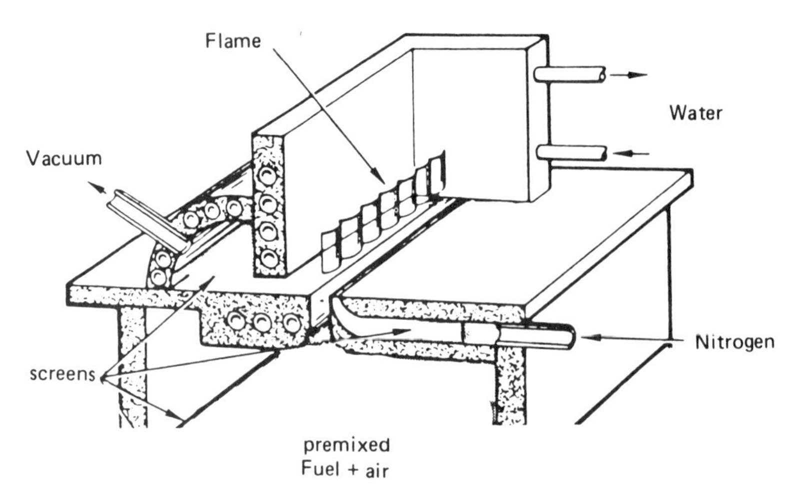

FIG. III-10 Slot Burner for Studying Cellular Flames. (After Markstein [1964], with modifications.)

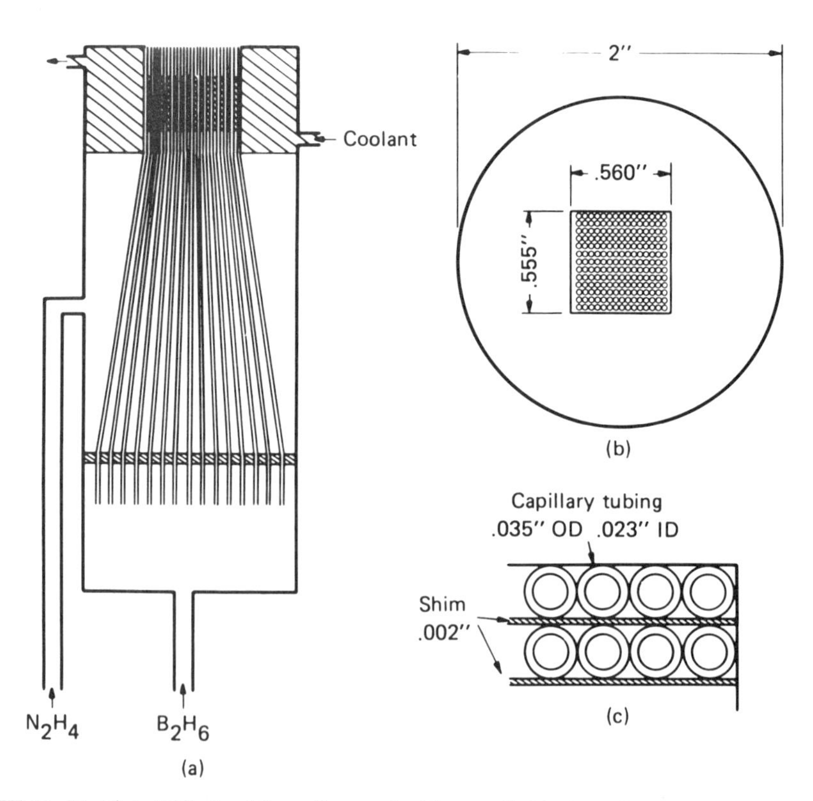

FIG. III-11 Multiple Diffusion Flame Burner for Hypergolic Fuels. (After Berl and Wilson [1961], with modifications.)

Eddy Burner

The eddy burner (Fristrom and Linevsky [1985]) (Figure III-12) is a device for simulating turbulent combustion acoustically. It consists of an orifice surrounded by a stabilizing flat flame. The orifice connects to an acoustic cavity driven by a loudspeaker. A pulse applied to the loudspeaker generates a toroidal eddy like a toy smoke ring generator. The loudspeaker input can be synchronized using a delay circuit. This allows pulse diagnostics to be synchronized for investigating the time behavior of the eddy in either single or continuous pulsing mode. If the orifice is circular, toroidal eddies are produced, but other geometries such as slots are also possible.

In the absence of drive, the combination of an orifice surrounded by a stabilizing flame produces particularly simple turbulent flames with uniform eddies (see Fig. II-8). This occurs because the orifice represents a single source of eddies whose size is controlled by the orifice and whose shedding rate is uniform.

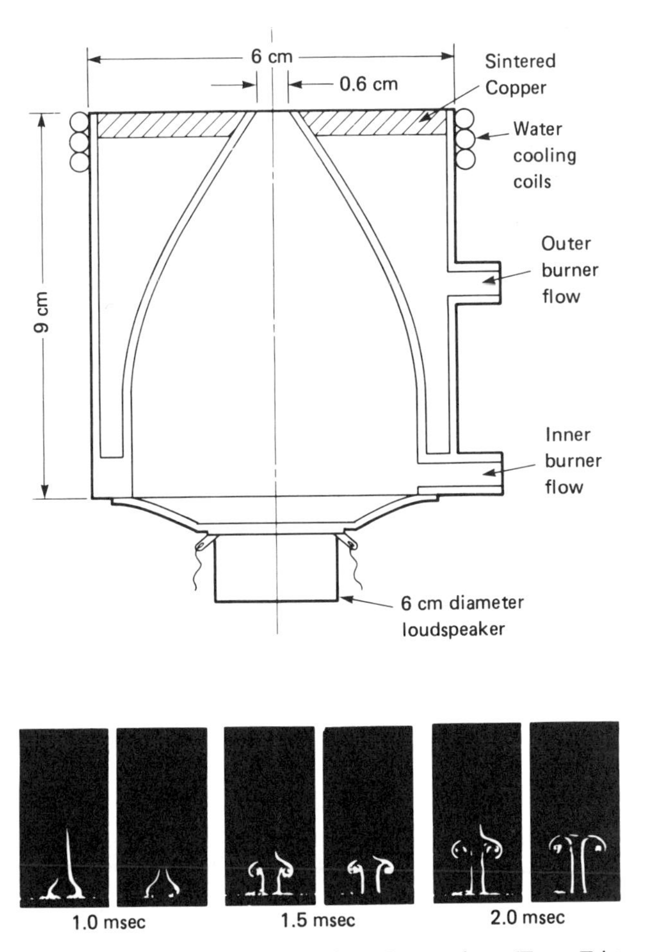

FIG. III-12 Coaxial Burner for Studying Flow–Flame Interactions. (From Fristrom, and Linevsky [1985].)

Curved Flame Burner

It is desirable in laser studies of flame structure to avoid excessive dispersion or "beam steering," as it is often called. To minimize this the flame front can be curved so that the beam enters the region of least density gradient. A burner designed by Stephenson [1978] for accomplishing this is shown in Figure III-6(g). It has a curved wall opposite the knife edge flame holder. The wall has several functions. The most important is that the gas flow is speeded at the edges. The flame adjusts to match the burning velocity with the flow velocity component normal to the flame front. The flame sheet appears pushed up at the ends, where gas flow is highest. The curvature of the flame allows the focused laser to probe the center of the flame with minimal disturbance.

Atomic Flame Burners

Flames can be produced using the atoms and radicals from electric discharges. The field was pioneered by Langmuir [1927] with his atomic hydrogen torch. This is used at atmospheric pressure for welding reactive metals. It has been largely supplanted for safety reasons by helium and argon electric arc "flames," called "heliarcs," in which excited electronic states rather than atoms carry the energy. Confined RF discharges with high-velocity streams are used for plasma spraying of high-temperature ceramics and metals. A laboratory-scale version called "Plasmatron"® is available commercially as a high-temperature source for atomic absorption analysis.

At reduced pressures (0.1–1 torr) the lifetimes of the atoms and radicals increase so that colorful flames can be formed with a wide variety of organic and halogenated compounds. Figure III-13 shows one type of such burner described by Gaydon [1974].

Flame Separation Burners

To facilitate the study of the inner cone of very rich flames, a burner that separated inner from outer cone was independently devised by Teclu [1891] and Smithels and Ingle [1892] (see Fig. III-14). This has been modified by Gaydon for spectroscopic studies [1974] using high-speed secondary air. Flame separation was discussed in Chapter II.

Stirred Reactors

The stirred reactor is a useful device for combustion studies (see Chapter II). The most recent version of the Longwell–Weiss reactor has toroidal geometry (Nenninger, Kridoiotis, Chomiak, Longwell, and Sarofim [1985]). This form gives better performance than the original spherical design. Figure II-18(A) gives the design and characteristics of the burner.

Microjet Burner

The microjet burner of Groeger and Fenn [1988] is a miniaturized stirred reactor that is small enough to be introduced into a molecular beam apparatus. The burner is useful for the study of intermediates and nonequilibrium combustion processes. Figure III-15 gives some design characteristics of the device.

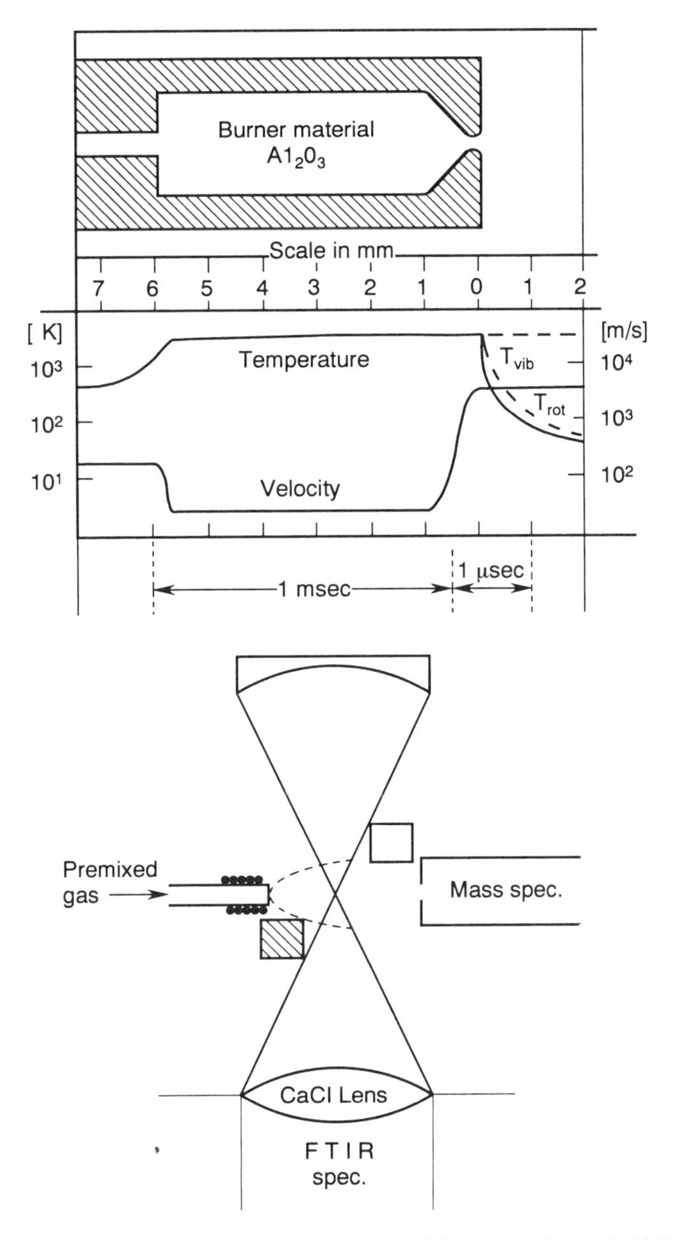

FIG. III-13 Schematic of microjet burner of Groeger and Fenn [1988].

Burner Housings

Because of the adverse effects of drafts and the desirability of studying flames at pressures other than atmospheric, a burner housing is usually required. It should provide a leak-tight chamber of suitable size with convenient access and visibility. One such transparent system is provided by a glass pipe cross or tee (Corning [1990]). They are

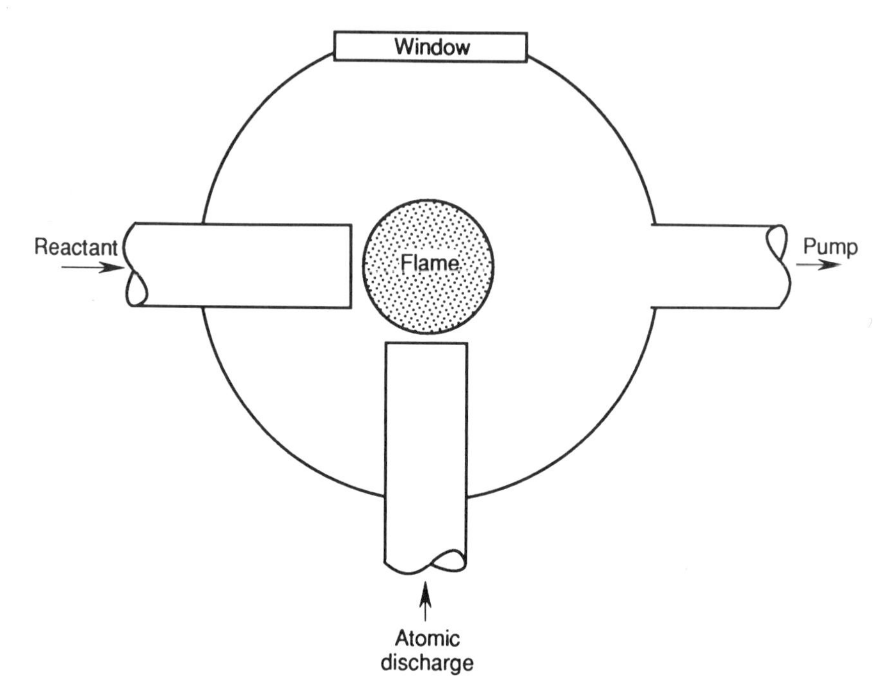

Fig. III-14 Laboratory setup for the production of "atomic flames."

available in standard sizes from 1 to 6 in. ID, together with many standard fittings and flanges. Larger sizes are available on special order. They can be connected using neoprene or Teflon gaskets or with O-rings. Housings can also be fabricated from round-bottomed flasks or bell jars. Because of shattering hazards they should be provided with a protective screen. Glass pipes are tempered glass and less hazardous but still should be used with caution. Metal housings can be fabricated.

Windows and access ports are necessary even in glass housings if photographic, optical, or cathetometer measurements are to be made. Windows can be of selected plate glass providing the flame does not impinge on them. Optical-quality Pyrex®, quartz, sapphire, and diamond windows are available commercially.

The exit gases are hot and require cooling. For small systems air cooling can be adequate, but water cooling is recommended. This should be done at the earliest convenient point beyond the flame to reduce the possibility of injury or fire.

Movement

In many studies it is necessary to move a probe or other device. For this one can use a shaft seal. Several satisfactory types are on the market. O-ring seals of any size can be fabricated from commercially available stock. Materials include buna rubber, nitrile rubber, silicone rubber, Teflon®, and Viton®. A wide variety of commercial movements are available for laser optic alignment. These can have as many as six degrees of freedom (three translations and three rotations) and are of high precision. Differential screw movements that can be reproducible within a millionth of an inch are

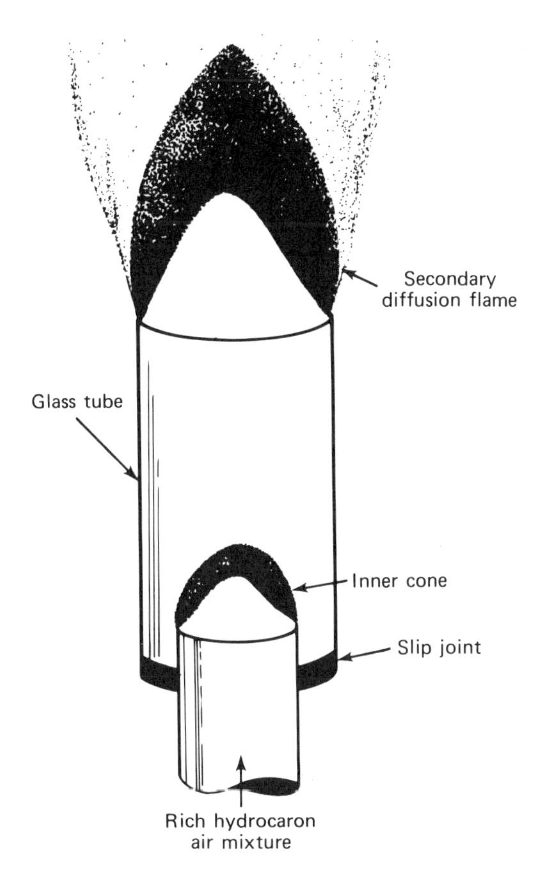

Fig. III-15 Teclu [1891]—Smithels–Ingle [1892] Burner for Flame Separation.

available commercially. Remote control can be obtained using stepping motors and piezoelectric drives. These devices are expensive and often overkill for flame work. Machinists' screw jacks and milling machine or lathe movements are recommended as rugged, reproducible, and inexpensive substitutes. Machinists' micrometers are also useful devices for measurements.

Pumps and Plumbing

Since flame studies are often made at reduced pressure, pumps play an important role in the combustion laboratory. They must be capable of handling the flow from the burner with enough excess capacity to allow choking flow across the exit orifice. This makes the pressure inside the burner independent of pump fluctuations. It must also resist the combustion gases, which may be both condensible and corrosive. In some cases cryogenic pumping may be used. An interesting technique was use by Frazier, Fristrom, and Wehner [1963] for the corrosive hydrogen–bromine flame. The HBr was reacted with ammonia to form involatile NH_4Br and excess Br_2 was condensed on a liquid nitrogen trap.

Pumps are characterized by two parameters, pressure range and pumping speed. These data are usually provided by the manufacturer in the form of a graph. Ordinary laboratory pumps (both rotary and piston) are constant-displacement devices and, therefore, show a roughly constant pumping speed over the upper part of their pressure range. The low-pressure limit is set by leakage and the vapor pressure of the oil. Piston pumps normally are useful down to about 20 torr, while rotary pumps are useful down to 10 μm (single stage) or 1 μm (double stage). The most generally useful pumps for flame work are the rotary type having a feature called "air ballast," in which air is injected into the second stage, sweeping out condensible gases. This allows the handling of condensible vapors as water, kerosene, and acetone. For high-speed pumping, steam ejectors or the Roots mechanical blower are also available.

For the lower pressure ranges, diffusion, turbo, ion, and cryostatic pumps are available (Guthrie [1963], Dushman [1962]), but flame studies are not usually undertaken below 1 torr because the required burner sizes and flow become impractically large.

One important consideration in pumping systems is the size of the piping. If it is too small, only a fraction of the pump capacity will be available. If continuum flow is involved (pipe diameter large compared with the mean free path), then the impedance offered by the line is proportional to its length and inversely proportional to the fourth power of the radius. Thus using Poiseuille's law the pumping speed S of a pipe is:

$$\dot{m} = \frac{\Pi M r^4 (P_1^2 - P_2^2)}{16 R T L \eta} \quad (3\text{–}3)$$

$$S_{\text{tube}} = \dot{m} \frac{RT}{M} \quad (3\text{–}4)$$

In these equations r is the radius, L the length (cm), P_1 and P_2 the up- and downstream pressures (dyn cm^{-2}), η the gas viscosity (g cm^{-1} sec^{-1}), $\dot{m}$ the mass flow rate (g sec^{-1}), A the orifice throat area (cm^2), γ the specific heat ratio C_p/C_v, M the molecular weight, T_i the upstream (stagnation) temperature (K), and R the gas constant (8.3×10^7 erg mole^{-1} K^{-1}).

Some constants for metering gases are collected in Table III–1.

Discussions of pumps and flow problems are given by Guthrie [1963] and Dushman [1962]. Ultrahigh-vacuum techniques are discussed by Roberts and Vanderslice [1966].

The required pump for a system can be estimated from the burner size, flame geometry, and maximum burning velocity with provision for a protective shroud flow. Total flow should then be multiplied by a factor of three to provide for choking exit flow so that the flame will not be influenced by pump fluctuations. It is wise to choose the largest pump that one can afford or house. This provides a safety factor for unforeseen requirements. Pumping speeds and flows add inversely.

Fixtures

Since these systems require vacuum lines, gas inlet lines, and water cooling lines, there is much plumbing involved. Standard iron, copper, and brass plumbing fittings are rugged, but tend to corrode. Therefore, stainless steel tubing and plumbing or Pyrex®

Table III–1 Gas Constants for Metering Devices

Gas[a]	M	γ	K	η
Acetylene	26.04	1.26	2.034	0.94
Air	28.95	1.40	2.00	1.71
Argon	39.94	1.67	1.805	2.10
CO_2	44.01	1.30	1.581	1.40
CO	28.01	1.40	2.034	1.66
Ethane	30.07	1.20	1.861	0.85
Ethylene	28.05	1.26	1.960	0.91
Helium	4.003	1.67	5.708	1.86
Hydrogen	2.016	1.40	7.61	0.84
Methane	16.04	1.31	2.629	1.03
Nitrogen	28.02	1.40	2.034	1.70
Oxygen	32.00	1.40	1.902	1.89
Propane	44.09	1.16	1.518	0.8

[a]The units of M are grams per mole; γ the ratio of specific heats is dimensionless; K is the constant in equation 3–5 multiplied by 0.0001; η is viscosity at 273 K (propane is at 293 K) in poises (g/cm sec) $\times 10^4$.

glass pipe is better. Metal lines require significant effort to install, and for this reason we recommend the use of plastic tubing and pipe when feasible. Plastic tubing is available in polyethylene, nylon, Teflon®, and PVC from 0.01 to 12 inches in diameter. These fittings offer the advantage of being flexible and noncorrosive. They are easier to install than metal fittings. Their major disadvantages are that most of them are limited to temperatures below the boiling point of water, and pressures below 100 pounds per square inch and above 10^{-5} torr. They are suitable for most flame work but not for high vacuum. The smaller polyethylene tubing is very flexible. This can be most advantageous. It can be cut with a knife, scissors, or saw. A variety of fittings are available for this type of tubing. They allow easy connection to each other or to standard metal and glass tubing or pipes. Two commercial types are Gyrolok® and Swagelok®. Others are also available. All appear satisfactory and are available in brass, stainless, Teflon® and nylon®. Connections can also be made using heat-shrinkable polyolefin or Teflon® tubing, available through many electronic supply stores. PVC pipe is easily glued. Flexible rubber and viton, as well as Teflon® and Teflon®-lined rubber tubing, are also useful.

Pressure and Flow Measurement and Control

Constant pressures and flows (better than one part in a thousand) are required for satisfactory flame structure studies. This is best accomplished using the electronic regulation systems now available commercially. These can be programmed and provide an electrical output. This is ideal for automated data gathering. These systems are expensive. Therefore, for the budget-minded experimenter the author suggests using calibrated orifices whose flow is proportional to the upstream pressure, which can be conveniently measured. Choking or critical flow requires a pressure ratio greater than 2.5. Under these conditions flow depends only on the orifice area and the upstream pressure and is independent of variations in downstream pressure. This isolates a burner located upstream of a critical orifice, making its pressure independent of minor pump fluctuations. If another critical orifice is used for the burner input, the flame will be as stable as the regulation of the inlet upstream pressure.

Pressure

Pressure regulation is generally required. Electronic flow and pressure regulators are excellent and offer an electrical signal for monitoring. They are, however, expensive, and for the budget-minded experimentalist some of the more primitive techniques are described. When a gas is available at a constant pressure higher than required for metering, it can be easily reduced by a needle valve. Common sources of constant pressure are the laboratory atmosphere or a cylinder of liquefied gas. Mercury, water, or oil bubblers can serve as atmospheric regulators. The atmosphere provides a good, but far from perfect, source of constant pressure. Day-to-day variations amount to several percent, and the rate of drift can be as high as several tenths of a percent per hour. Fluctuations of the order of a few hundredths of a percent can occur due to wind, air conditioning systems, and slamming doors. The pressure in a liquefied gas cylinder is constant if it is thermostatted. At high delivery rates the Joule–Thompson effect cools the incoming gas from a cylinder.

Excellent commercial pressure regulators are available. There are four general types: electronic regulators, spring-loaded bypass regulators, pancake regulators, and Cartesian divers. Electronic systems are best since they provide digital readout and computer access, but they tend to be expensive. A vacuum reference is preferable with bypass regulators to compensate for barometric fluctuations. The pancake regulator utilizes a flexible diaphragm and needle valve to control the upstream flow. Therefore, it does not spill material. This offers a significant advantage, but they can only be used over rather modest pressure ranges (up to 30 psi). The Cartesian diver uses as a reference the gas trapped in a float, which controls a needle valve. It is particularly useful at low pressures, since by filling the instrument with oil rather than mercury, precision regulation can be obtained at pressures as low as 1 cm of Hg. All of these regulators can be reproducible to one part in ten thousand if properly used.

Many devices are available for the measurement of pressure: manometers, diaphragm gauges, gas property gauges, and compression or McLeod devices.

Electronic pressure transducers are convenient, accurate, and provide an electrical output for automation or digital readout (AAAS [1992]). They are commercially available but expensive. Bourdon gauges of excellent accuracy are available and convenient. The mercury manometer can easily be read to a fraction of a millimeter. With precautions, a reproducibility of 0.001 cm is possible. The principal drawbacks are problems with spillage and toxicity. Open-end manometers require a knowledge of atmospheric pressure; so that closed-end manometers are more convenient. An order of magnitude improvement in precision can be attained by replacing mercury with oil. There are a number of problems, however, with solubility of gases in the oil, surface tension, etc. Another method of improving sensitivity is by tilting the tube. The practical limit of this is a ten-to twentyfold increase in sensitivity. The combination of these techniques allows measurement of pressure differences around atmospheric pressure to 2×10^{-3} torr. Even higher sensitivities are possible, but such measurements are often limited by the time constant of the system and variations in the reference pressure. One solution is Weinberg's ingenious pressure balance [1962].

Flow Measurement

The flow of gas can be measured by many methods. Commercial electronic mass flow controllers and meters are available with digital LED readouts. They are convenient and recommended. They are, however, expensive, and where economy is necessary, simpler flowmeter can be used. The most straightforward method consists of inserting a known impedance into the flow line and measuring the pressure drop across it. This impedance can take the form of an orifice or capillary.

A particularly useful flowmeter for flame structure work is the critical orifice (Anderson and Friedman [1949]). It offers high-accuracy flow isolation from downstream pressure fluctuation, linear dependence of flow on upstream pressure, and a low dependence of flow on temperature. The principal disadvantage is the requirement of a relatively high-pressure source of gas.

The governing equation for critical flow through an orifice is given in the following equation:

$$Q = CKAP \sqrt{\frac{T}{300}} \tag{3–5}$$

C is a constant factor called the *discharge coefficient,* which accounts empirically for the effects of boundary layer in the orifice. It is nearly unity for large orifices, but is in the range 0.8 to 0.95 for the orifices generally useful in flame studies. K contains all the constants for a given gas and the standard temperature and pressure. At other temperatures a temperature correction is required. Note that the equation gives flow of gas at the operational temperature. Table III–1 lists the constants necessary to use Eqs. (3–4 and 3–5) for various gases. P_1 is in atmospheres.

Flow is as constant as the regulation of temperature and pressure. Small orifices (ca. 0.006 cm) show deviations from the linear flow due to a variation in C with pressure. This does not prevent their use, but it makes calibration necessary.

Orifices can be operated below the critical pressure drop, but it is necessary to know both the upstream pressure and the pressure drop across the orifice. The governing equation is:

$$m^\circ = CA[2\rho(P_1 - P_2)]^{1/2} \tag{3–6}$$

where ρ is the density (g cm^{-3}) and the pressure drop is in dyn cm^{-1}. Noncritical orifices are used over a wider range, which is advantageous for low flows.

Large orifices are easily constructed by machining or drilling a hole of suitable size in a metal plate or shim stock. The orifice can be either sharp edged (length negligible compared with diameter) or contoured; the advantage of a contoured orifice is that its effective area lies closer to its geometric area. Sapphire watch jewels are available with contoured holes in diameters between 0.06 and 1 mm. They are inexpensive and can be mounted in Teflon® or soft metals with a swaging tool. Jewel and swaging tools are available at supply houses and some jewelers will mount to specifications for a nominal charge. Glasses for sealing sapphire jewels directly to quartz or Pyrex® are available from Corning. Sizes are designated in hundredths of a millimeter called an "MM." Thus a 0.06 mm diameter jewel is called a 6 MM jewel by the jewelry trade.

Small contoured orifices can be constructed from quartz tubing using the techniques for making sampling probes (Chapter VIII). Holes as small as 10^{-3} cm diameter can be made. Below this point capillaries are more satisfactory.

Needle valves are available with micrometer drives. These allow reproducible variable orifices for flow control and measurement. They often come with calibration factors, but it is wise to calibrate for the gas being used.

Capillaries are useful below flows of a microgram per second. Flow follows the Poiseuille law [Eq. 3–4]. Viscosity is temperature dependent, so it is often necessary to immerse the capillary in a constant temperature bath. Viscosity values are given in Table VI–1. If the capillary is coiled, as is convenient for large L, a term must be introduced to account for centrifugal acceleration of the gas. Flow depends on the radius of the coil. This correction has been worked out (Powell and Browne [1957]), and it is possible to make useful predictions.

Capillaries show a high dependence on temperature and downstream flow, but they are convenient for measuring small mass flows. Capillaries can be made from glass or steel tubing. Glass tubing is available with bores as small as 5×10^{-2} cm. Standard tubing is normally supplied in 4-ft lengths, but lengths up to 100 ft coiled for gas chromatography are available. Stainless steel hypodermic tubing is available with diameters between 0.1 and 0.001 in. Lengths as much as 100 ft are available.

For routine measurements rotameters are commercially available covering a wide range of gas flows. They are convenient instruments but have limited precision. Commercial home gas flow meters are satisfactory at atmospheric pressure but are somewhat clumsy.

Flow calibration is straightforward using a known volume and measuring the time and pressure rise. A few precautions must be observed: (1) Pressure must be low enough so that compressibility effects can be neglected; (2) temperature must be known; and (3) the volume must be isolated from the flowmeter by a critical orifice. For very low flows calibration can be made using the soap bubble technique, in which the rate of passage of a soap film through a calibrated burette is followed. This works well except with low-molecular-weight gases such as hydrogen or helium, which diffuse through the bubble, producing significant errors.

Problem Areas

Many special problems occur in flame studies including ignition, mixing, gas handling, corrosion, and flame instabilities. Some suggestions are collected in the following.

Ignition

Ignition is difficult in a closed vessel. One convenient technique is to use a small-capillary diffusion flame as an igniter. Start up with only oxidizer flowing through the burner, then introduce a *low* flow of fuel in the capillary and ignite. If the igniter is insulated from the housing, a laboratory tesla coil can be used to produce a spark that will ignite the diffusion flame. The main burner flame can then be established by introducing the fuel slowly until the proper flow is reached. *Be careful not to have a combustible mixture in the chamber initially.*

Mixing

Flame instabilities can result from poor mixing. Initial mixing is aided by bringing in fuel and oxidizer flows normal to one another, with the smaller flow along the overall flow direction. This introduces swirl and turbulent mixing. Mixing is also improved by passing the flows through a region packed with glass beads or stainless steel wool. This volume should be minimized because it increases the time constant for changing flow conditions and makes flame adjustments tedious.

The pressure level in burner chambers is set by the inlet flow and the pump off orifice size. A convenient arrangement is to have a large and a small valve in parallel. Major changes can be made with the large valve and fine adjustments with the small valve.

Gases with known, guaranteed purity are now available from several commercial sources. For suppliers see catalogs or compilations such as the yearly AAAS *Guide to Scientific Instruments* published as an issue of *Science*. It should be recognized gas purity is limited by contamination introduced by handling, and mistakes in handling can happen. Therefore, it is wise to analyze the delivered gas. Connections should be made with clean tubing. If the gas is a strong absorber, polyethylene or Teflon® tubing may be used, but even some of these products can show adsorption. Tubing should not be reused with other gases or gases of higher purity. In handling gases, remember that they can be dangerous. High-pressure cylinders should be restrained. Toxic gases should be stored in hoods, and the burned gases should be exhausted from the laboratory and treated if they offer a pollution hazard. Combustion gases are often corrosive, and one should use components that are compatible with the gases used and the exhaust gases expected. Teflon® and stainless steel are adequate for most combustion gases. Nickel or Monel® is required for chlorine. Fluorine requires specialized techniques and should only be handled by experienced personnel.

Handling Liquids and Solids

The handling of solids and liquids offers special problems. Vaporization is always an added problem. The metering of small amounts of liquids is difficult. Small syringes driven by a timing motor are available commercially. Capillary and orifice flowmeters can be employed, and special techniques have been devised for small flows of liquids (Calcote [1950]). It is usually better to vaporize whenever possible. Devices are available for metering powders, but the problems are even more severe, and vaporization is best where it is feasible.

Flame Instabilities

Movement can be a major problem in flames. They drift, flicker, and even whistle or roar. These difficulties can often be traced to poor flow regulation or flow instabilities, which are occasionally triggered by external vibration or noise. These are not new problems (witness the flickering of candles). Because of their complexity, solutions more often resemble witchcraft than science. The position of a flame front is affected by changes in flow, pressure, temperature, or composition. For small excursions there is an approximately linear relation between the changes of these variables and movement of the flame front. Thus, if it is desired to restrict flame front movement to one part in a thousand of its thickness, it is necessary to regulate these variables. Organ pipe oscillations of long burner tubes can be a problem. Flames can amplify in phase oscillations. By changing the effective tube length or inserting screens, the frequencies can sometimes be damped. Low-frequency pressure oscillations called Helmholtz resonances can occur when two volumes are connected through a flow impedance.

A general method for combating oscillations is to raise the frequency by reducing chamber and burner volumes and damping these higher frequencies with acoustic absorbers. Oscillations can be a major problem in open-tube burners. They are of less importance with screen stabilizer burners, as might be expected, since screens and packed chimneys are acoustic dampers.

Flow instabilities in open flames have two common sources: vibration-induced eddies and the Taylor instability of a hot gas column. The former is only a problem in laboratory burner systems under unusual circumstances. The inherent instability of a rising hot gas column is a more serious problem. This is responsible for the flickering observed with candle flames and in the tips of laboratory Bunsen flames. It is often quelled by adding a chimney, which captures the hot gas column before the instability develops.

IV

FLAME VISUALIZATION AND AERODYNAMICS

Characterizing flow in flames requires both overall visualization and a set of local profiles of: gas density, ρ; gas velocity, υ; and flow geometry. The local aerodynamic variables are related through the continuity equation, which expresses mass conservation:

$$\rho \upsilon A = \rho_0 \upsilon_0 = \dot{m} = \text{const.} \tag{4-1}$$

In this equation A is the area ratio; $\dot{m}$ the mass flow per unit area; υ is the velocity; υ_0 is the burning velocity; and ρ is the density.

It is necessary to visualize the overall flame geometry to establish the coordinate system for studies of flame microstructure. Flame geometry and microstructure are often considered in separate contexts, but they are related variables. In the one-dimensional systems considered in this book, geometry is usually expressed as a stream tube area ratio, $A = a/a_0$, which is a function of distance through the flame front. It can sometimes be inferred from overall flame shape, but direct measurements are more reliable. The relation between the idealized one-dimensional geometry used in flame structure analysis and real three-dimensional flames is discussed in Chapter IX, particularly in Figures IX-1 and IX-2. Pressure plays the role of an initial condition rather than a local variable because the changes across flame fronts are quantitatively negligible at the low Mach number of most flames. There are a few exceptions such as hydrogen–fluorine and diborane–oxygen flames, but even they only require minor pressure corrections. Density is an overall variable related to temperature and composition. Velocity is the most important aerodynamic variable because it provides the time–distance relationship for the flames.

Flame Visualization and Recording

Flow visualization has been a tool in aerodynamic research since the classic nineteenth century work of Toepler [1868] using flash Schlieren method. A multitude of methods have been developed for the measurement of instantaneous fields of scalars and vectors. Techniques include tomographic interferometry (Hesselink [1988]), planar laser-induced fluorescence (Hanson [1986]), laser speckle velocimetry (Adrian [1991]), molecular tracking velocimetry (Miles et al. [1989]), and particle image velocimetry (Adrian [1991]). These fields have been reviewed in the book of Merzkirch [1974], which can be supplemented by the compilations edited by Merzkirch [1982], Yang

[1985]. More recent reviews have been made by Settles [1985] and Adrian [1986; 1991]. Not all of these methods are useful for flame structure studies. In some cases the high temperatures and steep gradients limit accuracy. On the other hand, some methods for full field visualization are an overkill for characterizing the one-dimensional flow through laminar flame fronts. Nevertheless, this treasury of techniques deserves consideration as methods improve, are simplified, and become available commercially.

Lasers have become the illumination systems of choice, but Settles [1985] makes the point that in many applications, particularly Schlieren systems, lasers do not always provide the best light source because of coherent noise. He feels that many older visualization methods, such as stereoscopic viewing and sharp-focusing Schlieren, are underutilized. Weinberg's book [1963] and reviews [1982; 1986] (Schwar and Weinberg [1969]; Weinberg and Wong [1977]) provide excellent coverage of the special techniques applicable to combustion. Hanson [1986] has reviewed planar imaging techniques for combustion.

Flame geometry can be visualized using any of these techniques. The most common are direct photography, interferometry, Schlieren, and shadow visualization. Results are usually recorded photographically, although with increasing computing power the time is fast approaching when economic use can be made of digitized electronic imaging and image processing techniques. This would open up enormous data manipulation possibilities. Electronic recording techniques are being used, but they are still expensive for common laboratory use. Digitization can either be done directly using commercially available one- or two-dimensional optical multichannel analyzers (OMAs) and Vidicons, or indirectly using optical processing techniques (Yu [1982]) on photographic negatives or prints. Both techniques take a light image, intensify it, convert the pixel[1] intensities into digital form, which is collected on tape or disc for analysis, retrieval, or reproduction. Methods differ in the mode of initial detection of the light image. The OMA technique produces an electrical signal directly using photodiodes. The photographic method uses the time-honored photochemical reduction of a silver salt to free metal.

The OMA technique is by far the most rapid and most easily manipulated electronically, but at the time of writing this volume it offers lower resolution than photography. OMAs have around ten thousand pixels with eight levels of intensity. This compares with over a billion definable elements in a high-resolution 4 × 5 in. (10 × 12.5 cm) film using diffraction-limited lenses. The numbers of definable intensity levels are about equivalent. Being less expensive and more portable also gives the photographic process (followed by digital analysis) some economic advantages at the present time, but the situation is undergoing such dramatic development that any survey of the field would be obsolete before publication. Therefore, the researcher is advised to seek out techniques that are available locally, proven, and immediately useful. If they are adequate for the problem, use them. State-of-the-art techniques often require inordinate efforts to implement that can swallow the original problem and turn it into a development effort. In our own work we use initial photographic detection followed by digitization when necessary.

Another possibility (see Settles's review [1985]) is to use video equipment, which

1. A pixel (picture element) is the minimum resolvable area in an image.

is now mass produced and therefore economical. The results are stored on magnetic tape. Photographic data can be digitized using commercially available analog–to–digital equipment and stored on a small computer. The image manipulating and enhancing software, comparable to that used for photographic analysis, is available for personal computers, but none of the programs has been adapted directly for flame work. As an electronic detector, standard video equipment offers some of the advantages of the OMA at lower cost. It lies intermediate between the OMA and the photographic image enhancement method since the initial detector is electronic, but the recording is analog. Writing speed is modest (100 nanoseconds per pixel); minor modifications can widen its utility. For example, it might be used in place of a fast-scanning linear OMA. Because of the rapid development of the field, look at your resources first!

Flame Luminosity and Photography

The most conspicuous aspect of flames is their luminosity. They are chemiluminescent (Gaydon [1974]; Marodineamu and Boiteuz [1979]), and most of the visible light originates in the primary reaction zone. Radiating species include C_2, CH, HCO, and excited formaldehyde (OCH_2), whose band spectra can be resolved in low-pressure hydrocarbon flames. There is also a low-level continuum radiation in the blue attributed to the CO + O recombination (Gaydon [1974]). This peaks in the luminous region but extends over the entire equilibration zone. The hot gases also radiate thermally in infrared due to emissions from water and carbon dioxide. The sharp luminous zone provides a useful marker of the region of radical maximum since luminosity depends quadratically on the radical concentration. This occurs because most of the radiation in the visible is due to radical–radical bimolecular recombination. By contrast, the CO continuum, which depends linearly on radical concentration, outlines the radical equilibration region. Thus, visible flame radiation provides a good qualitative picture of the reaction regions and their relative radical concentrations.

Probe and thermocouple positions can be measured photographically, and this provides a convenient record of results. Precision of a few micrometers is feasible with good optics, but problems exist. Images are biased since the flames do not lie in a single focal plane (see Fig. IV-1). Another limitation is the refraction of the emitted light by the density gradients of the flame (Fig. IV-2). This varies inversely with pressure and

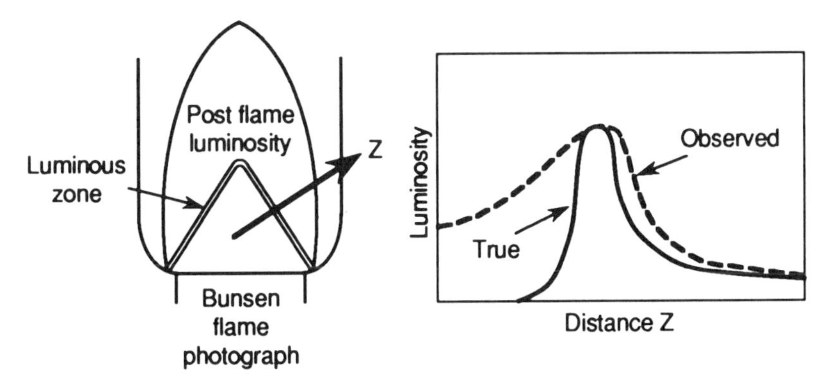

FIG. IV-1 Image distortion induced by out-of-plane light.

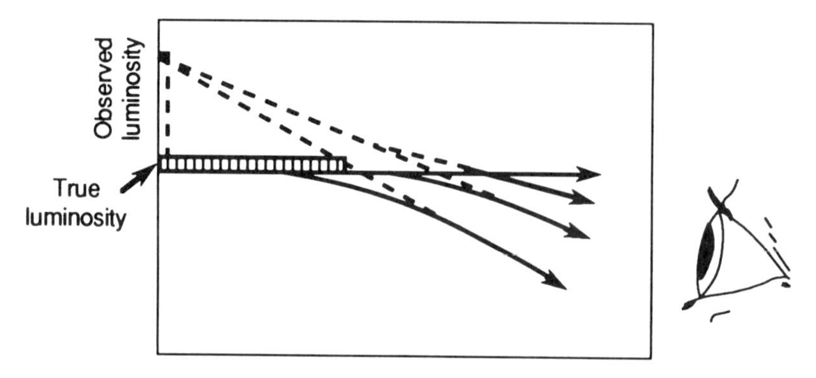

FIG. IV-2 Distortions of luminosity induced by density gradients.

becomes negligible below 0.2 atm. These effects can be minimized using sharp focusing with high-*f*-number lenses (i.e., large diameter with a high magnification ratio) and by blocking out the central cone of light. Luminous zone thicknesses above atmospheric pressure are dominated by this distortion. For purposes of establishing flame front geometry, most of these limitations can be minimized, and direct flame photography remains a popular technique that provides a convenient record of results. A comparison of several types of flame photography of Bunsen flames is given in Figure IV-3. In the not too distant future the new digitized photographic methods coupled with computer recording will open up new avenues for the analysis of flame photographic data.

In designing photographic systems one should consider: required field size, resolution, permissible distortion, object brightness, and exposure time limitations. These

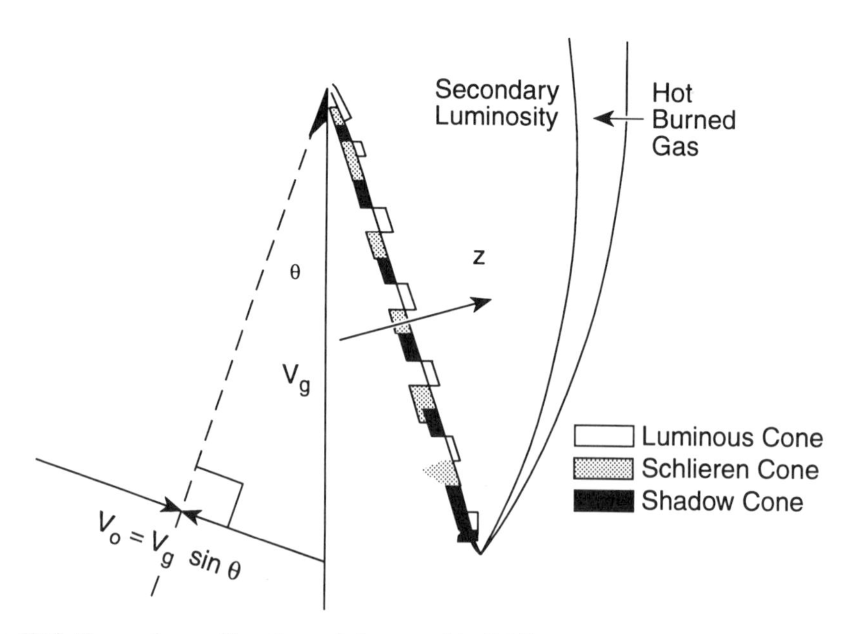

FIG. IV-3 Comparisons of locations of photographic, Schlieren, and shadow images of a Bunsen flame. (After Andersen and Fein, Univ Wisconsin Laboratory [1949].)

factors dictate optimum choice of lens, shutter, magnification ratio, film size, and speed.

A simple camera with manual adjustments is preferable to the modern automated models unless they are provided with manual override. Used camera stores usually offer a variety of choices in a range of prices. The features of a portrait camera are often convenient. One should use the largest film with the highest magnification; once the picture is on film, resolution for subsequent magnification is limited by film grain size. The quality of today's good camera lenses is high, but they should be checked for distortion around the edges of the image. This can be done by photographing graph paper under the expected conditions and examining the picture. This can provide reference for corrections. The lens size and *f* number should be chosen to provide an image of convenient brightness for visual inspection. Image brightness depends quadratically on the magnification ratio; so for high magnifications a high-speed (low-*f*-number) lens may be required for visual observation and manipulation. Most lenses operate best at *f* numbers between 4 and 8 because distortion increases with decreasing *f* number, while diffraction becomes a problem for *f* numbers above *f*8.

The lens should be chosen by considering the size of field to be covered. Most often, the field is small and an enlarged image is desired. In this case, it is good practice to reverse the lens unless it was specifically designed for photomicroscopy. This is the case because most camera lenses are designed for a large field with a small image. Thus if the required field is a few centimeters in diameter, a 35 mm camera lens is a good choice. With a large field, a 4 $\times$ 5 (in.) camera lens would be better.

Exposure time is controlled by the shutter. There are two common types: the compour shutter, which closes like an iris in a plane between the lenses, and the focal plane shutter, which is like a slot in a blind drawn across the film plane. The fastest compour shutters are usually 0.005 sec. There is no particular lower limit to the effective exposure time of a focal plane shutter since the sweep speed can be increased within reason and the width of the slit can be cut until diffraction becomes a problem. Practical limits lie below a microsecond. The focal plane shutter produces a distortion since each line on the film is exposed at a different time and one can get the "elongated baseball" type of distortion with moving events. Often this can be compensated by arranging the shutter direction so that the simultaneous events of interest lie along the shutter line. At their highest speeds, compour shutters expose the center of the image longer than the edges, so that effective exposure times are shorter than true open.

Most cameras have two synchronization contacts: (1) an instantaneous one that triggers at the moment of shutter initiation for electronic flash lamps; and (2) a trigger that is delayed by about twenty milliseconds for magnesium and zirconium flash lamps to compensate for their ignition delay time. The inverse problem of synchronizing a shutter with an external event can be accomplished using solenoid releases. The commercial devices have a least count time of about a millisecond. For more precise timing, external shutters such as Kerr cells or Faraday cells are available. Rapid streak and framing photographic recording methods have been developed that can record microsecond and shorter events (Rabek [1982]; Courtney-Pratt [1983]) using rotating drum and rotating prism cameras. For high framing rates, electronic methods can be superior to photographic recording methods.

A variety of films are available. Factors that should be considered are film speed and film resolution. These tend to be reciprocal elements because with proper development a grain requires a certain minimum number of quanta for exposure and this limit cannot be reduced. The highest speed with acceptable grain is the optimum choice. A major consideration for laboratory work is the mode of development. Conventional photographic film requires either a local darkroom or that pictures be sent out for development. This is inconvenient since one may wish to make adjustment according to the photographic results. The solution is to use self-developing films that require less than a minute to produce pictures. Most sizes, speeds, and color are available, with adapters for many cameras. In measuring film and prints, one should remember that film shrinkage may be neither isotropic nor linear.

Deflection–Diffraction Imaging

Flames may be visualized using the light deflected by the steep density gradients in flame fronts. The interpretation of the observed patterns of transmitted light is a complex exercise in physical and/or geometric optics. There are three limiting cases: (1) interferometry where the pattern deflections are proportional to density; (2) schlieren photography which is a dark field method where the image depends on the gradient of density; and (3) shadow photography where the deflections also depend on the gradient, but because of overlap the images outlined by the caustics are related to the second derivatives of density. These divisions are not rigid and intermediate systems are often used. Weinberg has written a definitive monograph [1963] and several reviews [1977; 1982] on the subject.

Interferometry

This technique utilizes the interference occurring between a coherent beam passed through the region of study and a comparison beam. When properly combined on a screen, the two waves interfere, producing an image outlining regions of differing optical path length (Fig. IV-4). This can be interpreted quantitatively in terms of density if the geometry is known (el Wakel [1976]). The technique is also discussed in connection with temperature determination from density measurements (Chapter V).

There are a number of optical arrangements for forming paired coherent beams, with one passing through the region of study and finally recombining them. Lens, mirrors, or gratings can be used. Weinberg [1963–1982] discusses many of these systems from the standpoint of combustion studies.

Schlieren Imaging

Schlieren imaging is a dark-field method that is most easily understood in terms of geometric optics. A beam of parallel light from a slit or point source is passed through a region of varying refractive index such as a flame and brought to a focus on a stop. This can be a knife edge, a wire, a small spot, or grid (Fig. IV-5). Regions containing varying refractive index will be deflected around the stop and can be brought to a focus, producing an image connecting regions of the same refractive index gradient. The stop allows index sign to be differentiated. If a knife edge is used with a slit source, only single direction gradients will be recorded. Depending on placement of the edge, the system can be made more or less sensitive. Horizontal/vertical deflection of either sign

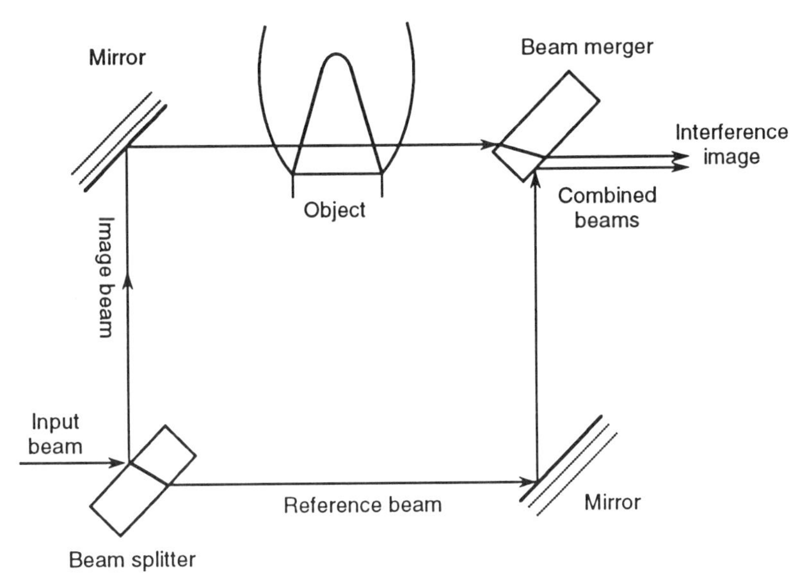

FIG. IV-4 Interferometry. Scheme of Mach Zehnder Interferometer.

can be isolated. These deflections can be interpreted quantitatively (Dixon-Lewis [1954]; Dixon-Lewis and Wilson [1951]). Deflection depends directly on index gradient. The usual Schlieren image of a flame locates a surface close to the region of initial temperature change because it measures absolute density gradient rather than fractional change (Weinberg [1963]). If a grating is used as a stop, a series of contours is produced that connect regions of equal gradient. This "Ronchi grating" system is useful for the quantitative work [Fig. IV-5(b)]. An interesting variant is the multiple source/stop "focusing Schlieren" of Kantrowitz and Trimpi [1950]. This system can be focused so that it is sensitive to gradients in a single plane.

Shadows and Moire Effect

Shadows are projected by a uniform beam of light passing through a flame. Either a divergent point source or parallel beams can be used. The sun or even a distant light bulb allows visualizing the steep gradients found in flames. Deflections in shadow photography are proportional to the gradient in density and a complex pattern results (Fig. IV-6). The surface visualized by a shadow picture is the caustic that marks the surface where the second derivative of density passes through zero. Weinberg [1963] has pointed out that because temperature is inversely proportional to density this occurs at a relatively low temperature in flame fronts. The deflection of single lines or grid images formed by parallel beams has been applied by Weinberg [1963–1982] to determine density and temperature profiles.

Full-field deflection mapping using diagonal parallel grids or Ronchi gratings is sometimes called the *Moire effect.* A recent example of such a study is that of Rau and Barzziv [1984].

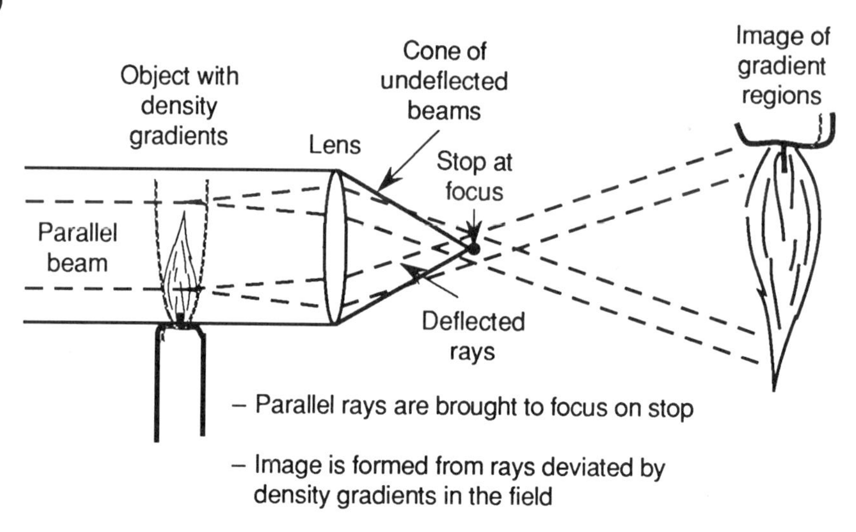

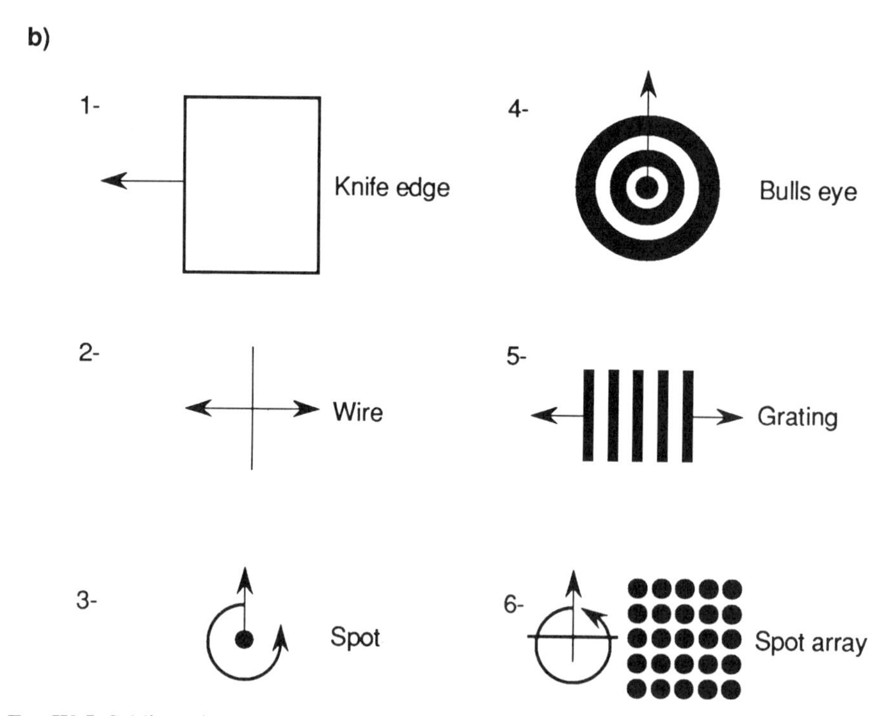

FIG. IV-5 Schlieren imaging. (a) Schematic of optical principles. (b) Source/stop geometries. Arrows mark directional sensitivity: (1) line source—knife edge stop (may be used at any angle); (2) Slit source—wire stop (visualized gradients normal to slit); (3) Point source—spot stop; (4) Point source—Bullseye "Ronchi" stop; (5) Line source—"Ronchi" grating; (6) Multiple point source—spot array for focusing Schlieren.

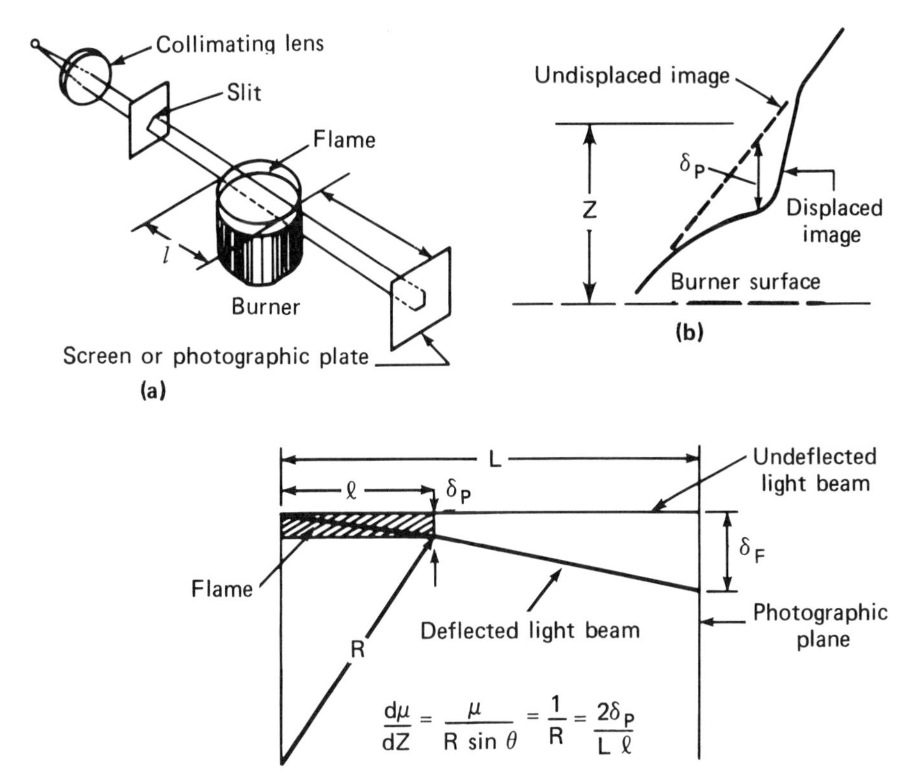

Fig. IV-6 Quantitative shadow. (After Weinberg, [1963] with modifications.) (a) Schematic with interposed diagonial slit to allow quantitative measurements of the image. (b) Image displacement analysis.

Holography

Holography[2] is the technique of recording multiple interferograms. This allows a three-dimensional reconstruction of the image. Transparent objects such as flames can be visualized using phase contrast techniques. Since interference requires coherent light, the usual source is a laser (see, for example, Fig. IV-7), but this is not an absolute necessity. Holography was developed by Gabor before the invention of lasers, and holograms that can be viewed in normal light are sold in novelty stores. They even grace stamps and magazine covers. However, for quantitative analysis, coherent monochromatic light is desirable.

A number of different techniques are employed for flame holography. These include (1) double exposure; (2) time lapse; (3) real time; and (4) time average. Some applications are discussed in the chapters on laser methods and temperature (Chapters VII and V).

2. Single-pass holography image is identical with a Mach Zehnder interferometric image. The difference and main advantage is that the hologram is obtained with a single path and path length errors are eliminated.

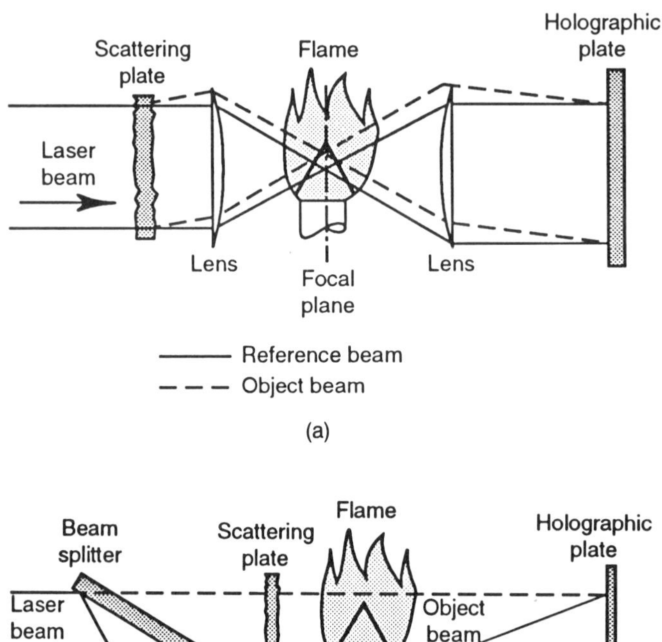

FIG. IV-7 Flame Holography. (After Merzkirch [1974].) (a) "In line " apparatus schematic; (b) "side-band" apparatus schematic.

The advantages of holography are high sensitivity and the three-dimensional visualization. Another advantage is the ability to average out optical path distortions common to test and reference beams. Disadvantages are expense and vibration sensitivity. Trolinger [1976] and Vest [1979, 1982] have reviewed the state of the art including the on-line production of holograms, heterodyne interferometry, and cine holography.

Tomography

Tomography is the technique for reconstructing sections through three-dimensional flows from an array of data taken at various viewing angles. The method, which is quite old, has recently been perfected for medical use in radiography. The subject has been reviewed by Vest [1982]. Deconvolution was a major problem prior to the recent and ongoing revolution in computing techniques. In the past only the cylindrically symmetric case using the classic Abel inversion was considered feasible. At present, almost any geometry appears feasible. Studies have been made of Bunsen flames (Faris and Byer [1988]) and diffusion flames (Chen and Goulard [1976]; Santoro, Semerjain, Emerson, and Goulard [1981]; Dosanji [1971]; Ravichandran [1988]). Tomography can be made using interferometry, holography, absorption, or beam deflection. The most sensitive of these is laser beam deflection, which Faris and Byer [1987] used to visualize jets expanding into a vacuum (Fig. IV-8).

Light Scattering Methods

Flames can be visualized by scattered light. The planar imaging of flames has been reviewed by Hanson [1986]. Scattering can occur from edges, small particles, or molecules. Diffraction is discussed in books on physical optics such as Wood's [1934] classic text.

Particulate Scattering

The light scattered from particles can be used to define regions in a flame and outline its geometry. Temperature regions are outlined because gas density changes reduce local particle concentration and hence scattered light. The particles can either be present in situ in the flame or be introduced as a tracer. If particles react or are formed in a flame, the reaction or nucleation surfaces are defined. Beyond this primitive method there is a class of velocity-sensing methods known as "pulsed light velocimetry" (PLV). These techniques allow the determination of local velocity in an illuminated region from the scattered light. The illuminated region can be a single focused point, or with planar illumination a complete field can be interrogated. The technique consists of photographically imaging the movement of particles using two or more short-duration flashes of light separated by known time interval(s). This multiple exposure image can be interrogated manually using the particle track method, in which velocity is derived by measuring the separation of paired images and applying the fundamental definition of velocity as the distance moved divided by the time interval. This was

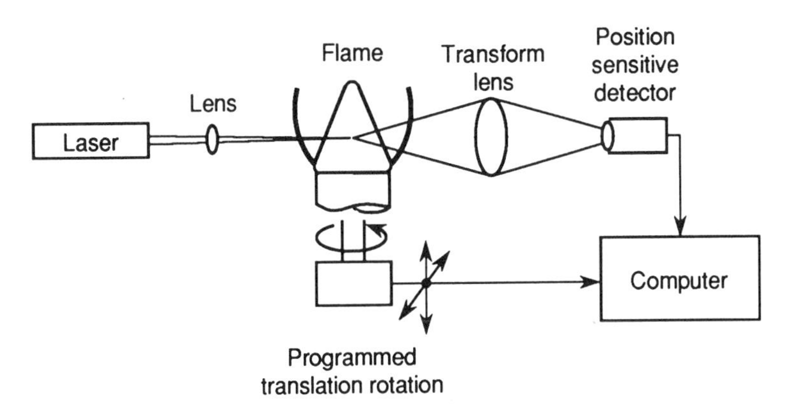

FIG. IV-8 Schematic of Apparatus for Laser Tomography of Flames. (After Faris and Byer [1987].)

applied by Fristrom, Avery, Prescott, and Mattuck [1954] to derive velocity profiles (see Fig. IV-9). This cumbersome manual method still finds some use in simple studies, but in more complex fields it has been supplanted by several methods that are automated and make use of computerized data reduction. The field has developed into a powerful diagnostic tool for studying complex flow fields. The area has been reviewed by Adrian [1991], who classifies pulsed light velocimetry (see Table IV–1), according to the markers used and the density of the images. Although the information is somewhat of an overkill for premixed laminar flame studies, they are of considerable importance for general combustion studies and are worth outlining here. Photochromic and fluorescent markers can also be used, but they have not found general use in flames because of the high temperatures involved. They are, however, useful in turbulence studies in cold gases and liquids. With respect to image density there are three modes: (1) the particle tracking mode, where the concentration of particles is so low that there is only a low probability of overlapping images and individual particle tracks can be interrogated; (2) the laser speckle velocity (LSV) mode, where particle concentrations are so high that the images of particles overlap in the image plane and the random phase differences between images create random interference patterns known as laser speckle; (3) in between these two limits lies the high-density PIV mode, where density is sufficiently high that every interrogation spot has many images but not so high that image overlap is probable. In this case velocity is determined by the separation of small groups of images. In some cases multiple coded pulses of light are use to improve the identification of groups. The methods merge into one another. Adrian [1991] gives a careful discussion of the relations, methods of analysis, and sources of error. A short discussion of the limiting methods is given in the section on local velocity measurements together with a discussion of particle introduction.

Rayleigh Scattering

This is the strongest molecular scattering process for normal molecules. It originates from the light interference associated with the inhomogeneities in gases due to statistical density fluctuations in regions only a few mean free paths across. It is density dependent, but since each molecular species has a characteristic cross section, it is also somewhat species dependent. An example from a study of eddy combustion is given in Figure III-12. The scattered light is diffracted, but not changed in wavelength. More details on Rayleigh scattering can be found in Chapter V on temperature and Chapter VIII on laser diagnostics.

Fluorescence

Fluorescence results from the excitation of molecular energy levels that have short radiative lifetimes and hence reradiate. This can be a very strong scattering process. Molecules that show fluorescence in the visible are generally radicals (molecules with an odd number of electrons). This usually requires excitation by an ultraviolet source.

Planar laser-induced fluorescence (PLIF) is a powerful diagnostic technique that lends itself to mapping. Hanson has reviewed the field [1986]. It has been used to make measurements of velocity in low temperature flows, both subsonic and supersonic. It has also been used for measurements of OH, NO, O_2, and Na concentrations and tem-

(a)

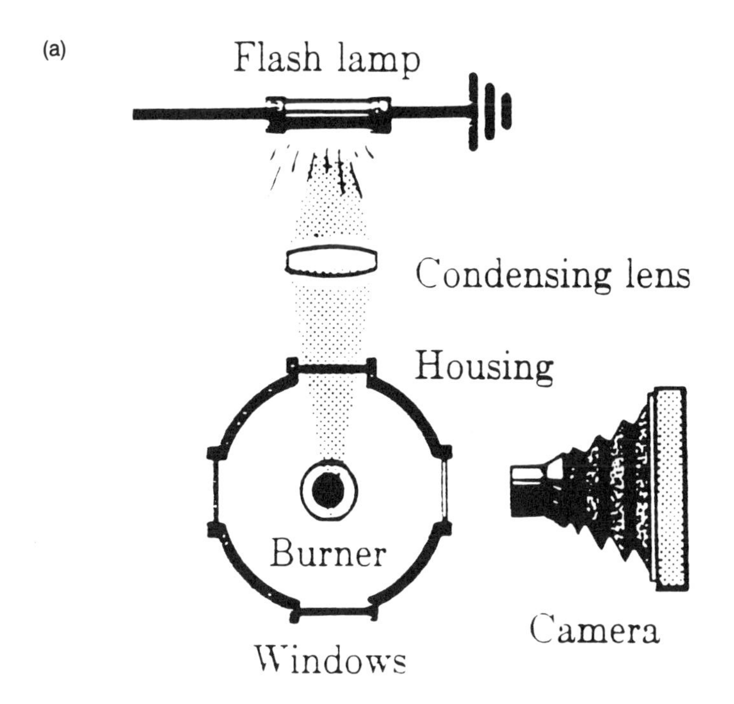

(b)

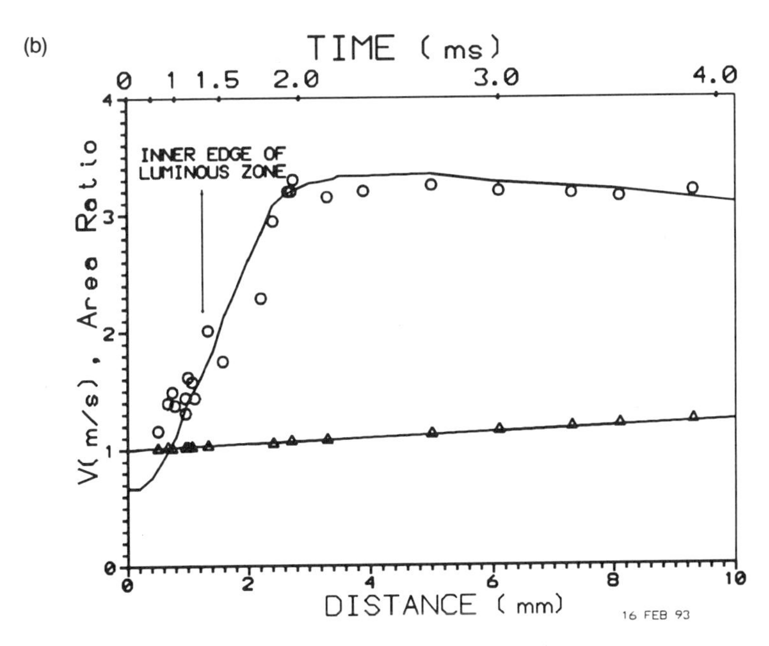

FIG. IV-9 Particle Track Studies. (a) Optical arrangement. (b) Measurements (circles) compared with $\upsilon = \upsilon_0(T/T_0)(M/M_0)$. (After Fristrom, Grunfelder, and Favin [1960].)

Table IV–1 Pulsed Light Velocimetry[a]

Density	Velocimetry	Comments
$N_1 \ll 1$	Low-density PIV	Images sparse in interrogation region
$N_1 \gg 1$	High-density PIV	Images dense in interrogation region
$N_2 \gg 1$	Speckle LSV	Images overlap and interfere

[a]Adrian's [1991] classification of pulsed light velocimetry using particles.

perature fields in flames (Kychakoff et al. [1984]). A schematic of a PLIF apparatus is shown in Figure IV-10. Some details are discussed in the chapters on concentration and laser diagnostics (Chaps. VI and VII).

Ramonography

Raman is the weakest of the first-order scattering processes (Lapp and Penney [1974]). Experimentally it occurs at any wavelength, but the strength of the scattering varies as the fourth power of the wavelength and quadratically with power level. Even in the visible region with focused laser excitation, the effect is weak. The technique is also discussed in the chapter on laser diagnostics. Raman scattering results when a molecule interacts with a light quantum absorbing (or emitting) its own vibrational quantum. The scattering process is nonlinear and mixes the two frequencies so that the scattered light is shifted by the characteristic molecular absorption. A downward shift is called *Stokes radiation,* while an upward shift is called *anti-Stokes.* The scattered light contains a shifted characteristic vibrational–rotational spectrum of the mol-

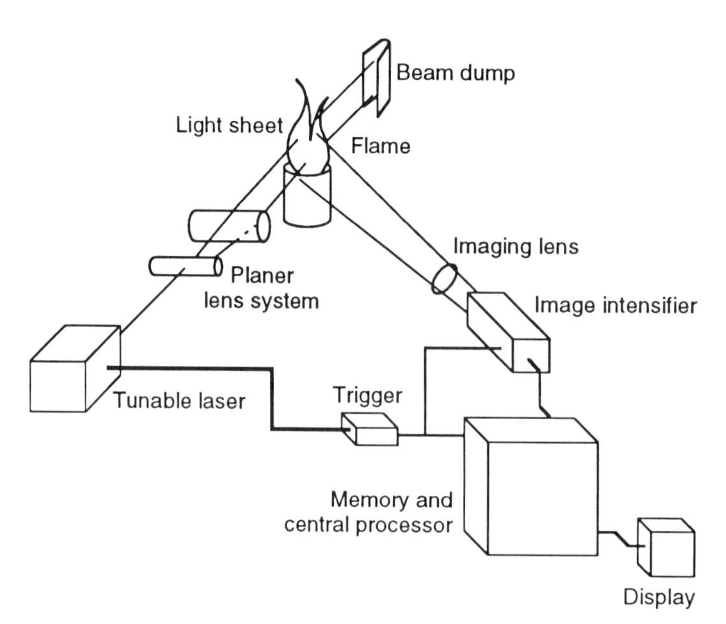

FIG. IV-10 Schematic of Planar Laser Fluorescence apparatus. (After Hanson [1986].)

ecule. This spectrum can provide information on both composition and temperature in flames. Where it is feasible to isolate characteristic portions of this spectrum, maps can be produced that outline the distribution of particular species (Hartley [1974]; Schefer et al. [1983]). These shifts are small, however, and isolation is difficult. The weakness of this effect has limited its use in mapping. Point measurements of local density, temperature, and composition are much easier. These measurements are discussed in their respective chapters (Chapters V and VI) and the chapter on laser diagnostics (Chapter VIII). Experimentally the major problem is the suppression of stray light from the incident beam and scattering from the Rayleigh processes and particles.

Stream Tube Geometry

The geometric aerodynamic variable in the continuity equation is (Eq. 4–1) stream tube cross section. As suggested in the preceding section on visualization, it is related to the overall flame geometry. In symmetric situations such as spherical or cylindrical flames, the geometry allows direct derivation of the stream tube function, but not all systems are so simple; so direct measurements are desirable. The area changes across the primary reaction zone in flames are of the order of the ratio of the flame front thickness to overall flame dimensions. A typical value would be 15% (see Fig. IV-9). In very large burners the change can be negligible, but it is always desirable to measure it directly. Area ratio affects velocity linearly (eq. 4–1). Its effects on the other variables are second order and are usually neglected in data analysis (see Chapter IX). If velocity, temperature and molecular weight (composition) measurements are available, area ratio can be inferred from the continuity equation and the perfect gas law (see Chapter IX).

Particle Streak Method

The most direct method of measuring stream tube area is the particle track technique previously described in the section on velocity. A long-duration light source (continuous or pulsed) is required. If pulsed single point particle injection is used, a single flash can be used to outline the streamlines. Area ratio can be inferred from measurements on a single streamline if the assumption of axial symmetry is used. It is better practice to use two or more streamlines. In an early study on flame structure, measurements on streamlines were used by Fristrom, Prescott, Newmann, and Avery [1953] to derive velocity profiles from which density and temperature profiles were inferred. This method is now obsolete because several undesirable assumptions are required and direct measurements are now possible.

Nusselt Number Method

An ingenious method of inferring the area ratio in flames was used by Peeters and Mahnen [1973]. They made use of hot wire measurements of Nusselt number in the cold flow jet of a flat flame burner. By combining this with an empirical relation between Nusselt number and Reynolds number, they were able to infer the change in velocity and hence infer area ratio change. Since the measurements were made in cold flow, a small error results from the effect of the flame front pressure drop on the flow. The area ratio is a slowly varying function; therefore, this error is probably not serious. A similar result could have been obtained by using the hot wire as a calibrated velocity probe.

Local Density Measurements

Density can also be derived from temperature and composition measurements through the equation of state:

$$\rho = \frac{PM_{av}}{RT} \tag{4-2}$$

$$M_{av} = \sum_{i=1}^{n} X_i M_i \tag{4-3}$$

In these equations ρ is density (g cm^{-3}); M is molecular weight (g/mole); P is pressure (atm); R is the molar gas constant (22,415 cm^3 atm mole^{-1}); T is temperature (K); X is mole fraction (dimensionless). The subscript zero designates inlet conditions; the subscript i is a species index; subscript av indicates an average.

Local density can also be derived from measurements of local velocity and stream tube cross section using the continuity equation (Eq. 4–1).

Experimental methods for measuring density include: (1) absorption by X-rays or alpha particles; (2) light deflection, absorption, or scattering, and measurement of the local speed of sound.

Since in flame studies density is commonly used to derive temperature or check consistency of measurements, discussion of these techniques will be deferred to Chapter V.

Local Velocity Measurements

Local velocity is the most important aerodynamic parameter for flame microstructure since it provides the time base. We have discussed some methods for velocity measurement in the section on visualization. Not all of these techniques are suitable for velocity profiles due to deficiencies in spatial resolution or inapplicability to regions of high temperature. Nevertheless, a number of direct techniques have been successfully used and will be discussed. Local velocity can also be derived from the continuity equation, providing stream tube geometry is measured or known from symmetry considerations, and the initial conditions (ρ_0, v_0) are specified.

Particle Tracer Methods

Particle tracking, one of the most common direct techniques for measuring velocities, is borrowed from aerodynamics. Small particles are introduced and their paths through the flame front are visualized as a function of time. This can be done by a number of methods: (1) the direct methods discussed in the section on visualization; (2) LDV (laser Doppler velocimetry); (3) laser speckle velocimetry; and (4) the "chirp" method. Particle methods suffer from some common problems: (1) Particles can affect flame propagation; therefore, the number and size of the particles must be small and they should be inert; (2) the boiling point of the material must be high enough so that the particles survive unless provisions are made to track the vapor cloud; (3) particles must be small to follow the velocity changes in flames (for example, the peak acceleration in a stoichiometric propane–oxygen flame at atmospheric pressure is $10^3 ms^{-2}$); (4) however, the particles must not be too small because of the thermomechanical effect (Waldman [1959]). This is the macroscopic analog of thermal diffusion. It is the force that a particle feels in a temperature gradient because the molecules on the hot side are more energetic than those on the colder side. In flames this sets a lower limit on the particle size that can be used for quantitative tracking studies.

One practical problem with measurements involving particles is the accumulation of material on windows and exposed optical elements. This is best avoided by minimizing the numbers of particles and by protecting optical elements using a buffer stream of gas.

Table IV–2 lists a number of particulate materials with their properties and an indication of the mode of application in flame studies.

Particle Generation and Sizing

It is desirable that particles be of a uniform, known size. If the source is a pulverized or precipitated solid, the material can be sized using sieves and/or any of the air elution separation methods. These techniques have been well developed by the aerosol community (Nieser, Wurster, and Haas [1981]; Henrich [1980]), and it is possible to select a narrow cut of particles with similar aerodynamic diameters. Particles are generally not spherical but, no matter what the shape, the important parameter is the ratio of weight to aerodynamic drag. Their aerodynamic diameter is the diameter of an equivalent sphere of the same density. Sized particles and commercial particle separators are available, but it should be remembered that particles tend to agglomerate so that,

Table IV–2 Physical Properties of Some Particulates Used in Flame Studies

Material	Melt (K)	Boils (K)	g/cm^3 at 298 K	Comments
Al	933	2325	2.7	Burns
NH_4Cl	—	800	1.53	Sublimes
$TiCl_4$	250	420	1.73	Forms TiO_2
ZnO	2225	—	5.47	Reduced
Al_2O_3	2275	2500	4.00	Good
Glass	675	—	2.24	Reflective
MgO	3025	3875	3.65	Good
SiO_2	2000	2500	2.65	Good
TiO_2	2400	$>$3000	3.84	Good
ZrO_2	3275	$>$4000	5.73	Good

unless precautions are taken, heavy composite particles will be formed and may skew the measurements. For this reason it is desirable to have a direct path from the particle sizing device into the flame. When this is not feasible, care should be taken to avoid the errors introduced by the larger particles. This can be done by "conditioning" the data using only those measurements associated with small particles.

Particles need not be initially solids; they can be introduced as dissolved solids in an aerosol mist that later evaporates, leaving the involatile particle. Another method is to produce particles by reaction, for example, $TiCl_4 + 2H_2O \rightarrow TiO_2 + 4HCl$. This produces a TiO_2 dust. Aerosol generation for introduction into flames is an integral part of the "atomic absorption" method (Mavrodineanu and Boiteux [1979]) long used by analytical chemists. They have devised many ingenious schemes for producing clouds of aerosols. These methods have been reviewed by Mavrodineanu [1967].

One reliable method of producing a monodispersal with uniform size and shape is by evaporation of droplets of salt solutions. This makes use of the instability of a small jet to the combination of an acoustic disturbance and the strong surface tension forces. Droplets of very uniform size can be produced using a thin stream of liquid acted upon by a single-frequency acoustic disturbance. Droplets can be produced either as a single stream or as a cloud. Commercial devices exist for producing and measuring such mists. If a suitable material is dissolved in the liquid and the solvent evaporates during transfer to or through the flame, one is left with uniform crystalline particles. In the interim before evaporation, the liquid droplet can serve as the aerodynamic tracer. This technique is particularly useful if it is desirable to minimize the number of particles and their interference with the flame. Commercial ink jet and powder jet typewriters offer components for making controllable powder and droplet generators. There is no lower limit to the minimum size produced since vanishingly small concentrations of salt can be used. However, in practice this can introduce problems associated with the vapor cloud formed by the solvent.

Particle Injection

There are almost as many methods of introducing particles as there are methods of producing them. The most common method is to have a reservoir of particles that can be introduced into the main flow. Some size control can be had by waiting for settling. Gaydon and Wolfhard [1979] recommend using a steel wool pad saturated with powder and tapped when particles are desired. These methods usually introduce more particulate than is desirable. For this reason it is common to restrict the particulate-laden flow to as small a region of the flame as possible, preferably in the plane where measurements are made. To a degree, this can be accomplished by suitable ducting. For example, Lewis and von Elbe [1987] "colored" the center of a tent-shaped flame to make sodium emission temperature maps (Chapter VII). We have used a dust-laden pipe cleaner stretched across the burner diameter as a line source of particles (Fristrom, Avery, Prescott, and Mattuck [1954]). Wibberly [1986] has devised an ingenious feeder. This principle can be extended so that only a single stream tube is injected. The limit is attained using the single droplet generator, which introduces a single particle in the flame at a given moment. This avoids the problems with the effect of the injected powder on the flame, since the path of a single particle is determined by the overall flow pattern and is only perturbed by local action of the particle itself.

Illumination Sources

Lasers offer very powerful light sources and are the illumination system of choice when they are available. Chapter VIII contains a discussion of the properties of some of the commercially available lasers. The light can be focused into a sheet using cylindrical lenses. The resulting light intensity can be high, and background scattered light can be almost completely eliminated by using a beam dump chamber. The principal disadvantage is the high cost. Unfortunately, the brightest visible lasers are pulsed and offer relatively low repetition rates (1–100 cps). The copper vapor laser, which can be operated as high as 10 kc, appears uniquely suited to flame-flow visualization. Continuous lasers are many orders of magnitude lower in intensity than pulsed lasers (for a given average power). They are suitable for some problems and can be modulated mechanically with shutters.

A convenient inexpensive source is the commercial flash bulb, which has a duration of about 0.02 sec and sufficient luminous flux to allow the visualization of 5 μm particles. For very slow flames where this time is not sufficient, several lamps can be used synchronized to follow one another. Modulation can be obtained by using a chopping wheel. It is convenient to use a lens system to focus a band of light on the chopping wheel slot and the slot on the burner axis. By using long-focal-length lenses, an approximate plane of light can be defined. With modern high-speed films, light intensity is not usually a major problem, and the brute-force light sources such as the ballisticly switched lamps (Fristrom [1953]) are no longer required. A high-pressure mercury arc modulated by the AC line voltage has been used (Weinberg [1963]). Commercial electronic flash lamps offer convenient powerful light sources. They can be synchronized with an external signal within a fraction of a microsecond.

Particle Modulation

Modulated particle injection can be accomplished using a single droplet or particle injector controlled by a repetitive pulse generator such as the commercially available ink jet printers. The virtue of the particle modulation method is that it minimizes the amount of material injected, keeping the system clean, minimizes stray light background, and makes the problem of interaction between flame and particles a moot question since the particle path depends on overall flow in the flame and negligibly on local irregularities induced by the particle. If single-particle injection is modulated, a single flash picture yields a series of images of successive particles along the same streamline with a separation determined by the particle modulation frequency.

Space Modulated Light—the "Chirp" Method

If the image of a grid is focused on the flame front (Fristrom, Jones, Schwar, and Weinberg [1971]; Cheng, Popovich, Robben, and Weinberg [1980]), a series of light pulses are scattered when the particle crosses the illuminated grid. Their time separation is related to the particle velocity. The grid should be aligned orthogonally to the flame front with its origin at a known point in the flame, and with a spacing significantly greater than the particle diameter. The resulting train of pulses can be interpreted in terms of a velocity profile, providing the absolute position and spacing of the grid are known. It yields a measurement of the velocity component along the direction of the grid. The method requires low particle densities to avoid confusing pulse trains from multiple particles. Since direct visualization is not involved, high-f-number collection optics with high-sensitivity detectors such as photomultipliers can be used.

The original application of this method by Fristrom, Schwar, Jones, and Weinberg [1971] used a small grid with uniformly varied spacing covering the range of expected particle diameters. The resultant output was a train of pulses. Hence the name *chirp method.* The modulation showed a minimum when the particle diameter matched the grid spacing. This provided a local measurement of both particle size and velocity.

Speckle Velocimetry

A relatively new (1988) and potentially very powerful velocity measuring technique is called *speckle velocimetry.* This is one of a class of velocity field techniques that were discussed in the section on visualization. The speckle method allows instantaneous measurement of complete two-dimensional velocity fields with spatial resolution in the image plane of around 1 mm and a precision of a few percent. Even movies can be made using pulsed lasers such as the copper vapor system, which can be operated at speeds as high as 20 kilocycles (Smith [1988]). This is an overkill for most flame studies but opens up new vistas for studying flame aerodynamics in complex time-dependent systems. Bibliographies on the technique are available in the reviews of Dudderar and Simkins [1987] and Adrian [1991]. The method is based on Young's classic optical experiment, which showed that the far-field diffraction pattern for a pair of small apertures illuminated by a coherent wave front is a pattern of parallel fringes (Fig. IV-11). Fringe separation depends on the ratio of the wavelength of the illuminating beam to the separation of the apertures. If more than one pair of apertures lie in the illuminating field, the pattern is strengthened, providing the pairs are equally separated and aligned. A pattern of this type can be produced from the photograph of a flowing field of particles illuminated by two (or more) flashes of light separated by a known time interval. Speed is obtained from the spacing of the fringe pattern, which is proportional to the separation of illuminating spots. The direction of the velocity vector is given by the angle of the fringe pattern. Local velocity is given by the separation of the particles divided by the time interval.

To obtain satisfactory fringes, the illuminating flash must be short enough so that the particle image is sharp, and the region interrogated must be homogeneous in velocity. These conditions can be conveniently met using a planar sheet of pulsed laser light for illumination and interrogating the resulting photograph with a laser beam reduced to the size required for satisfactory spatial resolution. As with all particle-seeding techniques, the results can be no better than the fidelity of the particles in following the flow fields. For this reason, particle size should be as uniform as possible and chosen small enough to follow the accelerations but large enough to avoid thermomechanical effects.

This is a very powerful technique, and the data accumulation potential is so massive that the realization of its full potential requires an automated readout system. This approach appears feasible, and systems for massive accumulation of flow field information in flames are being investigated (Smith [1988]).

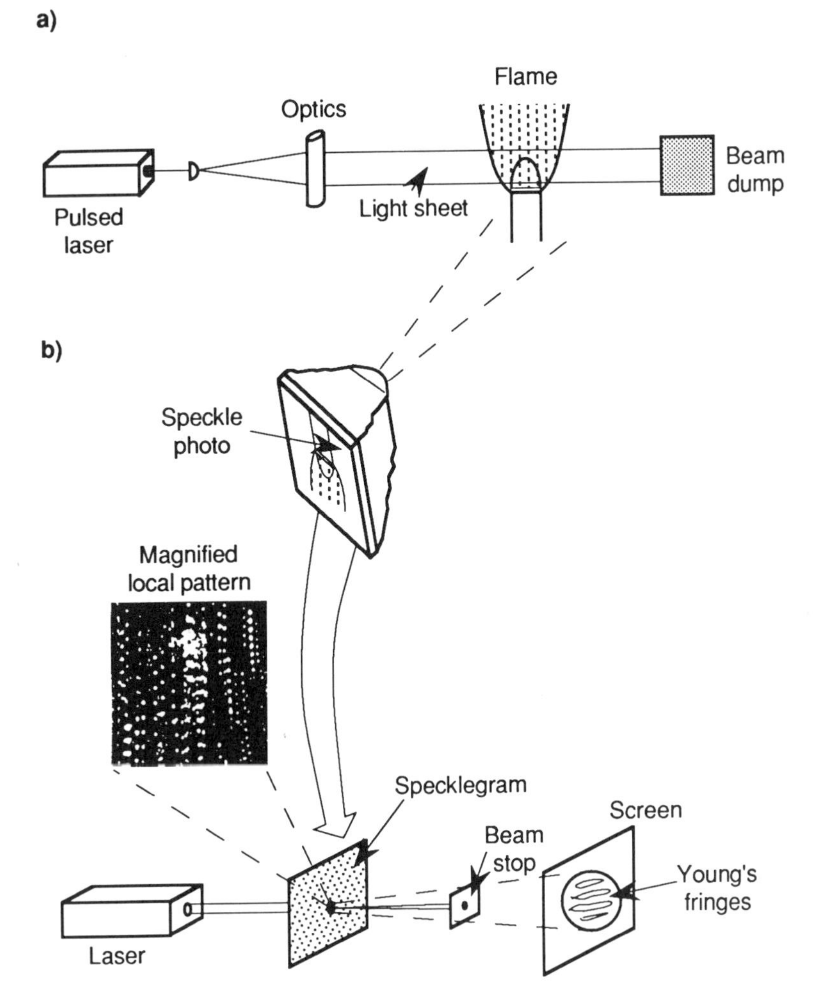

FIG. IV-11 Apparatus for Speckle Velocimetry: (a) Schematic for production of specklegrams. (b) Schematic of apparatus for interrogating specklegrams.

Laser Doppler Velocimetry or Anemometry

Laser Doppler velocimetry (LDV) can be considered as an analog to radar, where the frequencies are those of light rather than the radio region. This is a high-resolution method because wavelengths are measured in micrometers rather than centimeters, and spatial resolution is comparable to the wavelength used. The scattering agents are dust particles ranging in size between 0.2 and 1.5 micrometers in diameter. Such small numbers are required that normal room dust levels are often sufficient. The probe beam is the focal volume of one or two laser beams. In early LDV studies a single beam was used and the scattered Doppler-shifted light was beaten against a "reference beam." The difference frequency between the two beams detected by a photomultiplier is proportional to the velocity of the particle. This technique had the weakness that the output beat frequency depended on the angle between probe and detector. As a result it has been almost completely supplanted by the "dual beam" mode, where two equal intensity laser beams are crossed on a single focus point. This point is an ellipsoidal volume a few mean free paths in length and diameter. Interference between the two beams produces a "fringe" pattern of alternating high and low light intensity regions. When a particle crosses this pattern, it scatters the light in a sinusoidal pattern that maps the region of varying light intensity. The frequency of this scattered sinc wave is proportional to the velocity of the particle in the direction normal to the fringes and inversely proportional to the spacing of light fringes. Both components of velocity can be measured if a second orthogonal fringe pattern is established in the same volume. This requires a second wavelength of light to produce a separable signal.

To obtain clean results it is necessary to use high-intensity monochromatic light. Therefore, lasers offer the only practical light sources. A common choice is the blue and green argon ion laser. A powerful feature of LDV is the possibility of determining the direction of passage through the beam. This is accomplished by making the fringe pattern move through the volume, creating an artificial constant velocity that can be adjusted so that all particles will pass the fringe in the same direction. In the data reduction this constant velocity is subtracted out. Frequency shifting is commonly done either with Bragg cells (acoustic optic modulators) or rotating diffraction gratings (see Fig. IV-12).

The strength of scattered light and hence the signal depends on the angle between beam and detector, forward scattering being 100–1000 times greater than side or backscatter. Despite this backscatter is often the choice because this allows a common lens to be used for both incident and scattered light. Signals are small in combustion studies; so forward scattering is preferred whenever feasible.

Commercial versions of "dual-beam" LDV are available including computerized diagnostics suitable for small personal computers. There are two common designs; in one the fringe shifting is done with Bragg cells, while in the other the shifting is done using a rotating grating.

The source of the name *Doppler velocimetry* is obvious in the older single-beam mode, but less obvious in the dual-beam mode. The answer to that is that the two incident laser beams have different vector relationships to the particle trajectory such that the scattered lightly of the particle has a different Doppler shift for each beam. Analysis of Maxwell's equations indicates that the light scattered by each beam is shifted by half the Doppler frequency. When they are mixed on a nonlinear detector such as a photodetector, the difference output is at the Doppler frequency. This is analogous to the "beat" frequency between a tuning fork and a piano key that a piano tuner uses to adjust a key to true frequency. In electrical engineering jargon this is called "hetrodyning," because two shifted frequencies are mixed, as opposed to "homodyning," in the case of the single beam, where only the probe beam frequency has been shifted.

Homogeneous LDV Methods

Weinberg has developed an ingenious Schlieren/interferometric method for measuring the velocity of transparent objects. It employs the interference between Doppler-

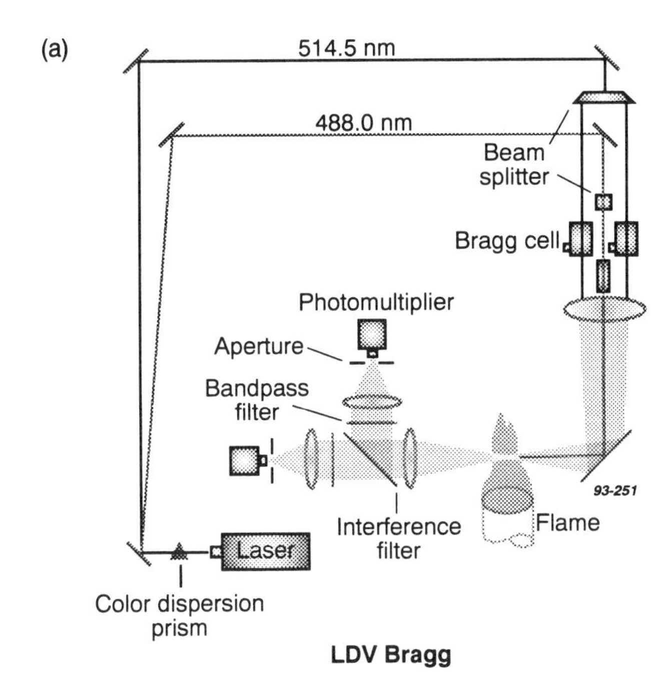

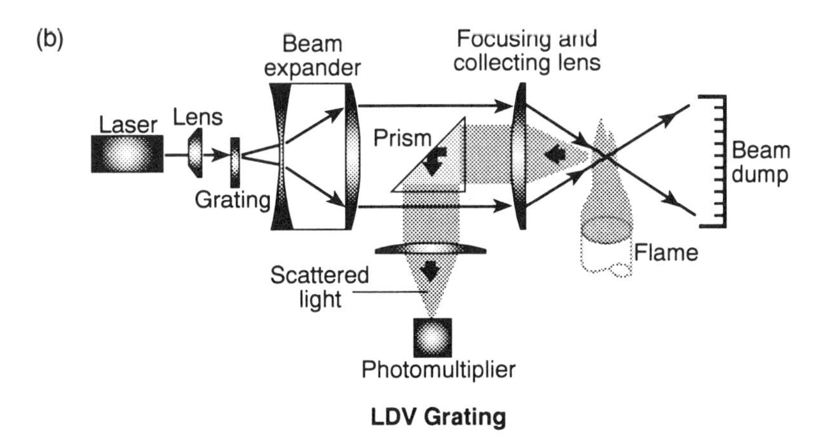

FIG. IV-12 Laser Doppler Velocimetry. (a) Dual-Beam Apparatus using Bragg cells. Apparatus uses two wavelengths to measure both components of velocity. (b) Dual-beam apparatus using grating splitter.

shifted Schlieren images and a reference beam (Schwar and Weinberg [1969]). The technique has been tested in the measurement of burning velocity of some flames (Weinberg and Wong [1975]).

The LDV technique can also be applied without particles using Rayleigh scattering or if the flame contains or is doped with a species that can be made to fluoresce or shows a strong Raman effect (Gustafson, McDaniel and Byler [1981]). As mentioned previously, Doppler measurements of low velocities are more difficult than high velocities, because the differences are smaller and the stability required of the reference signal must be that much greater. Miller and Hanson [1983] and Cheng, Zimmerman, and Miles [1983] have made velocity measurements in flow fields seeded with iodine and sodium, respectively. As with LDV, the Doppler shift between the moving fluorescing molecules and the undisturbed frequency is measured. In the past this had been done with supersonic streams, but improved experimental methods now allow the extension to low velocities. Raman/Doppler methods have been used by Gustafson et al. [1981] to study supersonic velocities, and the technique could in principle be refined to measure the low velocities of flames, thus removing the necessity of having a fluorescing species. Doping a flame with a fluorescing species is undesirable, although much of the problem could be avoided by using low concentrations or confining the injection to a single stream tube. If the Raman methods can be perfected, these problems disappear since most flame species are Raman active. Unfortunately the homogenous methods have not yet proven to be practical for routine use.

Pitot Probe Measurements

In principle, velocity is derivable from pitot probe measurements in flames. The pressure differences, however, are very low. It should be noted that pitot probe measurements differ from the pressure drop method for measuring burning velocity because the pitot probe is directed toward the flow and measures the dynamic pressure, while, in the pressure drop method, probes measure local static pressures. In a sense, in the pressure drop method the entire burner may be considered a pitot. In both cases, velocity is derived from the Bernoulli equation, which equates pressure drop with momentum change. However, such measurements have been made with quite acceptable precision, (Petersen and Emmons [1953]). Unfortunately several problems remain in interpreting the results quantitatively. The required probe sizes are small so that corrections must be made for boundary layer effects. In addition, the pressures are so low that even the hydrostatic head in a vertical probe heated by the flame gases can introduce a significant bias and error. A final difficulty is the long time constants associated with small tubes and low pressure differences. These can be long enough so that shifts in barometric pressure, air conditioning fans, and opening of doors produce significant errors in the reference pressure. The method has only been applied at atmospheric pressure, while most quantitative flame work is done at reduced pressures to improve spatial resolution. Pressure reduction introduces extra problems to an already difficult measurement. The measured pressures will be reduced proportionally, and provision for access and a reference pressure assume new dimensions. Despite these difficulties, the method is attractive because a probe similar to that used for sampling might be used, thus avoiding the problem of profile alignment.

Hot Wire Anemometry

Heat transfer from hot wires depends on the local velocity. This effect has been used to study jet flow velocities and measure turbulence levels. The field has been reviewed by Kovasznay [1954] and Comte-Bellot [1977]. Hot wire probes as small as 10^{-4} cm in diameter can be fabricated from Wollastin wire.[3] In principle the method might be applied in flames, but the high temperatures and steep gradients in composition and temperature would require elaborate corrections, and the method is limited to temperatures below the melting point of platinum.

Pulse Marking Methods (PTD Technique)

The flame gas itself can be used as a flow tracer for measuring velocity, providing a region can be marked. Hot wires have been used for marking in the pulsed wire method of Walker and Westenberg [1956]. This uses two hot wires, one pulsed to mark a gas region, and the other downstream as a detector. The advantage of this method over particles is that the inertial lag errors are inconsequential because of the small density difference. Another method is to heat the gas locally through laser absorption by some species in the mixture. This marks a parcel of gas by giving it a slightly higher temperature than its local surroundings. Detection can be by deflection optical techniques. The spatial resolution can be good if the laser is focused parallel to the flow and detection is made a short enough distance upstream since, as the flow proceeds, spatial resolution deteriorates due to enlargement of the region by conduction. The biasing due to thermal diffusivity in the temperature gradient could also prove a serious problem. The necessity of introducing an absorbing species might be a problem in some cases, although the amounts required are not large. Rose and Gupta [1985] in a test of the method used 500 ppm of NO_2 to dope the flame. In a variant of the PTD method, Sell [1984] studied jet flow velocity using amplitude measurements.

A small region can be marked by using laser ionization, providing an easily ionized species such as sodium is present. This technique was used by Schenck, Travis, Turk, and O'Haver [1982] using ionization probe detectors. Again, the spatial resolution of the method is limited by the region of the laser focus and by diffusion from the initial region of ionization. This can be optimized by focusing the laser beam normal to the flow. The resolution required for flame structure studies can be met if the detector is close to the source. Since the region affected by the technique is small, its effect on the overall flame flow is minuscule. The problem of requiring an easily ionized species in the flame gases can be overcome by working with very low concentrations or by confining the injection of tracer to a single stream tube. The method could be combined with the pulsed droplet injection technique to minimize effects of foreign species. The use of an alkali metal with a lower ionization potential and boiling point such as cesium would appear desirable.

Burning Velocity

Burning velocity is an important parameter used to characterize premixed laminar flames. Taken together with the inlet conditions it specifies the flame eigenvalue or

3. This is a very fine platinum wire imbedded in a silver sheath. The composite is formed by starting with a fine platinum wire, silver plating it, and drawing down the composite, reducing the silver and platinum diameters simultaneously. The platinum can be exposed by carefully etching away the silver with nitric acid.

mass flow per unit area. Experimentally burning velocity is the rate of propagation of a combustion wave through quiescent gas.

Any method for local velocity measurement could be used to establish burning velocity, but a number of special techniques have been developed. They are discussed in many combustion texts such as those of Strehlow [1983], Glassman [1977], Chomiak [1990]. Experimental methods can be divided into two general categories, propagation measurements and balanced flow measurements. The microstructure of the flames is almost the same in the two modes, differing only in the coordinate system used for viewing the flame front. In propagating methods, the flame coordinates move in the laboratory, while in balanced flow methods, they are fixed in laboratory coordinates. By contrast the flame geometries differ significantly between the two modes of propagation.

Methods for measuring burning velocity have been reviewed by Linnett [1953] and Andrews and Bradley [1972b]. In early reviews significant differences were found in measurements by the different techniques, and the question was raised whether burning velocity was truly a unique parameter. Most of these problems have been resolved. Two questions were involved: (1) Is burning velocity truly an invariant, or does its value vary from point to point along the flame front so that overall measurements simply represent an average? (2) What is the true relation between the ideal one-dimensional flame of theory and the experimental studies? These problems are discussed in Chapter X. The general conclusions are that in flames where stream tube expansion are moderate, burning velocity is invariant to a percent or two, providing the reference surface is the point of initial heat release rather than initial temperature rise. The cellular flames discussed in Markstein's book [1964] and mentioned in Chapter II should be excluded from this generalization.

The experimental methods can be divided into two general types: (1) propagation methods, where the flame passes through quiescent gas; (2) steady-state methods, where the flame is held stationary in laboratory coordinates.

Propagation Methods

Burning velocity measurements can be made by direct measurement of flame propagation. This requires locating the position of the flame front at two (or more) known times. Rapid visualization is required. Propagation velocity alone is not sufficient to define burning velocity. It is also necessary to know the flame front geometry and the flow constraints, and often the density change across the flame front.

Flame visualization has been discussed. Sensitive, relatively inexpensive video cameras are available, but most of the work in the literature has been done photographically. The speed and low luminosity of most flames makes direct photography difficult, but the use of high-speed Polaroid® films (up to ASA 20,000) has eased this problem. As a consequence most propagation measuring methods utilize Schlieren or shadow visualization with spark, flash lamp, or, more recently, laser sources. These can be pulsed and timed to a fraction of a microsecond. Continuous sources such as high-intensity mercury arcs can also be used with mechanical modulation of the light or with moving film such as a rotating drum camera.

Tubes

The earliest method for measuring flame propagation, used by Bunsen [1866] and Mallard and Le Chatelier [1883] (Chapter II), was to measure the time taken by a flame to propagate through a tube of known length. It is usual to measure propagation toward a closed end with the other open to the atmosphere. Propagation is thus into static gas. Unfortunately, there can be complications due to flow boundary layer and flame geometry (Jost [1946]; Lewis and von Elbe [1987]; Hibbard and Barnett [1959]). Therefore, in addition to time and distance, one must also visualize and measure the flame front area. Corrections are often made by using tubes of several diameters and using some extrapolation scheme. Recently the method has been refined by Fuller, Parks, and Fletcher [1969], who report excellent planar propagation in a vertical tube using an adaption of the opposed double-flame method of Raezer and Olsen [1962]. A diagram of the apparatus is given by Figure III–4.

Constant-Pressure Measurements

The propagation velocity at constant pressure of spark-ignited flames can be measured. What is measured is expansion velocity, which is significantly larger than burning velocity, since the gas expands during combustion and the flame kernel pushes the ambient gas ahead. The derivation of burning velocity from expansion requires a knowledge of the density change across the flame, together with a correction for finite flame front thickness. This latter is often made by extrapolating results to large diameters, assuming a model with constant flame thickness. The use of large flames where this correction would be negligible is impractical because of the onset of turbulence. When properly done, results from this method agree well with those from other methods.

Experimentally, the flame progress is usually visualized by shadow or Schlieren photography, although direct photography can also be used. It is necessary to isolate the combustible mixture from the surrounding atmosphere by using a slow-flowing jet, or a flexible confining method such as a soap bubble or slack bag. Flame kernels undergo complex initial propagation due to the effects of the initiating spark and establishment of the flame structure, but after the transients have decayed, the propagation approaches constancy. The initial geometry is toroidal due to the spark, but this quickly approaches spherical symmetry so that propagation can be deduced, assuming a simple constant flame structure. The apparatus is described in Chapter III (Fig. III-3).

Double Kernel Method

Many of the problems of the flame kernel technique can be avoided using an ingenious method suggested by Walker and developed by Raezer and Olsen [1962]. The essence of the method is to oppose a pair of simultaneously ignited flame kernels. As they expand toward one another, the flow of unburned gas is constrained along the line between their centers; the flames flatten and the two approaching flame surfaces approximate the flat flame geometry. Their relative movement asymptotically approaches twice the normal burning velocity. These burning velocities compare well with those derived by other reliable methods. The method combines the advantages of moderate sample size and single measurement. When combined with single flame kernel data, the final gas density can be inferred. The method has been extended to tubes (Fuller, Parks, and Fletcher [1969]) and bombs (Bradley and Hundy [1971]).

Combustion Bomb Method

If a flame is ignited in a vessel of constant volume, the initial behavior is the same as spark-ignited flames in open air. As burning continues, however, the pressure rises due to expansion. Initially, when the ratio of flame size to vessel volume is large, the rise is small but measurable and combustion can be followed through the pressure rise, or by photography of the growing spherical flame. The changing pressure increases the temperature through compression of the unburned gas. Consequently the interpretation is complicated, since the effects of changing pressure, temperature, and flame radius all require attention. Where the flame thickness can be neglected and gas ideality assumed, burning velocity can be calculated from the time rate of change of the flame diameter obtained from photography or hot wire anemometry (Bradley and Hundy [1971]). If only pressure rise data are available, a more complex interpretation is necessary (Yu, Law, and Wu [1986]). Satisfactory results can be obtained, but the number of corrections makes it more involved than direct methods. It does have the advantages of using minimal samples and providing burning velocities at several temperatures and pressures from a single run. A schematic of a typical apparatus is given in Figure III–3.

Steady-State Methods

If propagation is balanced by an opposing flow, a flame will stay fixed in laboratory coordinates. This is the principle of the Bunsen burner. At first glance this might appear as a precarious flow balance. However, in practice the flame adjusts its geometry to fit the flow and any flow higher than the propagation velocity (and below the blowoff velocity) will produce a steady flame of the Bunsen type. A discussion of flame stabilization is given in Chapter III. This behavior occurs because, no matter what the inlet flow profile, the flame adjusts its geometry so that the local component of inlet gas flow normal to the flame front is equal to the burning velocity (see Figs. II-4 and IV-3). Burning velocity is most reliably determined by measuring the local normal approach velocity directly using any of several methods to be described here. Alternatively the average overall burning velocity can be deduced from the overall flow given the total inlet flow and the flame surface. This assumes a uniform burning velocity and neglects edge effects. Some of the complexities that can arise in overall measurements can be visualized by referring to Figure III–6 which shows some of the myriad of flame geometries possible even with a simple symmetric tube. Special burners have been devised to produce flames with simple geometries where the boundary effects can be minimized. Critical discussions of the methods have been given by Linnett [1953]; Andrews and Bradley [1972b]; Jost [1946]; and Lewis and von Elbe [1982].

Flame Area Methods

As mentioned in the introduction, any method producing a laminar flame with measurable geometry can be used. This was pioneered on conical flames by Gouy [1879]. It is assumed that the burning velocity is constant along the flame front and normal to the visualized surface. It is also assumed that this surface lies at the surface of initial temperature rise. Corrections are required for edge effects and stream tube expansion in the flame front if the visualized surface lies at some other isotherm. The tip effect corrections are usually negligible. Most flame velocities in the literature have been determined using this method. In assessing literature on burning velocity measurements, one should examine the methods used because systematic errors as high as 50% can creep in due to visualization techniques, corrections for tip and boundary, and in rich flames by interdiffusion from the surrounding atmosphere. In the older literature, gross approximations such as measuring the flame height and assumed parabolic flow were used to minimize computation. Recent studies have been done with more care, but one should examine each piece on its own merits. For example, the early work of Jahn [1934] is probably as reliable as most modern work.

Flat Flames

The simplest geometry is the flat flame, which can be attained using screens, porous plates, or other flow-straightening methods discussed in Chapter III (Figs. III-7, and III-8) with a burner suggested by Edgerton and developed by Powling [1961]. If flow is adjusted so that the flame just departs from a flat profile, the burning velocity is quite simply the volume flow divided by the burner area. This balanced flow method can be applied by detecting flame liftoff either visually or by measuring heat transfer, the flame conductivity, or gas density.

By extracting heat from the flame using a water-cooled porous plate, flat flames can be stabilized at velocities substantially below the free space burning velocity. These flames have reduced enthalpy and are equivalent to precooled flames of equivalent final enthalpy. Such flames are very stable, and this type of burner is often used in flame structure studies. Burning velocity is obtained by determining the flow at the liftoff point, which is detected either visually or through measurement of heat transfer to the burner surface. At liftoff the heat transfer drops rapidly. This can be determined by measuring the temperature or the gradient at or near the burner surface or the temperature of the cooling water, providing a constant water flow is maintained.

Bertrand, Dussart, and Van Tiggelen [1979] showed that, if total flame electrical conductivity is measured as a function of burner flow, there is a sharp break in the current curve. This allows an accurate determination of the flow where the flame front first separates from the burner. This is identified with the burning velocity. The method gives very precise results and is useful for measuring small changes in burning velocity such as occur in inhibition studies.

Muller-Diethlefs and Weinberg [1979] use the change in density by Rayleigh scattering at the base of a flat flame as a function of flow to detect the flow at which the flame first begins to separate. This is also a precise method useful for measuring small changes in burning velocity.

Pressure Drop Method

The back pressure of a flame, due to the acceleration of the flame gases as they pass through the reaction zone, can be related to the burning velocity. Both the pressure difference between the burner and the surrounding atmosphere and the ratio of densities of the burnt and unburnt gases must be known. Measurement should be made across the flame front, and care should be taken to measure the static rather than the dynamic pressure as does the pitot probe (see the following). It was shown by Damkoehler as referenced by (Lewis and von Elbe [1988]), that the pressure difference between the unburnt gas and the final condition is[4]

$$v_0 = \sqrt{\frac{P_0 - P_\infty}{\rho_0(\rho_0/\rho_\infty - 1)}}$$

where ρ_0 and ρ_∞ are the initial and final densities and v_0 is the burning velocity.

For slow flames there is no difficulty in the large burners in locating a point to measure the pressure, but the difference is small. For fast flames, very small burners must be used, and there are practical difficulties in measuring the pressure without disturbing the flow of gases in the burner. The measured back pressure gives an average flame velocity over the flame front, including the dead-space area, and in this respect suffers from the same disadvantage as the total-area method. However, the form of averaging may be different.

Opposed Flame Methods

Many of the edge problems of the balanced flow methods can be solved by using the opposed double flame method of Yamaoka and Tuji [1984]. Here two flames are produced in opposed flow using a large cylindrical flame or other geometry. Because of the large size, area expansion is minimized, the perturbing effects of external atmospheres are eliminated, and the flow balance point can be accurately established. This method is particularly useful for slow flames near the extinction limit.

4. A correction for flame area expansion is also required.

V

FLAME TEMPERATURE MEASUREMENTS

Flames are associated with high temperatures and steep gradients. Temperature–distance data provide the key characteristic profile for flame structure studies, because most of the parameters used for analysis are temperature dependent. Figure V-1 provides a representative example.

The temperature–distance profile of an ideal one-dimensional flame would be a unique function of the inlet temperature, pressure, and composition of the incoming gas. In practice this is approximately true in the primary bimolecular reaction zone. Fortunately, this is the normal region of interest in flame structure studies and is most often measured. Outside this strongly coupled region the profile can be significantly distorted by area ratio variations, which change the relation between distance and residence time. These variations are quantitatively understood. In the approach zone no reaction occurs, energy is conserved, and the profile is a pure transport problem. Past the primary bimolecular reaction zone termolecular recombination takes over and the temperature–time relation becomes the invariant, providing energy is not lost to the walls.

These questions are discussed in more detail in Chapter IX in the section on one-dimensional behavior and zones in flame.

Local temperature is a well-defined variable in flames despite the conspicuous nonequilibrium processes associated with the flame luminosity. The residence time in most flames allows the establishment of a quantitatively valid Boltzmann equilibrium between internal and translational degrees of freedom to be maintained locally. This allows defining thermodynamic and transport properties and collision numbers. The nonequilibrium processes involve a quantitatively insignificant fraction of the molecules. This local equilibrium assumption is valid providing the logarithmic gradient of macroscopic variables (temperature, composition, density, etc.) is small compared with the mean free path of the gas. This is the fundamental requirement for the Chapman–Enskog treatment of kinetic theory (Chapman and Cowling [1939]) and is valid for flames in general. By contrast phenomena such as shock waves and detonations show significant deviations. Hirschfelder, Curtiss, and Bird [1964] give a very clear discussion of this point.

We shall assume that a single-valued parameter called the temperature exists and can be measured at every point in a flame. This controls molecular collisions and transport processes and allows the energy distribution of the *reactant* species to be calculated assuming Boltzmann equilibrium. The only known exceptions are the hydrogen–fluorine and hydrogen–chlorine flames, which are discussed in Chapter XI. This

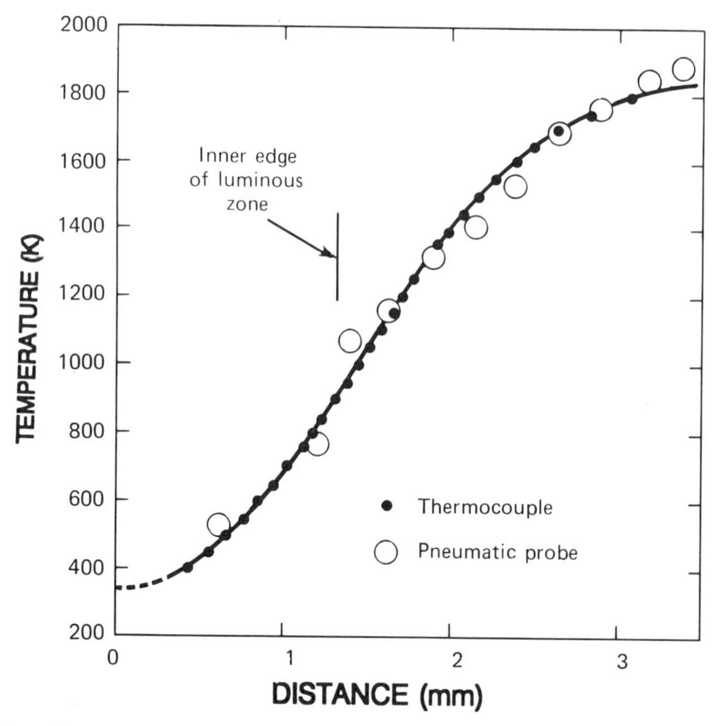

FIG. V-1 Experimental temperature profile comparing thermocouple and pneumatic probe measurements. (Data of Fristrom, Grunflelder, and Favin [1960].)

view should not blind one to the fact that flames contain many nonequilibrium populations that can offer vital clues on flame processes. The present chapter contains a discussion of methods for measuring this important quantity. From the aerodynamic point of view, the value of interest is the static temperature, that is, the temperature measured by a device moving with the stream. The distinction between this and the stagnation temperature (measured by a stationary device that brings the gas to rest adiabatically) is not important for most flames. This is the case because gas velocities are so low in most flames that the stream kinetic energy contributes a negligible part of the total energy. There are many methods of temperature measurement (see Table V–1), but not all of them are suitable for the high temperatures and steep gradients found in flames. The techniques can be classified as probe thermometry, radiation thermometry, density thermometry, sonic thermometry, and laser scattering thermometry.

Probe Thermometry

The most direct method of determining local temperature is to insert a thermometer probe. It must be small compared with the thickness of the flame front, and rugged enough to stand the high-temperature, corrosive flame environment. Thermocouples

Table V–1 Some Commercially Available Thermocouples Useful in Flame Studies[a]

Couple	EMF (V/K) × 10^6	Limit (K)	Comments
Chromel–Alumel®	12	1275	General use
Cu–Constantan®	14	400	Low temperatures
Fe-Constantan®	17	900	General use
Ir–Ir/0.4Rh	1.2	1800	Lean flames
Pt/Pt/0.1Rh	3.2	1550	Lean flames
Pt/.13Rh–Pt	3.6	1600	Lean flames
Pt/0.3Rh–Pt/0.06Rh	2.4	1800	Lean flames
W–Rh	4.1	2200	Rich flames
W–W/0.26Rh	4.8	2800	Rich flames
W/0.05Rh–W/0.26Rh	3	2800	Rich flames
W–Mo/0.5Rh		2500	Rich flames

[a]The output voltage is the nominal value over the range of use. It is given for comparison of couples *not for measurements*. The upper limit temperature is nominal.

are the most widely used thermal probes for flames (Friedman [1953]; Kaskan [1957]), but resistance thermometers (Gilbert and Lobdell [1953]) have also been used, and the recently developed optical fiber thermometers (Dils and Tichenor [1984]) show promise. We have even tried a quartz capillary/gallium thermometer (Fristrom and Grunfelder [unpublished]) and do not recommend it. This section covers the techniques and problems associated with probe thermometry. The discussion is specific to thermocouples, but much is also applicable to the other probes.

Thermocouple Measurements

The technique makes use of the thermoelectrical property of metals. If dissimilar conductors are connected with two junctions maintained at different temperatures, a potential is developed that is proportional to the difference in temperature. This temperature-dependent EMF is reproducible and is a function of the materials. The voltage is independent of the method of making the junction (wires may be welded, soldered, or simply twisted together) so long as good electrical contact is maintained, and *provided there is no appreciable temperature gradient across the joint.* The stringency of this latter requirement depends upon the junction size, the gradients, and the required precision. There are many high-temperature thermocouple pairs (Table V–2), but the most generally used in flame work is the Pt–Pt–10% Rh.

This technique offers high precision and an electrical output. Thermocouples as small as 10 μm in diameter can be made using Wollaston wire. High resolution is obtained with minimal disturbance of the flame.

Radiation losses are a major source of error above 1000 K, but even at this temperature with suitable corrections, temperatures reliable to 10–20 K can be obtained. Radiation corrections can be eliminated using the "null method," in which the thermocouple is heated electrically to balance the radiation loss (Wagner, Bonne, and Grewer [1960]). Hayhurst and Kittelson [1977] discuss some of the problems with this technique. Conduction losses can be minimized by aligning the thermocouple leads along the isotherms. Positions with respect to an absolute reference thermocouple can

Table V–2 Comparison among Methods of Determining Temperatures

Technique	Limit (K)	Effect	Cost
Thermocouple	2500	Yes	Low
Resistance	2500	Yes	Low
Aerodynamic	3000	Yes	Low
Pneumatic probe	3000	Yes	Low
Optical fiber	3000	Yes	Moderate
Pyrometry	None	Yes with additive	Low
Absorption	None	None	Low
Line shape	None	None	High
Interferometry	None	None	Moderate
Deflection	None	None	Low
Rayleigh	None	None	High
Raman	None	None	High
LIF	None	None	High
Cars	None	None	High

be established within 10 μm (Chapter III). Spatial resolution is limited by the size of the wire and the disturbances of the flame due to vibration and catalysis. Temperature differences as small as 0.1 K can be reliably measured with a spatial resolution of about 10 μm. Differential measurements yield reliable derivatives because the errors tend to cancel.

The apparatus consists of a suitably mounted thermocouple (see Fig. V-2) and a device for measuring the EMF. Techniques for fabrication of small noble metal couples are described in the Appendix to this chapter. A reference temperature is required at the second junction. This usually can be room temperature monitored by a reliable mercury thermometer since the required precision is modest. Where absolute measurements better than 1 K are required a standard junction is required. In most prac-

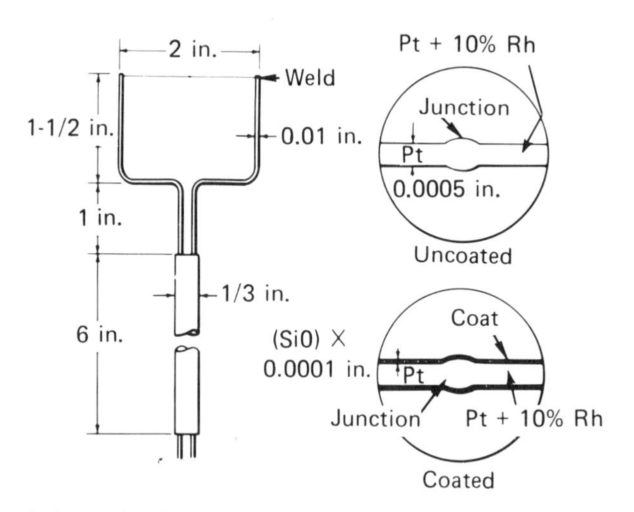

FIG. V-2 Details of thermocouple for temperature probing of flames.

tice a single noble metal junction can be used with the reference being furnished by the electric connections. The EMF should be measured with a potentiometer or more conveniently with an electronic millivoltmeter.

Resistance Thermometry

Resistance thermometry has been used to determine the maximum temperature of a low-pressure acetylene–oxygen flame (Gilbert and Lobdell [1953]). This technique makes use of the variation of resistance with temperature. A fine noble metal wire is used as the resistance element. Wires as fine as 10 μm in diameter have been used in hot wire anemometry (Chapter IV) and presumably could be applied to resistance thermometry in flames. In principle, resistance may be calculated from a wire's length and diameter, but in practice it is usual to calibrate the apparatus using a standard temperature bath. The experimental setup requires a source of constant current and a method of determining the wire resistance. For the wire sizes commonly used, the resistance is low and a Kelvin double bridge is used to minimize contact resistance errors. The problems of catalysis on the wire surface and evaporation of the wire mentioned in the literature (Shandross, Longwell, and Howard [1991]) could probably be eliminated by coating with silica (Kaskan [1957]) or rare earth oxide. As was the case with thermocouples, the accuracy of this method can approach a few degrees. The possibility of canceling the radiation correction through null measurements (Wagner, Bonne, and Grewer [1960]) would also improve the technique. Resolution as high as 0.1 mm can be obtained by careful alignment and tension mounting of the wire. The apparatus is simple, but the technique is less convenient than thermocouple thermometry, and gives averages over a line rather than at a point, as do thermocouples.

Many flame temperatures exceed thermocouple metal melting points. Raezer and Olsen [1960] developed an interesting method that avoids this limitation (Fig. V-3). It consists of intermittently inserting the thermocouple at a rate high enough so that the couple averages the insertion and withdrawal temperature. This ratio can be chosen at will, making the method applicable to any reasonable temperature. The technique is useful for measuring maximum flame temperatures of steady flames, but the movement disturbs the flame too much to make it useful for flame structure studies.

Another possible method for using a thermocouple above its melting point is to cool it by placing it in the throat of a sampling probe, where it registers a reduced temperature due to local adiabatic expansion. This technique has been explored at our

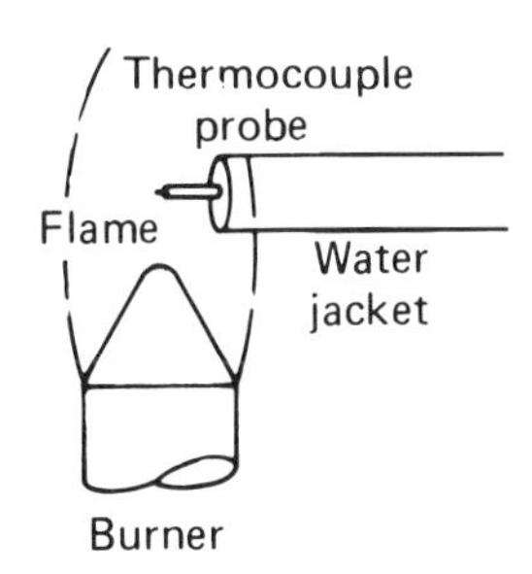

Fig. V-3 Schematic of the intermittent thermocouple of Raezer and Olsen [1960].

laboratory, and appears feasible. It requires calibration. Quartz, aluminum oxide, and sapphire probes appear suitable for such usage.

Sources of Error

A thermometer immersed in a gas stream may report a temperature differing from the true temperature for several reasons (Pollack [1984]). These effects can be divided into those that probes generally have on flames, and the errors peculiar to the technique employed. Generally the effects of probes on flames are reduced by minimizing the size of the thermometer. This is limited by practical problems of fabrication, fragility, and the heat transfer difficulty encountered when thermometers smaller than the gas mean free path are used. Disturbances by the probe can be classified as aerodynamic, thermal, or chemical, and are discussed in some detail with respect to sampling probes in Chapter VI.

Aerodynamically there are some differences between the actions of probe thermometers and sampling probes. One difference is the aerodynamic effect of a thermometer probe, which has a velocity-deficient wake that distorts the flame. This makes thermocouples read temperatures that are further downstream than their physical position. As a result, the device records a higher temperature than would be characteristic of the undisturbed stream. This shift is of the order of a few wire diameters. By contrast, sampling probes sample ahead of their physical position. This becomes significant for alignment with other profiles for the analysis of flame structure data. These problems are discussed in Chapters III, (see Fig. III-2) and IX.

Another major source of error is radical recombination on the thermocouple surface. Atoms and radicals are present in superequilibrium concentrations in flames. Metal surfaces catalyze recombination liberating heat. This can easily be detected because it produces a hysteresis in the temperature profile. The effect was used by Smeeton-Leah and Carpenter [1953] to estimate radical concentrations. The best method of minimizing this is to coat the thermocouple with a ceramic that has low catalytic activity. Two common coatings are silica and rare earth oxides. Techniques for coating are described in the Appendix to this chapter. The thermal effect of a probe as a heat sink is a function of the temperature difference between the stream and the thermometer. Reliable thermometry requires that these differences be small. This is not a problem if radiation losses are small, but they increase with the fourth power of the temperature.

Errors are reduced if the size of the thermocouple is decreased (up to the limit where its diameter becomes comparable with the mean free path) and the gas density and velocity are increased (up to about Mach 0.1, where the difference between static and stagnation temperature begins to be important). Conduction losses can be reduced in flat flames by aligning the support wires with the surfaces of constant temperature.

An order of magnitude calculation treating the thermometer as an energy sink due to radiation indicates that a 25 μm platinum thermocouple represents a sink which (in dimensionless form) is smaller than the sink of a 50 μm sampling probe. A better estimate of the interaction between flames and probe thermometers can be made making a balance between energy received by conduction from the gas and that lost by radiation and conduction along the supports. The following elementary analysis has been considered in more detail in the literature (Kaskan [1957]; Sato, Hashiba, Hasatani, Sugyima, and Kimura [1975]). For a spherical device of diameter d at a steady temperature T_c immersed in a gas of thermal conductivity λ temperature T_g ($T_g > T_c$), the heat transferred in per unit probe area may be approximated by $(2\lambda/d)(T_g - T_c)$ for a probe of small enough diameter so that its Reynolds number is much less than one. The heat lost by radiation to a wall at T_w is given by $\epsilon\sigma(T_c^4 - T_w^4)$, where ϵ is the emissivity of the probe and σ is the Stefan–Boltzmann constant. Equating these heat fluxes gives for the temperature error due to radiation

$$T_g - T_c = \frac{\epsilon\sigma d(T_e^4 - T_w^4)}{2\,\lambda} \tag{5-1}$$

In many cases ϵ and σ may not be known well enough for quantitative application of Eq. (5-1). But an upper limit in a typical case where $d = 2 \text{ x } 10^{-3}$ cm, $T_c = 2000$ K, $T_w = 300$ K, $\sigma = 10^4$ erg cm^{-1} K^{-1} sec^{-1}, $\lambda = 5.67 \times 10^{-5}$ erg cm^{-3} K^{-4} sec^{-1} would be estimated to be $T_g - T_c \approx 100$ K for a maximum $\epsilon = 1$.

A formulation for cylindrical geometries has been derived by Kaskan [1957] by estimating heat transfer using a Nusselt number–Reynolds number correlation.

$$T_g - T_c = \frac{1.25\epsilon\sigma d^{0.75}}{\lambda}\left(\frac{\eta}{\rho \upsilon}\right)(T_c^4 - T_w^4) \tag{5-2}$$

where η, ρ, and υ are gas viscosity, density, and velocity, respectively.

Wherc thc constants are not known, an effective calibration for a given thermometer can be determined by using a gas stream at a known temperature and measuring the resulting temperature. It is not always easy to provide a calibrated high-temperature gas stream; this is often estimated by assuming that the highest temperature measured corresponds to the adiabatic value. Since this is not always the case, it is desirable to confirm it by an independent method. It is also assumed that the thermal conductivity of the gas and the gas velocity are uniform over the region of application of Eq. (5-2). This is reasonable, since the correction is important only in the high-temperature region, where velocity and conduction variations are not great. The radiation correction remains one of the most serious sources of error in thermocouple measurements of flame temperatures. It can be minimized by using a sufficiently small device or by the "null method," in which the energy losses are balanced by electrical heating. The thermocouple absorbs heat from the flame if its temperature is lower than the flame gases and gives off heat to the flame if it is hotter. This results in an inflection in the temperature versus heating-current curve at the point at which the flame temperature is reached. The principle can be applied both to resistance thermometers and to thermocouples.

Since temperature measurements must be associated with a position, movement or vibration introduces errors. Induced vibrations can be a serious problem. Oscillation with amplitudes as great as half a millimeter and frequencies of the order of 50 Hz have been observed. They occur in regions of steep temperature gradient, and the driving mechanism probably is the alternating thermal expansion and contraction of the wire as it moves from one temperature region to another. This is intolerable and must be eliminated. The problem is best avoided by minimizing the length of fine thermocouple wire, using heavy, short support wires and drawing the thermocouple as taut as possible.

Thermometry by Measuring Gas Density

Gas density ρ can be directly related to temperature through an equation of state. The simplest of these is the ideal gas law.

$$T = \frac{PM}{\rho R} \tag{5–3}$$

Temperature can be derived using any method for measuring density or its derivatives providing the average molecular weight can be estimated with sufficient accuracy. This is quantitatively valid for low-density systems ($P < 3$ atm) and high temperature ($T > 300$ K). Most flame gases meet these conditions. There are a number of methods for measuring gas density, and several of them meet the temperature and spatial resolution requirements for flame studies.

Aerodynamic Measurements

Measurement of the aerodynamic profile allows derivation of the local density through the continuity constraint of the following equation:

$$\rho v a = const. \tag{5–4}$$

which in conjunction with Eq. (5-3) yields

$$T = T_o A \frac{Y}{Y_o} \frac{M}{M_o} \tag{5–5}$$

In these equations ρ is density (g cm^{-3}); v is velocity (cm sec^{-1}) and A is the stream tube area ratio (dimensionless). Velocity and stream tube area ratio measurements are discussed in Chapter IV.

Radiation Absorption Methods

Density can be determined by measuring the absorption of radiation by the total mixture or by a tracer species that does not enter into the flame reaction. Any spectral region can be used, but absorption coefficients increase with frequency. The ultraviolet and X-ray regions are particularly useful.

The most suitable transitions for tracers are those from the ground state to the first excited state. It is desirable that there be no other states with appreciable population at flame temperatures, since a correction must be made for the change of ground-state population with temperature. Mercury vapor provides a convenient tracer, although the possibility of catalysis must be considered. The 2537 Å resonance line in the quartz ultraviolet is the usual choice.

The absorption of oxygen in the Schumann region of the ultraviolet has been used to measure density in shock tubes (Camac [1961]). Vacuum spectrometry is required since air absorbs strongly in this region. This is a difficult technique. Tunable lasers can be doubled into this region, but where suitable lines are available, temperature measurements are better made by comparing two or more transitions as is described in the section on spectroscopic methods.

The absorption of alpha particles, neutrons, and electrons is proportional to gas density. Alpha particle absorption has been used to determine a final flame temperature (Shirodkar [1933]), and electron beam absorption has been used to map density in low-pressure wind tunnels (Schumacker and Gadamer [1958]), but neither of the methods has been used to study flame temperatures. Alpha particles and neutrons are scattered by nuclei, whereas electrons are scattered by the electron clouds. As a result, scattering is almost independent of molecular composition. This is advantageous for density measurements.

Flame temperature profiles have been determined using X-ray absorption (Mullaney [1958]), and the method is favored for studying the structure of detonation (Kistiakowsky and Kydd [1958]; Knight and Venable [1958]). X-ray absorption is a function of the total number of electrons and is, therefore, determined primarily by the number of atoms present and their atomic rather than their particular molecular arrangement. Absorption is thus directly proportional to the density. Common flame species have relatively low absorption, so that it is desirable—and usually necessary—to add a highly absorbing tracer gas such as xenon. The absorption of X-rays becomes proportional to the density of this tracer gas, and since the latter is unaffected by combustion, its density and the absorption of X-rays are proportional to the total density of the gas.

Soft X-rays are usually used for convenience and ease of shielding. They can be generated either with the use of pulse tubes or from radioisotopes. The former are convenient for studying transient systems such as shock or detonation waves. Suitable designs for flash X-rays are available (Knight and Venable [1958]) with microsecond duration and high power. Fe^{55}, on the other hand, provides a convenient stable source for steady-state measurements such as flames (Mullaney [1958]). An apparatus schematic using Geiger counter detection is given in Figure V-4. Other detectors are possible, such as electron multipliers and scintillation counters.

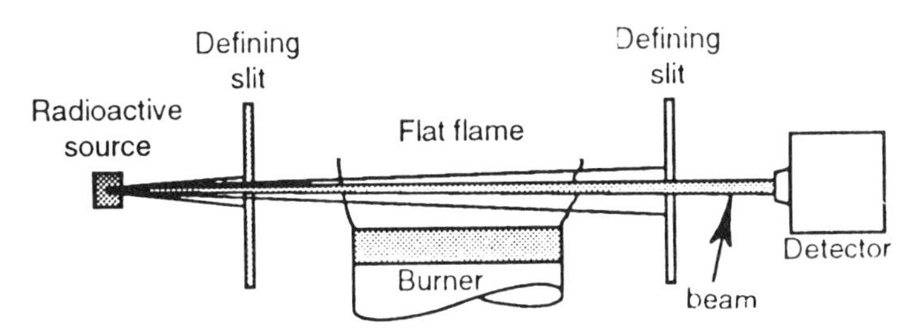

Fig. V-4 Schematic of apparatus for temperature profiles using X-ray absorption. (After Mullaney [1958]).

Density from Interferometry

Gas density can also be measured using the optical techniques of interferometry discussed in Chapter IV (Fig. IV-4). The velocity of light is a function of the density of the medium, so that beams passing through gases of differing density have different traversal times. If the beams were initially coherent, they will then be out of phase and interfere. To be coherent, the two beams must have the same source. This is usually accomplished with a half-silvered plate, which transmits part of a beam and reflects the rest. Interference does not occur between nonidentical sources, unless the phase of the sources can be controlled. This can be done in the radio or microwave region and in the optical region with lasers. This phase difference, or retardation, is proportional to the difference in density between the path taken by the measuring beam and that taken by the standard. The interference pattern can be interpreted quantitatively in terms of density by measuring the local fringe shift. The method has the advantage of optical techniques in that it does not disturb the system. Excellent precision can be obtained, although there are problems in defining the path length in the case of two-dimensional flames. The interpretation of the complex results from three-dimensional flames will be discussed in the section on tomography. Composition sensitivity is a second-order effect, since to a first approximation refractive index is proportional to the atomic volumes. Shock tube studies (Alpher and White [1959]) have shown that the simple relation between density and refractive index holds even in the presence of large concentrations of oxygen atoms and electrons. On the other hand, nitrogen atoms seriously disturb this relation. This means that ordinary combustion systems can be studied quantitatively by interferometry up to the temperatures at which appreciable dissociation of nitrogen occurs, namely 5000 K. Monochromatic radiation is not required, but greatly simplifies experimental adjustments.

Density Gradient Methods

The spatial derivatives of density can be integrated to derive a density profile and interpreted as temperature. Gradients can be conveniently determined using optical deflection methods. Strictly speaking, such Schlieren techniques measure $d\rho/dz$. Studies of flame geometry are often made using shadowgraph techniques discussed in Chapter IV. These have been discussed definitively by Weinberg [1963]. Other studies have been reported (Dixon-Lewis [1954]). Optical methods were discussed in Chapter IV, and laser techniques are discussed in Chapter VIII. An example of the technique is illustrated in Figure IV-6. If the gradient scale is determined and the initial density is known, the experimentally measured gradient curve can be integrated to yield a density curve. This can be interpreted in terms of temperature using the ideal gas law, providing local average molecular weight is known. Such studies have been primarily done on diluted flames, where the change in average molecular weight can safely be neglected.

In precision work, a second-order correction is subtracted to account for the actual parabolic path taken by the light in the flame. The density curve can be obtained by integrating the measured refractive index gradient curve. Temperature can be derived from density using the equation of state. The principal advantages of this technique are that the apparatus is simple, and the flame is not disturbed by the measurements.

The principal disadvantage is the necessity of composition and edge corrections. The use of a laser source simplifies the method.

Rose and Gupta [1985] have developed laser deflection methods that offer high sensitivity and precision measurements of density gradients. They have been used for temperature mapping through tomographic techniques. A short discussion of these methods with illustrations is given in Chapter IV, on flame visualization.

Sonic Velocity Methods

The temperature of a gas can be determined from the velocity of sound. In the case of ideal gases, the passage of a pressure disturbance such as sound is an adiabatic, reversible process, which travels with a velocity related to the compressibility (Hirschfelder, Curtiss, and Bird [1964] p. 728).

$$c = \sqrt{\left(\frac{\partial P}{\partial \rho}\right)_s} \qquad (5\text{–}6)$$

where the subscript s indicates constant entropy. Since an isentropic process in an ideal gas is characterized by the relation

$$\frac{P}{\rho^{\gamma}} = const. \qquad (5\text{–}7)$$

using the ideal gas law yields:

$$c = \sqrt{\gamma P/\rho} = \sqrt{\frac{\gamma RT}{M}} \qquad (5\text{–}8)$$

In these equations c is the sonic velocity and γ is the ratio of specific heats. c is the high-frequency or frozen sound velocity, that is, the velocity at a fixed gas composition. Since flame gases approach ideality, Eq. (5-8) should provide a reasonable approximation. This method has also been used to measure the temperature in electric arcs (Suits [1941]).

To apply this technique to the measurement of flame temperature profiles, it is necessary to know both the average molecular weight M and the average ratio of specific heats γ. This information can be obtained from a composition profile, from the assumption that γ is constant throughout the flame front, or from some other experiment that determines this ratio. The heat capacities are the so-called "frozen" values. Sound waves can be visualized optically since they induce small density gradients. For structural measurements spatial resolution can be a problem, but this does not limit the method for determining final flame temperature. Sound disturbs flames. This can be avoided by using a single pulse since a sonic pulse transmits no disturbance ahead of itself.

The Pneumatic Probe Method

Another sonic method is the pneumatic probe (Moore [1958]). This consists of two orifices placed in series. The second orifice has an area at least three times that of the first so that both can be pumped in series under choking conditions. For flames the probe must be thermally resistant and for structural measurements the orifice must be small and offer minimal aerodynamic disturbance. A quartz probe similar to that used in sampling studies (see Chapters VI and VII) is suitable. If the pressure drop across an orifice is sufficiently great, the gas velocity through the throat is sonic and the flow depends only on the upstream pressure, temperature, molecular weight, and specific heat ratio with a minor Reynolds number correction for boundary-layer effects. This was discussed in Chapter IV. If two orifices are connected in series so that critical flow occurs through both, then it is easy to show that the temperatures and pressures upstream of orifices 1 and 2 are related by

$$T_1 = KT_2(P_1/P_2)^2 \qquad (5\text{–}9)$$

where K is (assuming constant molecular weight and specific heat ratio) a constant determined by calibration. Thus if T is the unknown temperature at the entrance to a suitable probe orifice, it may be obtained from a measurement of T_2, P_1, and P_2.

The pneumatic probe provides method for temperature measurement that can be related to composition when the same probe is used for both. Although calibration is required for quantitative work, the results are not influenced by the external environment and no radiation correction is needed. The calibration constant is a function of the Reynolds number of the orifice. For a very small probe this effect varies with temperature. It is not necessary to make the calibrations at high temperatures, however, since the Reynolds number can be simulated by varying molecular weight and viscosity, rather than the temperature. This is important since it is difficult to provide calibration temperatures above 1500 K.

It is necessary that both orifices operate in the continuum flow regime and that radical and atom concentrations are small, since they recombine before entering the second orifice. This changes the molecular weight and ratio of specific heats. For convenient operation it is necessary to minimize the volume between the two orifices since the equilibration time is proportional to this volume. Precision better than 1% is required in the pressure measurements. Diaphragm gauges and mercury manometers are commonly used. The mass spectrometer provides a convenient pressure measuring device if composition studies are being made—the total pressure being the sum of the partial pressures. An experimental setup is shown in Figure V-5 (Fristrom, Grunfelder, and Favin [1961]). Some representative results are given in Figure V-1.

Errors associated with radical recombination in the probe can be avoided by injecting a known flux of a radical scavenging gas as is done in the probe studies described in Chapter VI. In this case composition measurements provide the ratio between probe flux and injected flux.

The advantage of using the pneumatic probe is that it allows a probe simultaneously to measure temperature and composition. This avoids difficulties in alignment of profiles.

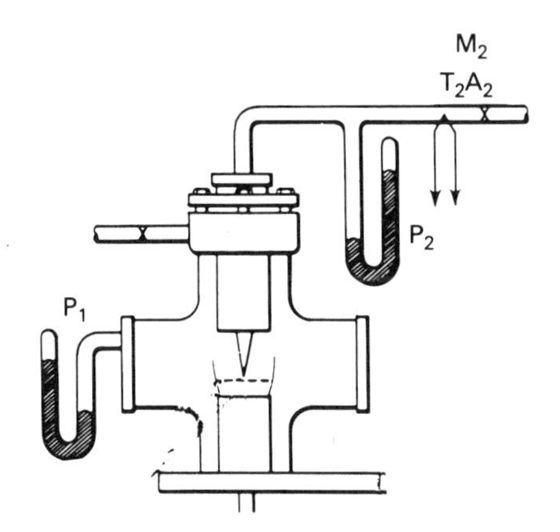

FIG. V-5 Pneumatic probe apparatus for temperature measurement (Fristrom and Grunfelder [1961]).

Temperature from Molecular Beam Velocity Measurements

Temperature can be inferred from beam velocity and Mach number measurements in systems where molecular beam inlet sampling is used. This technique is described in Chapter VII, p. 189.

Radiation Thermometry

Measurement of electromagnetic radiation provides a versatile method for determining flame temperatures. Techniques range from simple optical pyrometry to sophisticated spectroscopy and may utilize any wavelength from the X-ray to the microwave region. These techniques are nonperturbing and have no high- temperature limit. The principal disadvantages lie in their poor spatial resolution and the fact that they give an average over the line of sight.

Viewed from the molecular standpoint, the energy of a system is distributed among the molecules in quantized energy levels. If the molecules are far enough apart so that interactions are negligible, as is the case with gases at moderate pressures, the energy can be divided into a translational part, which can be considered classically, and an internal part, which is quantized. The quantization of the internal molecular energies results in light being absorbed and emitted only at particular wavelengths that are functions of the molecular parameters. If interactions between molecules become strong, as is the case with very dense gases, liquids, and solids, then the emission and absorption lines are broadened, and in the limit coalesce into the continuum black-body distribution.

If the internal energy states are in equilibrium among themselves, they will be distributed according to the classical Maxwell–Boltzmann law (the difference between classical and quantum statistics being negligible at flame temperatures). Thus, the probability that a molecule will be found in a particular state E is proportional among other factors to $\exp(-E/RT)$, where the temperature T is identified with the ordinary translational temperature if the internal and translational degrees of freedom are also in equilibrium. The intensity of emission or absorption for a given transition between internal states is proportional to this "Boltzmann factor," $\exp(-E/RT)$, so that in principle a single absolute measurement of absorption or emission for a known transition should allow T to be determined. In practice, because of the difficulty in determining emissivity, it is more convenient to make two (or more) measurements at different wavelengths and derive temperature from the intensity versus energy plot. This method offers the added advantage that deviations from a Maxwellian distribution and dependence of absorption–emissivity on wavelength can be detected. Many methods have been devised for making such measurements (Penner [1959]; Forsythe [1941]), and a few of these will be described. Most systems consist of three elements: (1) a flame with suitable geometry so that homogeneous regions can be optically probed; (2) an optical system consisting of a monochromator and a detector for measuring intensities; and (3) a comparison emitter.

Optical Pyrometry

The simplest radiation method is optical pyrometry. Temperature is measured by comparing flame emission with a light source at a known temperature. In commercial pyrometers, the filament of an incandescent lamp is superimposed optically on the image of the flame region being studied. Filament temperature is adjusted by changing the applied voltage, and when the two images are at the same temperature, they will tend to merge. The equivalence point can be determined visually or by use of a photocell with a precision of a few degrees. Tungsten or carbon filaments are used for general pyrometry, and they are calibrated against a black body whose temperature is measured by some auxiliary method. Commercial instruments require minor modifications to measure flame structure. The method can be directly applied to flames containing solids, although a correction should be made for the dependence of emission on particle size. The problems with coal dust flames are discussed by Tichenor, Mitchell, Hencken, and Niksa [1984]. Flames usually do not absorb (or emit) significantly in the visible region; so in the absence of solids they must be colored with a material that emits in the region being studied (e.g., salts of sodium, lithium, iron, iodine, etc.). Since the spectrum of most additives consists of discrete lines, this technique is commonly called the *line reversal* method. With sodium salts added, for example, the familiar yellow *D*-line doublet is involved. When viewed with the comparison source, these lines appear brighter or darker than the background depending on whether the comparison source temperature is lower or higher than that of the sodium-containing flame. When the two temperatures are the same, the lines tend to merge. At the reversal point the temperature of the unknown and comparison emitters are equivalent. These additives are foreign to the flame system, and the question of disturbance arises. However, if the amounts required are small, the problem is usually not serious. The most important limitation of colored flames is that emission is negligible below 1700 K for most additives. In an experimental setup it is necessary to provide a flat geometry and add the emitter to a uniform section of the flame—otherwise some average temperature will be measured. A schematic for these measurements is given in Figure V-6. Spatial resolution of the order of 1 mm is attained, and the afterburning region of flames can be studied using this technique (Gaydon and Wolfhard [1979]).

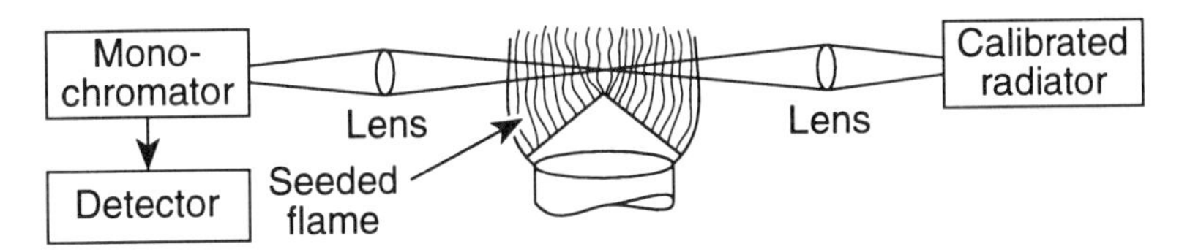

FIG. V-6 Apparatus for flame photometry using line reversal.

In a study of flame temperatures using sodium as a seed, Lapp and Rich [1963] pointed out that if the lamp temperature is higher than the flame, three measurements can be made: (1) lamp emission, (2) transmission of the lamp through the flame, and (3) the emission of the seed molecules. This allows determination of both the temperature and the concentration of the seed molecules.

If two spectral regions are compared, it is not necessary to know the emissivity, since the "grey body" assumption (i.e., that it is the same in the two regions) is usually sufficient. Measurements are usually made in the infrared, where emission intensity is high. Some care must also be taken to avoid problems of averaging, optical depth line broadening, etc., since finite spectral regions rather than spectral lines are being compared.

Another convenient technique is the two-path method, which consists of measuring the intensities of radiation for one and two traversals using a mirror. This is essentially a reversal method in which the source is used as its own background, and is particularly useful if the radiation varies with time. To apply this method it is necessary to know the reflectivity of the mirror and to assume that emissivity varies slowly with wavelength. Resolution comparable to other optical methods can be obtained, although the double traversal complicates the problem of maintaining the narrow light beam required for high spatial resolution. The main uncertainties associated with the density of emission in the region chosen are the assumption of constant emissivity, and the effects of heterogeneities in the optical path. Errors can amount to hundreds of degrees, although careful work can hold them down to a few degrees.

Optical Probes

A recent addition to probe thermometry is the optical fiber probe (Dils and Tichenor [1984]). The sensor for this method is a quartz or sapphire optical fiber coated with an iridium film which in turn is coated with evaporated Al_2O_3. The film acts as a "black body" for the fiber which leads the emitted light to the detector which is a two-color pyrometer. The output is coupled to a microcomputer which provides for convenient data reduction and output. High frequency response (50kc) and precise temperature determination are claimed for this instrument (Fig. V-7). By using infrared the low temperature limit can be reduced to 370K. The currently available commercial instruments employ two millimeter sapphire rod sensors which are coarse for structural measurements, but it clearly has a good potential for flame structural studies. There

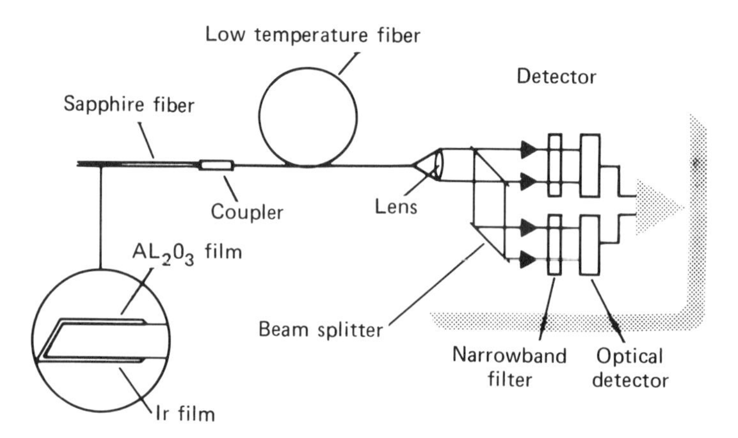

FIG. V-7 Schematic of apparatus for optical fiber pyrometry (Dills and Tichenor [1984]).

would appear to be no fundamental reason for not using finer fibers (there is, of course, a wavelength cutoff frequency).

Particulate Thermometry

Particle emission in flames can also be used for thermometry. Care is required in interpretation of results because small-particle emission is biased. Inert particles lie below the carrier gas temperature, while reacting particles show higher temperatures. In denser systems the effect of particles on the flame requires consideration including such factors as catalysis, radiation loss, etc. If the particles react, their heat generation becomes a major factor. For these reasons, particle thermometry has centered on the study of the combustion of particles. The use of a two-color system with the grey body assumption is reliable for individual particles, and they can be followed as a function of time and position with a suitable optical system. In principle, single-particle thermometry with inert particles could be used as a thermal probe for structural measurements, but no such studies appear to have been attempted. There have, however, been successful studies of coal particle combustion temperatures by Tichenor, Mitchell, Hencken, and Niksa [1984].

Line Intensity Methods

The most accurate and detailed temperature information can be obtained from the measurement of relative intensities of a known series of spectral lines. If a system is in thermal equilibrium, a plot of the logarithm of the quotient of the line intensity by its transition probability versus the initial energy level yields a straight line whose slope is $1/T$. This is a very sensitive test for equilibrium. Care must be exercised to avoid spurious results. Such studies give the temperature associated with a particular species, indicating if it is in thermal equilibrium, and often provide clues as to the reasons for nonequilibrium behavior. At present nonequilibrium cases are too complicated for quantitative interpretation, and their consideration must be deferred until the simpler equilibrium situation is understood.

In competent hands, this method is capable of high precision and great delicacy. However, spatial resolution is limited, and the method is restricted on the low-temperature side. The rotational fine structure of vibrational–electronic bands provides a particularly favorable case for such an analysis, since the lines are spaced closely enough so that the problem of changes of sensitivity with wavelength is minimized. Transition probabilities are available for some triatomic and larger molecules with high symmetry such as benzene. However, this is not generally the case for polyatomic molecules (Herzberg [1945]), which have complex spectra. However, most of the important combustion species have been analyzed. The OH radical has been a particular favorite, since it has a well-known, easily accessible spectrum in the near ultraviolet. Other species that have been used are CH, Li, Fe, I_2, and HCl. An experimental difficulty associated with such studies is that high-resolution spectrometers must be used to resolve the lines. This is expensive equipment that many combustion laboratories do not possess. Only weak emission (and therefore absorption) lines should be studied, since otherwise the intensity relationships are distorted by "self-absorption," and equilibrium may not be established.

Studies can also be made using absorption (Wagner, Bonne, and Grewer [1960]) rather than emission spectra. OH rotational lines are a favorite target. The techniques are the same as those used for absorption spectroscopy studies of composition measurements. These are described in Chapters VI and VIII. Using iodine as a tracer, it is possible to make studies at relatively low temperatures (Beck, Chen, Uyehara, Winans, and Myers [1955]), but this should be used with caution since I_2 is a powerful flame inhibitor that enters the flame reactions.

The advent of tunable lasers, diodes in the infrared, and dye lasers in the visible (and doubled into the ultraviolet) has made these measurements much more convenient, albeit more expensive.

Laser temperature studies can be made either by absorption or by fluorescent emission (Kichakiff, Hanson, and Bowie [1984]). Since absorption studies are line-of-sight measurements, the problem of path length and boundary interferences must be considered. These are usually solvable. One elegant solution consists of Stark modulating the absorption line using a second crossed laser beam (Farrow and Rahn [1981]). The technique is described in Chapter VIII. Resolution can be high since the intersection of two lasers can be defined at their focal points within a few wavelengths. The method requires that the field strength of the crossed laser beam be sufficiently high to move the absorbing level out of the probing laser's frequency. This requires only modest laser power. The result is a modulation of the absorbing signal at the repetition rate of the crossed laser beam. Single-point studies are usually combined with concentration measurements, and in some cases temperature and composition fields can be mapped using laser fluorescence of a pair of adjoining rotational lines (Chan and Daily [1980]).

Line Shape Methods

If the mechanism of line broadening is understood, it is possible to derive temperature from the line shape. Broadening can stem from many different factors, however, and caution should be exercised in interpreting such data. In the optical region the principal source of line broadening is the Doppler effect, due to the random distribution of velocity components parallel to the radiation. Self-absorption broadening is a complicating factor that can usually be corrected (Rea and Hanson [1983]). This type of broadening can be directly related to temperature. On the other hand, in the X-ray region the principal source of broadening is the natural linewidth, which is independent of temperature. An early study was made of rocket exhaust gases, where spatial resolution problems were minimal (Strong, Bundy, and Larson [1949]). From the shape of the sodium *D* doublet, it was possible to deduce both temperature and gas velocity. Velocities were comparable to molecular velocities and this made the line asymmetric. The advent of tunable lasers has made general line shape studies feasible. There is adequate change in shape with adjacent rotational lines for reasonable temperature sensitivity. The problem arises with interpretation. The line broadening mechanisms are complex. The general mechanisms for line broadening are known (Doppler broadening, collisional broadening, and Stark broadening), but present theories require empirical parameters for fitting. This means that for quantitative interpretation a fitting parameter must be determined experimentally. Once this is established, flame temperatures can be measured. Rea and Hanson [1983] have demonstrated the feasibility of the technique using a doubled ring laser. Reliable measurements are feasible, although the apparatus is complex, and depend on the choice of the Voigt parameter, which at present is in some doubt. The method may find application in conjunction with other more detailed studies, which would justify the complex experimental setup. Laser methods are discussed in Chapter VIII.

Laser Scattering Methods

Convenient high-powered lasers have made a variety of new techniques feasible for temperature measurements in flames. Characteristically these are rapid noninvasive light-scattering methods. We will discuss them as a group with reference to temperature determination. These methods are used not only for temperature, but also for composition and velocity measurements. Experimental details are covered in Chapter VIII which covers laser diagnostics. With the exception of Rayleigh scattering, where temperature is derived from density, the methods are spectroscopic in nature, and temperature is derived from relative line intensity measurements.

Laser methods can be used to define point regions, lines, or areas, depending on the geometry of illumination. Point measurements are the most sensitive, but they are the slowest, and they lack spatial correlation. Sensitivity depends on illumination power (and hence the laser), but this is limited by electric breakdown of the gas. For this reason it is advantageous to stretch the laser pulse to as long a period as is compatible with freezing the information. For normal flames this period exceeds 10 microseconds which is a time long compared with normal laser pulses.

If point measurements are made, the resolution normal to the laser beam axis is determined by the sharpness of focus of the laser beam. At best this is of the width of a few wavelengths. In the direction along the axis, resolution is determined by the collection optics. The criticality of these limitations depends on the geometry of the experiment. In linear and area scattering studies, experimental resolution depends on the imaging optics. This can be high, but the price paid for this resolution is a reduction of illumination intensity.

Three limiting types of scattering can be distinguished, and will be discussed in the following sections.

Rayleigh Scattering

In Rayleigh scattering the light is elastically scattered by the inhomogeneities in refractive index that occur on a distance scale corresponding to the mean free path in the gas. This occurs at every wavelength, but strength increases with frequency. It is the strongest and most easily implemented of the single-photon scattering methods so it is often combined with more complex methods as a monitor and to provide auxiliary information. The scattering is proportional to local density, which in turn can be interpreted as a temperature using the ideal gas law assuming that the local average molecular weight is known or can be considered approximately constant (Smith [1978]; Robben [1976]). Unfortunately scattering depends on species so that strictly speaking composition and sensitivity coefficients are required. However, in cases where there is a dominant diluent or the coefficients for reactants and products are close enough, it can provide a useful first-order measurement.

Resonant Fluorescent Scattering

In resonant fluorescent scattering the exciting frequency is coincident with an energy level of the molecule. This is usually limited to radical species, and the required excitation lies in the ultraviolet. The scattered lines are sharp and characteristic of the molecule. Intensity depends on the lifetime of the excited state.

For quantitative studies one must either work under saturated conditions or know the local composition and quenching efficiencies of the components. When only visualization and temperature are required, these requirements can be relaxed by making relative measurements on adjacent lines whose initial states can be assumed to have similar quenching properties. Usually rotational transitions are used (Crosley [1980]). Temperature maps of flames have been obtained using these techniques (Chapter VI). The technique is only applicable to a restricted number of molecules with currently existing sources. Several, however, are common radical species in flames. The excitation lines usually lie in the ultraviolet; so the experiment requires a tunable visible laser that can be frequency doubled. Another variant useful for low temperatures is two-line atomic fluorescence using traces of indium (Omenetto, Browner, and Winefordner [1972]; Haraguchi et al. [1977]).

Raman Scattering

Raman scattering is a nonresonant method where the exciting frequency can be any wavelength, but the scattered frequency is re-emitted with the addition or subtraction of energies characteristic of the irradiated molecule. It can be considered as a transposition of the infrared rotational–vibrational energy levels around the exciting frequency; however, the selection rules differ from those in the infrared (Lapp and Penney [1974]). The most important difference is that homonuclear diatomic levels are active; so it is possible to study diatomic gases such as hydrogen and oxygen.

Vibrational bands and rotational fine structure are used in estimating temperature (Lapp and Penney [1974], Burlbaw [1983]). Three general methods are used. (1) Two bands or lines are chosen and their relative heights measured. This is essentially two-color spectrometry; unfortunately the optimal choice of line pairs can change with temperature. (2) A second technique is to compare Stokes and anti-Stokes bands or lines. This technique is useful where only limited spectral resolution is available. (3) The most powerful method is band shape matching, where the vibrational-rotational pattern is matched (see Fig. VIII-6). In favorable cases, a precision of a few degrees may be obtained over a range of several thousand degrees. There is no particular upper limit of temperature.

This yields high precision and is useful over a wide temperature range. Unfortunately it requires a detailed knowledge of the spectrum and separation of interfering lines. The required information is available for the diatomic and a few of the triatomic species, but this is adequate for most flame work. The sensitivity of the method is best for light species, particularly hydrogen, but because of the weakness of the Raman effect, measurements are limited to flames near atmospheric pressure or higher. Sensitivity also depends on the fourth power of the exciting frequency; so blue and ultraviolet laser sources are favored. The method is discussed in more detail in Chapter VIII.

Multiphoton Scattering

In multiphoton scattering measurements molecules interact with more than one photon at a time acting as a nonlinear transducer mixing sum and/or difference frequencies that are resonant with energy levels in their irradiated molecule. The most common of these is CARS (Coherent Anti-Stokes Raman Spectroscopy), which has been used for temperature measurements in flames (Farrow, Lucht, Flower, and Palmer [1986]). The result is a resonant Raman spectrum, in which the sum or difference frequency of the irradiating laser beams coincide with one of the molecular levels. This level is emitted as a third laser beam. Because the scattering is coherent and directed, it can be many orders of magnitude stronger than incoherent scattering. The process is particularly useful where there is a background of emitted light, as in combustion systems with radiating particles such as sooty flames. By accepting a small angle, background radiation can be reduced to negligible proportions. As with other multiphoton processes, the technique is experimentally complex, since two sources must be aligned and one laser must be tuned in frequency. For systems with high background, however, the complexity is worth the cost. Other multiphoton processes have been demonstrated (see Chapters VI and VIII), but none is in common use for measuring temperature.

Comparisons among Methods for Determining Temperature Profiles with Some Value Judgments

A number of methods for determining temperature profiles are available (Table V–1). Each possesses advantages, but the most useful ones for detailed structure studies have been the thermocouple and the inclined slit methods. There have been several comparisons of techniques (Figs. V-8 and V-9) (Wagner, Bonne, and Grewer [1960]; Kichakiff, Hanson, and Howe [1984]; Robben [1976]; Fristrom and Westenberg [1965]; Ray [1981]). Thermocouple measurements have been compared with aerodynamic determinations, with the pneumatic probe method, with spectroscopic measurements—both in absorption and emission, and with the inclined slit technique. They have been compared with calculated adiabatic flame temperatures. It is heartening to find that the agreement among methods is generally satisfactory.

The thermocouple method combines high precision and spatial resolution with ease of measurement. The three principal problems are the emissivity correction, the spatial shift due to aerodynamic wake, and the possibility of catalytic recombination on the thermocouple surface. The upper temperature is limited by the materials used, but in many systems this is not a problem.

The inclined slit method offers the advantages of reasonable precision and spatial resolution, ease of measurement, and the virtue of not disturbing the flame. It provides an absolute position reference in the flame front and is amenable to instantaneous measurements. The sensitivity of the technique drops off with temperature, but otherwise there is no particular upper limit. Difficulties include moderately complicated data reduction and the requirements of local average molecular weight information. The derivation of temperature from aerodynamic measurement allows only limited precision and spatial resolution so that it is principally useful as a cross check on the more precise measurements and to establish reference coordinate surfaces.

Pneumatic probe measurements offer modest precision and spatial resolution and the advantage of connecting composition and temperature data. A radiation correction is not required. Problems include the effects of the probe on the flame and a dropoff of sensitivity with temperature. The upper limit for its use is set by the material of fabrication (usually quartz or sapphire) and problems with correcting results for the changes brought about by radical recombination in the probe. For small probes there may be Reynolds number effects to consider.

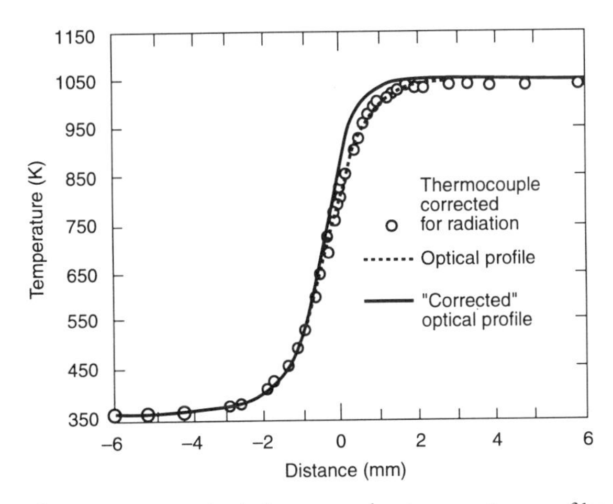

FIG. V-8 Comparisons among methods for measuring temperature profiles: Inclined slit and thermocouple (Dixon-Lewis and Isles [1960]).

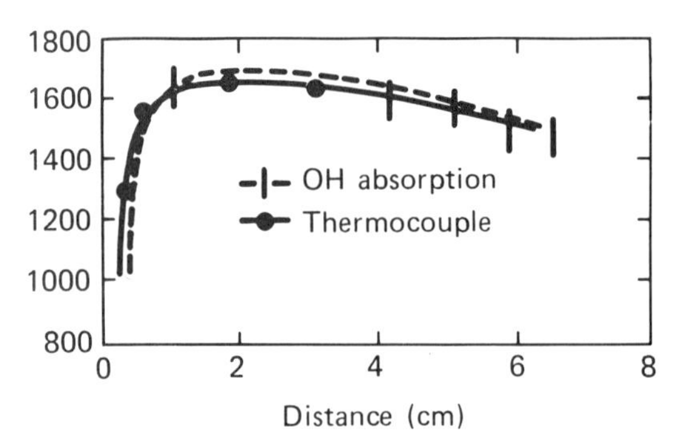

FIG. V-9 Comparisons among methods for measuring temperature profile: Spectroscopic absorption and thermocouple (Wagner, Bonne, and Grewer [1960]).

Emissivity and spectroscopic measurements offer good precision but limited spatial resolution. They have the advantage of not disturbing the flame, although optical pyrometry requires introducing an emitter and some problems may arise. The principal problems are nonequilibrium emission and the restriction to relatively high-temperature regions.

X-ray absorption measurements offer moderate precision and spatial resolution. Its use has been limited because the apparatus is not common in combustion laboratories and may involve safety problems.

Interferometry offers good precision and spatial resolution and the advantage of not disturbing the system. It is difficult to apply in very steep gradients, but the technique is well complemented by inclined slit measurements. This type of measurement has been held back by the cost and complexity of the equipment, although recently interferometers of modest cost have appeared. The use of laser sources simplifies the technique. Holography (see Chapters IV and VIII) has been used by Reuss [1983] to measure the temperature field in a radially symmetric flames.

Appendix The Construction of Small Thermocouples

Local temperature measurement in flames can be made with small thermocouples. They are available commercially (Omega Corp). Their fabrication is tedious, but the technique can be mastered. Two operations are involved: welding of the junction and coating of the final couple to reduce catalytic errors. The couples are generally made of noble metals (Pt–Pt–10% Rh or 13% Rh) or (Ir–Ir/Rh). In low-temperature work Chromel–Alumel couples can be used. These wires are available commercially. Chromel-alumel and platinum-platinum/rhodium can readily be made as small as 0.001 in. in diameter with reasonable bead size. Below this diameter, the wires are so fragile that a strong breath will break them, and they are difficult to manipulate. Iridium couples are stronger and stiffer and can easily be made as small as 0.0005 in. in diameter.

The two methods which have been used to make junctions are electric welding and flame welding (soldered junctions are unacceptable because of the high temperature). Electric welding should be used on non-noble metals to avoid oxidation. Flame weld-

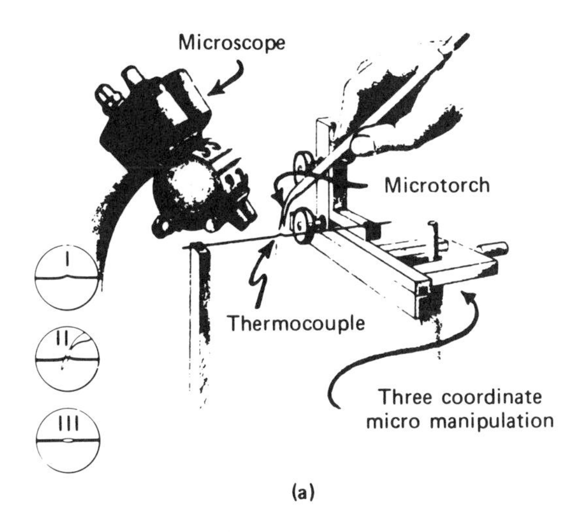

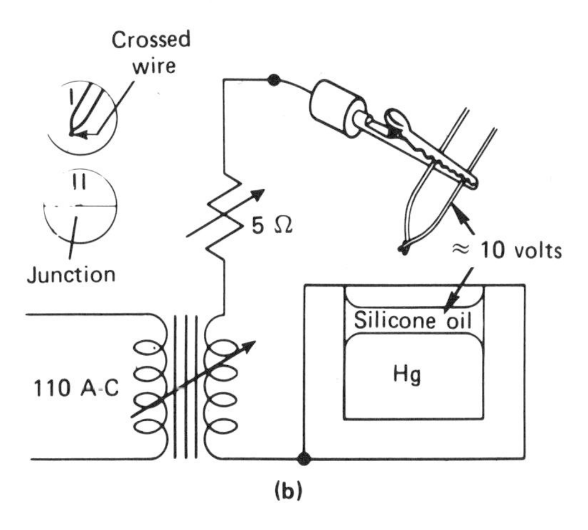

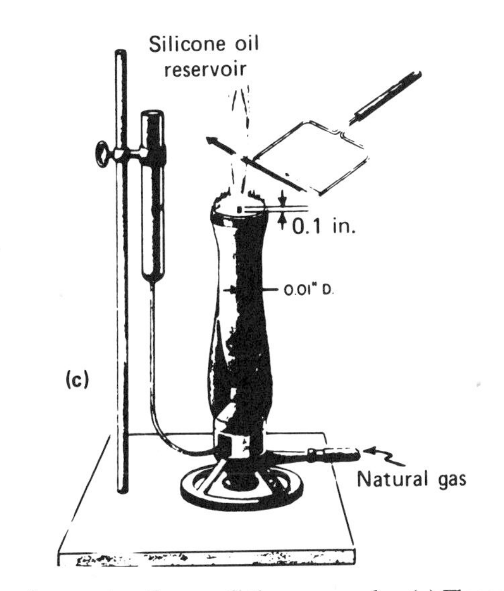

Fig. V-10 Techniques for constructing small thermocouples. (a) Flame welding; (b) Electrical welding; (c) Silica plating in flame.

ing is convenient for noble metal couples. Simple welding techniques can be used on wires as small as 0.03 mm, and techniques for welding small couples are discussed by Fristrom and Westenberg [1965] and Stover [1960].

To flame weld small noble metal couples it is necessary to use a micromanipulator to butt the two wires together with just sufficient compression so that the wires bow slightly. When the wires are held in this position a microtorch is momentarily applied to the junction to heat it just to the melting point of the platinum. At this time the wires move slightly and the flame should be immediately withdrawn or the couple will be burned. A satisfactory couple will have a bead that is no more than 50% greater in diameter than the original wire. The junction should be strong enough to stand a *light* tug with the fingers.

The couple can now be mounted on a suitable support by welding. Since the support wires are larger than the thermocouple, it can be wound onto the support and welded. The thermocouple can then be drawn taut by bending the support frame *carefully.*

The final operation is coating. This is most easily done by "flame plating" with silica (Kaskan [1957]). This technique consists of passing the couple slowly (about 1 sec per pass) through a flame containing particles of silica. These particles deposit in a homogeneous coat on the couple if the flame temperature is adjusted to the proper point (1860°C). This technique was introduced by Kaskan [1957] and studied by Mason and Theby [1984]. The thickness of the coating can be controlled by passing the couple through the flame several times and checking the thickness under a microscope after each pass (shown in Fig. V-10). The recommended silicone compound is dimethyl siloxane, but other silicone oils (e.g., DC 703) are satisfactory. Under 400 power magnification the coating should appear like translucent, fine porcelain with no pin holes. Other resistant oxide coatings have been tested by Cookson, Dunham, and Kilham [1964] and Kent [1970]. Shandross, Longwell, and Howard [1991] investigated mixed Yt_2O_3–BeO coating for use with fuel-rich systems. Methods of improving time constants to allow measurements of fluctuating temperatures in turbulent flames have been discussed by Bradley, Liau, and Missaghi [1989] and Cambray [1984]. The time resolution of thermocouples was compared with Rayleigh scattering by Chandran, Komerath, Grisson, Jagoda, and Strahle [1985].

VI

COMPOSITION MEASUREMENTS IN FLAMES

Composition is the intensive chemical variable in flames. This is often presented as a family of profiles giving composition as a function of distance through the flame front. Figure VI-1 gives an example of one such experimental study. Other examples can be found in Chapters XI and XII, which describe the various flame chemistries.

The three composition variables associated with flames are discussed in Chapter IX, which covers the analysis of flame data. They are:

1. Concentration, a scalar measuring the amount of a species per unit volume;
2. Flux, a vector measuring the amount of a species passing a unit area in a unit time. In flames this vector is pseudoscalar taking on positive or negative values depending on whether the flux is directed with the flow or against it[1];
3. Rate, a scalar measure of the rate of change with time. It is the spatial derivative of flux or the temporal derivative of concentration.

The basic experimental variable is concentration. Its units are amount per unit volume. Amount can be expressed as: mass, partial pressure, moles, or number of molecules. Volume can be in English, CGS metric, or SI metric. The old-fashioned CGS metric system with gram moles is used here rather than the newer SI system. This is simply a matter of the conservatism of the author combined with the fact that a substantial part of the literature is in the older units. There are no quantitative differences between the systems; so this should offer no difficulty to the reader. Fluxes and rates involve time in addition to composition units. In this book and most scientific work it is given in seconds. Some relations among units are given in Table A-1 of the Appendix A at the end of the book.

From the viewpoint of kinetic theory all of the species in flames are considered

1. Flames are three dimensional so that in principle the description of their flux requires tensor notation. This is not the case for steady-state flames because there must be a propagation direction and flame gradients must be small normal to this direction to maintain steady state. Off-diagonal elements of the tensor can be treated as perturbations on a dominant vector. This breaks down in time-dependent situations such as ignition or extinction and may or may not be a factor in turbulent flames.

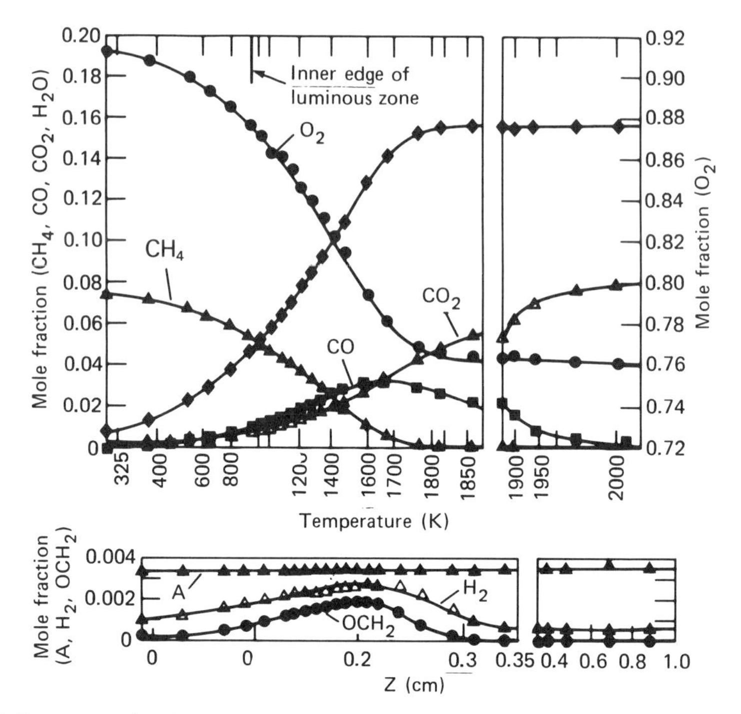

FIG. VI-1 Representative Composition Profile Using a Quartz Microprobe with Mass Spectral Analysis. (0.08 CH_4–0.92 O_2 at 0.1 atm). (Replottted from data of Fristrom, Grunfelder, and Favin [1960].)

molecules since they undergo collision, energy exchange, and reaction. However, it is convenient to subdivide the category of molecule further. Common divisions are: atoms, molecules, radicals, and ions.[2]

2. Atoms are the monatomic molecules. They can be normal molecules with completely paired spins as are the noble gasses, He, Ar, etc. If they have unpaired electrons, as do H•,O:, and Cl•, they are radicals. To emphasize the importance of the unpaired spins they are often indicated by dots (or dashes), but in many cases these indicators are left off. Finally, if they are charged, as for example, are H^+ and Cl^-, they are called ions.

Molecules in the restrictive sense includes all uncharged species with paired electron spins. This includes polyatomic species, diatomics, and atoms such as the inert gases, which have closed shells of electrons. Radicals are molecules with one or more unpaired spins. They should not be confused with ions (see below), which are charged. Most ions are also radicals with unpaired spins. It should be noted that although all molecules with an odd number of electrons are radicals, molecules with an even number of electrons can also be radicals if two electrons are unpaired. Atomic and molecular oxygen and the ground state of CH_2 are examples of such diradicals. Radicals include polyatomic molecules such as CH_3, diatomic molecules such as OH, and atoms such as H. They are usually more reactive than spin-paired molecules because unpaired electrons promote low-activation-exchange reactions. This makes them important in flame reactions. Most radicals are transient, being stable only in high-temperature equilibria, but a few exist under normal STP [Standard Temperature (298 K); pressure (1 atm)] conditions. As might be expected, these are oxidizers (i.e., electron acceptors). Common stable radicals include oxygen and many oxides of nitrogen and the halogens. Ions are molecules possessing

Species may also be classified according to their function in the reaction scheme. They can be reactants, intermediates, or products. Reactants can be subdivided according to whether the species is a single-decomposition reactant such as ozone or acetylene, a fuel such as hydrogen or methane, or an oxidizer such as chlorine or oxygen. Any species whose concentration passes through a maximum in a flame can be called an intermediate, although the term is sometimes reserved for chain reaction radicals. Intermediates can be molecules, radicals, or atoms. Products can also include radicals, molecules, and atoms. It is worth remembering that some species will play differing roles, depending on the flame. A species can play more than one role in flames. For example, fuels such as hydrogen, carbon monoxide, and acetylene can act as fuels and still appear as a final product. Excess oxygen acts similarly in fuel-lean flames.

There are two general approaches to determining local composition: sampling and in situ analysis. Each offers advantages. Some comparisons among methods are given at the end of this chapter. Sampling offers the advantage that any analytical method may be applied at leisure but it can disturb the flame and introduces quenching problems with reactive species. In situ analysis has the advantage of not disturbing the flame, with the principal disadvantage being that it requires the use of several different techniques to provide a complete analysis.

This chapter offers an introduction to the field of composition measurements in flames. It includes both the measurements and the interpretation of the results. Microprobe sampling and analytical techniques are covered. Chapter VII extends the discussion of sampling to molecular beam inlet probing and provides a more detailed analysis of flow in probes and sources of errors. Chapter VIII covers the field of laser diagnostics. These techniques apply not only to determination of composition, but also temperature and velocity.

Probe Sampling

Sampling requires the withdrawal of a representative local sample of flame gases. The sample can be sent directly into the analytical instrument or to a sampling bottle for future analysis. Two methods have been developed for flame studies: microprobing and molecular beam inlet probing. With microprobes the transfer time to the point of analysis is relatively long, so that reactive species such as atoms and radicals show up as product species that are stable under STP conditions. For example, a sample of flame gases that contained H, O, and OH would only show the products of recombination, H_2, O_2, and H_2O.

---

a net charge. Again they can be poly-, di-, or monatomic. Ions are even more reactive than radicals, but their concentration in flames (ca. 10^{-7} MF) is so low that their reactions contribute only a small fraction of the conversion reactions, which are dominated by radical–molecule and radical–radical reactions. Molecules and radicals will be the principal topic, taking the view that ions and other trace species, while interesting in their own right, are not required for a quantitative description of flames.

By contrast with molecular beam inlet, the sample is transferred under collisionless flow conditions to the detector, which is usually a mass spectrometer. Under these conditions the population of labile molecules is frozen, and atoms and radicals can be analyzed directly. The sample is adiabatically expanded to a low local temperature so that they are usually in the ground rotational state. The vibrational states are frozen close to their high-temperature inlet values. This presents some problems for mass spectrometry because the cracking pattern of a molecule depends on its vibrational state. Usually these problems are minor and can be surmounted. Probe sampling problems are explored in more detail in Chapter VII. The present chapter will center on microprobe sample transfer and analysis together with a few general remarks on flame sampling problems.

Three questions must be answered to assess the reliability of microprobing: (1) What effect does the process have on the flame? (2) Does the sample represent local concentration or flux or something intermediate? (3) Is the reaction quenched, or can the quenching process be reliably modeled so that the composition can be inferred?

Flame Disturbance

The flame disturbance of small quartz microprobes is generally minor, although there can be problems with flame attachment near extinction limits. This is because they withdraw small samples, have small probe angles, and do not extract large amounts of heat from the flame. By contrast molecular beam inlet probes offer significant perturbation due to heat losses, radical recombination, and flow disturbance because of the wide angle required for beam formation. In choosing a method the strengths and weaknesses of the two methods should be weighed. The ideal is to combine the two methods to provide cross checks.

Sample Interpretation

It was shown in early studies by Westenberg, Raezer, and Fristrom [1957] that if a probe inlet is sonic that local concentration is sampled. This occurs because both stream and diffusion velocities in flames are small compared with the sonic velocity at the probe throat. Local concentration in the sampling region is maintained by lateral diffusion in the flame. The more detailed modeling discussed in Chapter VII sustains this view.

Two major questions are involved in interpreting probe samples: (1) What is the aerodynamic history of the gas sampled? and (2) What is the postsampling reaction history and ultimate fate of the reactive species? Stable species offer no problem. Early studies by Fristrom, Prescott, and Grunfelder [1957] showed that they pass through heated quartz tubes with negligible decomposition or reaction under conditions that simulate flame sampling. Sampling results for major species are independent of sampling conditions below a critical pressure drop (Fig. VI-2), but results can be biased by a percent or two in regions of high radical concentration by the addition of recombination products.

By contrast radical species that are unstable under sampling conditions must appear as recombination products. The question is, what is the effect of the chain reac-

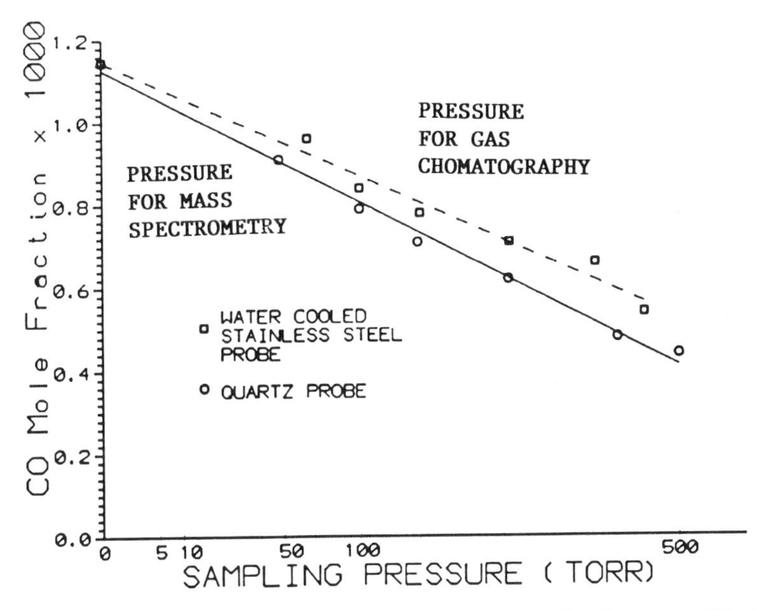

Fig. VI-2 Effect of probe sampling pressure level on quenching of carbon monoxide. Data of Schonung and Hanson [1981] replotted against square root of sampling pressure level. For trace quantities, where concentrations are of the same level as the radical species, a low sampling pressure is desirable.

tions they introduce have on the compositions of the stable species? In the worst case they could affect the concentrations of major species through chain reactions.

Fristrom [1983] commented that since a temperature drop will not quench low-activation-energy radical reactions, radicals must be recombined before radical–molecule chain reactions become significant. In practice, wall recombination appears to minimize such biases, although ESR studies show that some atoms and radicals survive sampling and travel significant distances through quartz sampling probes (see discussion in the section on ESR probing).

Sample Collection and Transfer

One of the advantages of microprobe sampling relative to molecular beam inlet systems lies in the ability to collect a sample and transfer it. This widens the scope of instruments that can be used for analysis and allows use of the optimum analytical methods for each species. It removes the necessity of having a dedicated instrument. The other side of this coin is that the sample must be taken, preserved, and transferred. The problems associated with adsorption of the sample on inlet tubes and walls are added. This is particularly severe with polar species, the most notorious of which is water. When an on-line instrument is available, the problem can be solved by running the system to equilibrate the walls with the incoming sample. If such a system is allowed to flow until constant analytical results are obtained, continuity considerations dictate that the sample analyzed be identical with the inlet sample. The time required to attain equilibrium depends on the flow and the material of the walls. The least absorbent material is virgin Teflon, but some block Teflon is porous and not suitable for vacuum systems. PVC and polyethylene tubing are also useful and convenient. They should be outgassed initially to remove traces of polymerizing agents. Glass is intermediate, and metals are the least satisfactory. Of the metals, stainless steel and nickel appear the least adsorbent and most resistant to corrosion. For the halogens either monel or nickel are recommended. Precautions are necessary in setting up the sampling flow. For example, the inlet to most mass spectrometers is effusive, while the probe samples are taken in continuum flow. To provide an unbiased sample for the mass spectrometer it is necessary to make the portion of the sample diverted effusively into the spectrometer a small fraction of the total flow. The flow in the sampling line should be continuous.

Where batch samples are required, sampling bulbs of glass, metal, or polymer can be used. Care must be exercised that the vessels are clean and that adsorption does not bias the results. A few general rules should be observed. The sample bulb should be equilibrated with the sample by allowing the sample to flow through as long as is practical before collection. The larger the sample and the higher the pressure, the smaller will be contamination and adsorption problems. Unfortunately, for effective quenching, samples should be taken at a low pressure (< 0.01 of the flame pressure). The sample can be recompressed, but care should be taken to avoid condensation. Each sampling problem should be considered separately. Samples should be compared using differing conditions. One can only assume that a representative sample is obtained if constancy is obtained using reasonable variation of sample pressure level, time, and size of sample. Temperature should also be varied if condensation is a problem. Recompression is necessary for gas chromatography since it is often necessary to inject the sample at several atmospheres. In such recompression it is necessary to heat the sample above the boiling point of the least volatile constituent. Water requires particular care since it is often a major constituent. This can be avoided by absorbing or reacting the water and measuring it separately, but this adds a step that is often inconvenient.

Analytical Methods

With batch sampling any convenient analytical technique can be applied. A combination of several techniques can provide the best information since there are wide variations in sensitivity and selectivity. Often the choice will be dictated by local availability—whatever works and is most convenient is the best choice. The discussion will touch on some of the strong and weak points of some common analytical methods, which are compared in Table VI–1. This is too large a subject for comprehensive treatment in these few pages, so the interested reader is referred to treatises by Tine [1961], Bamford and Tipper [1969], Rabeck [1982], and Demtroder [1981].

Orsatt Analysis

This is a primitive method that dates back to the beginning of chemistry as a science. It is a very general technique that consists of measuring the partial pressure of a gaseous

Table VI–1 Comparisons among Analytical Techniques

Technique	Applicability	Cost	Character
Orsatt	Volatile	Low	Simple
Mass spectra	General	Low–high	
Magnetic defl.	General	Moderate	Compact
Double focus	General	High	High resol.
Time of flight	General	Moderate	Speed
Quadrapole MS	General	Moderate	Sens.
Ion trap	High mass	Low	Mass
Fourier trans.	General	High	Resolve
ESR	Stable radicals	Moderate	Radical
Abs. spect. (AS)	General	Low–High	
Microwave AS	Polar molecules	Moderate	Select
Infrared AS	Polyatomic fuels	Moderate	
Visible AS	O_3, Br_2, NO_2	Moderate	
Ultraviolet	General	Moderate	Sens.
X-ray emission	Elements	Moderate	
NMR	Quadrapole	High	Select.
Chromatography	Complex mixtures	Moderate	Separ.
Gas	General	Moderate	Separ.
Liquid	High mo. wt.	Moderate	Special
Polarography	Peroxy compounds	Moderate	Liquids
Thermal cond.	Simple mixtures	Low	Separ.

constituent after eliminating the other species. Species isolation can be done using several methods: freezing out constituents, adsorbing them, or reacting them. The detector can be any pressure measuring device. This method has been applied in flame studies by Friedman and Cyphers [1955] and Frazier, Fristrom, and Wehner [1963]. It has fallen into disuse since chromatographs and mass spectrometers have become commonly available because manual operation made complex analysis too slow and manpower-expensive. With the advent of computer-guided systems these problems could be solved and, properly applied, it might find wider applicability.

Gas Chromatography

Gas chromatography is one of the most widely applied methods. the field has been reviewed in several books such as those of Kaiser [1963], Ettre [1979], and Jennings [1992]. It is basically a technique for separation by differential adsorption followed by suitable detection. Separation is accomplished by forcing a slug of sample to flow through a tube filled or lined with adsorbent. Since each species possesses a different adsorption characteristic, the arrival times at the detector will differ. Ideally, the species will be separated and arrive as a resolved "peak" whose area (the product of response by the dwell time) is proportional to the amount of the species in the original sample slug. Strongly adsorbed materials are delayed by the adsorbent while the lightly adsorbed materials pass through close to stream velocity. The relative delay time can be related to the absorbance and the width of a "peak" to the diffusion of the species in the carrier. "Columns" have been tailored for many analytical problems. By combining several columns with temperature and flow programming, one can separate

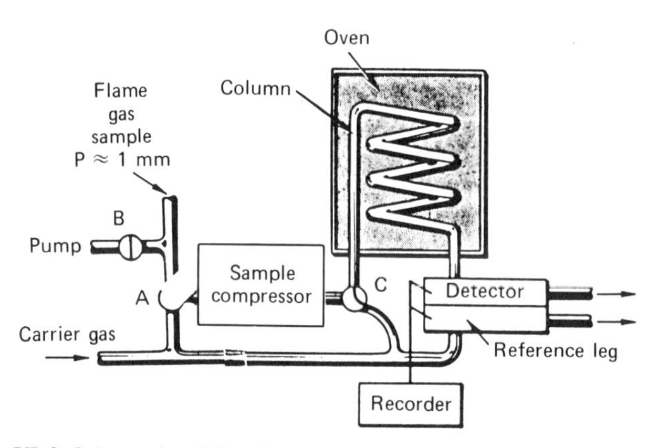

FIG. VI-3 Schematic of Gas Chromatograph with Sample Compressor.

mixtures of almost any complexity. One must know ahead of time what constituents to expect so that column and detector can be matched to the analysis. This offers no problem in routine analysis, but systems where one does not know what species to expect can offer unpleasant surprises. A major difficulty is that identification of compounds depends on arrival time at the detector. The analysis depends, of course, on correct identification of the species. For this reason it is desirable to use an analytical method that gives some clue as to the identity of the species passing through to confirm that it indeed is the one expected. The combination of a gas chromatograph with mass spectrometer detection is considered to be one of the most powerful analytical tools. Other detectors, such as thermal conductivity gages, ion gages, beta detectors, and hydrogen flame conductivity detectors, offer special convenience or sensitivity for some applications. Several detectors can be used in sequence so that most analytical problems in flames can be resolved. Chromatography can be considered to be a standard method for flame analysis. One major problem has been that of sample injection. Chromatography is essentially a batch analysis method. For flame analyses it is necessary to collect a sample under low-pressure conditions, where the flame gases can be quenched, and then compress the sample to over an atmosphere for injection into the chromatograph. Modern instruments are so sensitive that they allow samples to be injected directly. Figure VI-3 shows a schematic for using a chromatograph to batch sampling flames.

Absorption Spectroscopy

Absorption spectroscopy offers a convenient analytical tool, and many texts such as Pungor [1967] and Demtroder [1981] discuss its theory and practice. The variable measured is the loss of intensity of a beam of radiation due to sample insertion. For an absorbing species present at a partial pressure $P_i = PX_i$, the relation between incident radiation intensity I_0 and that detected, I, after passage through a sample of thickness L is given by Beer's Law

$$I = I_0 \exp(-\alpha_i P_i L)$$

where α_i is the absorption coefficient, a function of temperature.

From the above equation it is clear that I_0 gives the absorption per unit length for the pure substance at unit pressure. In general it is necessary to deal with single absorption lines of low strength, or Beer's Law may not be obeyed. For species more complex than diatomic molecules, the coefficient must be determined empirically.

Any region of the electromagnetic spectrum may be used for analysis, but the infrared and the ultraviolet have been the most useful.

Mass Spectrometry

Mass spectrometry is one of the most useful analytical methods for flame studies. It is based on the observation that electron bombardment of a molecule produces a distribution of ions that are characteristic of the molecule and the electron energy. One of the virtues of the method is that it provides an analytical method with comparable sensitivity for all flame species. Electron bombardment produces every positive ion that can be formed by removing an electron or breaking a bond and removing an electron. Few negative ions are formed because most molecules to not have stable levels for electron capture. Those that do (e.g., O_2^-) also form positive ions so mass spectrometry is primarily concerned with positive ions. The relative abundance of the fragment ions is a reproducible function of the electron beam energy and the strength of the bond that had to be broken.

Magnetic Deflection Mass Spectrometry

The original mass spectrograph of Aston separated the ions using their differential deflection under crossed electric and magnetic fields. Either the collector position, the magnetic field, or the electric field can be changed to verify the mass-to-charge ratio of the ion detected (Fig. VI-4). This instrument is manufactured commercially by several companies with mass resolutions varying from one in four in helium leak detectors

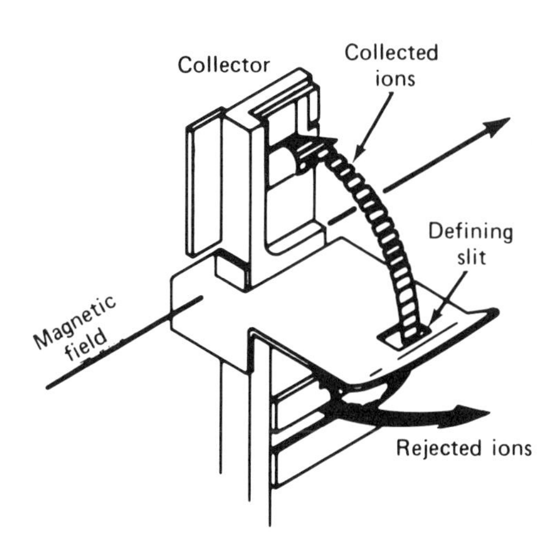

FIG. VI-4 Schematic of Magnetic Deflection Mass Spectrometry.

through the common single-deflection instrument, which gives unit mass resolution up to mass 50 through the double-focusing instruments used for isotope work, which can separate one mass unit in ten thousand at mass 100. High mass resolution is obtained with a sacrifice of convenience, speed, and sensitivity, so one should use a system with resolution matched to the problem. For flames single unit mass resolution up to the highest species considered is convenient, but not necessary. Resolution beyond this is superfluous unless one can reach the plateau where the mass differences between nitrogen, ethylene, and carbon monoxide at mass 28 become separable. Since N_2 is 28.0134, CO is 28.01, and C_2H_4 is 28.054, ethylene can be separated using 1/1000 resolution, but the separating of CO from N_2 requires 1/10,000. Such resolving power is available in commercial double-focusing instruments and recently in the better quadrupole instruments.

Time-of-Flight Mass Spectrometry

In time-of-flight (TOF) mass spectrometry, initial ionization is by a pulsed electron beam. Laser ionization can also be used and has advantages of high selectivity. In this instrument a short ionization pulse (a few hundred nanoseconds) forms a bunch of electrons that are then pulse accelerated by a grid and allowed to pass down a flight tube to a detector (Fig. VI-5).

The instrument makes use of the principle that an electric field imparts kinetic energy proportional only to an ion's charge, and because of differing masses each ion will have a characteristic terminal velocity. If the accelerated ions are allowed to fly down a drift tube and impinge on a detector, a time-varying signal will be obtained from the single ionization burst. Ion arrival times are proportional to the square root of their masses.

Commercial instruments operate at repetition rates between three and one hundred kilohertz. Increased frequency allows increased sensitivity and time resolution. Lower frequency allows study of ultrahigh masses.

The original display for the instrument was an oscilloscope. This offered only lim-

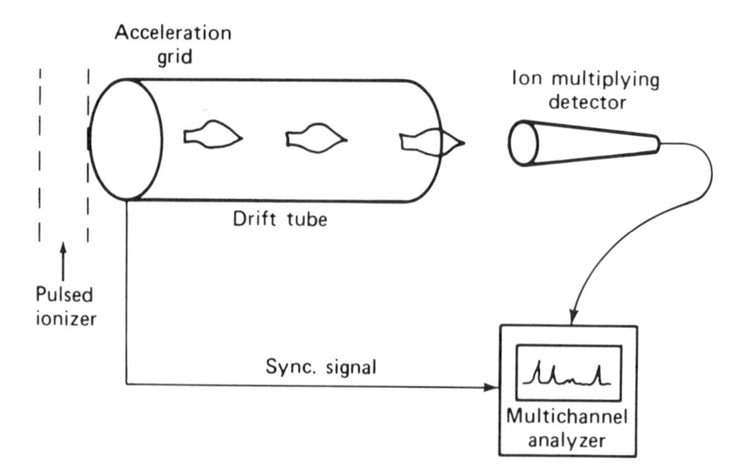

FIG. VI-5 Schematic of Time-of-Flight Mass Spectrometry.

ited sensitivity and range; so timed gates were introduced that separated signals for a given mass-to-charge ratio. This allowed integration and greatly increased sensitivity and dynamic range. This was at the sacrifice of much information, since only gated signals could be studied. With the advent of large-scale memory chips and high-speed buffers, it has become possible to digitize and collect the complete spectrum of and record for computer analysis as many as a dozen cycles (more with added memory). Signal-averaging algorithms allow data handling either in real time or as desired. This makes the TOF a very powerful instrument since it can obtain complete spectra at a 25 msec rate with the time of a given peak known to within a fraction of a nanosecond (K. Lincoln, NASA Ames [1987, private communication]). This is ideal for studying time-dependent phenomena, and the temporal resolution is well matched to transient flames.

Quadrupole Mass Spectrometry

One of the most sensitive and convenient mass spectrometers is the quadrupole. It is one of the most common instruments found in combustion laboratories. This was called a "mass filter" by its inventor Prof. H. Paul of the University of Bonn, Germany. It uses AC and DC fields arranged in quadrature and operated so that there is only a single open mass-to-charge trajectory through the axis of the quadrupole fields for a given choice of the AC and DC fields (see Fig. VI-6). This can be done using practical fields, frequencies, and sizes of quadrupoles. Very satisfactory commercial instruments are on the market. They offer high sensitivity and good mass resolution. They have the interesting property of offering the same mass resolution irrespective of mass number. Usually the instrument is adjusted to resolve unit masses, but this can be increased by a factor of over 500 in special instruments. The RF frequency required varies with mass number so that when a wide range of masses is required, as in flame studies, more than one RF head is required and the system must be tuned in each mass range. The inclusion of masses one and four is particularly inconvenient. They can be

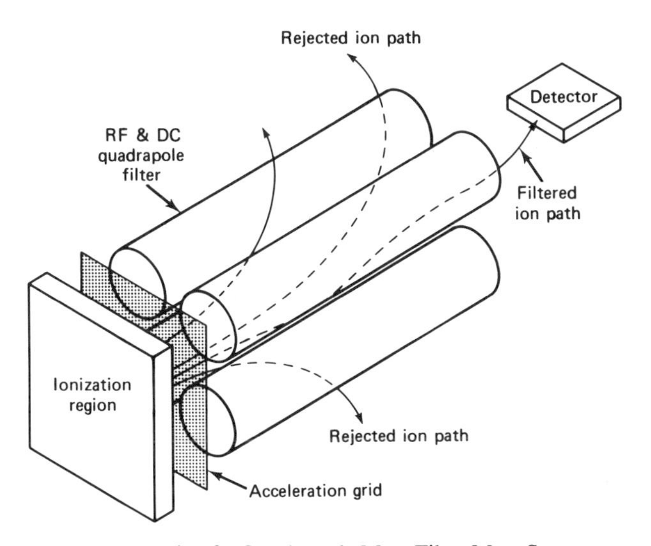

FIG. VI-6 Schematic of a Quadrupole Mass Filter Mass Spectrometer.

scanned across moderate mass ranges (two to one) at moderate frequencies (a few hundred Hertz using a single mass setting of the quadrupole voltages.

Fourier Transform Mass Spectrometry

A recent entrant in the field of mass spectrometry is the Fourier Transform Mass Spectrometer (FTMS) (Wilkins and Gross [1981]). This is an ion cyclotron operating in the time domain. The new instrument makes use of superconductor magnets whose high magnetic field raises the resonance frequency of the ion to a convenient detection range. The technique is to "chirp" the drive periodically as ions are made and injected. The signal is detected in the time domain as a wave train whose Fourier transform is the mass spectrum desired. The availability of microprocessors that can record and store this mass of information and software for handling Fourier transforms with minimal effort have made this a promising and powerful technique. It combines high sensitivity with high mass resolution because information is collected on all ion oscillations simultaneously and the attainable mass resolution is limited only by the number of repetitions.

Other Mass Spectrometers

A number of mass spectrometer designs have been tried such as the ion linear accelerator and the field ionization source time-of-flight spectrometer. A recent entry is the ion trap mass spectrometer described in a review article by Cooks, Glish, McLuckey, and Kaiser [1991]. It is a quadrapole instrument in which the field is formed using three electrodes. Often the ionization is external to the instrument. It is particularly useful for high mass numbers. It is conceivable that this instrument might be useful in looking at flame ions or other tasks. It is one of the most compact instruments yet produced. One should keep an open mind and judge each new instrument on its own merits.

Analysis of Mass Spectral Data

Many modern mass spectrometers include built-in computerized analytical capabilities including a mass spectral library for common species. Computerized data handling is a marvelous labor-saving tool, but one should be cautioned that library sensitivities and cracking patterns are average ones and true quantitative analysis can only be obtained by daily calibration and updating the coefficients used in the analysis. The following remarks are intended to acquaint the neophyte with the principles behind mass spectral analysis. The field is discussed in introductory texts but a formal introduction through one of the short courses available through the American Chemical Society or a local university would be a good investment for someone entering the field.

Mass spectral data are obtained in the form of a spectrum, which gives the distribution of ion flux as a function of the mass-to-charge ratio. Mass-to-charge ratio is usually expressed as the ratio of molecular weight to electronic charge and referred to as a mass number. If standard inlet conditions are used, the output at any given mass number is proportional to the number of ions formed, which in turn is proportional to the inlet concentration (partial pressure of the species in question). Under the conditions of high electron beam energy, which are usually used to obtain high sensitivity, all of the ions that can be made by breaking bonds are formed in varying degree. No ions are formed by rearrangement of the original molecular bonds, or recombination. For example, the spectrum of water yields ions at mass numbers 18 (H_2O^+), 17 (OH^+), 16 (O^+), and 1 (H^+), not 32 (O_2^+) or 2 (H_2^+). In addition, all of the possible ions appear in their normal isotopic abundance, as well as some doubly ionized species.

These patterns and relative intensities have been cataloged for standardized conditions in a number of sources. A particularly useful compilation is that of the Standard Reference Data Program of the National Institutes of Standards and Technology (SRDP/NIST), which offers a set of discs suitable for personal computers (Lias [1992], Stein [1992]). Other reliable sources are also available, such as the eight peak index of the Royal Society [1983]. The field is served by a monthly journal, the *International Journal of Mass Spectrometry and Ion Physics*.

The probability of interference of the spectrum of one species with another is rather high. Fortunately the contributions of each species to a given mass number are additive, so that it is possible to derive the partial pressures, even in these cases. As the complexity of a mixture is increased, the problem of interference grows.

Flame samples differ from many analyses discussed in the literature because of the substantial amounts of oxygen and oxygen-containing compounds involved. "Carburized filaments" are undesirable because of the reaction of oxygen with the carbides to form spurious carbon monoxide. This problem is best avoided by using a rhenium filament, which forms no stable carbides, or by running with a clean tungsten filament.

In analyzing a spectrum it is necessary to make some reasonable choice of expected species. All species of quantitative importance should be included, although the smaller the number chosen, the easier the analysis. A common approximation that is valid for fuel-lean hydrocarbon flames is to assume no hydrocarbons higher than in the fuel species, no oxygenated compounds except formaldehyde (possibly methyl alcohol), and no nitrogen compounds other than elemental nitrogen (the absence of NO should be confirmed). This choice should be checked for consistency by analyzing some typical spectra completely showing that all of the observed peaks are accounted for quantitatively. Having established the qualitative makeup of the mixtures to be analyzed, a characteristic mass peak is assigned to each species. This should preferably be its strongest mass peak and not interfered with by other species. Where it is possible to assign unique mass peaks that no other species of the mixture possesses, the analysis is trivial since the partial pressure is the peak height (corrected for background) multiplied by the sensitivity factor obtained by calibrating the instrument. The mole fraction is then easily obtained as the partial pressure of the species divided by the sum of all of the partial pressures.

In general, more than one species contributes to the height h of a given mass peak. Fortunately the contributions of each species are additive, so that the observed peak height is the sum of the product of the partial pressure P_i of each contributing species multiplied by its respective sensitivity factor $(K_i)^p$. Thus

$$h_p = \sum_{i=1}^{n} P_i (K_i)^p$$

One such equation can be written for each characteristic mass peak measured, that is, one for each species expected. This constitutes N linear algebraic equations with N unknown partial pressures that can be solved by straightforward means. With computers and programs available today this is a simple problem of matrix inversion.

It is good practice to take a full spectrum and check that the experimental values can be reproduced within reasonable limits of error. Major discrepancies indicate the presence of a species not considered in the analysis.

If the number of species is large, the resulting spectrum becomes very complex and the foregoing type of analysis—although always sufficient in theory—will not yield the desired accuracy. The most direct solution of this problem is the separation of the sample. A number of commercial systems combine gas chromatography with mass spectrometry.

Special Probe Methods

Several special probing methods have been developed for the measurement of radical concentrations. They include scavenger probes, ESR probes, optical probes, and tracer techniques.

Scavenger Probe Sampling

Scavenger probe sampling combines microprobe sampling with chemical scavenging. This allows the determination of radical concentrations using microprobe techniques. It has been applied to flames by Fristrom [1963] and Volponi, McLean, Fristrom, and Munir [1986]. It avoids the limitations of scavenger studies by isolating the scavenging reactions from the flame. The advantages of the technique are: (1) spatial resolution; (2) radical concentrations being determined absolutely relative to a known stable species; and (3) smaller flame disturbance than molecular beam inlets.

The method illustrated in Figure VI-7 consists of introducing a scavenger molecule into the probe just beyond the sampling orifice. It reacts rapidly with radicals to produce identifiable products. If the secondary radicals are unreactive under probe conditions, they will disappear principally by recombination with each other. Scavengers that react with a single radical offer a problem because the exchange reactions will funnel all radicals into the scavenged species. In this case the scavenger measurement represents the total radical concentration rather than an individual scavenged species. For quantitative studies, the decompression, mixing, and scavenger reaction rates must be rapid compared with any other reactions the radical can undergo, and the scavenger reaction must produce a unique product. Of the exchange reactions in the Hydrogen–Oxygen system, the worst case is $OH + O = H + O_2$. It occurs on every tenth collision between O and OH, independently of temperature. Excess scavenger must be mixed into the sample rapidly compared with this rate (i.e., before many $OH + O$ collisions occur). In most flames, O and OH concentrations are well below 1% so that scavenger should be mixed on a short time scale compared with 1000 collisions per molecule. This can be accomplished by injecting within a fraction of a millimeter of the tip. A

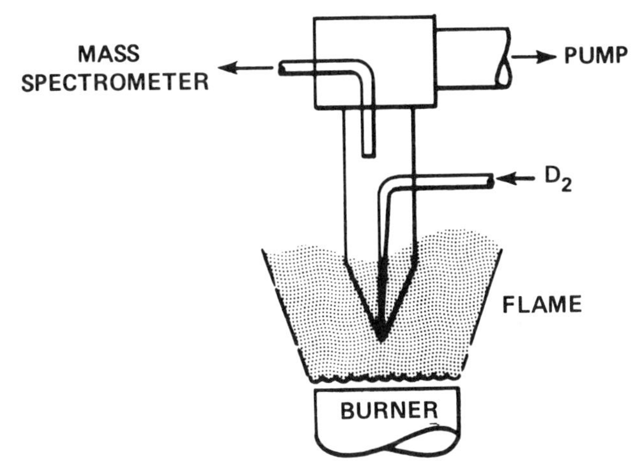

FIG. VI-7 Schematic of Scavenger Microprobe Flow Arrangement using Deuterium as a Scavenger. (After Volponi, McLean, Fristrom, and Munir [1986].)

satisfactory scavenger must react rapidly and quantitatively with all flame radicals. The author feels that these conditions can be met using D_2 and I_2 and perhaps by other systems. They react with most of the radicals found in C–H–O flames at rates that should allow scavenging of the radicals before exchange reaction biases occur.

Scavenger fluxes should be comparable to or larger than the sampling flux. As long as mixing is sufficiently rapid, scavenger quenching is favored by higher pressures and temperatures, since this reduces the importance of wall reactions. Such a probe should collect unbiased, quantitative samples of both stable and radical species. This is an attractive possibility because of the simplicity of microprobe sampling, and because flame disturbances can be minimized. Quantitative measurement of flame radical species only requires calibration with stable species. This is important because the quantitative determination of absolute radical concentrations is often difficult. In rich flames where complex organic radicals occur, gas chromatography and mass spectrometry are required. Iodine as a scavenger yields easily separable and identifiable organic radical derivatives. Problems that occur are wall exchange of water and DI and possible reactions of iodine atoms with oxygen.

Electron Spin Resonance Probing

Since labile atoms and radicals have unpaired electrons, they are paramagnetic. They show absorption in the microwave region due to transitions between the Zeeman levels when placed in a magnetic field. Many common radicals such as H, O, N, OH, halogen atoms, etc., have been observed in the gas phase in this way, and commercial spectrometers are now available with sufficient sensitivity to detect low-pressure samples (0.1–5 Torr).

Electron spin resonance (ESR) was used by Azatyan, Panilov, and Nalbandyan [1961] for radical detection by allowing a flame to burn inside the resonant cavity of

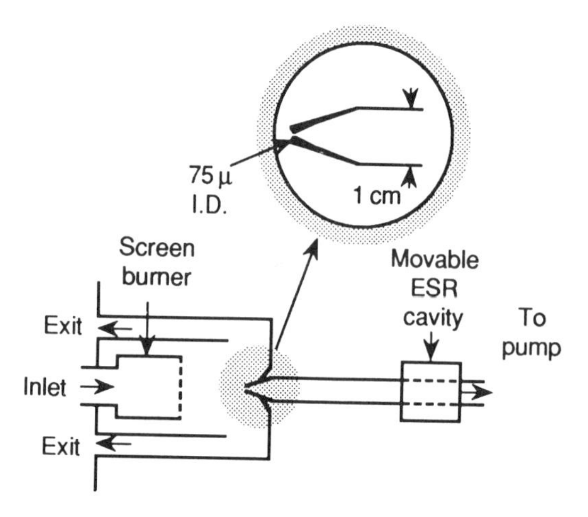

FIG. VI-8 Electron Spin Resonance Spectrometry for Flame Sampling. (Fristrom and Westenberg [1963].)

the spectrometer. There are problems in deriving structure from such experiments. The tomographic techniques of nuclear magnetic resonance imaging might be applied but have not yet been developed for use in flames. A simpler approach combining probe sampling with ESR spectroscopy to measure absolute atom concentration profiles in flames was developed by Westenberg and Fristrom [1965]. Using the apparatus shown in Figure VI-8, the gas samples withdrawn from the flame zone were pumped directly through the ESR detecting cavity. An important feature of the experiment was the ability to move the cavity and its associated magnet along the sampling flow tube. In this way the decay of atoms along the flow tube could be corrected. Loss in the probe tip appears negligible. H and O atom profiles were measured, and OH could have been measured using a different cavity. For such studies the flame should contain known amounts of some stable radical such as O_2 or NO to act as a calibrating reference species. Bregdon and Kardirgan [1981] studied hydrogen atoms. Sochet's group in France has extended the technique to the study of halogen atoms and hydroxyl and peroxy radicals (Carlier et al. [1984]; Pauwels et al. [1982]). Ksandropolo, Sagindykof, Kudaibergernov, and Mansurov [1975] were able to freeze out and detect peroxy radicals in a propane flame and showed that adsorbed fuel species could bias the results in rich flames. Herron and Peterson [1991] studied sampled hydrogen atoms behind a sonic orifice using a simple atomic absorption technique.

An analogous technique has been developed by Wagner's group at Göttingen using laser rather than microwave radiation, but it has not been applied to flames.

Rate and Tracer Studies of Radical Concentrations

In the early days of flame structure studies radical concentrations were deduced from the observed rates of disappearance of stable species. This technique is mostly of historic interest now that direct measurements of radical concentrations are feasible. The procedure is the reverse of obtaining rate constants from measured rate data. The reaction of CO was used by Fristrom and Westenberg [1961] to estimate OH concentration in a methane flame.

Where no convenient reaction is available, a tracer species can be introduced. A number of such reactions have been used by Fenimore and Jones [1958]. For example, deuterium and heavy water are convenient additives for the study of H concentration using the rate of formation of HD. The same tracers might also be used to study the concentration of OH through the formation of HDO, but this reaction also proceeds rapidly on glass and metal walls, and it is difficult to obtain meaningful concentration measurements. Nitrous oxide has been used as a tracer in flames to measure the concentration of hydrogen and oxygen atoms. Some caution must be exercised in using such tracer techniques. It must be established that in the flame being studied the assumed reactions are dominant and reverse reactions are negligible. In the region where the reactions can be assumed to be in partial equilibrium, the radical concentrations can be calculated (Chapter X).

Optical Probes

Flame gases may be examined spectroscopically either in situ or by using conventional analytical techniques on a probe withdrawn sample. An interesting variant is the sampling technique developed by Schonung and Hanson [1982] for the study of turbulent flame fluctuations and illustrated in Figure VI-9. It uses the sampling line as an absorption cell.

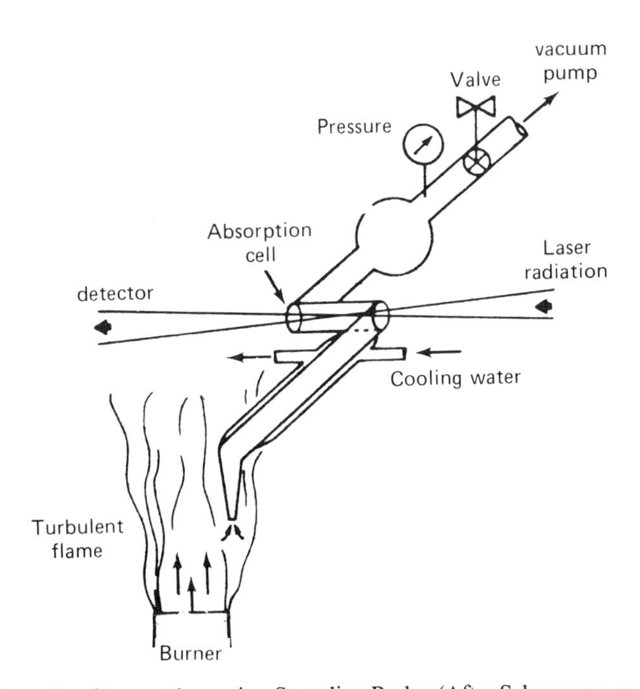

Fig. VI-9 Schematic of Laser Absorption Sampling Probe. (After Schoenung and Hanson [1982].)

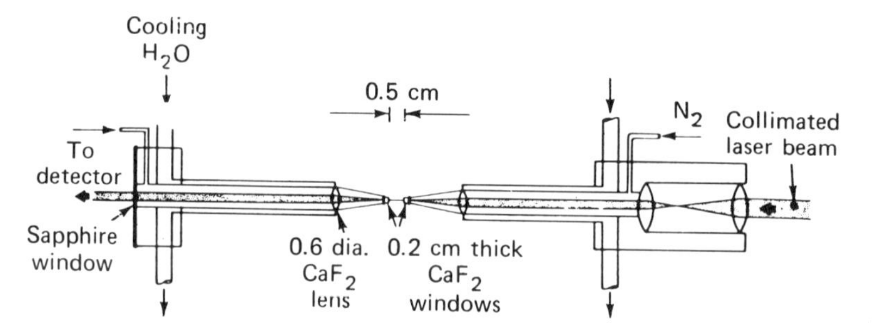

FIG. VI-10 Schematic of Light Guide Probe. (After Schoenung and Hanson [1982].)

Another ingenious optical probe was also developed by Schoenung and Hanson [1982] for absorption studies and is shown in Figure VI-10. This probe allows easy variation of the path length so that optimal absorption can be obtained and end effects corrected for. It is also useful for species that absorb so strongly that, in normal path lengths, absorption is too complete to allow analysis.

In many flame studies the path length required is inconveniently long. This difficulty can be overcome by using multiple-pass optics that reflect the beam through the region a number of times, allowing an order of magnitude increase in sensitivity. This technique is illustrated in Figure VI-11.

Laser Methods for In Situ Analysis

The direct analysis of flame gases using optical methods has become feasible, sensitive and powerful with the advent of lasers. This does not disturb the flame and thus avoids

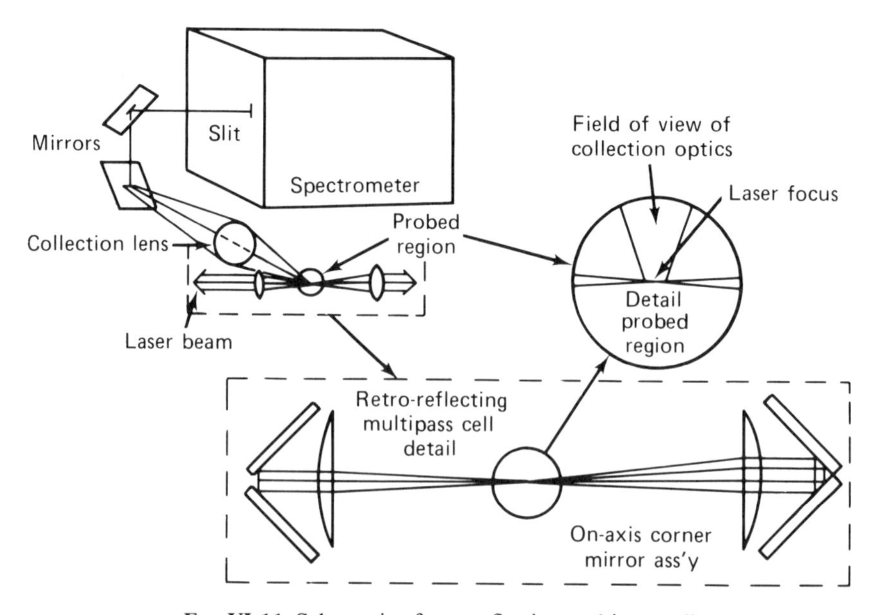

FIG. VI-11 Schematic of retroreflecting multipass cell.

many of the vexing problems of probe interference. In addition, it is possible to measure local temperatures, individual state populations, and radicals that are difficult to sample. It is even possible to measure species populations that are not in Boltzmann equilibrium. This power and selectivity does come with a price. The cost of the equipment is significantly higher than that of probe sampling techniques that accomplish the same objective. The methods are so selective that it is often necessary to employ several methods to obtain the complete analysis required for flame studies. This can make the alignment of profiles a significant problem. In many cases one may be overwhelmed with detailed information that is difficult to interpret quantitatively, and some of which may be irrelevant to the problems at hand. Nevertheless, in situ analysis by optical techniques offers a powerful set of diagnostic tools.

There are three general optical methods: (1) absorption, (2) scattering, and (3) emission. Absorption is a line-of-sight method unless some auxiliary technique such as Stark modulation is employed. This requires planar geometry or the use of tomography to provide local concentrations, and these factors may limit spatial resolution. Stark modulation and tomography are discussed in the chapter on laser methods (Chapter VIII). The choice of spectral region depends on the species. As a general rule polyatomic species such as fuel molecules are best studied in the infrared region, where they usually have rich spectra. Homonuclear diatomic species such as nitrogen and oxygen must be studied in the ultraviolet region because they have no absorptions in the infrared or visible. This can offer problems because of interference from the surrounding atmosphere. Some atoms and radicals have absorptions in the visible, but most of them lie in the ultraviolet. Again problems can occur due to atmospheric interference. The absorption chosen must be of reasonable strength, neither so weak that absorption cannot be detected nor so strong that the flame is opaque at the chosen wavelength. This problem can be mitigated by use of the optical probe described in the previous section.

In scattering process the incoming light is absorbed by the molecule and then reemitted. In Rayleigh scattering the light is reemitted elastically without change in wavelength. In Raman scattering the emitted light is either shifted upward (Stokes lines) or downward (anti-Stokes lines) by the addition (or loss) of a quantum from the molecule's vibrational–rotational energy. In fluorescent scattering the incoming wavelength is resonant with one of the transitions in the molecule, and an excited state is formed that then reradiates in a short period (10^{-4}–10^{-9} sec) depending on the nature of the transition.

Since molecules are nonlinear oscillators, they can interact with light in even more complex ways, leading to a host of multiple quantum processes where more than one quantum is absorbed. If there is a suitable transition in the molecule, these waves can be added and subtracted so that emitted light can significantly shift. These processes and their use in flame studies are cataloged and discussed in Chapter VIII where laser diagnostics are covered.

Either conventional or laser light sources can be used. Lasers are usually more convenient because of their selectivity, intensity, and ease of collimation. For many of the weaker scattering processes they are the only practical source. Most conventional sources require the use of a monochromator, the exception being low-pressure discharges used for atomic absorption of the elements. The virtue of conventional sources is their cheapness relative to lasers and in some cases their compact size. The advent of lasers has opened a wealth of opportunities for studying flame structure without interference. These techniques are discussed in more detail in Chapter VIII and will only be outlined here. Two general types of measurements are made: absorption and scattering. Absorption measurements are line of sight and require a flat flame or the use of tomographic techniques. By contrast, scattering measurements can provide point measurements if the laser is brought to a focus, or map areas if planar illumination is used. Mapping techniques were outlined in Chapter IV on flame visualization. One difficulty with laser spectroscopic methods comes from their most useful aspect, namely their specificity. A different wavelength and often a separate experiment are required for each species. By contrast, a mass spectral analysis simultaneously covers all species. In other cases where a single process is being studied the specificity is desirable and necessary. The lack of flame disturbance is usually well worth paying for.

Absorption Spectroscopy

Absorption spectroscopy is one of the most powerful analytical methods. Nonlaser methods were mentioned earlier as an analytical method. Tunable lasers have greatly improved the sensitivity and convenience of absorption spectroscopic study of flames. In the infrared region most are diodes, while in the visible most are fluorescent dye lasers, either pumped by a flash lamp or another laser. By doubling or mixing with another laser, wavelengths well into the ultraviolet can be reached. Laser sources are discussed in Chapter VIII. To study a particular species one must choose a spectral region where it absorbs but is not overlaid by some other absorber. Commonly the ultraviolet region is used for the study of radical species while the infrared region is used for studying fuels and fuel fragments. Absorption spectroscopy is by nature a line-of-sight method; so the achievable spatial resolution depends not only on the characteristics of the probing beam, but its dispersion by flame density gradients. For quantitative work it is necessary to know the effective path length and correct for end effects.

One technique for defining position is Stark modulation, using a crossed laser beam, as developed by Farrow and Rahn [1981]. This makes essentially point measurements possible since the intersection of two laser beams defines a small volume. The technique is illustrated in Figure VI-12 and further discussed in Chapter VIII. This technique could be used in any spectral region, but has proven particularly useful in the infrared for studying fuel breakdown in flames.

Laser Scattering Methods

A number of laser scattering methods are used in composition studies. One of them, Rayleigh scattering, has little species specificity. As a consequence it is used more for density than for composition studies (Smith [1978]). Through density it provides a useful temperature probe (Chapters V and VIII). Fluorescent scattering is particularly useful for radicals; Raman scattering is useful for species present above 1 percent mole fraction. Below 1 percent, the experimental difficulties make work inconvenient unless some special technique such as Resonance Raman is used. The third alternative is a collection of multiphoton processes such as Coherent Anti-Stokes Raman Scat-

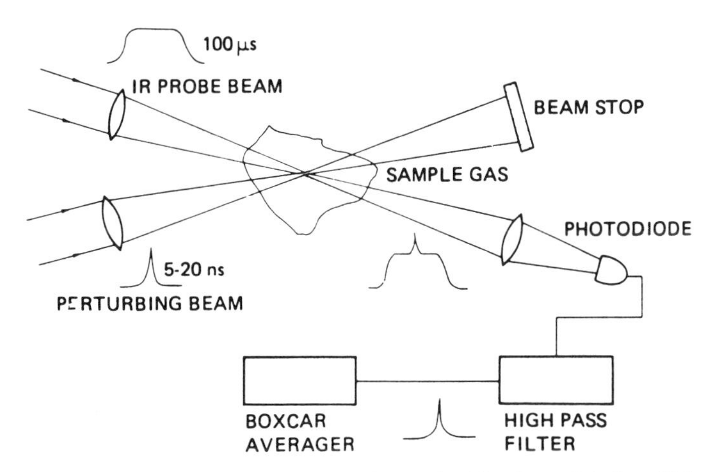

FIG. VI-12 Spatially resolved IR Spectroscopy using optical Stark modulation. (After Farrow and Rahn [1981].)

tering (CARS), Photogalvanic Spectroscopy, Inverse Raman, and Raman Inverse Kerr Effect Spectroscopy (RIKES). The simplest and most generally useful method has been Spontaneous Raman. CARS is useful in systems having high light backgrounds, such as flames with particles.

The two-photon photogalvanic effect has been used to measure hydrogen and oxygen atom distributions in flames. The photogalvanic effect is the ionization by inlet light (Goldsmith [1982]). It is detected by putting collection electrodes across a flame. Two-photon fluorescence was used by Goldsmith and Anderson [1985] to study atom concentrations in flames. In this case the detection is made using a monochromator and photomultiplier.

The multiphoton processes are useful for species such as atoms whose absorptions lie in the ultraviolet, where flames are opaque.

Flame Emission and Photometry

The emissions of flames provide valuable insights into flame processes. This is important in many contexts, but it is only of peripheral interest in the analysis of major flame processes since visible emissions stem from minor side reactions. Flame emissions are chemiluminescent and require radical–radical reactions to provide sufficient energy to radiate in the visible. The area will only be mentioned briefly in this book. Nonetheless there have been many interesting studies of the structure of flame emissions such as those of Wohl and Welty [1955] and Vanpee, Vidaud, and Cashin [1974], and there is a substantial body of literature.

Flame photometry provides a standard method for analyzing metals either in emission or using atomic absorption. These interesting applications, which lie outside the scope of this book, have been discussed in the book of Mavrodineanu and Boiteux [1965].

It is difficult to correlate species concentrations with emission because the intensity depends not only on concentration and temperature, but also on the mode of excitation. The problems have been clarified in the cases of certain flame emission bands and lines connected with the alkali metals by Bulewicz, James, and Sugden [1955, 1956], and this has been used to make flame structure studies. The technique consists of introducing trace quantities of an alkali metal or one of its compounds. Under flame conditions the metal will enter into equilibria with the radicals H and OH so that the concentration of free metal is reduced. This results in a reduction of the observed emissivity relative to what would be expected from the inlet concentration of the metal and the temperature. The amount of this decrease can be deduced from the temperature, the equilibrium constant of the metal salt, and the observed luminosity. Alkali hydrides and hydroxides also emit, and their concentrations can in principle be determined by emission (or absorption) measurements of their band spectra.

Certain chemiluminescent emissions (Table VI–2) have been identified with the presence of radicals. Parts of the continuous emission background found in many flames have been correlated with particular atom–molecule reactions. Examples of this behavior are the reaction of oxygen atoms with CO and NO by Clyne and Thrush [1961] and the reaction of hydrogen atoms with NO by Padley and Sugden [1959].

Some caution should be used in applying these methods, but properly applied they can be quite useful.

Charged Species

Charged particles were first associated with flames by Volta in the 1700s at Padua, when he demonstrated that electrostatically charged bodies could be discharged by brushing them with a flame. Many advances have been made since this qualitative observation, but the source and role of ions in flames is still a subject of debate. It is well established that ions are produced in the primary reaction zones in considerable excess over the concentrations to be expected from chemical equilibrium considerations, and hence must be produced by some nonequilibrium process such as radical reactions. The ions are nonthermal and associated with CH fuels since neither H_2–O_2 or CO–O_2 flames show significant ion concentrations. One candidate is $CH + O = HCO^+ + e^-$.

Flames are electrically neutral overall, and separation of charge is negligible. They can be considered to be a dilute plasma with a fractional concentration around 10^{-7}. The negative particles are predominantly electrons, while the positive particles are generally complex ions. This is to be expected, because few molecules possess energy levels stable enough for an added electron (i.e., show a positive electron affinity), whereas the loss of an electron to form a positive ion rarely decreases molecular stability unless the electron is a bonding one. Experimentally, the association of individual ion peaks with a particular species is a difficult problem, although the fact that they are charged means that they can be determined in very low concentrations.

Table VI–2 Radical Concentrations from Flame Photometry

Radical(s)	Dominant Equilibrium	Wavelengths	Species Measured
H[a]	$LiOH + H \rightleftharpoons Li + H_2O$	6708 Å	Li
H[b]	$Na + HCl \rightleftharpoons NaCl + H$	5890, 5896 Å	Na
H[c]	$Cu + H + M \rightleftharpoons CuH + M^*$	4288 Å	CuH (0,0)
H[d]	$H + NO \rightarrow HNO^*$	Red continuum	HNO^*
O[d]	$O + NO \rightarrow NO_2^*$	4500–5200 Å Green continuum	NO_2^*
O[e]	$O + CO \rightarrow CO_2^*$	3200–3900 Å Blue continuum	CO_2^*
OH[f](?)	$CuOH + M^* \rightleftharpoons Cu + OH + M$	5350–5550	CuOH
	$MnOH + M^* \rightleftharpoons Mn + OH + M$	6150–6250 Å	MnOH
OH[g]		Continuum 4500 Å	Na
H;OH[h]	$H + H + M \rightarrow M^* + H_2$ $H + OH + M \rightarrow M^* + H_2O$	Various resonance lines	Na, Tl, Ag, Pb, Fe

[a]The deficiency of emission compared with equilibrium values is measured. This is usually obtained by comparison with the sodium emission at the same point. The defect on emission is identified with the drop in Li concentration due to formation of LiOH. The H-atom concentration can be obtained from

$$[H] = \frac{K_T[H_2O][Li]}{[LiOH]} \approx \frac{K_T[H_2O]I(Li)}{\delta I(Li)}$$

where K_T is the equilibrium constant for the reaction, I(Li) is the observed intensity of the species in the parentheses, $\delta\ I$(Li) is the defect in emissivity of the species in the parentheses, and brackets indicate concentration.

[b]The drop in emission upon the addition of small amounts of Cl_2 is measured. This defect is identified with the drop in Na concentration due to the formation of NaCl. H-atom concentration is obtained from the equation

$$[H] = \frac{K_T[HCl][Na]}{[NaCl]} = \frac{K_T[HCl]I(Na)}{\delta I(Na)}$$

[c]The emission of CuH is measured at the band head. Its intensity with a small temperature correction is proportional to H-atom concentration. Absolute concentrations are obtained by calibration, using the known H-atom concentration in the equilibrium flame region.

[d]The intensity of the red (green) emission is proportional to the product of [H], [O], and [NO]. Since the NO is regenerated and remains essentially constant, the emission is approximately proportional to the atoms.

[e]Intensity of emission proportional to the product [O][CO].

[f]The measured emission is proportional to the concentration of CuOH, which is in turn proportional to the concentration of OH radical. This interpretation is now considered doubtful.

[g]The continuum emission due to Na has been correlated with OH concentration. Absolute concentrations are obtained by calibrating the emission, using the known concentrations in the equilibrium flame gases.

[h]The chemiluminescence of these metals has been correlated with the three-body combination processes. The data are too complicated for use in determining radical concentrations.

Electrical Conduction in Flames

If electrodes are inserted into a flame and a voltage applied, current passes. The voltage–current behavior can be represented by an equation of the form:

$$E = A[(a/d)i + Bi^2] \tag{6–1}$$

where a is the area of the electrodes and d is the distance between them, A is a constant called the specific resistance of the flame gases (typically about 10^{-6} cm^{-1} ohm^{-1}), and B is a constant characteristic of the electrodes.

In cases where electrons are not emitted by the cathode, B will be about 10^{14} ohms cm^{-2} $amps^{-1}$. The linear (in current) term in this equation represents the potential drop in the bulk of the gas and is analogous to Ohm's law in ordinary conductors. The positive charge is carried by molecular ions whose masses are of the order of 10^{-24} g, while the negative charge is carried by electrons, which are some thousands of times lighter. In addition, the collision cross section of electrons is much smaller, and this is reflected in a difference in the mean free paths of the two carriers. This means that the velocity the electrons attain between collisions is much higher that of the ions. Since the specific ion conductances are proportional to these velocities, it can be seen that electrons carry the major part of the current, the ions being responsible for less than 1 percent.

The quadratic term represents the potential drop near the electrodes, most of which occurs in the vicinity of the negative electrode. This asymmetry is a result of the great difference in mass of the two charge carriers. Providing electrons are not generated at the negative electrode, the current in this region must be carried predominantly by positive ions. Since they have low mobility, there is a large potential drop in this region. The potential drop in the neighborhood of the positive electrode results from the buildup of positive space charge due to the disappearance of electrons.

If only the ohmic part of the potential drop characteristic of the gas phase is considered, an estimate can be made of the concentration of charged species provided ion mobilities are known. Ion mobility is defined as the average velocity of an ion in a unit electric field. Positive ions in flames have mobilities of a few centimeters per second per volt, while that of the electron is about 2500. The current carried by these two conductors is, considering the linear term of Eq. (6-1), given by

$$i = aneG(k_+ + k_-) \approx aneGk_- \qquad (6\text{-}2)$$

In this equation, G is the voltage gradient (V/cm), k_+ and k_- are the mobilities of the positive ion and the electron, respectively, e is the electron charge, and n is the concentration of positive or negative carriers (they are equal for singly charged ions, since the flame is electrically neutral overall).

The concentration of charged particles n may be obtained from Eq. (6-2) by measuring or knowing the other quantities. In a typical case i/aG is 10 ohm cm and $n = 2.5 \times 10^9$ ions per cm. This is of the order of 10^{-10} mole fraction in the burned gas region. Higher concentrations are found in the reaction zone.

Ion Mass Spectrometry

The best technique for identifying flame ions is direct mass spectrometry. Quantitative studies of ion concentration profiles in low-pressure flames were initiated by Deckers and Van Tiggelen [1959], Knewstubb and Sugden [1959], Bulewicz and Padley [1963], and Calcote [1963].

The apparatus used is a mass spectrometer without an ionizer. An orifice and a set of focusing electrodes define the sample of ions. Care must be taken in the design of the sampling inlet and pumping system. It is desirable to operate the sampling inlet in continuum flow so that the sample is not biased by the probe boundary layer, which is much cooler than the bulk of the flame. However, it is also necessary to maintain Knudsen flow conditions inside the spectrometer (mean free path large compared with the apparatus) to avoid producing spurious ions.

Any type of mass spectrometer can be used. Individual ion profiles have been measured. Such information could play an important role in understanding the kinetics of ion reactions in flames, but as yet they have not been subject to quantitative analysis probably because of the difficulties with handling the coupled ambipolar diffusion of ions and electrons.

Langmuir Probe

The Langmuir probe was one of the earliest devices for measuring ion concentrations. It provides a measure of ion or electron concentration and effective electron temperature (Calcote and King [1955]; Calcote [1963]). A probe consists of a large-area and a small-area electrode (see Fig. VI-13). Current is limited by the arrival of ions (or electrons) at the surface of the small electrode and is proportional to its area. The method is analogous to polarography in electrolytic solutions. If the small electrode is positive, the current is limited by, and proportional to, the electron concentration, while if the small electrode is negative, the current is limited by, and proportional to, the positive ion concentration. The ratio between the areas of the small and large electrode must be especially small for electron detection because of the high mobility of the electron. The technique has been criticized because of flame disturbances, but useful results can be obtained.

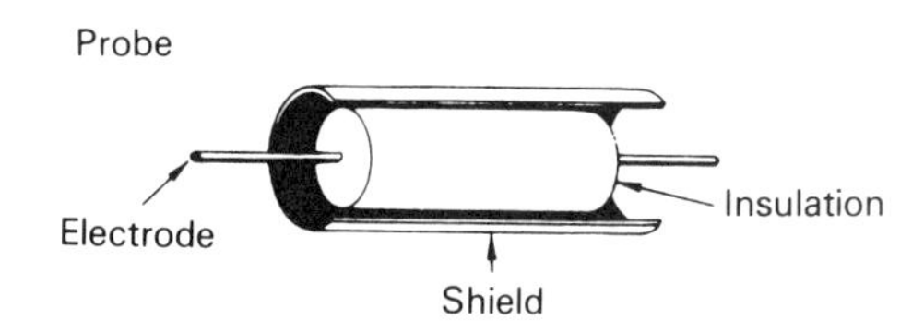

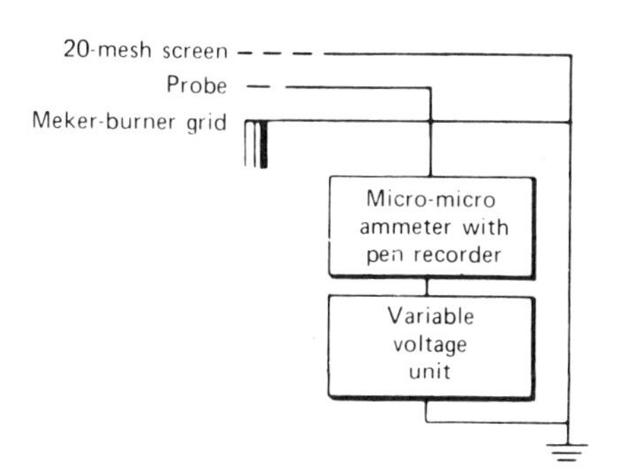

Fig. VI-13 Langmuir Probe for Studying Ions in Flames (After Calcote [1963].)

Radio Frequency Absorption

If an electric field is applied across a flame, the ions and electrons are accelerated along the field. In an alternating field the particles will oscillate between the electrodes with an amplitude proportional to the applied voltage and inversely proportional to the frequency. Electrons have higher mobilities than the ions; so there is a difference in the frequency response of electrons and ions. As the frequency is increased, the amplitude of the ion response drops off. At some point the ions move through less than a mean free path during the cycle. Since energy can only be transferred from the field by collisions, this means that although the ions oscillate in high-frequency fields, they do not dissipate energy. For most ions this occurs at a few tens of megacycles. This is convenient because it means that impedance measurements made on flames above this frequency region will depend only on the electron concentration. The fact that flames absorb energy above the megacycle region is evidence that electrons are the major negative carriers.

Two experimental techniques have been used. One employs the 30 MHz region with a resonant loop for deducing electron concentration. The second is in the gigahertz region, where the wavelength is a few centimeters. Here resonant cavities or waveguides are used instead of resonant loops. The cavity technique (Fig. VI-14) is analogous to the resonant loop method. In the waveguide method the microwaves are treated as an optical beam and the attenuation resulting from interposing the flame in the microwave beam is measured. If the measured attenuation is β(dB/cm), then the conductivity of the flame is given by

$$\sigma = \frac{\beta c}{17.4\pi} \tag{6–3}$$

where σ is the conductivity (ohms^{-1}) and c is the velocity of light (3×10^{10} cm/sec).

The conductivity can be related to the concentration of electrons by

$$n = \frac{\sigma m(\omega^2 + \omega_e^2)}{\omega_e e^2} \tag{6–4}$$

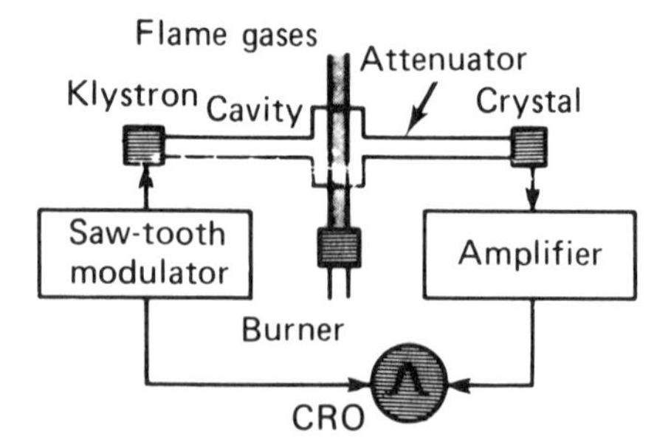

FIG. VI-14 Electron Concentrations in Flames Using Microwave Absorption (3 GHZ). (After Bulewicz and Sugden [1958].)

where e is the electronic charge, $= (4.8\times10^{-10}$ esu); m the mass of the electron (9.1×10^{-28} g); ω the angular frequency of the microwave field (sec^{-1}); and ω_e the angular collision frequency of electrons (sec^{-1}).

Photographic Method

A photographic method has been used by Weinberg [1962] to determine the spatial distribution of ions in flames. At ordinary pressures, high ion velocities can be induced by fields well below electrical breakdown. If a uniform normal field is used, an ion will strike the withdrawal electrode at a position differing only slightly from its origin. By use of a matrix of electrodes in contact with a photographic paper or film, the charge density distribution in a flame can be determined. A fine matrix can be made by casting a wire brush in plastic and machining it into a flat plate. This serves the dual purpose of protecting the photographic paper from light and of increasing sensitivity by concentrating local current to a point. The resulting picture has the appearance of a half-tone print.

Measurements of Charge Production Rates

When detailed electron and individual ion profiles become available, it should be possible to determine the net rates of production of ionic species in the same manner as is used to determine the production rates of stable species. The equations are similar to those for neutral species, (Chapter IX), differing only in the substitution of ambipolar for ordinary diffusion coefficients. This is necessary because the electron diffuses rapidly compared with the positive ions and the two are coupled through electrostatic forces.

Overall charge generation rate can be measured by placing a flame of known geometry in a "saturation" electric field which extracts the ions as rapidly as they are formed. This approach has been applied by Ward and Weinberg [1962] to the study of flat diffusion flames. The apparatus consists of a burner, a set of electrodes, and current and voltage measuring devices. There are complications from geometry, perturbing effects of the ion wind, and in some flames where secondary ionization occurs before saturation. The technique appears worth exploiting in simple systems where quantitative interpretation should be possible.

Comparisons and Value Judgments

No single method for measuring compositions in flames is ideal. Some measurements are easily made and require less sophisticated equipment. Some have limited spatial resolution, some disturb the flame, some are only sensitive enough for major species, some apply only to radicals, some only to stable species.

Table VI-3 provides some comparisons among methods. Each study has special requirements; so one should keep an open mind and use the most convenient method consistent with obtaining the required results. Where it is possible, more than one technique should be used to allow an estimate of the true precision of the results.

Table VI–3 Comparisons among Analytical Methods Used in Flame Studies

Technique	Comments
Microprobes	Sensitivity, selectivity with chromatography
Scavenger probes	Detects radicals
Molecular beam inlet	Best for radicals
Absorption spectroscopy	Line of sight
Conventional	Cheap
Laser	Good but expensive
Laser–Stark	Point measurements
Laser scattering	Proportional to cross section
Particles	Comparison
Atomic fluorescence	Strongest
Fluorescence	Good for radicals
Rayleigh (N_2 488 nm)	No selectivity
Rotational Raman N_2	Major species
Vibrational Raman N_2	Major species
Multiphoton	Special uses
Fluorescence	Far UV
CARS	Luminous flames
Optogalvanic	Atoms and radicals
Tomography, holography	Mapping techniques

In general, optical methods give the least flame disturbance, but a combination of several different optical techniques is required to make complete flame studies. By contrast, molecular beams and microprobes, which can use chromatography and mass spectrometry, allow one to study all species. They are useful for validating and calibrating optical methods. All line-of-sight methods require corrections for nonuniformities and end effects. An optical probe that allows variation of path length is excellent for this. Laser fluorescence methods have high sensitivity for radicals in the 10–100 ppm range. Where complex mixtures are involved almost the only method is microprobe sampling followed by chromatography. Many of the multiphoton techniques look promising, but all are difficult and at present experimental. Radicals do not survive microprobing unless scavenging is used. Molecular beam mass spectrometry is one of the best overall techniques, but the disturbance of the flames often requires considering the probe as part of the flame system for quantitative analysis. For this molecular beam inlets or laser fluorescence is best. For trace species in the ppm range there is no substitute for sampling with chromatography and/or mass spectrometry. Complex mixtures require chromatographic separation or the positive identification offered by spectroscopy or laser scattering. The choice is always a judgment call based on requirements and available equipment. Table VI-3 provides some perspective on the problem. This represents the opinions of the author. It should be remembered that, in addition to personal prejudice, this table also represents the time of writing. Many of these estimates will change, usually for the better.

The agreement between differing methods is encouraging (Figs. VI-2 and VI-15). Not only do most probe methods agree with laser optical methods, but also with detailed modeling. Despite these positive indications, one should always be wary and check composition measurements by independent methods and against detailed models whenever possible.

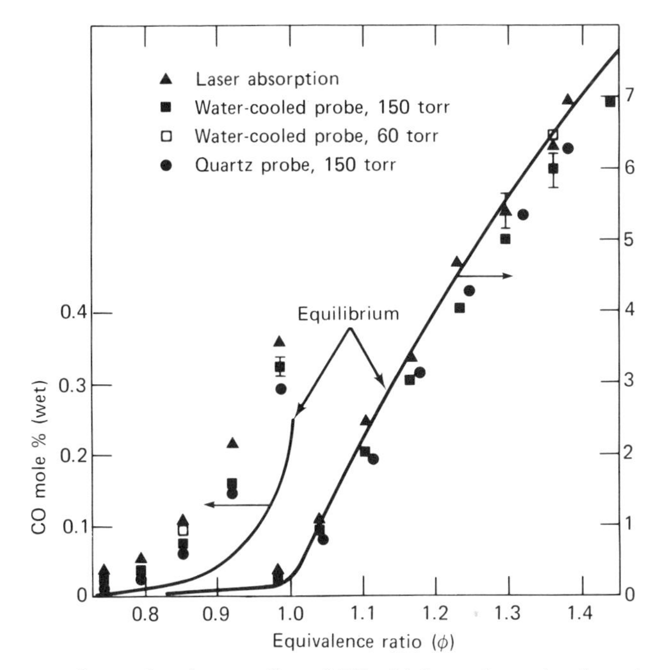

Fig. VI-15 Comparisons of probe sampling of NO with laser absorption in various methane–air flames. (Falcone, Hanson, and Kruger [1979].)

Appendix The Construction of Microprobes

Quartz tubing is the easiest material for the fabrication of microprobes. It is available in a wide range of sizes and possesses excellent mechanical and thermal properties. Sizes up to 1 cm can be easily worked using ordinary glassblowing equipment (oxygen–propane or natural gas flame). A discussion of quartz handling can be found in Strong [1944]. Two precautions should be observed in working with the material: glasses should be worn to protect the eyes, and a well-ventilated room should be used. The operations involved in pulling microprobes are illustrated in Figure VI-16.

To make the taper, it will be found that gravity furnishes a much more uniform pull than the hand of a nervous laboratory technician. To ensure circular symmetry, it is desirable to use two torches and rotate the tubing slowly (30 rpm). The taper and wall thickness can be controlled by judicious heating and selection of the pulling weight. Thin-walled tubing can also be fabricated using these techniques.

When the tube has been reduced below 0.5 mm o.d., the drawing weight should be reduced, a microtorch or pair of microtorches substituted for the glass blowing torch, and the process viewed with a low-power microscope. Glasses need not be worn for this operation since the low aperture and glass of the microscope protect the eyes. The procedure is essentially the same as used in rough drawing except that more care must be taken. The taper should be drawn below the expected orifice size.

The drawn tube is notched with a broken edge of fine ceramic at a point such that the hole will be 2–3 times the required final orifice diameter. This operation requires some practice since it is difficult to judge wall thicknesses. The tube is broken off by

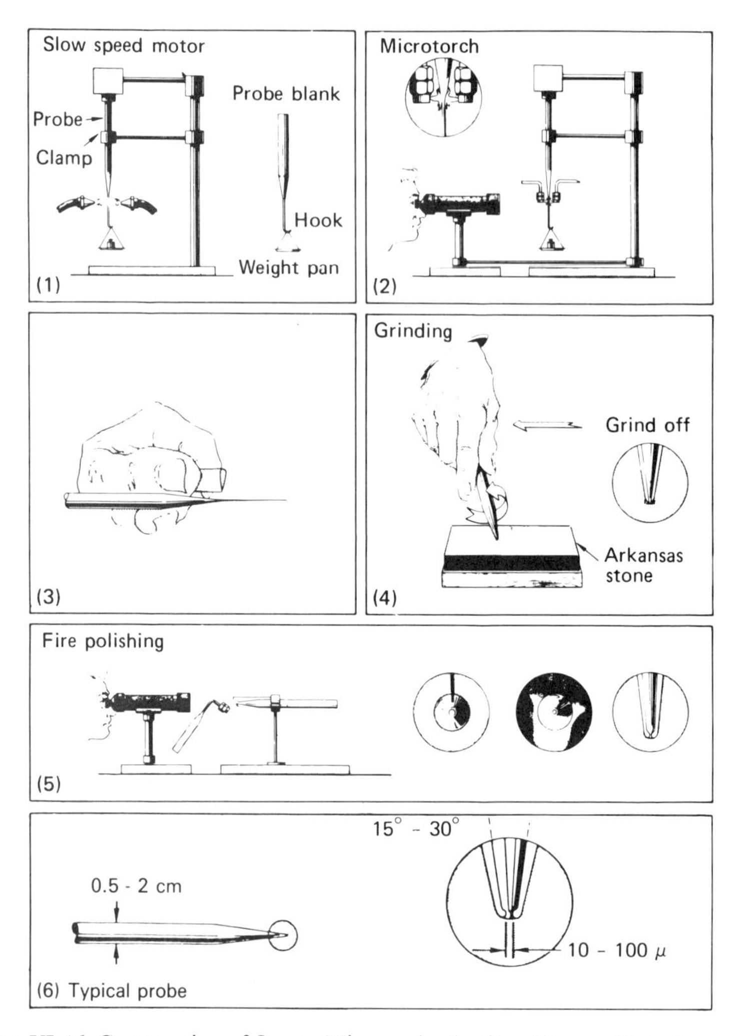

Fig. VI-16 Construction of Quartz Microprobes by GlassBlowing Techniques.

applying pressure with a fine quartz fiber. The cut should be examined to make sure it is clean and normal to the axis of the probe.

To allow the formation of a properly shaped orifice it is desirable to grind some of the excess wall from the tip of the tube. This can be done by hand with a fine Arkansas stone using xylene, water, or some other lubricant, or (with *caution*) on a high-speed, fine carborundum wheel such as a Dremel™ tool. The latter method is much more rapid but requires some practice. The resulting tip should be examined under a microscope to see if it is uniform. Before one proceeds, the tube should be thoroughly cleaned with acid and rinsed with distilled water.

The final orifice is formed by fire polishing under a microscope. With careful manipulation, the orifice diameter can be controlled to 20 μm. A wire of the diameter desired is mounted in the microscope field for comparison with the probe tip in Figure VI-16. The microflame is then slowly brought in contact with the tip. As the quartz softens, the dark orifice surrounded by glowing quartz will be seen to contract slowly. When the proper diameter is reached, the flame is withdrawn and the orifice freezes rapidly through radiation cooling. Its size should be checked by a traveling microscope and/or by its flow rate. If it appears satisfactory, it should be cleaned with nitric or hydrofluoric acid and stored in distilled water.

Quartz is not suitable for high-temperature–high-pressure flames where the increased heat transfer can heat it above its fusing point. Such flames must be studied with either a cooled probe or a high-temperature ceramic probe. Several ceramics are available in tubular form. They are of two classes: (1) high-temperature porcelains—trade names include Mullite™ and Coors™; and (2) aluminum oxide—trade names include Luclaox™ from General Electric Corp., Alumina™ from Coors, and Synthetic sapphire. The aluminum oxides will stand over 2200 K flames at atmospheric pressure. Unfortunately these materials are more difficult to work than quartz because they have relatively sharp melting points and show little glassy behavior. Probes can be fabricated using mandril casting techniques.

VII

PROBE SAMPLING FROM COMBUSTION SYSTEMS

OWEN I. SMITH

Department of Chemical Engineering
University of California, Los Angeles

Two approaches may be taken in the experimental study of the details of flame structure. One involves flame sampling, for example, the physical extraction of flame gases for subsequent analysis. The other involves in situ analysis, primarily by optical means. Sampling necessarily leads to some degree of distortion of the chemical composition of the flame at and around the point of sampling, but allows the use of a wide variety of powerful techniques in sample analysis, the most popular of which have historically been gas chromatography and mass spectrometry. In situ methods have the advantage of being essentially nonobtrusive, but, as seen in Chapter VI, a given technique is usually less general, often applying to at most only a few of the species present in the flame.

Flame structure measurements shed light on a number of important aspects of combustion. These include thermodynamics, kinetics and reaction mechanisms, transport, emissions of toxic pollutants, etc. The choice of a diagnostic method or methods for experimental characterization of flame structure depends largely on the use to which the measurements are to be put. In many cases, sampling and in situ methods can be used interchangeably. In other cases, such as in the determination of elementary rate parameters for reactions involving radicals, sampling methods should be used with caution. Here, definitive measurements are usually performed by a nonobtrusive optical method (and usually in a system a good deal less chemically and hydrodynamically complicated than a flame). In still other cases, in situ methods may not be feasible at all. The characterization of trace levels of complex organic pollutants such as dioxins formed in the incineration of wastes is one area of great current interest to which this situation pertains.

This chapter deals with the perturbations to flame structure induced by probe sampling. We will review the factors influencing the degree of perturbation as they are currently known. Techniques that can be used to minimize sampling perturbations or to account quantitatively for their effects will also be discussed.

Probe Types: A Qualitative Description

Probes used in the study of flame structure are most often fabricated from quartz or water-cooled stainless steel, although other materials with good high-temperature mechanical properties such as various ceramics, platinum, and iridium are used occasionally. Two basic geometries are used that differ in aspect ratio. The microprobe is a very low-aspect-ratio probe. A version suitable for use with many laboratory-scale systems is easily made by drawing the end of a piece of quartz tubing to a point (see Appendix to Chapter 6). Microprobes are not always small. Designs utilized in the evaluation of utility boilers and hazardous waste incinerators may be over two meters in length and several cm in diameter (Tine [1961]). These are fabricated from metals for structural reasons. Water cooling is usually required.

Where the goal is to sample gaseous species (as opposed to particles), sufficient pumping downstream to insure sonic conditions at the orifice is nearly always used. The required pressure ratio is typically somewhat greater than that required to choke a simple orifice due to viscous losses in the narrow entrance section. For small probes, a pressure ratio of at least 5 should be used. Because convective heat transfer is an important mechanism for quenching chemical reactions in microprobes, the body is often water cooled, even in quartz probes where this is usually not required structurally. This is especially important if the body of the probe is situated within the flame.

Many variations of the basic microprobe have been developed for special applications. Two of the more important are the dilution microprobe (Ranade, Werle, and Wasan [1976]) used in the sampling of particulate streams and the scavenging microprobe (Volponi, McLean, Fristrom, and Munir [1986]) used in the sampling of very reactive species. The dilution probe incorporates a diluent gas injected through the probe wall just downstream of the sampling orifice. This is done to lower the particulate number density upon which the rate of coagulation of the smallest particles is highly dependent. In the scavenging probe a radical scavenging gas such as D_2 is injected into the diverging region of the probe. The object here is to form more stable species which are not reactive at higher temperatures which may exist downstream and can survive collisions with the probe walls.

By comparison with the microprobe, the probes used with molecular beam sampling systems tend to have much higher aspect ratios. These probes are usually drawn from quartz tubing, and may be of a simple conical shape or may involve compound angles. Unlike the microprobe, molecular beam probes diverge over their entire length; so the density and temperature of the sampled gas decrease continuously downstream of the orifice. The rate of decrease depends on the divergence of the nozzle, the pressure ratio across the orifice, and the thermodynamic properties of the sampled gas. In sampling from low-pressure sources the expansion can be made rapid enough so that the original vibrational state distribution of the sampled gas is preserved. The rotational state distribution, on the other hand, is usually nearly fully accommodated. Typical rotational and translational temperatures at the point of free molecular flow are usually around 20 K. In molecular beam–mass spectrometer systems, the central core of this expanding jet, which presumably has not undergone collisions with the probe wall, is passed via another nozzle through one or more additional stages of differential pumping, and then directly into the mass spectrometer ion source.

If the quenching properties described in the previous paragraph are to be realized in practice, it is important to keep the edge of the sampling orifice sharp. Ideally the wall thickness of the orifice should be substantially less than one orifice diameter. The most effective way to accomplish this in practice is first to draw the probe so that the interior surface of the tip is as sharp as possible. Then the exterior surface of the tip is carefully ground away perpendicular to the probe axis to create the orifice. Finally the blunt tip resulting from the previous step is sharpened by grinding the exterior wall near the orifice.

Perturbation Mechanisms

It is convenient to separate the phenomena leading to probe-induced distortions to flame structure into two categories, depending on the location of the perturbation relative to the sampling orifice. In this chapter we will refer to those processes acting upstream of the orifice as external distortions and to those acting downstream as internal distortions. Mechanisms leading to internal distortions can include homogeneous reactions (including particle coagulation), heterogeneous reactions on the interior walls of the probe, and the production of concentration gradients by the hydrodynamics of the expansion. External distortions are much more difficult to characterize. The mechanisms leading to distortions here are basically the same as for internal distortions; however, the importance of each now depends not only on the probe material, geometry, and orifice size, but also on the nature of the combustion system being sampled. Factors that are recognized as being important include the role of radical transport in propagation of the flame, the chemistry of the species responsible for chain branching, and the degree to which the flame is stabilized within the combustor.

In comparison with microprobes, the more massive molecular beam probes show a much greater tendency to induce external distortions. On the other hand, such internal distortions as exist in molecular beam probes can usually be corrected for with a reasonable degree of confidence by the use of well-founded correlations. With the conventional microprobe, gross internal distortion to the concentration of radicals is obviously unavoidable. In fact, the goal here is to maximize the rate of destruction of those radicals participating in the dominant branching reactions. If this can be done rapidly enough, say by catalyzed recombination on the probe walls, then the concentrations of the stable species will be preserved. Of course, catalyzed recombination results in the same products as are formed homogeneously in the flame, but the radical concentrations are seldom large enough to make a measurable difference.

External Distortions

External distortions have been treated analytically by several authors including: Stepowski, Puechberty, and Cottereau [1981]; Smith and Chandler [1981]; and Yi and Knuth [1986]). These perturbations have also been characterized experimentally in one-dimensional premixed flames by Stepowski, Puechberty, and Cottereau [1981] and Smith and Chandler [1986]. Less direct experimental measurements of external distortions have been made by Biordi, Lazzara, and Papp [1974] and Cattolica, Yoon, and Knuth [1982]. Because external distortions are much more severe in molecular beam probes, most past work has been focused on these.

In one-dimensional premixed flames there appear to be two major factors leading to external distortions. The first is the tendency of the probe surface to catalyze the recombination of flame radicals. Since the upstream diffusion of these radicals, principally O, H, and OH in hydrocarbon flames, is an important flame propagation mechanism, anything that interrupts this can significantly change flame structure (Hayhurst, Kittelson, and Telford [1973]). The second is less obvious, and is related to the suction at the sampling orifice. This affects the axial velocity field upstream of the orifice, resulting in strong radial gradients beginning a few orifice diameters upstream of the probe tip.

Surface-Catalyzed Radical Destruction

The effect of surface-catalyzed radical destruction on flame structure has been directly observed experimentally for both OH (Stepowski, Puechberty, and Cottereau [1981]) and CN (Smith and Chandler [1986]). Figure VII-1 depicts the perturbation of CN concentration in a weakly stabilized low-pressure hydrogen–oxygen premixed flame doped with 1.3% HCN. The sampling probe used here is constructed of quartz, with a 60° included angle and a 0.5 mm orifice diameter. CN radical concentrations were determined by laser-induced fluorescence. The spatial resolution is about 0.4 orifice diameters axially and about 1.4 diameters radially. The CN concentration profile for the unperturbed flame, measured without the probe present and normalized by the peak value (estimated to be of the order of 100 ppm), is depicted on the right-hand side of the Figure VII-2. The decrease in CN concentration near the probe walls is obvious, particularly when sampling upstream of the unperturbed peak. CN concentrations at the entrance to the orifice are up to 50% lower than those at corresponding locations

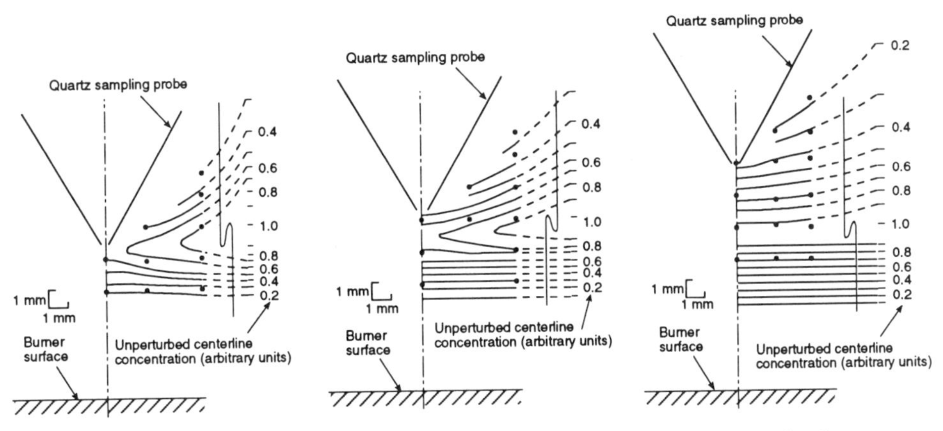

FIG. VII-1 Normalized concentration contours for CN in a 3.32 kPa H_2–O_2–Ar flat flame doped with HCN at three burner-probe distances. The unperturbed CN profile (normalized to peak) is shown on the right. Measurements were made by LIF using R(4) through R(12) 0–0 lines in the B-X transition. (After Smith and Chandler [1986].)

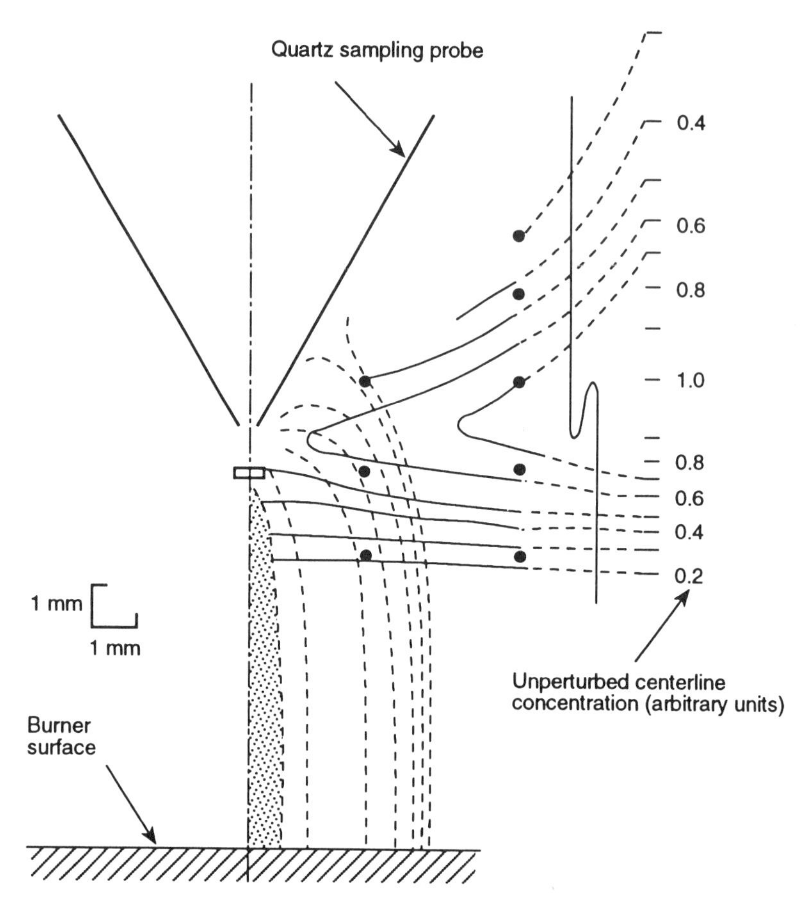

FIG. VII-2 Relation between streamlines and normalized CN concentration contours for gas sampled by a molecular beam probe from a low-pressure H_2–O_2–Ar flame doped with HCN. While much of the sample is drawn from a highly distorted region near the probe surface, the shaded portion passed to the mass spectrometer for analysis is relatively free of such distortion. The probe volume of the laser-induced fluorescence system illustrated by the rectangle upstream of the orifice is too large to distinguish perturbations induced by suction.

in the unperturbed flame. These perturbations were present whether or not suction was applied to the orifice; however, as will be seen later, the radial resolution of the measurements are not sufficient to resolve the details of distortions caused by suction at the orifice. Surprisingly, CN rotational temperature measurements showed no detectable (>100 K) thermal perturbation, even just upstream of the orifice.

Much the same behavior has been observed for OH in low pressure propane–oxygen flat flames (Stepowski, Puechberty, and Cottereau [1981]). LIF measurements indicate that the molecular beam style probe depletes the OH concentration upstream of the sampling orifice when sampling from the preheat or reaction zones. The degree of depletion decreases with increasing separation of the burner and sampling probe until the net convective and diffusive flux of OH is in the downstream direction. Here (e.g., far downstream in the reaction zone and in the postflame zone) no significant depletion is observed. This is in agreement with previous studies in the postflame region (Smith and Chandler [1986]). Since OH typically peaks farthest downstream of the major chain carriers in hydrocarbon flames, this suggests that the perturbation is caused by reduction of the upstream diffusive flux of OH, and probably H and O as well, due to reaction on the probe wall. This hypothesis is not inconsistent with the results depicted in Figure VII-1, although for CN the distortion along the centerline clearly extends to the downstream side of the peak. Numerical calculations indicate that the distortion observed here could well be caused by perturbations to the H and OH profiles, which influence the rate of formation of CN from HCN and peak much farther downstream.

The studies cited indicate that probe measurements of radical species in low-pressure flat flames can be significantly in error when sampling upstream of the point where the net flux of *any* of the major chain carriers is against the flow. However, the maximum distortions observed will likely not exceed a factor of two, even for a very unstable radical peaking early in the reaction zone at relatively low concentration. For mechanistic studies such qualitative information can still be very useful. In systems such as flow reactors, where upstream radical diffusion is less important, probe distortions by this mechanism should not be significant.

The degree of perturbation due to heterogeneous radical destruction and upstream diffusion blockage is known to be quite dependent on the geometry of the sampling probe. For molecular beam probes of large included angle (80–100°) and small orifice diameters, gross distortions of both radical and stable species concentrations have been observed (Hayhurst, Kittelson, and Telford [1973]). Where very large distortions are present, it is usually possible to observe visible flame attachment. To complicate matters further, the degree to which upstream radical diffusion may be reduced without resulting in flame attachment is highly dependent on the system itself. For high-pressure, strongly stabilized flames, relatively blunt (80° included angle) probes have been used successfully (Cattolica, Yoon, and Knuth [1982]). For very weakly stabilized flames, flame attachment has been known to occur even with uncooled quartz microprobes (Smith and Thorne [1986]). In these flames, the rates of radical production and recombination are so closely balanced that virtually any additional mechanism for the destruction of chain carriers can extinguish the flame locally. Unfortunately, it is in these weakly stabilized flames that the flame structure is most sensitive to the chemistry.

Mathematical models can be of at least qualitative assistance in specifying probe geometry. For the purpose of dealing with distortions caused by heterogeneous reactions, the approach taken by Smith [1981] is probably the most helpful. The model consists of an incompressible potential flow over an infinite cone aligned with the flow and with a point sink imbedded in the tip. The resulting streamlines are depicted in Figure VII-2. The stagnation streamline, which divides the sampled gas from the unsampled, intersects the probe wall about 2.5 mm downstream of the orifice. Clearly, a large fraction of the sampled gas is pulled through the boundary layer beginning at the stagnation point and developing in the upstream direction toward the sampling orifice. For microprobes, the gas eventually analyzed will be representative of the entire sample. In molecular beam systems, only the core of this flow (represented by the shaded area in Fig. VII-2) is analyzed. Application of a standard Nusselt number correlation for stagnation point flows (Kays [1966]) results in nondimensional temperature perturbations as are displayed in Figure VII-3. Here T_o, T_w, and T_s are the unperturbed gas, probe wall, and sampled gas temperatures, respectively. As can be seen, blunt probes can result in a mean sample temperature almost completely accommodated to that of the probe, especially at low pressures. As a result, it is advisable to utilize a material of low thermal conductivity and emissivity for molecular beam probes, and to maintain a minimum wall thickness between the orifice and the stagnation point.

If heterogeneous radical reactions are assumed to be very fast so that the concentration field near the wall is governed by mass transfer, then an analogous calculation could be performed using the equivalent Sherwood number correlation. Recognizing that even quartz is highly catalytic at flame temperatures (Scheffer [1982]), it is clear that the probe angle should be kept as small as possible. As for heat transfer, this holds especially for low-pressure flames. Unfortunately, we will see later that relatively large probe angles are needed to avoid internal distortions in molecular beam probes.

Figure VII-3 indicates that thermal distortions should be very small for uncooled microprobes. Experimentally determined temperature perturbations for these probes are somewhat larger. Decreases of 65–200 K have been observed just upstream of the orifice by Biordi, Lazzara, and Papp [1974] and Bowman [1977]. This is reasonable since for probes of very small internal angle, the finite size of the sampling orifice will start to become important. This is not accounted for in the model, and will tend to increase the calculated perturbations. Nevertheless, it is generally accepted that for most premixed flames, stable species in relatively large concentrations can be sampled by microprobes without significant distortion.

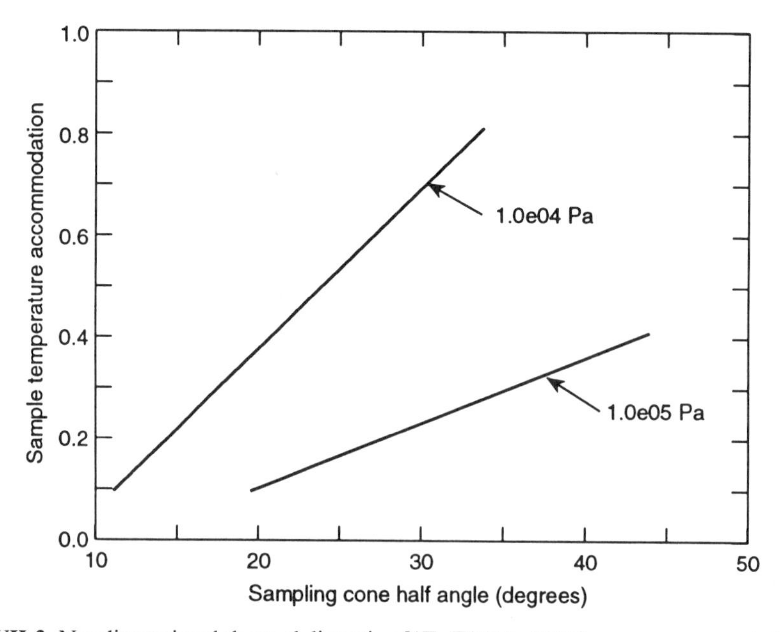

FIG. VII-3 Nondimensional thermal distortion $[(T_S-T_0)/(T_W-T_0)]$ for mean gas sample from a stoichiometric methane–oxygen–argon flame at source pressures of 10^4 and 10^5 Pa. The sampling orifrice diameter is 0.25 mm.

Effect of Suction

For some time it has been recognized that gas samples analyzed in molecular beam–mass spectrometer studies of flame structure are representative of the unperturbed flame at a position somewhat upstream of the sampling orifice. This phenomenon is referred to as *apparent profile shift.* Apparent shifts of various magnitudes are described in the literature, ranging from 2 orifice diameters by Stepowski, Puechberty, and Cottereau [1981] to 5 orifice diameters by Biordi, Lazzara, and Papp [1974] and Catollica, Yoon, and Knuth [1982]. Such corrections apply to all of the species samples, stable and radical alike. Experience has shown that, except very near the burner surface, an upstream shift of the experimentally determined profiles results in good agreement between molecular beam–mass spectrometer, microprobe, and optical measurements for stable species.

The numerical calculations of Yi and Knuth [1986] address the phenomena responsible for this profile shift. As before, the model used consists of an infinite cone oriented parallel to the flow. In this case, a sink of finite size (disk sink) is used, and near the orifice compressibility is accounted for. Comparison with the earlier model by Smith [1981] shows that compressibility is important only very near the orifice. The ratio of the centerline axial velocity to the (sonic) velocity at the orifice is still less than 0.1 at one diameter upstream of the orifice. The finite size of the sink plays a role at greater distances (up to 5 diameters from the orifice axially in both directions, and 10 diameters radially), but this still does not change the location of the stagnation point significantly, at least for the 80° included angle probe studied.

Figure VII-4 depicts the streamlines and concentrations contours computed for a probe immersed in a binary mixture with an artificially generated concentration gradient. The flow is nonreactive, the concentration gradient being supported by a numerical source at a reference plane located 2.5 diameters upstream of the orifice and a numerical sink located 7.5 diameters downstream. The unperturbed gradient is depicted at the top of the figure. In the vicinity of the orifice, perturbation of the concentration gradient by the acceleration of the core flow and the reversed flow along the wall is obvious. The net result is that the portion of the sample along the centerline, which is what is analyzed in molecular beam–mass spectrometer experiments, corresponds to the unperturbed concentration at a point 1–2 orifice diameters upstream. It should also be noted that the mean sample composition, the important property in microprobe sampling, is much closer to that of the undistorted mixture. Hence, profile shifts are not expected to be significant with the microprobe.

When mass convection dominates the conditions at the orifice, for example,

$$\frac{1}{Re\,Sc} = \frac{d/a_o}{d^2/D_{AB}} > 1$$

the computed apparent profile shift δ is correlated by

$$\frac{\delta}{d} = 0.19\,(Re\,Sc)^{1/2}. \tag{7–1}$$

Here d represents the orifice hydraulic diameter, a_o the sonic velocity at the orifice and D_{AB} the diffusion coefficient at the orifice. The ratio of the hydraulic (d) and geometric

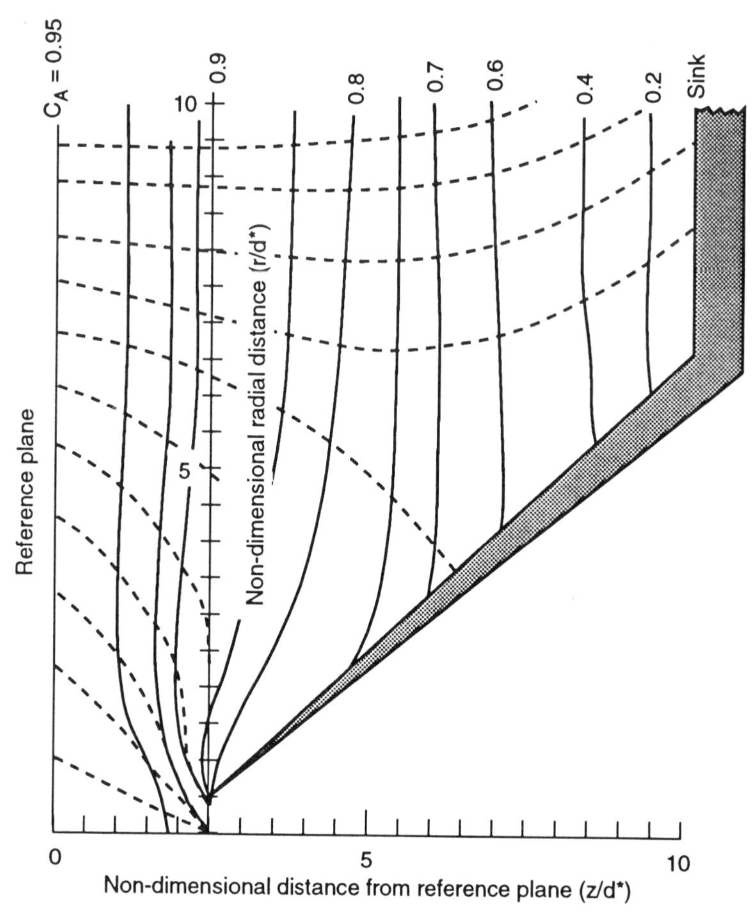

FIG. VII-4 Numerically generated streamlines (—) and concentration contours (---) for a molecular beam probe immersed in a nonreacting binary mixture. A concentration gradient is artificially maintained by a source ($C_A = 0.95$) at the reference plane located 2.5 orifice diameters upstream of the sampling point, and a concentration contour at the orifice center (—•—) indicates that the gas analyzed by a mass spectrometer is effectively extracted about one orifice diameter upstream of the sampling point. (Adapted from Yi and Knuth [1986].)

(d^*) diameter is related to the orifice discharge coefficient (D) by

$$\frac{d}{d^*} \alpha \sqrt{D}$$

The reciprocal of the *(Re Sc)* product can then be thought of as the ratio of the characteristic convective time to the characteristic diffusion time.

Theory predicts that for reactive flows the apparent shift should have the same dependence on the *(Re Sc)* product, but that the leading coefficient will increase slightly. For very reactive flows, such that the ratio of the characteristic convective time to the characteristic reaction time is greater than or of order 1, mass convection will no longer dominate and Eq. (7-1) will not be valid. Nevertheless, the correlation has shown good agreement with OH profile shifts identified experimentally in the post-flame zone.

Internal Distortions

Internal distortions take place after the sample has entered the orifice. They result either from a failure to quench the chemical reactions taking place in the flame or from various gasdynamic mechanisms. Chemical reactions may be quenched by lowering the concentration of the reactants or lowering the temperature. If the concentrations of stable species only are of interest, radical removal can be effectively used to quench reactions. Since all combustion reactions with low to moderate activation energies involve the participation of radicals, if these species can be effectively removed, higher temperatures within the probe can be tolerated. Radicals may be removed either heterogeneously or homogeneously. In the latter case, the process may be accelerated by the introduction of a radical scavenging gas. In either case, the rate of radical removal is likely to be limited by mixing, so that radical removal by itself is seldom fast enough to be a satisfactory quenching mechanism. For this reason, most probes are designed so that the sampled gas initially undergoes a rapid expansion, resulting in a very high cooling rate.

In molecular beam probes, this expansion is the only mechanism for quenching. For microprobes, a combination of aerodynamic and convective cooling with radical removal is used. In molecular beam probes the gasdynamics of the expansion can also lead to distortions by inducing radial concentration gradients. These can take place by a variety of mechanisms, all of which are of a physical rather than a chemical nature. Such distortions are unimportant in microprobes because once the expansion stops, diffusive transport quickly eliminates any radial gradients that might be present.

Quenching in Microprobes

A schematic representation of the temperature and Mach number profiles within a microprobe is given in Figure VII-5. Provided that the flow at the orifice is choked, a supersonic expansion will occur within the probe tip. At some point within the probe, viscous losses result in a transition back to subsonic flow. This usually occurs within the probe body, but for small-diameter probes can occur within the diverging entrance section. Here the sample is heated by compression, and may reach temperatures at which chemical reactions again become important. Because both convective cooling and wall radical loss are relatively slow, it is desirable to delay the onset of subsonic flow as long as possible. This can be done by minimizing viscous losses, for example, by using as large a diameter orifice and entrance section as practical and locating all bends or sudden changes in cross-section as far from the tip as possible. Thus, one key to satisfactory performance of a microprobe is how far down the probe supersonic flow can be maintained, and how much radical loss and convective cooling occurs within this interval.

The internal aerodynamics of microprobes have been investigated both theoretically and experimentally by Colket and co-workers [1982]. They used a one-dimen-

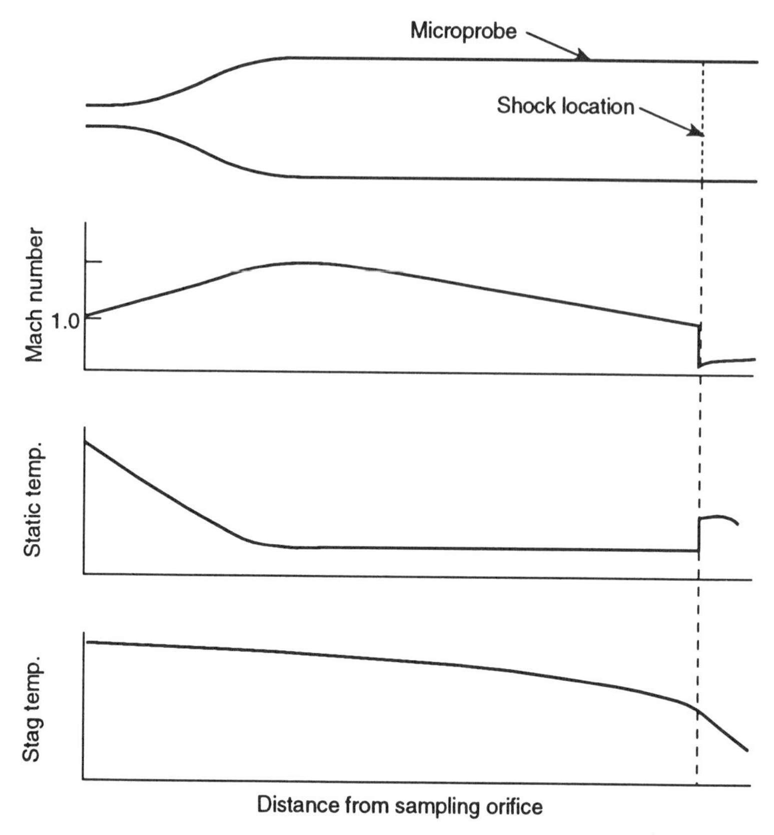

Fig. VII-5 Sketch depicting Mach number and static and stagnation temperatures in the supersonic expansion region in a microprobe. (Adapted from Colket et al [1982].)

sional gasdynamic model to relate the change in Mach number with axial position to the effects of convective heat transfer, viscous losses (through a skin friction coefficient), and probe geometry. In evaluating probe designs, they endeavored to achieve the following conditions:

1. The sample static temperature is reduced to below 1000 K at a rate of at least 108 K/m in the supersonic expansion;
2. The resulting static temperature is maintained until the stagnation temperature is reduced to less than 1000 K by convective heat transfer;
3. Convective heat transfer is rapid enough to effect the reduction in stagnation temperature in less than 1 msec.

Calculations indicate that these conditions can be met by microprobes with relatively large-diameter orifices and entrance sections, provided that pumping is adequate to maintain a low enough ratio of back to ambient pressures. The magnitude of the pressure ratio required is predicted to be highly dependent on the dimensions of the entrance section, since as the channel size decreases viscous drag increases more rapidly than convective heat transfer (Colket et al. [1982]). Thus, when sampling from identical ambient conditions, the smaller the probe diameter, the lower the back pressure required to achieve quenching. For very small orifices, such as those commonly used with quartz microprobes sampling laboratory flames, it may not be possible to maintain supersonic flow throughout the entrance section regardless of the back pressure. Finally, theory and experiment both indicate that lower back pressure ratios are required for lower ambient pressures (e.g., as the ambient pressure is reduced, the required back pressure must be reduced by a proportionally larger fraction).

While our understanding of internal distortions in microprobes from a theoretical perspective is far from complete, these devices have seen a great deal of use. Experience indicates that it is, in most cases, not difficult effectively to quench the reactions of the major stable species found in combustion systems. However, a few stable species are known to form homogeneously within microprobes and, when they are present in very low concentrations within the flame itself, this can lead to significant distortion. The classic example is the oxidation of NO by the peroxy radical to form NO_2 within the probe (Amin [1977]; Kramlich and Malte [1978]; and Johnson, Smith, and Mulcahy [1979]). In the relatively cool expanding jet, significant amounts of the peroxy radical are formed by a variety of fast gas-phase reactions.

$$\mathrm{H} + \mathrm{O_2} + M = \mathrm{HO_2} + M$$

$$\mathrm{OH} + \mathrm{OH} = \mathrm{HO_2} + \mathrm{H}$$

The peroxy radical is relatively stable below about 800 K, and is destroyed slowly both homogeneously and heterogeneously. Hence, concentrations of order 10 ppm persist within the probe long after the major chain carriers have been destroyed. The peroxy radial reacts with NO to form NO_2, which is thermally stable below about 1100 K.

$$\mathrm{HO_2} + \mathrm{NO} = \mathrm{NO_2} + \mathrm{OH}$$

Due to the slightly negative temperature coefficient of the reaction between HO_2 and NO (≈ -500 cal/mole), it can be quite rapid even at low temperatures.

Quenching in Molecular Beam Probes

If a small length-to-diameter ratio is maintained in the orifice channel of a molecular beam probe, viscous effects will not be important. If relatively large internal angles are utilized, the expansion downstream of the orifice can usually be approximated as a free jet. This greatly simplifies theoretical analysis of aerodynamic quenching. Unlike microprobes, molecular beam probes must be designed to quench reactions involving very reactive radicals. This, coupled with the relative simplicity of the flowfield, has stimulated considerable work in this area. As a result, our understanding of internal distortions in molecular beam probes is on a much more quantitative basis than that for microprobes.

Chemical relaxations in expanding jets have been discussed by Knuth [1972, 1973] and, from a somewhat different point of view, by Young [1975]. Knuth combined the sudden freezing model of Bray [1961] with the freezing point criterion of Phinney [1964] to obtain a chemical relaxation equation for an elementary reaction with an

Arrhenius rate coefficient. His expression for the chemical freezing point is given in terms of the local velocity U and temperature T at a nondimensional axial distance x/d downstream from the sampling orifice, the stagnation sound speed a_o, and two dimensionless parameters $\tilde{E}$ and $\tilde{\tau}$.

$$\tilde{E}\, e^{\tilde{E}} \frac{U}{a_o} \frac{\partial T}{\partial(x/d)} \approx -0.5\, T\, \tilde{\tau} \tag{7-2}$$

For most reactions, the dimensionless thermodynamic parameter

$$\tilde{E} = \frac{E' + K'(E'' - E')}{RT} \tag{7-3}$$

is the more important. Here E' and E'' represent the activation energies of the forward and reverse reactions, respectively, and, for the reaction

$$\sum_\alpha \nu'_\alpha \mathrm{A} \rightarrow \sum_\alpha \nu''_\alpha \mathrm{A},$$

K' is given by

$$K' = \frac{\sum_\alpha (\nu'_\alpha \nu_\alpha / \chi_\alpha)}{\sum_\alpha (\nu_\alpha^2 / \chi_\alpha)} + \frac{\sum_\alpha (\nu_\alpha^3 / \chi_\alpha^2)}{\left(\sum_\alpha (\nu_\alpha^2 / \chi_\alpha)\right)^2} \tag{7-4}$$

In Eq. (7-4), χ_α is the mole fraction of species α at the point of sampling and $\nu'_\alpha \equiv \nu''_\alpha - \nu'_\alpha$. The second dimensionless term in Eq. (7-2) is the scaling parameter,

$$\tilde{\tau} = \frac{\nu_s\, d}{a_o} \tag{7-5}$$

Here ν_s is a constant for a given isentrope, and is given in terms of the reaction relaxation time at constant enthalpy and pressure $\tau_{h,p}$ by

$$\nu_s \approx \frac{e^{\tilde{E}}}{\tau_{h,p}}. \tag{7-6}$$

In Eq. (7-2), the effect of nozzle geometry is embedded in the temperature derivative, while that of orifice diameter enters principally through the scaling parameter. In molecular beam probes, these are the primary design variables influencing the point of quenching. Ideally, one would like to quench chemical reactions *at* the sampling orifice, as opposed to downstream in the expansion. This condition will be realized (Knuth [1973]) for

$$\tilde{\tau} \leq 4(\gamma - 1)\, \tilde{E}_o e^{\frac{\gamma+1}{2}\tilde{E}_o} \tag{7-7}$$

where the subscript (o) denotes the source conditions. From Eqs. (7–5)–(7–7) the maximum desirable orifice diameter is given by

$$d \approx a_o (\tau_{h,p})_o 4(\gamma - 1) \tilde{E}_o e^{\frac{\gamma - 1}{2} \tilde{E}_o}. \tag{7–8}$$

Where the orifice diameter is no greater than that specified in Equation (7–8), chemical relaxations are no longer a primary concern in specification of the probe internal angle. Pumping requirements also favor small orifice diameters, and this often presents fabrication problems in quartz nozzles, where it becomes increasingly difficult to maintain small orifice length to diameter ratios as d decreases.

Gasdynamic Distortions

Sample composition can be distorted by various physical mechanisms that induce radial gradients within the expanding jet. These gasdynamic distortions occur in both molecular beam and microprobes, but because the resulting radial inhomogeneities are quickly eliminated by diffusive mixing when the expansion stops, they are important only in the former. In molecular beam probes the extent of composition distortion depends not only on the sampling probe, but also on the flow elements downstream in the expansion. One aspect of proper molecular beam sampling system design is to reduce the magnitude of these distortions to negligible levels or, where this is not possible, to be able to account for them quantitatively. In most cases, an additional design goal is to achieve maximum beam density at the point where the chemical analysis is conducted. Maximum beam density leads to maximum sensitivity since the response of most analytical instruments is proportional to the number density.

Apparatus representative of molecular beam sampling systems is diagrammed in Figure VII-6. A supersonic beam is formed by means of an expansion through an orifice in the sampling probe. The central portion of the resulting jet then passes through another orifice (skimmer orifice) into a collimating chamber, and thence into the detection chamber. The collimated beam is usually analyzed by means of a mass spectrometer. Occasionally, other diagnostic techniques such as microwave spectroscopy are used.

Knuth [1972] has shown that for the simplified model of a molecular beam system depicted in Figure VII-6 (spherically symmetric flow originating at the source orifice), the number density at the detector (n_d) is approximately related to that at the skimmer entrance (n_1) by

$$n_d \approx n_1 \frac{A_1}{\pi L_{1d}^2} S_1^2. \tag{7–9}$$

S_1 is the speed ratio at the skimmer entrance (the ratio of the hydrodynamic velocity at the skimmer entrance to the most probable random speed normal to the hydrodynamic velocity at the same location), A_1 is the cross-sectional area of the skimmer orifice and L_{1d} is the skimmer–ionizer distance. Equation (7–9) is valid for $S_1 > 1$ and $\xi_{max} > \pi/2$, conditions that are easily satisfied by most designs. The objective of maximizing number density at the detector motivates a design that produces large speed ratios at the skimmer entrance and keeps the skimmer–ionizer distance as small as

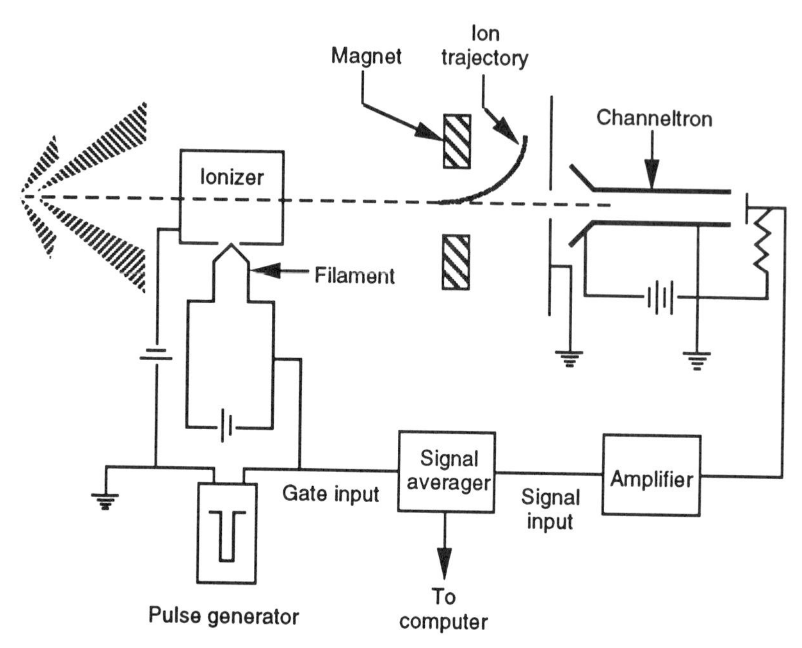

FIG. VII-6 Schematic diagram illustrating geometric parameters of importance in the design of molecular beam sampling systems.

possible. As we shall see subsequently, both of these requirements can conflict with the need to understand internal gasdynamic distortions quantitatively.

Three major physical distortion mechanisms have been identified in molecular beam systems.

1. Pressure diffusion;
2. Skimmer interference;
3. Mach number focusing.

Pumping requirements and elimination of skimmer interference are primary considerations in skimmer design. Design parameters include orifice diameter, skimmer geometry, and sampling–skimmer orifice separation, as well as source chamber dimensions. Processes that may initiate hydrodynamic composition distortions occur in the expanding jet between the sampling nozzle and skimmer, and are intimately connected with skimmer design. These lead to Mach number focusing and skimmer interference. Pressure diffusion can also occur here but is more dependent on sampling nozzle design. Each of these processes has been reviewed by Knuth [1972, 1973]. Our discussion is extracted primarily from these sources, to which the reader is referred for more details. In view of the requirements imposed by sampling nozzle design, it is seldom (if ever) possible to eliminate each of these hydrodynamic distortions by design. Skimmer interference is the least understood in a quantitative sense, followed by pressure diffusion. Elimination of these effects is a primary concern.

Composition distortion by pressure diffusion can occur in the region just downstream of the sampling orifice and across the barrel shock separating the jet from the surrounding gas. The effect of pressure diffusion is correlated by the Reynolds number based in sampling conditions and orifice diameter (Re_o). Knuth [1972] recommends selection of sampling orifice diameter such that $Re_o \geq 10^3$ to avoid pressure diffusion near the source orifice, and distortions across the barrel shock. $1/Re_o(P_o/P_s)^{1/2} \geq 0.1$ These criteria can be satisfied simultaneously only for fairly large pressure ratios ($>10^4$). For pressure ratios in the range usually used for flame sampling (10^4–10^5) they constrain the sampling orifice diameter rather severely.

Where these requirements dictate an orifice diameter larger than that specified by the criterion for quenching at the orifice (Eq. 7–8), it is probably best to be satisfied with a quenching point slightly downstream. In this case, an appropriate model of the expansion must be adopted and recourse made to Eq. (7–2) to determine the quenching distance. In evaluation of Eq. (7–2) (using a given d), relatively simple results are obtained for the asymptotic models (e.g., free jet or isentropic channel flow) described by Knuth [1973] and Smith, Wang, Tseregounis, and Westbrook [1983], respectively. For a given nozzle angle, the calculations of Sherman [1963] provide some guidance in choice of model. Table VII–1 lists the ratio of free jet density at a hypothetical conical surface of half-angle θ to that on the centerline. For $\theta = 60°$ the free jet is an excellent model for the expansion, while for $\theta = 30°$, it is a poor description.

Skimmer interference is a catch-all term covering a number of hydrodynamic processes that result in decreased beam density and mean velocity, increased velocity distribution width, and composition distortion. Figure VII-7 displays a schematic of the shock structure in the source chamber. Skimmer interference results if the standoff shock caused by interaction of the sampling chamber wall with the jet is not situated far enough downstream of the skimmer orifice, or if the shock structure becomes detached from the skimmer tip. Other causes of skimmer interference are molecules reflected from both external and internal skimmer walls and oblique shocks formed inside the skimmer.

Recalling (from Eq. 7–9) that operation with high speed ratio leads to high beam density, and noting that for a free jet prior to translational freezing (Fisher and Knuth [1969]) S_1 increases with increasing separation of the sampling–skimmer orifices

$$S_1 \;\alpha\; \left(\frac{x_1}{d}\right)^{\gamma-1}$$

it is clear that the sampling and skimmer orifices should be separated by as large a distance as possible without inducing skimmer interference. For a given x_w, the nondimensional shock standoff distance $[(x_w - x_s)/x_w]$ decreases with increasing pressure ratio (P_o/P_s) (c.f. Figure 6 of Knuth [1972]). Since skimmer interference-free operation

Table VII–1 Ratio of Free Jet Density at Half Angle Θ to That on the Centerline ρ/ρ_1 as a Function of Θ

Θ	**20**	**30**	**40**	**50**	**60**	**70**
ρ/ρ_1	0.79	0.58	0.37	0.19	0.08	0.02

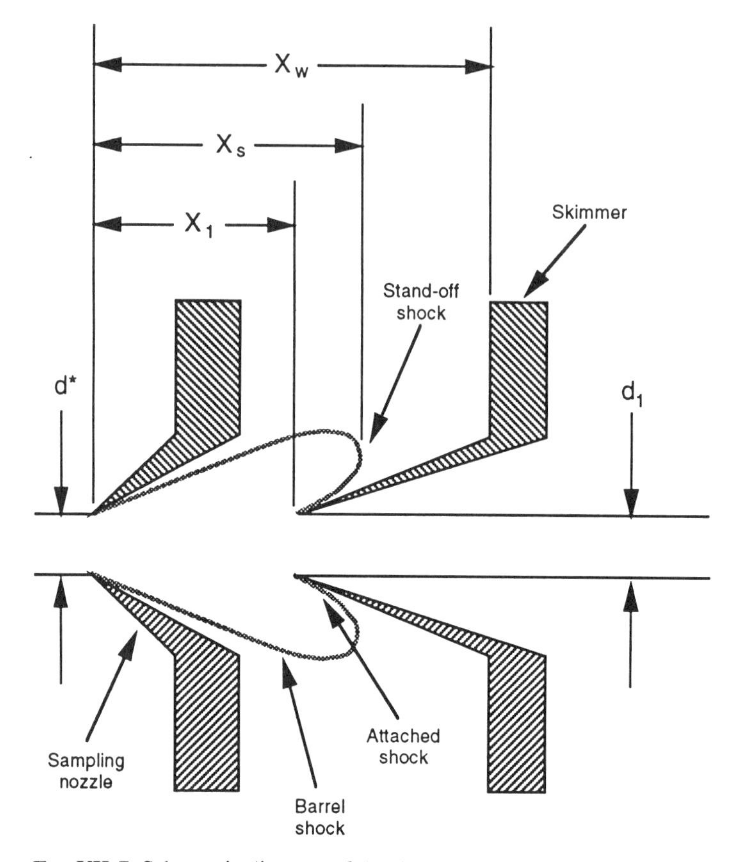

FIG. VII-7 Schematic diagram of shock structure in the source chamber.

requires that the standoff shock be located downstream of the skimmer tip, large pressure ratios (on the order of 10^4–10^5) are usually needed. Nevertheless, for most designs one still finds the base of the skimmer located far enough from the sampling orifice so that

$$\frac{x_w}{d} \geq 1.3\left(\frac{P_o}{P_s}\right)^{1/2},$$

in which case the location of the standoff shock is independent of the chamber dimension (x_w) and is given by

$$\frac{x_s}{d} = 0.67\left(\frac{P_o}{P_s}\right)^{1/2}. \quad (7\text{–}10)$$

For this case Knuth [1973] recommends x_1 satisfy

$$x_w - x_1 \geq x_w - \frac{x_s}{2}. \quad (7\text{–}11)$$

This equation represents a compromise between the need to maximize beam density at the detector and the need to avoid skimmer interference. Although Eq. (7–10) was developed for a free jet, our experience indicates that it predicts the standoff shock location fairly well, even in rather confined jets ($\theta = 30°$). This can easily be verified by observing beam intensity while increasing x_1, in which case the onset of skimmer interference from the standoff shock is evidenced by a dramatic decrease in intensity. It also points out the need of provision for adjustment of x_1 with the system in operation.

Skimmer half-angle (δ) should be selected to keep the shock structure attached to the skimmer tip and minimize reflections from external and internal surfaces flame sampling applications, $\delta < 45°$ is usually sufficient to keep the attached shock in place. For small angles, external scattering increases exponentially with δ while scattering from internally reflected molecules decreases. Knuth suggests an internal skimmer angle

$$\delta = 3 \sin^{-1} \frac{1}{M_1} \tag{7–12}$$

as a reasonable compromise; however, we have successfully studied mixture compositions using an internal angle of 16° and skimmer mach numbers as large as 20 (Smith, Wang, Tseregounis and Westbrook [1983]).

Skimmer orifice size is dictated by considerations of beam intensity (Eq. 7–9), skimmer interference, and pumping requirements in the collimating chamber. Disregarding the latter two requirements for the moment, analysis of a more general form of Eq. (7–9) (c.f. Eq. 7 of Knuth [1973]) indicates that orifice diameters greater than

$$d_1 = x_1 \frac{4}{S_1} \tag{7–13}$$

lead to negligible improvement in beam intensity. Skimmer interference for relatively large orifice diameters is correlated by the ratio of the Knudsen number based on the number density at the skimmer entrance and the source temperature to the Mach number at the skimmer Kn_{01}/M_1). Fisher and Knuth [1969] have found that the mean velocity in the beam is practically unaffected if this ratio is greater than 0.3–0.4. Since Kn_{01} is inversely proportional to d_1, this may limit the maximum useful orifice diameter. More frequently, background scattering in the collimating chamber will limit orifice size. This condition will be avoided if the number density in the collimating chamber is small enough so

$$n_c \, Q_{A_c} \, L_{12} \leq 1. \tag{7–14}$$

Here Q_{A_c} is the collision cross section of species A in the collimating chamber and L_{12} is the distance from the skimmer tip to the detection chamber. n_c is related to the skimmer conditions and the collimating chamber pumping speed (S_c) by

$$n_c = \frac{n_1 \, U_1 \, \pi \, d_1^2}{S_c}$$

where U_1 represents the hydrodynamic velocity at the skimmer entrance.

Mach number focusing results from the fact that for a mixture, the speed ratios of the lighter components are generally smaller than those of the heavier. This results in enrichment of the beam in species of higher molecular weight than the mean and corresponding depletion in those lighter. The governing parameter is the enrichment factor (α), defined for a mixture of minor species A in major species B (usually the diluent) as

$$\alpha_A \equiv \frac{(n_A/n_B)_d}{(n_A/n_B)_l} .$$

As before, the subscript d denotes conditions at the detector and l denotes those at the skimmer. For a binary mixture consisting of a small amount of species A in B, α can be expressed in terms of speed ratios at the skimmer entrance by

$$\alpha_A = \frac{1 - e^{-s_{1A}^2 \xi 2}}{1 - e^{s_{1B}^2 \xi 2}} \qquad (7\text{–}15)$$

where ξ is as defined in Figure VII-6. It is seldom possible simultaneously to satisfy the design requirements outlined previously and to achieve large enough speed ratios so that $S^2\xi > 1$. In this case, Mach number focusing corrections must be applied for all species differing substantially from the mean molecular weight of the mixture. Procedures for calculation of the required speed ratios are reasonably well established for cases of both continuum and free molecule flow at the skimmer. For diatomic major species subsets of the former case exist, depending on whether rotational degrees of freedom are frozen at the skimmer. For a free jet expansion, the appropriate equations have been given by Sharma et al. [1965] for monatomic major species, and by Yoon and Knuth [1979] for diatomic (or linear polyatomic) major species. Some aspects of the latter analysis have been extended to an isentropic channel expansion by Smith et al. [1982].

Since avoidance of Mach number focusing is not a primary goal of system design, we will not address the calculation of speed ratios in any detail. Nevertheless, we should point out that the speed ratios are functions of both the source temperature and density, each of which changes abruptly through the flame. Caution should be exercised in the use of experimentally determined enrichment ratios unless the these can be measured under the conditions applicable to the actual flame experiments. For stable species, this problem may be avoided by introducing a calibration mixture as an effusive beam rather than a supersonic beam. The effusive source should be located downstream of the skimmer. The Mach number focusing corrections are then replaced by a relatively simple effusive flow correction. For effusive flow, the mixture composition at the detector is related to that in the bulk gas by

$$(\chi_A)_d = (M_A/\bar{M})^{1/2} (\chi_A)_o$$

where the detector and bulk mole fractions are denoted by subscripts d and o respectively, and $\bar{M}$ represents the mean molecular weight of the bulk mixture.

Summary of Distortion Mechanisms

Experiment and theory both suggest that external distortions from microprobes are minimal when sampling reasonably well-stabilized premixed flames. Thermal and catalytic disturbances are small because of the low aspect ratio of these probes. Suction will result in radial concentration gradients near the orifice; however, the radial nonuniformity of the sampled gas tends to cancel when the area weighted mean is taken, leading to negligible apparent profile shift.

Molecular beam probes can produce very significant external distortions, both through catalysis of radical recombination and the effects of suction. To minimize catalysis and thermal effects, the probe should utilize as small an internal angle as is consistent with the system being sampled and internal distortion mechanisms. As the source pressure is raised, probes of progressively larger aspect ratio may be safely used. For systems in which heat conduction and radical diffusion play a role, a relatively noncatalytic material with low thermal conductivity and emissivity should be used, and the wall thickness near the orifice should be kept as small as possible. If an alternative is available, molecular beam sampling should not be used on systems where transport plays a dominant role. This specifically applies to diffusion flames. Conversely, they can be used with a much greater degree of confidence where transport is unimportant, as in a well-designed flow reactor. For molecular beam–mass spectrometer systems, where only the central core of the sample is analyzed, flat flame profiles must be shifted upstream to correct for distortions due to suction. Where mass convection dominates conditions at the orifice, Eq. (7–1) may be used to determine the magnitude of the shift.

Microprobes are subject to major composition distortions due to failure to quench chemical reactions adequately. Although scavenging microprobes may yield some information regarding radical concentrations, the sampling of such species with conventional microprobes is out of the question. It is now well established that internal probe reactions can lead to distortions for some of the stable species found in flames in relatively low concentrations. NO_2 and H_2O_2 provide examples of such species. For microprobes with small orifice diameters and entrance sections, it may be impossible to achieve enough convective cooling during the supersonic expansion to quench reactions of these species after subsonic flow is restored. Nevertheless, where comparisons have been made of major species concentrations obtained by microprobe sampling and more trustworthy techniques, agreement has usually been quite satisfactory. Where large discrepancies have been observed, such as in very weakly stabilized flames, external distortions have been responsible.

Molecular beam probes should be designed to minimize the effects of pressure diffusion and skimmer interference within the jet. This can usually be done without too much difficulty. The quenching of chemical reactions, particularly radical–radical reactions with little activation energy, presents a more difficult problem. These reactions cannot be quenched at the orifice, and in extreme cases may persist until the free molecular flow regime is reached. If the internal flow field is well approximated by a free jet, this will occur rapidly; however, the large probe internal angles necessary may well result in unacceptably large external distortions. Fortunately, high-pressure systems, which are difficult to quench internally, also are less susceptible to external distortion. The converse applies for low-pressure systems. For atmospheric-pressure, highly stabilized flat flames, probe internal angles of up to 80° have been used successfully. For low-pressure, less highly stabilized flames, probe angles as low as 40° may be necessary.

If pressure diffusion and skimmer interference are to be avoided, it will usually be impossible to achieve large enough speed ratios to neglect Mach number focusing for at least some of the species found in hydrocarbon flames. Fortunately, reliable enrichment ratio correlations are available for several idealized internal flow fields.

Mach number focusing corrections depend strongly on the temperature of the sampled gas and the flow field regime at the skimmer (continuum or free molecular). For this reason cold flow calibrations can lead to systematic errors. The following methodology, which has been found to give reasonable quantitative results for a number of combustion systems, is recommended.

1. Perform cold flow calibrations (detector response vs. number density at the detector) using an effusive beam;
2. Conduct the experiment (e.g., measure the number densities at the detector);
3. Using an appropriate model of the internal flow field to calculate enrichment ratios, correct the measured number densities at the detector for Mach number focusing to obtain number densities at the sampling orifice; This will require the sample temperature, which may have to be obtained independently;
4. Correct these values for apparent profile shift to obtain number densities in the combustor.

Special Techniques in Molecular Beam Sampling

In the past 20 years, molecular beam sampling followed by mass spectrometric analysis has been utilized to study a wide variety of combustion phenomena in steady systems. Examples include laminar premixed flames (Peeters and Mahnen [1973]; Biordi, Lazzara, and Papp [1973]), sooting flames (Bittner and Howard [1982]), and rocket exhausts (Goshgarian and Solomon [1972]). The relations already presented or referenced in this chapter are sufficient to specify the design of quantitative sampling systems for such applications. However, molecular beam sampling is by no means limited to obtaining species profiles in steady systems.

Temperature profiles may be obtained by examining the moments of the velocity distribution within the beam, and under certain circumstances real-time species concentrations can be obtained for unsteady systems. In both of these cases, additional considerations enter into the design of the sampling system. In the following sections we will outline how such measurements are performed. In the interests of brevity and in keeping with previous portion of this chapter, the experimental work leading to the development of many of the necessary gasdynamic correlations used will be largely neglected. Specific literature citations are provided for the additional relations needed. The basis for these correlations can be found in these, and in references cited therein. In this sense, the following sections are best considered as a guide to the applicable literature.

Temperature Measurement by Molecular Beam Sampling

Molecular beam sampling may be used to measure flame temperature as well as composition (Cattolica, Yoon, and Knuth [1982]). For an adiabatic expansion performing no work on the surroundings

$$h_o + \frac{U_o^2}{2} = h_d + \frac{U_d^2}{2}. \tag{7-16}$$

As before, the subscripts o and d denote conditions at the source and detector, respectively, U is the hydrodynamic velocity, and h is the specific enthalpy of the mixture (including rotational and vibrational as well as translational energy, if applicable). During the expansion, energy is transferred out of the various modes into directed translational motion upstream of their respective freezing surfaces. Downstream of the freezing point for a given mode, energy is no longer transferred to or from the mode. Taking for simplicity the case of a monatomic diluent B in large excess,

$$h \approx \frac{5}{2}\frac{R}{M_B}(T - T_o)$$

so that if the hydrodynamic velocity at the source can be neglected with respect to that at the detector

$$T_o = T_{trans,\infty} + \frac{M_B U_d^2}{5R}. \tag{7-17}$$

Here $T_{trans,\infty}$ represents the temperature at the point in the expansion where the translational degrees of freedom freeze. This point is correlated in terms of the translational relaxation time and source conditions by Eq. 11 of Young, Rodgers, and Knuth [1970]. Once the freezing point is determined, it is a simple matter to extract the Mach number at the freezing point and the temperature from this. For a free jet expansion, Eqs. 13b of Knuth [1973] and 9 of Smith, Wang, Tseregounis, and Westbrook [1983] are appropriate. The freezing point is itself a function of the source temperature, however, so that Eq. (7–17) must be solved iteratively. A suitable starting point is

$$T_o \approx \frac{M_B U_d^2}{5R}.$$

For a diatomic major species, rotational freezing makes the calculations much more complex. Fortunately, vibrational degrees of freedom usually freeze immediately so that their presence can be neglected. Since the preceding analysis for a monatomic gas will apply downstream of the rotational freezing point, the usual procedure is to modify the source conditions in the correlation of translational freezing point to reflect rotational relaxation. Source temperature, pressure and orifice diameter for an effective monatomic gas are given by Eqs. 3.10–3.12 of Yoon and Knuth [1979].

Solution of Eq. (7–17) for the source temperature requires measurement of the mean velocity downstream of the translational freezing point. A variety of time-of-flight (TOF) techniques can be used to determine the velocity distribution (Lucas, Petterson, Hurlbut, and Oppenheim [1984]; Alcay and Knuth [1969]), which must be measured downstream of the translational freezing point. One method involved mechanically chopping the beam and using the mass spectrometer to acquire the TOF signal (Goshgarian and Solomon [1972]). A simpler alternative apparatus is illustrated in Figure VII-8. This method eliminates the need to account for the flight time of ions within the mass filter and complexities involved with the mechanical chopper. It also provides better control over the beam pulse width and duty cycle. The apparatus consists of a pulsed ionizer, permanent magnet and electron multiplier along with associated electronics. Short pulses (1–10 μsec) of ions and metastable species are created by periodically biasing the filament supply from near ground to ≈ -100 V. The ionizer surfaces are held slightly below ground so that no electrons enter the ionizing volume in the absence of the bias pulse. Ions are deflected from the resulting pulsed beam by an electrostatic or magnetic field, leaving only the metastable species. These are not affected by either the detector or ionizer bias voltages, and are also produced in much larger quantity than the ions. The metastable species are detected by an electron multiplier located on the beam axis of a distance L from the ionizing volume. The detector output is amplified, and the results in the time domain averaged over several cycles. If the collimating chamber is large enough to provide reasonable flight times, higher signal strengths can be obtained by conducting the measurements there. Otherwise, the apparatus can be located in the detection chamber.

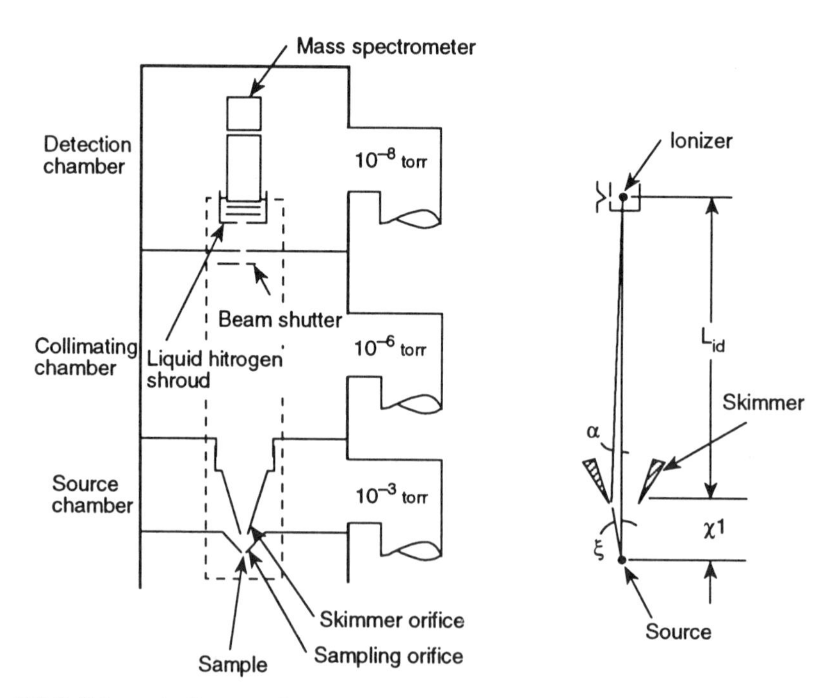

FIG. VII-8 Schematic diagram of an electronic arrangement for measurement of speed distribution in a molecular beam. The apparatus may be in place anywhere downstream of the translational freezing point.

The time-dependent detector output $I(t)$ is related to the metastable number density at the detector $N(t)$ by the convolution

$$I(t) = \int_0^t g_d(\lambda)\, N(t - \lambda)\, d\lambda \tag{7–18}$$

where $g_d(t)$ is a dynamic response function of the detector and amplification electronics usually expressed in terms of a characteristic time τ as

$$g_d(t) \; \alpha \; e^{-t/\tau}$$

The number density at the detector is given by

$$N(t) = n \int_0^t A(\lambda)\, \rho(t - \lambda)\, d\lambda \tag{7–19}$$

where n is the molecular number density at the ionizer and $A(t)$ is the gate function for the pulsed ionizer. Taking first moments and normalizing, we have

$$\eta_1\, [\rho(t)] = \eta_1[I(t)] - \eta_1[g_d(t)] - \eta_1[A(t)] \tag{7–20}$$

where η_1 denotes the first (normalized) moment. The mean velocity is then given by

$$\bar{U}_d = \frac{L}{\eta_1\, [\rho(t)]}. \tag{7–21}$$

For a given set of detector and ionizer parameters, the last two terms in Eq. (7–20) will be constants. They can be evaluated for a known source temperature (usually room temperature) using measured $\eta_1[I(t)]$ along with Eqs. (7–17) and (7–21).

The procedure outlined have results in temperature profiles for one-dimensional premixed flames that show good agreement with thermocouple measurements in the postflame region. Here, typical differences between the two methods are about 50 K, usually within the uncertainty of the thermocouple measurements. The TOF temperature profiles show the same apparent profile shift as the molecular beam species profiles. When the TOF profiles are shifted in accordance with Eq. (7–1), good agreement is also obtained in the region of large thermal gradients.

Molecular beam systems designed to produce good species data will usually also produce good TOF temperature data. The accuracy of the measurements depends on the accuracy with which the mean speed downstream of the translation freezing point can be determined. The length of the flight path is critical here. It must be long enough so that the flight time at the highest temperatures of interest is large compared with both the digitization interval and the width of the electron pulse that produces the metastable species. For a pulse width of 10 μsec, duty cycle of 1 kHz, and a digitization rate of 200 kHz, a flight path of 25 cm will result in an uncertainty of about ± 25 K at typical postflame temperatures.

Molecular Beam Sampling from Unsteady Systems

Molecular beam sampling from systems in which the thermodynamic state varies with time is much more difficult than for steady systems. As before, the usual goal of molecular beam measurements is to follow species concentrations, now as functions of time. The temporal variation of temperature plays a major role in determining the detector output, and must be known to obtain concentration measurements. Where beam chopping in used, the $T(t)$ can be determined by a much more detailed analysis of the kind presented in the previous section. Alternately, the temperature (and pressure if applicable) can be followed by other means, or calculated from an idealized model of the process (Young, Rodgers, and Knuth [1970]; Young, Wang, Rodgers, and Knuth [1970]). Examples of unsteady systems amenable to study in this manner include ignition, flame propagation through stratified media (including flame extinguishment), and internal combustion (Slone and Ratcliff [1983]).

Unsteady systems introduce the following additional complications:

1. Short sampling intervals are usually required to provide adequate time resolution, resulting in decreased sensitivity.

2. Variation of source pressure with time will affect the pressure ratio (p_o/p_s) and may induce skimmer interference [see Eqs. (7–10) and (7–11).

3. The flight time of molecules from the sampling source to the ionizer is usually not negligible with respect to the time scale of the process.

If the process of interest can be reproduced with a reasonably short duty cycle, the sampling time can be extended by averaging over many cycles. Even for major species this is usually desirable. To achieve good signal-to-noise ratios, averaging over several hundred cycles is often necessary. This is an obvious consideration in designing the system to be sampled.

The latter two complications have bearing on the design of the sampling system. Systems exhibiting cyclic pressure variations may induce very large changes in the pressure ratio. The conductance of the sampling orifice will result in a phase difference between the source and sampling chamber pressures. If changes in pressure ratio are too large, some compromise in sampling–skimmer orifice distance will be necessary to minimize skimmer interference if the entire cycle is to be sampled (Young, Wang, Rodgers, and Knuth [1970]). This will entail some loss in accuracy of the calibration.

Complications induced by finite flight times are perhaps best illustrated by example. Consider a laminar deflagration propagating past the sampling orifice normal to its axis. This might occur in the study of ignition processes, for example. For simplicity, we assume a dilute mixture so that the mixture molecular weight and specific heat ratio are nearly constant. If the source Reynolds number is $\geq 10^3$, the slip velocity between molecules of differing molecular weight will be negligible; that is, they will all possess the same average velocity $\bar{U}$, given by

$$\bar{U} = \left(\frac{2\bar{\gamma}}{\gamma - 1} \frac{R}{M} T_o \right)^{1/2}. \tag{7–22}$$

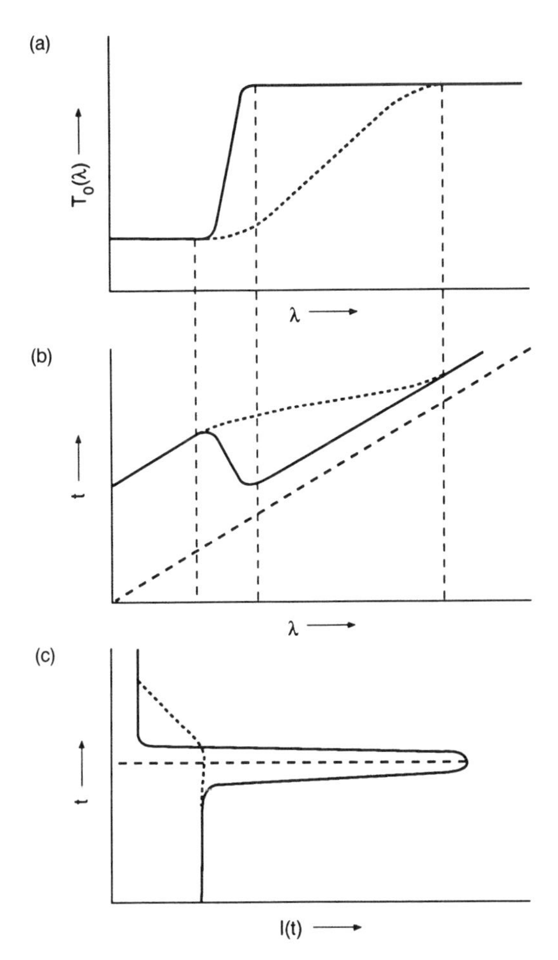

Fig. VII-9 Sketches illustrating the relation between the thermodynamic state of the sample and mass spectrometer signal for two unsteady systems with differing temperature gradients. (a) Variation of source temperature with time; (b) Relation between the time of disturbance at the sampling orifice (λ) and the time at which it is evidenced at the detector (t) for the cases depicted in (a); (c) Mass spectrometer signal (I) resulting from a reaction product for the cases presented in (a) and (b).

This would normally be the case for a properly designed system, since the same criterion ($Re_o \geq 10^3$) applies to the elimination of pressure diffusion in the initial stage of the expansion.

For this process, the variation of temperature with time at the sampling orifice might be as given in Figure VII-9(a). Where Eq. (7–22) applies, a change in concentration at the sampling orifice occurring at λ is evidenced at the detector at time t, given by

$$\tau = \lambda + \frac{L}{U} \tag{7–23}$$

Here L is the distance between the translational freezing surface ($\approx$ the sampling orifice) and the detector. The complication, which influences system design, occurs because $\bar{U}$ is a function of T_o, which is itself a function of time. For very steep temperature gradients, this may cause molecules sampled later in the cycle to arrive at the detector earlier. This phenomenon is termed *beam overrun.*

The relationships between t and λ for the temperature profiles of Figure VII-9(a) are sketched in Figure VII-9(b). For the steeper temperature gradient, molecules sampled at three different times will arrive at the detector simultaneously. Further, this situation will pertain to the most interesting part of the cycle, that is, that where the heat release occurs. The detector signal resulting from this behavior is depicted for a reaction product in Figure VII-9(c). For the case where λ is a multivalued function of t, signal deconvolution will be extremely difficult. In theory, this could be accomplished by repetition of the experiment at several different flight distances L. In practice, however, it is best to reconfigure the system to eliminate this behavior.

To do this we require the slope in Figure VII-9(b) to be always non-negative. The required relation is obtained from Eqs. (7–22) and (7–24),

$$\frac{\partial T}{\partial \lambda} = 1 - \left(\frac{\bar{M}}{2C_p T_o^3}\right)^{1/2} L \frac{\partial T_o}{\partial \lambda} \geq 0. \tag{7–24}$$

Beam overrun is most likely to occur in the preheat zone, where the temperature gradient has nearly reached its maximum value yet the temperature itself is still fairly low. For a given flame, the inequality (7–24) may theoretically always be met by decreasing L. Often this is not practical, in which case an effort should be made to lower $dT_o/d\lambda$, either by decreasing the flame front velocity or increasing its thickness. Low pressures and large diluent concentrations will obviously help.

The pressure dependence of the temporal temperature derivative is approximated for a laminar flame by

$$\frac{\partial T_o}{\partial \lambda} = \frac{\Delta T}{\partial} S_u \alpha P_o^{n-1} \tag{7–25}$$

where ΔT represents the temperature increase through the flame, δ the flame thickness, S_u the burning velocity, and n the global reaction order. For methane–oxygen flames, the global order is approximately 2. For 10% methane and a pressure of 0.1 atm, the laminar flame thickness, flame speed, and adiabatic flame temperature are 0.2 cm., 55 cm/sec, and 2215 K, respectively. Equation (7–24) predicts that the maximum tolerable flight path under these conditions would be about 35 cm, so that this constraint is not too severe for low-pressure flames.

In some situations beam overrun is almost unavoidable. Figure VII-10 shows the result of a molecular beam measurement of the transient argon signal in a flash-ignited $NO_2/CH_4/O_2/Ar$ mixture initially at 300 K (Young, Rodgers, Cullian, and Knuth [1971]). The measured source temperature [corresponding to Fig. VII-9(a)] is shown in one vertical plane. The temporal number distribution of number density at the detector has been calculated by a considerably more detailed treatment than that outlined here, and is shown as a function of the source time at the bottom. The projection on the remaining vertical plane represents the arrival time of the distribution peak at

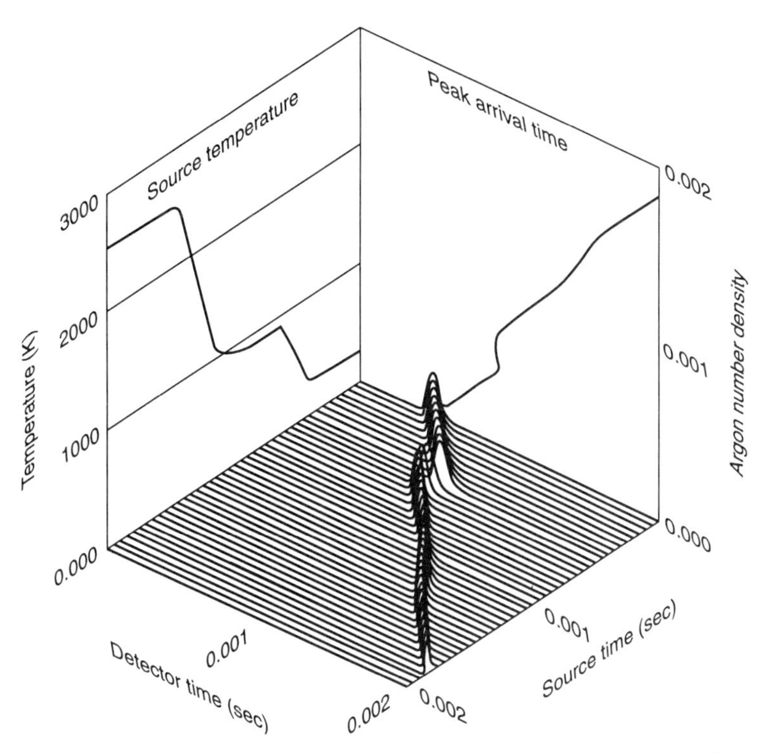

FIG. VII-10 Beam overrun for argon in a flash-ignited $NO_2/CH_4/O_2/Ar$ mixture. The measured temperature history (the left-hand vertical plane) shows an initial temperature rise due to flash heating followed by an induction period and further heating due to combustion. The relation between the source time and detection time [as in Fig. VII-9(a)] is shown in the right-hand vertical plane. The bottom plane shows the temporal distribution of argon number density at the detector. (From Lucas, Petterson, Hurlbut, and Oppenheim [1984].)

the detector [corresponding to Fig. VII-9(b)]. The mass spectrometer signal for unchopped operation [Fig. VII-9(c)] would correspond to the number distributions in terms of detector time summed over the discrete intervals in source time. Beam overrun is clearly indicated in the peak arrival time projection. As expected, this occurs in the early stage of heating by photolysis. The temperature increase resulting from combustion (1000–2600 K), while resulting in a steep thermal gradient, does not lead to beam overrun because of the higher initial temperature.

In the event of significant slip velocities at the translational freezing point, Eq. (7–22 cannot be utilized, and the analysis becomes much more complicated. The mean speeds of individual species at their respective translational freezing points must then be calculated by the procedures outlined in the discussion of TOF temperature measurement. Since one needs $\bar{U}(T_o)$ for each species, the process is quite tedious and is best avoided by increasing the source Reynolds number. More extensive theoretical treatments that account for the effect of beam chopping are available (Goshgarian and Soloman [1972]; Young, Rodgers, Cullian, and Knuth [1971]). Where the signal is intense enough to allow use of a chopped beam, transient temperature measurements can also be obtained by molecular beam methods.

VIII
LASER DIAGNOSTICS

The advent of lasers has made a myriad of new techniques for studying flame structure practical. They offer high selectivity and sensitivity and the prime advantage of being nonintrusive. The field is enormous; so this short chapter can only outline the application of laser methods to the study of flames. A list of books and review articles is given in Table VIII–1. This chapter will begin with a short discussion of lasers and their availability as of this writing. (The reader should be cautioned that the field of optical devices is undergoing rapid development and there are many new devices being introduced which have great potential for this research and may supplant current devices.) A laser is an optical device that makes use of the negative absorption coefficient a media possesses when there is a population inversion of one or more energy levels. The population inversion required for lasing is absolute. The excited-state population is larger than the ground state. This requires an external energy source to be maintained as a steady state. The source is usually called a "pump." This can be by an electric discharge, by photolysis using a flash lamp, by chemical reaction, by aerodynamic methods or using another laser. Under such conditions a beam matching the wavelength of inversion is amplified rather than absorbed. The effect is proportional to path length and therefore is increased by multiple passes through the medium. Most lasers employ an optical cavity defined by mirrors (Fig. VIII-1), which provide the boundary conditions for the beam and produce coherence and directivity. However, they are not essential since any beam passing through an inverted medium will be amplified.

The output of a laser is monochromatic, directed, and phase locked. The precision of these conditions depends on the quality of the cavity. The linewidths of commercial laser outputs varies from a few angstroms for dye lasers to a few hertz out of 10^{15} Hz for the ring lasers, which are used for frequency standards and line shape studies. The directivity of lasers varies but is of the order of a few milliradians. This can be greatly improved by careful attention to the figuring of the optics, but this is rarely necessary for combustion studies. The phase coherence of lasers depends on the quality and length of the mirror cavities relative to the wavelength. It must be uniform on a small scale compared with the wavelength to avoid reflections which would produce destructive interference and destroy lasing. For commercial lasers it must be a small fraction of a wavelength.

The nonequilibrium populations found in flames and most electric discharges are only partial and therefore do not support lasing. This rarity of complete inversion and the absence of reliable predictive criterion for the conditions makes the search for new

Table VIII–1 Review Articles on Laser Diagnostics

Topic	References[1]
COMBUSTION	A-TBMP&T,C,Ec,L,L&P,PW&B,S&S
COMPOSITION	C,D&T,H&L
DEGENERATE FOUR WAVE MIXING	F&R
DETECTORS, PROCESSORS	C&L,He,P
FLUORESCENCE	K-H,K,L,S&S,W,ZO&W
HOLOGRAPHY	S
LASERS	C&B,M&D,M
RAMAN	Har,LP,L&P(eds)
RAYLEIGH	PW&B
SPECTROSCOPY	D,D&T,Ea,F&W,H&L,K,M,W
TEMPERATURE	C,D&T,E&W,L,L&P,L&P(eds),ZO&W
VELOCITY	A,DM&S,D&W
FLOW VISUALIZATION	A,DM&S,Han,Hes,HS&P

[1]The letters represent the initials of the author's last name in the references given below.

Adrian, R. J., "Particle-Imaging Techniques for Experimental Fluid Mechanics," *Ann Rev. Fluid Mech.* **23,** 261 (1991).

Attal-Trétout, B., Bouchardy, P., Magre, P., Péalat, M. and Taran, J. P., "CARS in Combustion: Prospects and Problems," *Appl. Phys. B.,* **50,** 445 (1990).

Chang, R. K. and Long, M. B., "Optical Multichannel Detection," in *Light Scattering in Solids,* p. 179, Cardinam, N. and Guntherodt, G. (eds), Springer-Verlag, NY (1982).

Chiang, F. P., and Reid, G. T. (eds), *Optics and Lasers in Engineering,* **9,** pp. 161, Elsevier, NY (1988).

Crosley, D. R., *Laser Probes in Combustion Chemistry* ACS #34, American Chemical Soc., Washington, DC (1980).

Demtroder, W., *Laser Spectroscopy,* Springer-Verlag, NY (1981).

Druet, S. A. and Taran, J.-P.E., "CARS Spectroscopy," *Prog. Quantum Electron.* **7,** 1 (1981).

Dudderar, T. D., Meynart, R., and Simpkins, P. G., "Full Field Laser Metrology for Fluid Flow Measurement," in *Optics and Lasers in Engineering,* **9,** pp. 161–325, Chiang, F. P. and Reid, G. T. (eds), Elsevier, NY (1988).

Durst, F., Melling, A. and Whitelaw, J., *Principles and Practice of Laser-Doppler Anemometry,* Academic Press, NY (1976).

Easley, G. L., *Coherent Raman Spectroscopy,* Pergamon Press, Oxford (1981).

Eckbreth, A., "Recent Advances in Laser Diagnostics for Temperature and Species Concentration in Combustion," *Eighteenth Symposium (International) on Combustion,* p. 1471, The Combustion Institute, Pittsburgh, PA (1981).

Elder, M. L., and Winefordner, J. D., "Temperature Measurements in Flames: A Review," *Anal. At. Spectrosc.,* **6,** 293 (1983).

Farrow, R. L. and Rakestraw, D. J., "Detection of Trace Molecular Species Using Degenerate Four-Wave Mixing," *Science* **257,** 1894 (1992).

Fraser, I. M., and Winefordner, J. D., "Laser-excited Atomic Fluorescence Spectroscopy," *Anal. Chem.* **43,** 1693 (1971).

Hanson, R. K. "Combustion Diagnostics: Planar Imaging Techniques," *Twenty first Symposium (International) on Combustion,* p. 1677, The Combustion Institute, Pittsburgh, PA (1986).

Hanson, R. K., Seitzman, J. M., and Paul, P. H. "Planar Laser-Fluorescence Imaging of Combustion Gases," *Appl. Phys. B.,* **50,** 441 (1990).

Harris, T. D., and Lytle, F. E. "Analytical Applications of Laser Absorption and Emission Spectrosocpy, in *Ultrasensitive Laser Spectroscopy,* p. 369, Kliger, D. S. (ed), Academic Press, NY (1983).

Harvey, A. B. (ed), *Chemical Applications of Nonlinear Raman Spectroscopy,* Academic Press, NY (1981).

Hesselink, L, "Digital Image Processing in Flow Visualization," *Ann Rev. Fluid Mech.* **20,** 421 (1988).

continued

Table VIII–1 Review Articles on Laser Diagnostics (continued)

Klainer, S. M. (ed), "Special issue devoted to Fluorescence Spectroscopy, *Opt. Eng.* **20,** 507 (1983).

Kohse-Höninghaus, K., "Quantitative Laser-Induced Fluoresence: Some Recent Developments in Combustion Diagnostics," *Appl. Phys. B,* **50,** 455 (1990).

Lapp, M., "Flame Temperatures from Vibrational Raman Scattering," in *Laser Raman Gas Diagnostics,* p. 107, Lapp, M. and Penney, C. M. (eds), Plenum Press, NY (1974).

Lapp, M. and Penney, C. M., "Raman Measurements in Flames," in *Advances in Infrared and Raman Spectroscopy,* p. 204, Clark, R. J. H., and Hester, R. E. (eds), Hayden and Son ltd., London (1977).

Lauterborn, W., and Vogel, A., "Modern Optical Techniques in Fluid Mechanics," *Ann Rev. Fluid Mech.* **16,** 223 (1984).

Lucht, R. P. "Applications of Laser-induced Fluorescence Spectroscopy for Combustion and Plasma Diagnostics," in *Laser Spectroscopy and its Applications,* p. 623, Radziemski, L. J., Solarz, R. W., and Paisner, J. A. (eds), Marcel Dekker, NY (1987).

Maitland, A. and Dunn, M., *Laser Physics,* N. Holland Pub. Co. Amsterdam-London (1969).

Murray, J. R. "Lasers for Spectroscopy," in *Laser Spectroscopy and Its Applications,* p. 91, Radziemski, L. J., Solarz, R. W., and Paisner, J. A. (eds), Marcel Dekker, NY (1987).

Penner, S. S., Wang, C. P., and Bahadori, M. Y., "Laser Diagnostics Applied to Combustion systems," *Twentieth Symposium (International) on Combustion,* p. 1194, The Combustion Institute, Pittsburgh, PA (1985).

Pratt, W. K., *Digital Image Processing,* Wiley-Interscience, NY (1978).

Radziemski, L. J., Solarz, R. W., and Paisner, J. A. (eds), *Laser Spectroscopy and Its Applications,* Marcel Dekker, NY (1987).

Schofield, K., and Steinberg, M., "Review of Quantitative Atomic and Molecular Fluorescence in the Study of Detailed Combustion Processes," *Opt. Eng.* **20,** 501 (1981).

Smith, H. M. *Principles of Holography,* Wiley, NY (1975).

Valenti, J. J., "Laser Raman Techniques," in *Laser Spectroscopy and Its Applications,* p. 597, Radziemski, L. J., Solarz, R. W., and Paisner, J. A. (eds), Marcel Dekker, NY (1987).

Wehry, E. L. (ed), *Modern Fluorescence Spectroscopy* (two vols), Plenum Press, NY (1976).

Zizak, G., Omenetto, N., and Winefordner, J. D., "Laser-excited Atomic Fluorescence Techniques for Temperature Measurements in Flames," *Opt. Eng.* **23,** 749 (1984).

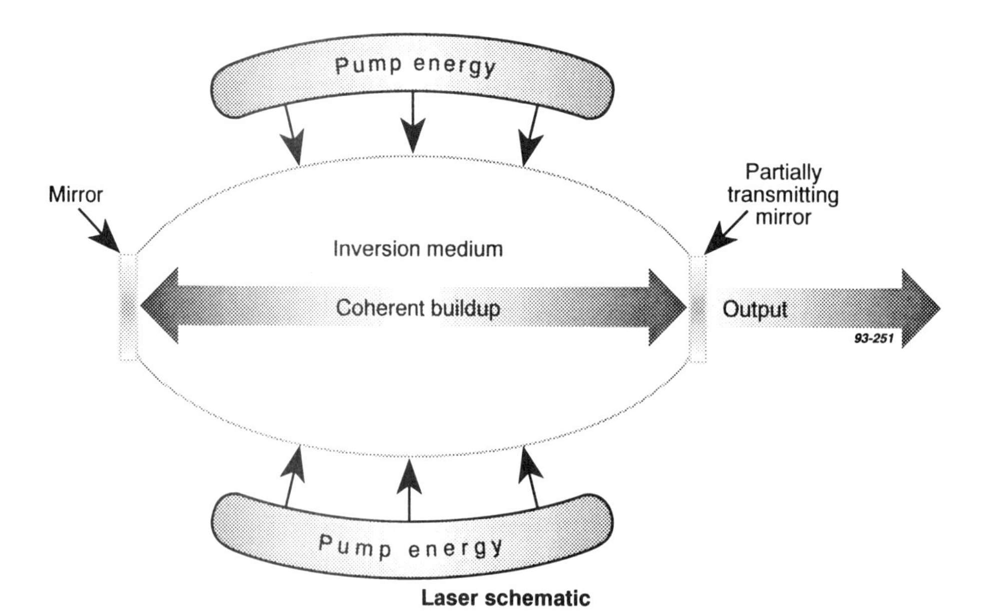

FIG. VIII-1 Schematic of laser operation.

Table VIII–2 Representative Types of Commercially Available Gas Lasers

Medium	Wavelength (μm)	CW Power (W)	Comments
CW lasers			
Argon	0.275–0.529	0.1–50	300 types
Argon ion	0.455–0.497	4–750	100 types
Ar/Kr	0.454–0.676	20	7 types
CO_2	10.6	1–10^4	200 types tunable
CO	5.2–5.9	1–20	20 types tunable
DF	3.6–4	1–1000	6 types
He/Cd	0.325–0.442	0.001–5	40 types
He/Ne	0.633	0.001–0.07	300 types
HF	2.6–3	2–1000	10 types
Kr	0.338–0.799	0.1–6	40 types
Xe/He	2–4	0.1	2 tunable

Medium	Wave Length (μm)	Pulse (J)	Rep Rate (S^{-1})	Comments
Pulsed lasers				
ArF	0.193	0.000001–0.8	2–1000	60 types
Ba	1.5–2.55	0.001	10000	1 type
CO_2	10.6	0.1–3000	1–5000	150 types
Cu	0.511–0.578	0.001–0.02	6000–20000	20 types
Au	0.628	0.0001–0.001	20000	5 types
KrF	0.248	0.00001–1	1–100	70 types
N_2	0.337	0.0001–0.01	1–100	12 types
XeCl	0.308	0.01–2	10–1000	50 types
Xe/He	0.351	0.01–200	10–200	50 types

lasers a continuing treasure hunt. Tables VIII–2–VIII–4 give some properties of the myriad of commercial lasers listed in the 1992 buyer's guide of *Laser Focus World.*

Types of Lasers

Lasers are classified as continuous or pulsed. The medium can be gaseous, solid state, or dyes dissolved in liquids. The energy pump can be a flash lamp, an electric discharge, a chemical reaction, or another laser beam.

Table VIII–3 Representative Semiconductor Lasers

Medium	Wavelength (μm)	Comments
CW lasers		
Laser diodes	0.43–1550	250 types of various wavelength and power
GaAlAs	0.81–0.85	5 types, 1–1000 mW output
InGaAsP	1.27–1.33	7 types, 0.1–2 mW output
InGaSbP	1.27–1.33	4 types, 0.1–2 mW output
Pulsed lasers		
Diodes	0.67–25	Over 100 diodes with rep. rates up to 10^9 Hz
InGaAsP	1.06, 1.27–1.33, 1.52–1.57	10000 Hz

Table VIII–4 Representative Solid-State Lasers

Medium	Wavelength (μm)	Power (W)	Comments
CW lasers			
Er:YAG	2.9	2	
Nd:YAG	1.06	0.001–50	300 types easily doubled and quadrupled
Ti-Sapphire	0.67–1	1	Tunable

Medium	Wave Length (μm)	Energy (J)	Rep Rate (Hz)	Comments
Pulsed solid-state lasers				
Alexandrite	0.2–2	0.01–250	30	Tunable
Holmium	2.1	2	20	
Nd:YAG	1.06	.0001–.2	20	500 types, doubles and quadruples
Ruby	0.69	0.1–30	0.07–5	50 types
Ti-Sapphire	0.1–0.95	1–100	0.1–1	Tunable

Continuous versus CW Lasers

The helium–neon and the CO_2 lasers are examples of continuous systems. The neodymium:YAG (Nd:YAG) and the copper vapor lasers are examples of pulsed lasers. Pulse rates vary from a few hertz for the Nd:YAG through a few hundred hertz for the nitrogen laser to ten kilohertz for the copper vapor laser. Power levels from microwatts for the tiny solid-state diodes through the giant terajoule lasers used for fusion studies.

State

The state of the medium determines the power density because the higher the density the higher the power level possible. However, the mechanisms for depopulating excited states usually increase with density. In solids, however, the atoms are held in place, and this inhibits deexcitation. By contrast, in gases deexcitation is slowed by the low density and long time between collision. Dyes dissolved in liquids offer the principal examples of lasers employing a liquid medium where the density is high but the molecules are still mobile.

Pumps

The original ruby laser is an example of a flash-lamp-pumped system. The CO_2 laser is an example of an electric-discharge-pumped system. The hydrogen–fluorine laser is an example of a laser pumped by a chemical reaction. There is even a flame-pumped laser (Linevsky and Carabetta [1974]), which uses the carbon disulfide flame with oxygen or N_2O. Flame lasers are rare because the most common flames produce water, which is one of the most effective molecules for relaxing nonequilibrium vibrational populations. There are aerodynamic lasers where the inversion is maintained by rapid adiabatic expansion cooling of hot flame gases using a wind tunnel. Finally, there is a large class of tiny solid-state diodes in which the population inversion is maintained by high-voltage electrons crossing semiconductor boundaries. This may be considered the solid-state analog of an electric discharge. A laser can be pumped by the beam from another laser, and this technique is commonly used for tunable dye lasers.

Most systems lase at only a few lines. Some of them, such as the alexandrite or Ti-sapphire, can be tuned over modest ranges. Some solid-state lasers are tunable electrically. The most common high-powered tunable systems are dyes that fluoresce over a wide band of frequencies. By using a grating mirror and/or an etalon it is possible to force much of the fluorescence energy into a single lasing line that can be tuned over the range of fluorescence. Several dyes covering a range of frequencies are available (see Fig. VIII-2). A different high-powered tunable system is the solid-state Alexandrite laser, which can be electrically tuned over a reasonable range. The free electron laser employs the interaction between relativistic electrons and a varying magnetic field. This can produce lasing over a wide range of frequencies from microwave to the ultraviolet. The free electron system is, however, a major facility rather than a commercial instrument.

One of the most versatile methods for obtaining tunable laser output is to use a dye laser. Many dyes fluoresce over a wide range of wavelengths when excited by visible or ultraviolet light. By using a light-dispersing element such as a prism, grating, or etalon, it is possible to force the wide dye output into a single lasing wavelength, which can be tuned over the range of the dye. Pumping can be done by flash lamp or by another laser. Common choices are the nitrogen laser and the doubled Neodymium:YAG lasers. The coverage available by tunable dye lasers is indicated in Figure VIII-2. The extension of these ranges by frequency doubling using commercially available birefringent crystals is also indicated in the figure. Even wider ranges into the far ultraviolet are possible by tripling and quadrupling laser outputs using a pulsed gas jet such as argon, but these systems are research programs in themselves. Commercial equipment will no doubt eventually become available.

Many of the experiments of interest in combustion require visible and ultraviolet light. Therefore, lasers are often doubled, tripled, or even quadrupled. This is done by using a nonlinear optical device, often a birefringent crystal. Frequency multiplying requires the highest power, but it is much more efficient in the optical region than in

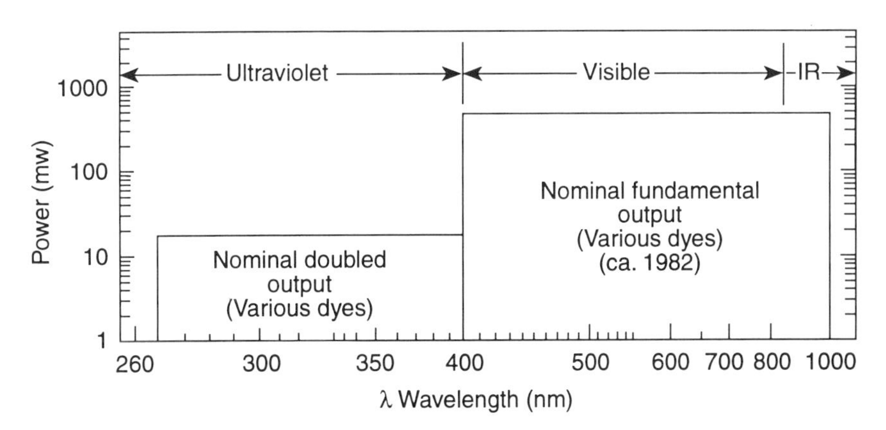

FIG. VIII-2 Tuning Range and Coverage of some Dye Lasers and their Extension by Doubling.

Table VIII–5 Optical Detectors

Detectors	Wavelength (μm)	Sensitivity (W)	Remarks
Photomultipliers	0.5–1	10^{-18} @ 1 Hz	Engstrom [1980]
GaAs (77 K)	1–5	2×10^{-12} @ 1 Hz, 1 cm^2	
HgCdTe (77 K)	10.6	5×10^{-11} @ 1 Hz, 1 cm^2	

the microwave region. Optical efficiencies can exceed 90% with good optics. The most common source is the workhorse Nd:YAG laser in the Q switching mode. In Q switching the power of the laser is built up over many cycles and triggered using a dye mirror, which breaks down at a critical intensity, dumping the stored energy in a single pulse.

Detectors for laser experiments are listed in Table VIII–5.

Laser Diagnostic Methods

Lasers are used as diagnostic tools in flames in a multitude of modes, and new techniques are still being introduced. Some of the geometries for using laser beams are shown in Figure VIII-2. This chapter can only provide brief outlines of some of the methods. Table VIII–6 lists some techniques that have been applied to combustion problems as of this writing. They may be classified as absorption methods and scattering methods, and these major methods may be further subdivided. Schematics of some of the common laser diagnostic methods are given in Figure VIII-3. The techniques will be outlined to aid in understanding the distinctions among them. For more definitive treatments the reader is referred to literature surveys and books listed in Table VIII–1.

High power levels are desirable in flame studies because all of the phenomena depend strongly on power level. For this reason most studies employ pulsed lasers where instantaneous power is high. This approach is limited by the electrical breakdown of the flame gases, and some high-powered lasers have been designed to generate longer pulses, providing higher total energy levels without exceeding breakdown. This desirable approach is generally not available commercially, but such facilities are

Table VIII–6 Cross Sections of Various Scattering Modes[a]

Type of Scattering	Cross Section	Comments
10 μm particles (ME)	10^{-7}	
0.1 μm particles (ME)	10^{-13}	
Atomic fluorescence	10^{-13}–10^{-18}	Strong, visible
Molecular fluorescence	10^{-19}–10^{-24}	Simple molecules
Rayleigh	10^{-27}	N_2, 488 nm
Rotational Raman	10^{-29}	N_2, sum of all lines, 488 nm
Rotational Raman	6×10^{-31}	N_2 single strong line, 488 nm
Vibrational Raman	5×10^{-31}	N_2, strokes, Q branch, 488 nm

[a]After Lapp and Penney [1974].

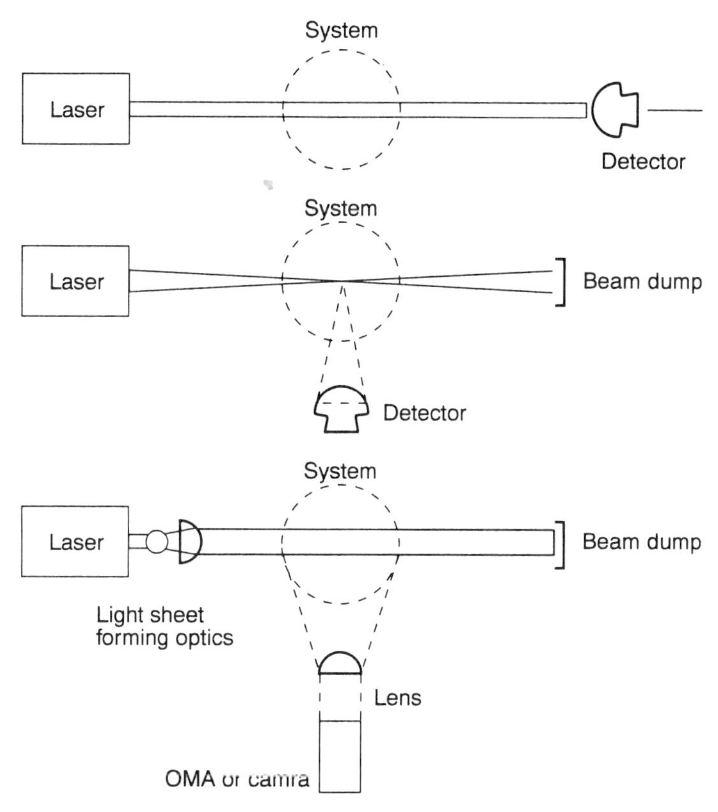

Fig. VIII-3 Some Geometries used in Laser Combustion Studies. (a) Direct Beam Absorption; (b) Focused Beam for Scattering; (c) Planar Imaging.

available at the Sandia National Laboratory, Livermore, California, which will host collaborators on projects of mutual interest in combustion.

Most lasing transitions are in the infrared region with relatively few in the visible or near ultraviolet. This is the case because the levels that can be easily inverted are vibrational states and because the lifetimes of states varies roughly with wavelength. Exceptions are the Free Electron Laser (FEL), which requires a large facility and the X-ray laser, which requires a nuclear detonation for pumping. These are not suitable for the usual combustion laboratory.

Absorption Methods

Lasers offer an ideal monochromatic source for spectroscopy. For individual lines it is necessary to tune to the line center carefully to avoid statistical errors in absorption. If scanning methods are used, resolution is determined by the scanning rate. General methods are listed in the references to Table VIII–1. Because of the high available power multiple reflection techniques (Fig. VI-11) are possible. This greatly increases sensitivity. Alden et al. [1983] used Raman to analyze fire gases.

The sensitivity of absorption spectroscopy can be improved by using laser derivative spectroscopy, which operates on the wings of lines. The improvement can exceed 1000. The ultimate in sensitivity is probably obtained using intercavity absorption, in which the system becomes part of the laser cavity. It is difficult to obtain spatial resolution using this technique; so it has not been widely applied to flame structure studies. Special techniques such as laser acoustic spectroscopy (West, Barrett, Siebert, and Reddy [1983]) and Stark modulated absorption spectroscopy (Rahn, Farrow, Koszykowski, and Mattern [1980]) are discussed in the section on mixed techniques.

Light can be absorbed by a molecule to excite one or more modes of internal energy: rotational, vibrational, or electronic. These modes are quantized and furnish valuable tools for identifying molecules and their state of excitation. Characteristically rotational levels lie between the microwave region and the far infrared; vibrational levels lie between the infrared and near-visible region; electronic levels lie between the visible and ultraviolet. Pure rotational spectroscopy is described in the standard text of Townes and Schawlow [1982]. Vibrational Infrared Spectroscopy is described in the authoritative text of Wilson, Decius, and Cross [1981]. The general field is described by Herzberg [1945,1950,1966,1971] in his classic series of books on the spectroscopy of atoms and diatomic and polyatomic molecules. Some reviews and books on laser spectroscopic methods are collected in Table VIII–1. Methods include: laser resonance absorption, laser derivative Spectroscopy, and Fourier Transform Spectroscopy. Other possibilities include:(1) Laser-induced fluorescence (LIF), also called Laser-Induced Chemiluminescence (LIC), which is useful for the study of radicals.

Traditional absorption spectroscopy requires a line-of-sight path with corrections for end effects. Resolution is generally limited by the planarity of the flame being studied. Lasers provide higher intensity sources and allow better discrimination, but the techniques are essentially those using the older light sources. Conventional techniques are described in earlier chapters (V and VI). One method of avoiding line-of-sight problems is to Stark modulate using a crossed laser beam. This technique, developed by Rahn and Farrow [1989], is described in Figure VI-12.

Mixed Techniques

There are a number of experimental techniques that require lasers to become feasible but involve other techniques. They will be discussed here. Lasers also play a decisive role in improving and simplifying optical imaging using, planar visualization PLIF (Hanson [1986]), holography, and tomography (Druet and Taran [1981]), as well as the quantitative studies of optical deflections reviewed by Weinberg [1982]. These methods are also discussed in the chapters on visualization and aerodynamic measurements (Chapter V) and temperature (Chapter VI).

For the techniques under consideration the important property may be the high powers, the monochromaticity, or the coherence, but the laser is usually the key. Methods include: (1) optical interference and deflection techniques, (2) Laser-induced ionization (ELIB), which allows local elemental analysis [Schenck, Travis, Turk, and O'Haver [1982]), (3) optogalvanic studies of radical by Goldsmith [1982], (4) Laser Acoustic Spectroscopy (LAS) (West, Barrett, Siebert, and Reddy [1983]), and (5) Optical Stark modulated Spectroscopy (OSMS) by Rahn, Farrow, Koszykowski, and Mattern [1980].

Optical Methods

Because of their monochromaticity, phase coherence, and low divergence, lasers are ideal sources for any optical experiment. They transform interferometry from a facility into a simple student experiment. Shadow and Schlieren photography become almost trivial. The application of these techniques to combustion studies is discussed in the chapters on visualization (Chapter III) and temperature (Chapter IV).

Simple lenses can be employed with lasers because the beam diameters require very small *f* numbers. Corrections are not required for chromatic aberration. As Weinberg has pointed out, lasers can be used to produce interference lenses of almost any size photographically. These can correct for window imperfections in combustion systems.

Holography, Tomography, and the techniques for Planar Imaging are discussed in the chapter on visualization (Chapter IV). A schematic for a typical holographic experiment is given in Figure IV-7. A schematic for laser tomography is shown in Figure IV-8.

Laser Optogalvanic Spectroscopy

Laser optogalvanic spectroscopy (LOGS) is an interesting mixed spectroscopic technique that has proved useful for detecting radicals and atoms in flames (Goldsmith [1989]). A schematic of the apparatus is given in Figure VIII-8. This is a multiphoton technique employing high-intensity visible radiation. Generally three or four visible photons are required to ionize the atom. Multiphoton absorption is required both because the far-ultraviolet lasers necessary for the processes are not generally available and because flames are opaque in this region. The crossed-beam optical setup such as is used for Stark modulated spectroscopy (see Fig. VI-12) is modified by adding a collector electrode in the burned gas region and focusing the laser on the region to be studied. The atom is detected by the increase in conductivity by the electrons produced by the laser ionization when the resonant frequency is crossed. Modulation techniques can improve the sensitivity. The technique is applicable to most radical species. It is interesting to note that in studying H and O atoms Goldsmith [1982] found significant interference by the lines of nitric oxide.

Laser Optical Stark Modulation

The electric fields in a focused laser beam are sufficiently high to be able to shift rotational and vibrational levels outside their respective linewidths. This procedure has been used by Rahn et al. [1980] in an ingenious method of improving the spatial resolution and sensitivity of laser absorption spectroscopy. The technique consists of crossing the probe beam with a high-intensity Stark field beam. The intersection of the two beams defines a very small volume in the flame. By modulating the Stark beam the spectroscopic signal is modulated and can be detected using phase-sensitive detection. The method has been demonstrated in infrared absorption spectroscopy of flames and was discussed in Chapter VI and illustrated in Figure VI-12.

Laser-Acoustic Spectroscopy

Laser-acoustic spectroscopy (LAS) is a sensitive spectroscopic technique in which absorption is detected by applying pulsed monochromatic light on the sample at a known frequency and phase. Absorption is detected by the sound that is induced in the sample. Phase-sensitive detection can be used, and the sensitivity is high. This is a quite general technique since all that is required is a suitable absorption. Tunable lasers offer the ideal source for flame structure studies since the laser can be focused. The spatial resolution can be a few wavelengths of the laser frequency. It is necessary to know the spectrum of the molecule under study and the temperature. Since any absorption could lead to a signal, confusion can occur in complex mixtures.

The technique has been applied to measure local sound speed in flames from which temperature is derived. The method uses laser deflection to detect the passage of the wave produced by a second focused laser (Cool, Bernstein, Song, and Goodwin [1989]). Velocity measurements have been made by similar techniques by detecting the locally heated area passing through the flame front. A schematic of a typical apparatus is given in Figure IV-14. Laser ionization has been used in a similar manner. These techniques are described in the chapter on velocity determination (Chapter IV).

Emission from Laser-Induced Breakdown

Lasers are often used as plasma sources for elemental metal analyses (Cremers and Radziemski [1987]). Commercial spectrometers with photomultipliers fed by monochromators tuned to the elemental emission lines are available. Multiplexing for more than one element simultaneously can be done by spacing detectors at the focus points for the elemental lines dispersed by a spectrometer. This technique has been used in flames by Schenck et al [1982] and colleagues at Sandia. It possesses the spatial resolution typical of laser focus. Both metallic and nonmetallic elements show detectable emissions so that relatively complete elemental analyses are possible.

Scattering Methods

The most productive use of lasers in combustion studies have been the scattering techniques. Table VIII–6 the relative orders of magnitude for the various types of scattering. Some of the techniques that have been developed are listed in Table VIII–7. The relations among vibration–rotation levels used in combustion diagnostics is illustrated in Figure VIII-4. These were difficult or impossible using conventional light sources, but have been made practical by the high power available using lasers. First-order scattering phenomena are discussed first. Second-order scattering in gases is forbidden by symmetry conditions; so the third-order and higher processes are discussed in the final section.

Classical Summary of Nonlinear Optics

Eckbreth [1981] pointed out that a classical picture is adequate to visualize the basis of the physical phenomena used in laser diagnostics, and the following paraphrases his arguments. Readers interested in more detail are referred to texts such as that of Mait-

Table VIII–7 Acronyms of Some Laser Diagnostic Processes with Recent References

Name	Acronym	References (1985–1992)[a]
Crossed-beam phase-matched CARS	BOXCARS	Eckbreth [1978]
Coherent anti-Stokes (Stokes) Raman spectroscopy	CARS	Druet and Taran [1981]; Eckbreth et al. [1982]
Coherent Raman spectroscopy	CRS	Easley [1981]
Emission (from) laser breakdown	ELIB	Cremers and Radziemeski [1987]
Fourier transform infrared spectroscopy	FTIR	Chen [1987]
Higher-order Raman anti-Stokes spectroscopy	HORAS	Harvey [1981]
Higher-order Raman Stokes spectroscopy	HORSES	Harvey [1981]
Inverse Raman spectroscopy	IRS	Daigreault, Morris, and Schneggenburger [1983]
Laser absorption spectroscopy	LAS	Radziemski, Solarz, and Paisner [1987]
Laser resonance absorption spectroscopy	LRA	Radziemski, Solarz, and Paisner [1987]
Laser deflection imaging	LDI	Faris and Byer [1988]
Laser deflection velocimetry	LDV	Durst, Melling, and Whitelaw [1976]
Laser derivative spectroscopy	LDS	Demtroder [1981]
Laser-induced fluorescence	LIF	Lucht [1987]
Holography		Smith [1975]; Caulfield and Sun [1970]
Laser Moire interferometry	LM	Rau and Barzziv [1984]
Laser shadowgraph	LS	Weinberg [1982]
Multiphoton fluorescence spectroscopy	MFS	Goldsmith [1985]
Mie scattering	LMS	Adrian [1991]
Optogalvanic spectroscopy	OGS	Goldsmith, Miller, Anderson, and Williams [1991]
Optically modulated Spark spectroscopy	OMSS	Rahn et al. [1980]
Particle imaging techniques	PIT	Adrian [1991]
Planar laser-induced Fluorescence	PLIF	Hanson [1986]
Raman inverse Kerr effect spectroscopy	RIKES	Valenti [1987]
Rayleigh scattering	RS	Smith [1978]
Spontaneous Raman spectroscopy	SRS	Easley [1981]
Spontaneous resonant Raman spectroscopy	SRRS	Valenti [1987]
Stimulated Raman gain spectroscopy	SRGS	Valenti [1987]

[a]For references prior to 1985, see Lapp and Penney [1974 and 1977], Crosley [1980], and Penner, Wang, and Bahadori [1984].

land and Dunn [1969]. It should be recognized that the full treatment requires consideration of quantum mechanics.

Optical phenomena are governed by Maxwell's equations. When approximately monochromatic field components are involved, the electric field and polarization susceptibility can be expanded in terms of Fourier components. Induced polarization can be expressed as a power series.

The higher-order susceptibilities are nonlinear and become important at high power levels. The linear term is described by the refractive index, which can have both real and imaginary parts, which is commonly modeled by an oscillating charge. The refractive index and its gradients are responsible for optical interference and deflection

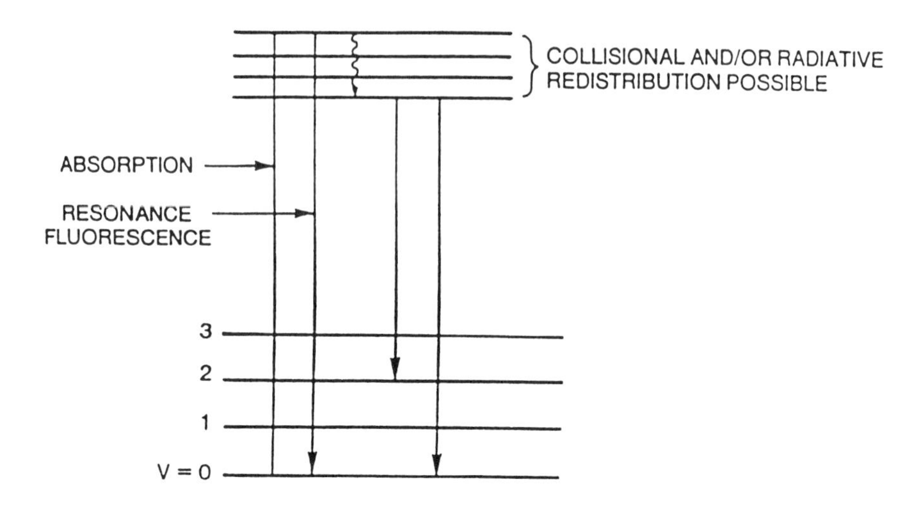

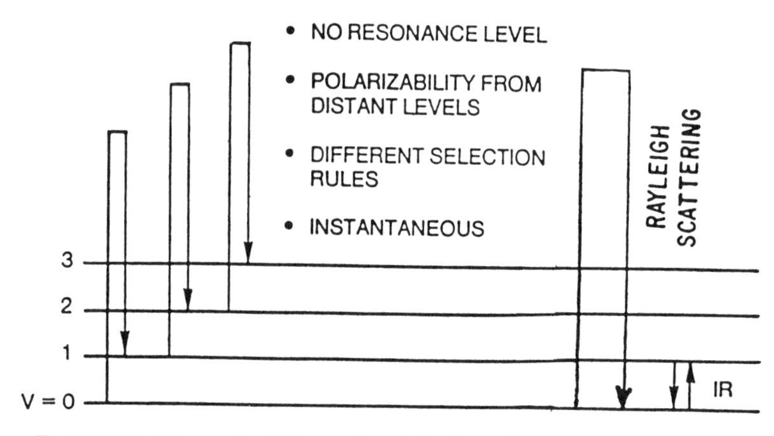

Fig. VIII-4 Schematic of Energy Levels in Scattering Processes.

methods for visualizing flames. Raman and Rayleigh scattering arise from polarization induced through the linear susceptibility. Rayleigh scattering is the term that has the same frequency as the incident frequency. Polarizability of a molecule depends on the molecular configuration, and therefore the induced term is modulated by the internal motions of the molecule. This leads to Raman frequencies that are shifted by the rotational and vibrational levels of the illuminated molecule. This can be interpreted as a difference frequency between the incident wave and the internal motions of the molecule.

Successive higher-order polarizations are weaker by a factor that is proportional to the ratio of the incident field to the intermolecular field. Intermolecular fields are of the order of 3×10^{10} V/m so that even for laser intensities of 10^{13} W/m^2 the ratio of successive orders is down by a factor of a thousand. Since the second-order susceptibilities are zero because of symmetry considerations, the first higher-order effects such as CARS appear in third order and are down by a factor of a million as compared with the first-order Raman scattering. Table VIII-6 gives the relative orders expected for some of these processes.

Particulate Scattering (Mie–Tyndal)

The scattering of light by particles has been known since Tyndall's studies in the nineteenth century (Partington [1961]). It is a diffraction phenomenon that shows a complex angular dependence on particle size because of interference between waves from various points around the circumference of the particle. The quantitative description of this, which is due to Mie, can be found in texts such as Durst, Melling, and Whitelaw [1976] and the review article of Adrian [1991]. In premixed laminar flames the principal use of this phenomena is in velocity determinations using entrained particles. Methods for particle tracing of stream tubes are discussed in Chapter III. More direct use of the scattering processes is made in Laser Doppler Velocimetry (LDV) and Speckle Velocimetry (SV) (Table VIII-7). The experimental methods as applied to combustion are described in this chapter. LDV makes use of the Doppler shift induced in the scattered light by the moving particles. The beat frequency between a reference beam and the scattered beam is detected by mixing them on a photomultiplier. LDV systems are commercially available and in common use. A schematic of a typical apparatus is given in Figure IV-12. Speckle velocimetry (SV) uses a photograph of a particle field illuminated by flashes at periodic intervals. This forms multiple images of particles displaced by the flow during the flash lamp interval. If such a photograph is illuminated by coherent light, the related particle images form parallel Young's fringes in the far field. The particle speed is proportional to the fringe separation divided by the illuminating flash interval. The direction of flow is defined by the angle of the fringes. A schematic of a typical apparatus is given in Figure IV-11. The experimental methods used in these studies are outlined in Chapter IV.

Rayleigh Scattering

Rayleigh scattering is elastic and hence is not shifted in frequency. It results from the statistical density inhomogeneities in fluids that occur at distances of the order of the mean free path. It depends on the refractive index of the gas and is, therefore, not very sensitive to the specific molecule involved. As a consequence, it is useful for overall density determinations, but not concentration. There is a significant difference between the scattering of large and small molecules, particularly hydrogen. Therefore, at least a semiquantitative knowledge of composition changes and molecular Rayleigh scattering cross sections is required to interpret the data quantitatively.

Because the scattered light is not shifted, Rayleigh scattering is interfered with by particulate scattering and any source of stray light. The techniques for the use of Rayleigh scattering to infer density and derive temperatures are discussed in Chapter V. A schematic is given in Figure VIII-5. Some results are given in Figure V-10.

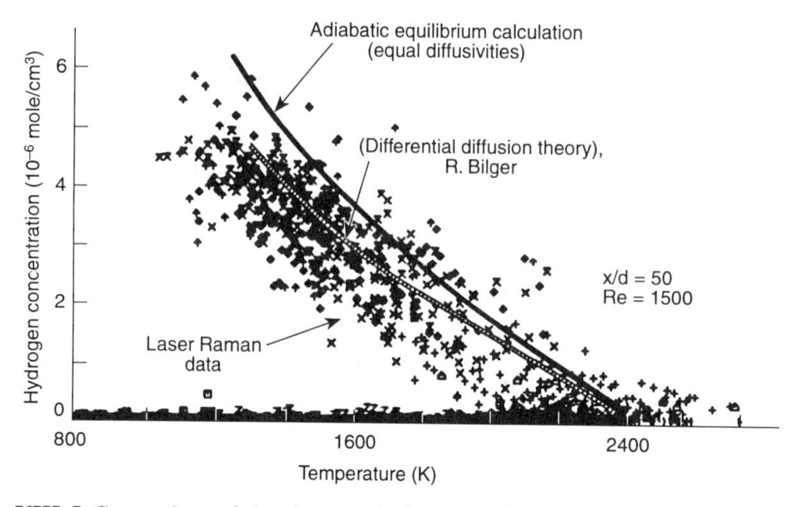

FIG. VIII-5 Comparison of simultaneous hydrogen and temperature data taken on a turbulent hydrogen flame in air at a variety of radial positions fifty tip diameters downstream of the nozzle. (After Drake, Lapp, Penney, Warshaw and Gerhold [1981].)

Laser-Induced Fluorescence

Laser-induced fluorescence (LIF) involves laser absorption in which the interrogating beam is resonant with an internal triplet energy level. If the energy is absorbed and then reradiated, the processes is called *fluorescence.* If the initial state is singlet, the characteristic time for fluorescence is 10^{-8} to 10^{-9} sec. If the initial state is a doublet π the characteristic time ranges from 10^{-4} to 10^{-7} sec (see Fig. VIII-4). Fluorescence is in competition with collisional quenching and internal conversions to another internal state. The relative rates of these processes controls the usefulness of the technique. If the internal conversion is to a long-lived triplet state, it may be populated highly enough to lase. Elementary discussions of these processes are found in most texts on physical chemistry.

The experimental arrangement (Fig. VIII-3) consists of illuminating the region to be studied and viewing it at an off-axis angle. The system can use: (1) focused point illumination with photomultiplier detection, (2) linear illumination with linear Optical Multichannel Analyzer (OMA) detection, or (3) planar illumination with a two-dimensional OMA. The arrangement used depends on whether the purpose is point measurements, linear measurements, or planar visualization.

To obtain quantitative results the molecular spectrum must be known and it must have absorptions accessible to tunable lasers. The Einstein coefficient for radiative decay must be known. Finally the efficiency of the processes must be established since it is in competition with nonradiative transfer, collisional deactivation, and in some cases even chemical reaction. These restrictions limit quantitative applicability of the method, but for many systems it offers a highly sensitive method, particularly for the study of radicals. Sensitivity can be as high as parts per million. A number of techniques exist for evaluating quenching. One method is to use a laser pulse short compared with the quenching time and to measure the quenching directly. This of course requires a high response time. Another approach is called *saturated fluorescence* (Crosley [1980, 1984]) in which the incident laser intensity is made so large that the absorption and stimulated emission are large compared with collisional quenching. Under these circumstances the quenching rate can be ignored.

The scattering is strong enough to allow its use in the planar visualization of flames (Hanson [1986], Cattolica [1984]). A schematic for a typical experiment is shown in Figure IV-4.

Spontaneous Raman Scattering

Raman scattering is the first-order inelastic scattering that results from the interaction between the incoming light and the internal energy levels in the molecules. The scattered light is shifted either up (anti-Stokes) or down (Stokes) by the quantized energy levels within the molecule. The result is essentially a vibration–rotation spectrum shifted from the infrared into the visible region by the addition of the illuminating light. The selection rules differ between infrared and Raman so that the spectra, although similar, are not identical. The most significant difference is that homopolar diatomic molecules, which do not emit or absorb in the infrared, do so in Raman. This is important for combustion studies because it allows such species as nitrogen, oxygen, and the halogens to be studied. Raman scattering cross sections scale with the fourth power of the illuminating frequency. For this reason visible and ultraviolet sources are favored, even though the effect occurs in principle at any frequency. The application of the technique to combustion using laser methods was developed in the late 1960s and early 1970s, and there have been several reviews (Lapp and Penney ([1974]; Penner, Wang, and Bahadori [1984]; Hanson [1986]; Crosley [1980]; Eckbreth [1981]). A schematic of an experimental arrangement is given in Figure VIII-5.

One of the virtues of Spontaneous Raman Scattering (SRS) is that there is no direct interference from scattered light of the exciting laser. To utilize this advantage fully, however, requires isolation using double monochromators. It is also not affected by quenching, in contrast with laser fluorescence. Many species can be monitored simultaneously, and in air flames nitrogen offers an ideal calibration species. The Raman cross sections for most common diatomic species are well known (Durst, Melling, and Whitelaw [1976]). However, it should be remembered that the detailed spectrum of even such simple molecules as water have not been completely mapped.

SRS is excellent for temperature and concentration measurements of major species. The sensitivity of the technique for temperature measurement can be seen in Figure VIII-6, which illustrates the use of both vibrational and rotational SRS. It is useful down to a few parts per thousand. A good example of such studies in flames is the study on propane by Kaiser, Rothschild, and Lavoie [1983], illustrated in Figure XII-32.

Higher-Order Scattering Processes

There are many higher-order multiphoton scattering processes. The most useful higher-order process at present is Coherent Anti-Stokes Raman Spectroscopy (CARS). It is third order in susceptibility. The technique utilizes two laser beams, a

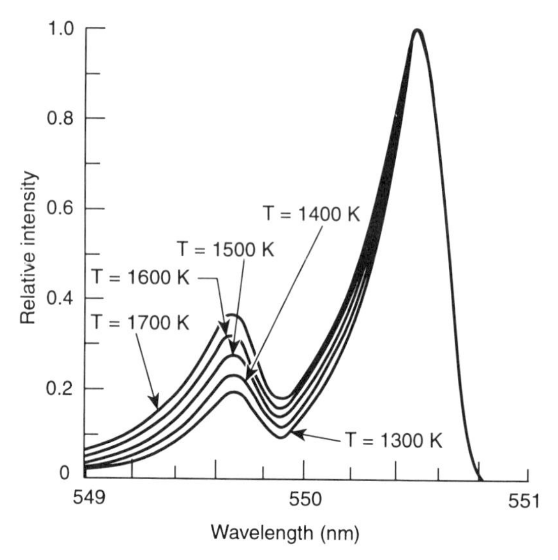

Fig. VIII-6 Temperature sensitivity of Stokes vibrational Raman band structure as calculated by Lapp [1974]. Nitrogen is an example with unresolved rotational fine structure. The peak intensity of the ground state band is normalized to unity to emphasize the temperature sensitivity of the contour. The spectra of figures 6 and 7 are computed using similar monochometer slit widths (0.16nm FWHM). The differences are due to the differences in molecular weight between the two species. Rotational energy levels and hence line separations depend inversely on molecular weight.

pump, and a Stokes probe. They must be carefully aligned to interact through the third-order media. If there is a Raman-active resonance Ω_v that is matched by the difference frequency $\Omega_v = \Omega_1 - \Omega_2$, a coherent new laser beam at the Raman-shifted frequency $\Omega_3 = 2\Omega_1 - \Omega_2$. Phase matching is required for efficient generation by parallel alignment of the beams. This is difficult in combustion studies because of density gradients, and this factor limits the spatial resolution. The difficulty is reduced by using crossed-beam phase matching, where high spatial resolution is possible. This arrangement is called *BOXCARS.*

The big advantage of CARS is that a coherent beam output is obtained so that it is possible to reduce interference by background light simply by separating the detector from the system. Collection efficiency is high. It is a relatively strong process, being several orders of magnitude higher than equivalent spontaneous Raman emission.

Although there is no threshold power level for CARS, a high intensity pulse is required to obtain significant output. The usual choice for the pump laser is the ruby laser or a frequency-doubled Neodymium:YAG laser. The Stokes laser is usually a tunable dye laser pumped by part of the output from the pump laser. The geometry for CARS and BOXCARS scattering is shown in Figure VIII-7.

Flame Images and Processes from Pointwise Measurements

Simultaneous pointwise Raman temperature and composition data can be accumulated as histograms which can be combined to provide two or three dimensional

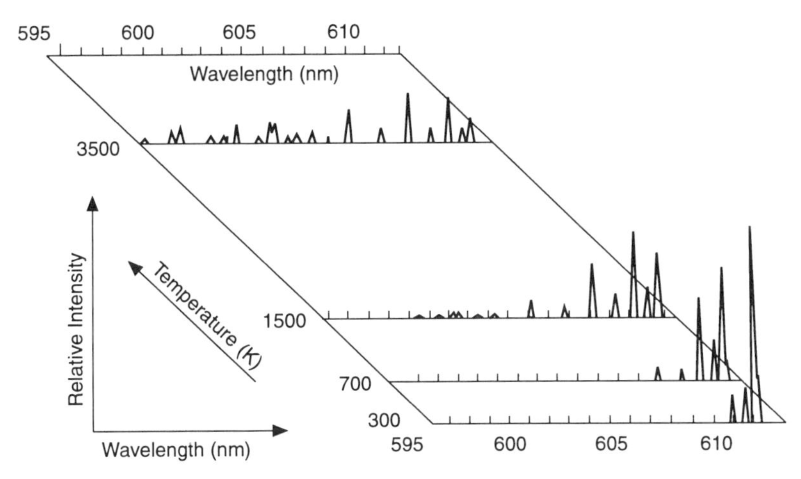

Fig. VIII-7 Temperature sensitivity of Stokes vibrational band structure as calculated by Lapp [1974]. Hydrogen is an example with resolved rotational fine structure. The spectra of figures 6 and 7 are computed using similar monochometer slit widths (0.16nm FWHM). The differences are due to the differences in molecular weight between the two species. Rotational energy levels and hence line separations depend inversely on molecular weight.

images of turbulent flames (Lapp, Drake, Penney and Pitz [1988]). This powerful tool can approximate the probability density functions employed in most theories. For example, Drake, Lapp, Penney, Warshaw and Gerhold [1981] have compared measured temperature and compositions with local equilibrium calculations (see Fig. VIII-5). They attribute the observed deviations to the differential diffusion of fuel hydrogen in this turbulent diffusion flame.

Stimulated Raman Gain/Loss Spectroscopy

There are other third-order processes besides CARS. Many of these have been used in combustion studies (Table VI-2). In stimulated gain/loss spectroscopy a pump beam produces a gain or loss in a crossed pump beam at the Stokes or anti-Stokes frequency. The advantage of this technique over CARS is that it does not require phase matching. The spontaneous Raman spectrum is produced. The fractional change in combustion situations is small (10^{-4}–10^{-5}). It has the disadvantage that only one frequency is interrogated at a time. This presents no difficulty in the study of steady-state flames, but for time-dependent systems such as turbulent flames it becomes a problem. The technique is somewhat more sensitive than CARS (Penner, Wang, and Bahadori [1984]). The experimental arrangement for this and other crossed-beam processes is given in Figure VIII-3.

Raman-Induced Kerr Effect

The Raman-induced Kerr effect (RIKE) is a third-order processes. The experimental system is similar to that of CARS or SRG/LS, consisting of crossed pump and tunable probe laser beams. The pump laser rotates the polarization in the probe beam. As a

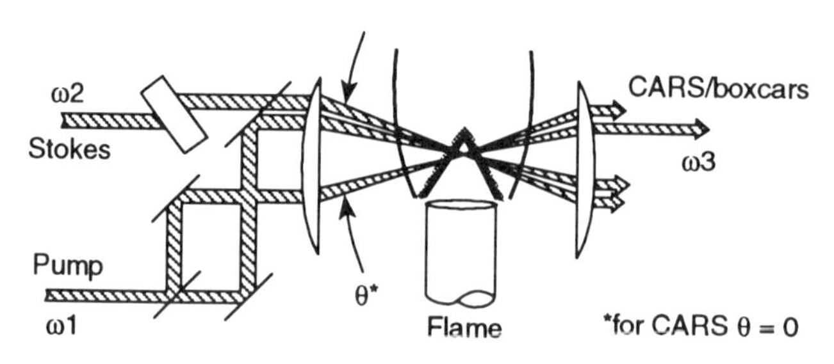

FIG. VIII-8 Schematic of CARS Apparatus.

result the Raman spectra produces changes in intensity, as seen by a detector with a polarizing filter. Eckbreth [1981] indicates that it may be of limited use in combustion studies, but the question remains open.

Comparisons and Comments

Laser diagnostic methods have become a standard for flame structure studies because they are noninvasive, offer both species and state selectivity, and allow characterization of nonequilibrium populations.

All of the techniques of Table VIII–7 have shown promise in one area or another, but the researcher must evaluate the merits and problems of the technique relative to his problem.

There are, of course, weaknesses in laser methods: (1) A number of methods must be combined to provide a complete flame description; (2) some of the techniques are difficult and costly; (3) resolution is usually limited to about 1 mm^3, although in principle resolution should approach a few wavelengths; (3) flames may be opaque to the desired wavelengths requiring the use of multiphoton techniques; (4) absolute calibrations can be difficult; (5) sensitivity of some of the techniques is limited. Taken as a whole, however, lasers offer a family of powerful, useful tools for flame studies. Techniques that have been explored are summarized in Table VIII-7, which gives references.

The agreement between laser and probe methods of measuring temperature and concentration is generally good, and laser methods often allow the detection of probe interference problems. It should be remembered that to apply most of the laser methods it is necessary to know the spectra of the molecules involved. In multicomponent systems this can become a confusing factor. Temperatures can be determined within a few degrees under optimal conditions, but in flames 10–20 K is more common.

Lasers have made the optical methods of interferometry and shadow/Schlieren practical for flame studies of local density. In addition it has opened up a myriad of visualization possibilities using Planar Laser-Induced Fluorescence (PLIF), holography, and tomography. These techniques are also discussed in the chapter on visualization (Chapter IV).

Laser absorption spectroscopy is a powerful, versatile technique, but it requires either a planar flame or a path-defining technique such as Optical Stark Modulation

(OMSS) or optical probes. Crossed laser modulation techniques such as OSM, RIKES, OHRIKES, Laser Derivative Spectroscopy, Stimulated Raman Gain/Loss Spectroscopy, and Optogalvanic Spectroscopy greatly increase the sensitivity and versatility of absorption methods for studying flames. Sensitivities for trace species are typically 10^{-4} to 10^{-5} mole fraction at atmospheric pressure. Using intercavity absorption increases the sensitivity even more, but this technique does not possess the resolution required for most flame structure studies.

Rayleigh scattering is the strongest of the scattering processes. It is mainly useful for visualization and density determinations and requires some knowledge of local composition to provide quantitative determinations. The principal interferences come from particulate material.

Spontaneous Raman Spectroscopy offers an excellent technique for studying temperature and minor species down to around 10^{-4} mole fraction at atmospheric pressure. Raman studies are commonly photon limited. As with all scattering techniques, sensitivity is proportional to pressure so reduced pressure studies are difficult, but on the other hand this favors high-pressure studies. Rotational Raman is significantly stronger than vibrational Raman (Table VIII–5), but it is difficult to work with because shifts are so small that interferences from the stimulating beam present a problem. In some cases these can be surmounted using high-dispersion spectrometers and narrow line filters such as iodine vapor. The technique is difficult to use in sooting or highly luminous flames.

Laser induced fluorescence is the method of choice for trace radical species. It allows detection in the 10–100 ppm region. Quenching is the most common difficulty in obtaining quantitative measurements. There are a number of techniques for determining quenching rates, but the problem is best handled by avoiding it using saturation spectroscopy.

CARS/BOXCARS is the method of choice in sooting, particulate, or high-luminosity environments. Its principal drawbacks are experimental difficulty and extreme sensitivity to beam alignment.

LDV gives high-precision velocity determinations (Drain [1980]), but spatial resolution is limited to about 1 mm. It should be remembered that although precision is high, the absolute accuracy is limited by the particulate tracers used. In flames this can be a severe limitation since accuracy is limited for large particles because of acceleration errors with large particles, while small particles introduce errors in flame fronts because of the thermomechanical effect.

Optical deflection techniques appear to offer a universal laser method that in principle allows determination of velocity, temperature, and composition all using the same technique. At present the utility resolution and sensitivity of this family of methods remains to be assessed for flame structure studies.

Acknowledgments

The author appreciates the contributions which were made by Drs. M. Lapp and P. Witze of Sandia National Laboratory, Livermore, and by Prof. R. Lucht of the University of Illinois in updating the coverage of this chapter.

IX

THE ANALYSIS OF FLAME DATA

Earlier chapters have covered the experimental methods for studying the microstructure of flames. It is now appropriate to set forth the mathematical apparatus required to reduce the experimental data and derive fluxes and rates. This is done using the equations of change which we will usually call the *flame equations*. The first rigorous formulation based on kinetic theory is due to Hirschfelder, Curtiss, and Bird [1964]. It is the basis for the present formulation, which was developed by Westenberg and Fristrom [1960, 1961, 1965]. Other treatments for the reduction of experimental data have been given by Biordi, Lazzara, and Papp [1975c] and Hastie [1975]. Solutions for modeling have been described by Dixon-Lewis [1967, 1972], Warnatz [1977, 1981], Oran and Boris [1981], and Peters and Warnatz [1982]. The equations will be summarized here and made specific to data analysis. The full understanding of these equations requires a general knowledge of chemistry and physics including: thermodynamics, fluid dynamics, chemical kinetics, heat and mass transfer, and some understanding of kinetic theory. For their manipulation some familiarity with "stiff" differential equations is desirable. There is not sufficient space in this book to provide a proper treatment of these diverse areas. Each of them is worth a book in its own right, and indeed such treatises can be found in most technical libraries.

Results will be presented here in what is hoped will be a useable form. A minimum of background material is presented in order to conserve space. Those interested in details are referred to the literature. The book of Gardiner [1984] is particularly useful because it covers various aspects of the combustion problem, each presented by an authority in the field. In the area of transport phenomena and heat and mass transfer, the books of Bird, Stewart, and Lightfoot [1960] and Rossler [1986] provide comprehensive treatments. In thermodynamics the classic book of Lewis and Randall [1961] is a standard. The most comprehensive coverage in combustion theory is the book by Williams [1985] and the Russian contribution by Zeldovich, Barenblatt, Librivich, and Makhviladze [1985]. In numerical methods Peters and Warnatz [1982] have edited a workshop with many approaches. The old flame structure book by Fristrom and Westenberg [1965] contains a useful summary of background material.

The critical difference between flames and ordinary flow systems is the strong interaction of the transport processes with reaction and flow. As a consequence the analysis of experimental data requires an associated family of transport, thermodynamic, and physical parameters. Methods for obtaining the required auxiliary physical, thermodynamic and transport data can be found in the critical monograph of Reid, Prausnitz, and Poling [1987]. The discussion will be opened with short summaries of the relevant experimental variables and the transport processes.

Summary of Experimental Variables

In considering the equations for flame structure analysis it is desirable to know which system parameters must be known or measured and identify which dependent variables are required. The input system parameters are the pressure, the inlet mass flow rate, initial gas composition, and temperature. If heat is extracted by the burner, the system is nonadiabatic, and this loss must be measured to establish the inlet enthalpy. Alternatively the temperature gradient at the surface of heat extraction or final temperature can be measured. The local experimental variables that describe the flame are the temperature, density, velocity, area ratio, and a set of species concentrations. The maximum number of variables that could be measured is $s + 4$, where s is the number of species. These are not all independent, since there are always the continuity equation and an equation of state that relate variables. This reduces the required number to $s + 2$. In addition, there is always another equation relating the concentration variables. If mole fractions X_i or mass fraction, f_i, are measured

$$\sum_{i=1}^{n} X_i = 1 \quad \text{and} \quad f_i = X_i M_i/\overline{M} \tag{9-1}$$

or if absolute concentrations N_i are used, then

$$\rho = \sum_{i=1}^{n} N_i M_i = \sum_{i=1}^{n} f_i$$

The element and energy conservation equations can also be used to reduce the required number of experimental profiles, but they are usually reserved for consistency checks (Westenberg and Fristrom [1960]).

The required $s + 1$ dependent variables may be chosen in various ways depending on flame geometry and available experimental techniques. Table IX–1 relates some of the possibilities. Distance is the usual experimental independent variable for flame structure measurements, but results may be expressed using any convenient variable. For example, temperature is a convenient independent variable for theoretical purposes (see Fig. IX-1). In this case temperature gradient becomes a dependent variable, and distance is obtained by integration. For propagating flames such as spark-ignited flame kernels, time is the experimental independent variable. In flames of ideal geometry such as a true flat flame or spherical flame, area ratio may be calculated from geometric considerations. Otherwise it must be measured (see Fig. X-2). It may also be noted that if concentration is obtained in dimensionless fractional units, the density is usually derived from the equation of state using the initial conditions with local temperature and the molecular weight derived from the composition. On the other hand, if concentrations are obtained in absolute units, say, N_i (moles/cm^3), then density is implicit and temperature is redundant. Table IX–2 gives the relations between common units of composition and rate.

Table IX–1 Sets of Variables Sufficient for Characterizing Common Experimental Flames

Flame System	Independent Variable	Distance z	Time t	Velocity v	Area Ratio A	Density ρ	Temp T	Concentration X_i, N_i
		Dependent Variables						
Flat	z		Calc.	Calc.	A	Calc.	T	$X_i; i = 1, 2, \ldots$
Flat	z		Calc.	v	Calc.	Calc.	T	$X_i; i = 1, 2, \ldots$
Flat	z		Calc.	Calc.	A	Calc.	Calc.	$N_i; i = 1, 2, \ldots$
Spherical	z		Calc.	Calc.	Calc.	Calc.	T	$X_i; i = 1, 2, \ldots$
Conical	z			v	A	Calc.	Calc.	$X_i; i = 1, 2, \ldots$
Expanding flame kernels	t	Calc.		Calc.	Calc.	Calc.	T	$X_i; i = 1, 2, \ldots$
Theory	T	$\int\left(\frac{dT}{dz}\right)$	Calc.	Calc.	Const.	Calc.		$X_i; i = 1, 2, \ldots$

Each line of the table gives the symbol for the variables that must be measured directly by some technique. The others are then obtained by calculation.

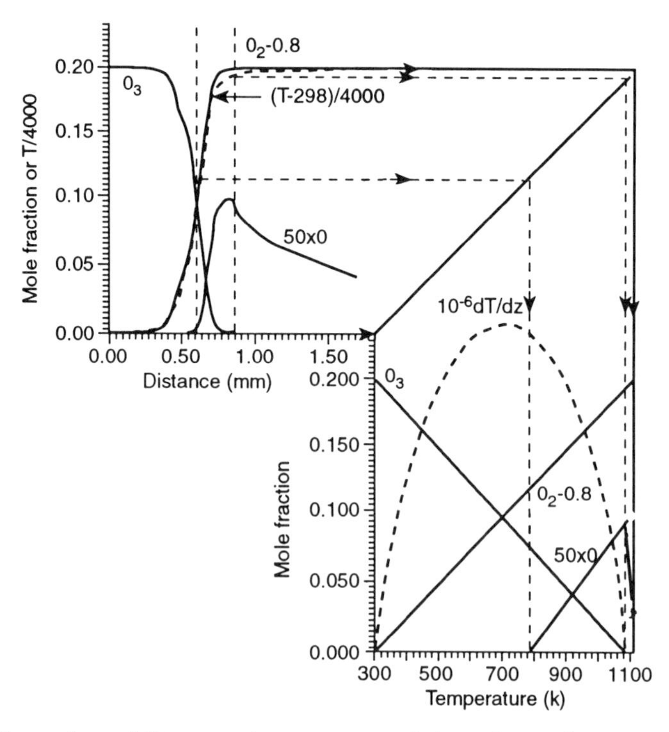

FIG. IX-1 Comparison of distance and temperature as independent variables of presenting flame structure data. (Ozone Data of Warnatz [1978], replotted.)

Table IX–2 Conversion Factors for Some Concentration Units

Given a Quantity in the Units Below	Multiply by Table Value to Convert to the Units to the Left	$cm^3\ mol^{-1}$	$L\ mol^{-1}$	$m^3\ mol^{-3}$	$cm^3\ molecule^{-1}$	$(mm\ Hg)^{-1}$
$cm^3\ mol^{-1}$		1	10^3	10^6	6.023×10^{23}	$62.40 \times 10^3\ T$
$L\ mol^{-1}$		10^{-3}	1	10^3	6.023×10^{20}	$62.40\ T$
$m^3\ mol^{-1}$		10^{-6}	10^{-3}	1	6.023×10^{17}	$62.40 \times 10^{-3}\ T$
$cm^3\ molecule^{-1}$		0.1660×10^{-23}	0.1660×10^{-20}	0.1660×10^{-17}	1	$10.36 \times 10^{-20}\ T$
$(mm\ Hg)^{-1}$		$16.03 \times 10^{-6}\ T^{-1}$	$16.03 \times 10^{-3} T^{-1}$	$16.03\ T^{-1}$	$96.53 \times 10^{17} T^{-1}$	1

Transport Processes in Flames

The dominant transport processes in flames are diffusion, thermal conduction, and thermal diffusion, effects of viscosity and inverse thermal diffusion being negligible (Hirschfelder, Curtiss, and Bird [1964]). The processes are related because they are all a consequence of the transport of some physical property through the gas by molecules under the influence of a gradient of an intensive property. On a molecular level they are all functions of binary collisions between molecules. *Ordinary* or *molecular diffusion* is the transfer of one species relative to the others under the influence of a concentration gradient. Viscosity is the transfer of momentum by a velocity gradient; thermal conductivity is the transfer of energy by a temperature gradient. Cross-product processes also occur. The only one of importance in flames is thermal diffusion, which is the transfer of a species relative to the others as a result of a temperature gradient. It is a consequence of the temperature dependence of diffusion. Even mass flow can be considered as a process of transport of mass under a pressure gradient. In flames, however, flow is more conveniently treated separately using fluid dynamics.

Each transport process has a corresponding coefficient. Tables of these coefficients for some common combustion species are given in Appendix A-1 at the end of the book. When experimental values are not available, they can be derived with reasonable reliability from the molecular constants given in Table A–1 using formulae given in Appendix at the end of the chapter. The velocity of diffusivity, which is used in the analysis, is obtained by multiplying the transport coefficient (normalized to the dimensions of diffusivity [$cm^2 \, sec^{-1}$]) by the logarithmic gradient of the appropriate intensive property [Eq. (9–3)]. In this equation, Y is the intensive property, D_t is the corresponding transport coefficient, V_t is the transport velocity, and z is distance.

$$V_t = \frac{D_t}{Y}\frac{dY}{dz} = D_t \, d(\log(Y))/dz \tag{9–2}$$

The flux of a species then becomes the product of the concentration and a characteristic velocity, the sum its individual species velocity and the mass velocity. To bring out the relations among the transport process and gain an understanding of the coefficients, it is convenient to normalize the coefficients to the dimensions of diffusion (cm^2/sec). One often talks about the diffusion of heat. This is the transport of heat and should not be confused with thermal diffusion, which transports species. One can also talk of the diffusion of momentum, molecular diffusion, etc. They can be combined as dimensionless ratios. The most common of these are: the Lewis number $\lambda/(\rho C_p D) \approx 1.24$, which is the ratio of thermal diffusivity to molecular diffusivity; the Schmidt number $\eta/(\rho D) \approx 0.83$, which is the ratio of viscous diffusivity to molecular diffusivity, and the Prandtl number $\eta C_p/\lambda \approx 0.67$, which is the ratio of viscous diffusivity to thermal diffusivity. The theoretical values that were derived from kinetic theory provide a reasonable approximation for simple molecules (Hirschfelder, Curtiss, and Bird [1964]). It should be observed that there many other permutations since the ratio of any two dimensionless numbers is also a dimensionless number.

Only a limited number of direct measurements of transport coefficients of even simple one- and two-component systems are available over the temperature range required for flames. As a consequence it is necessary to resort to "rigorous" kinetic theory for estimates of both the primary coefficients and the multicomponent parameters required for flames. This is a complex but relatively well understood field. It is based on the formulation of kinetic theory developed independently by Enskog [1922] and Chapman [1923] and presented in the book by Chapman and Cowling [1939]. The extension to multicomponent systems was made by Hirschfelder, Curtiss, and Bird [1964], and the extension to polyatomic, polar species and thermal diffusion by Monchick, Mason, and their collaborators [1961a, b]. The most comprehensive treatment of methods for deriving these parameters both using kinetic theory and by more empirical methods is given in the treatise of Reid, Prausnitz, and Poling [1987]. This source is recommended both for its extensive compilations and critical discussion of methods for estimating parameters. For reliable, up-to-date information the Standard Reference Data Program (SRDP) of the National Institutes of Science and Technology (NIST) is the most extensive and reliable source. Their continuing critical reviews and publication of data make their information the data of choice whenever it is available. They publish compilations and a journal and distribute tapes and discs suitable for both PCs and mainframe computers (see references in Appendix IX-A at the end of the chapter).

Fortunately flames offer a relatively simple environment for estimates of transport coefficients. Pressures are moderate, temperatures are high, and binary trace approximations are often quantitatively useful for mixtures. A discussion of some approximate methods that are appropriate for flame studies are given in Appendix IX-A.

Analyzing Flame Data

Flame data are usually analyzed using the previously mentioned flame equations. They are straightforward in principle but complex in practice. The basic constraints are:

1. Conservation energy;
2. Conservation of mass and further the individual conservation of each elemental species;
3. Differential equations for energy transport;
4. Differential equations for molecular transport; and
5. Differential equations for species transformation.

Taken together with appropriate physical, thermodynamic, transport, and rate parameters, these equations provide a complete flame description. Two methods may be used:

1. The synthetic approach, in which the flame equations are solved in their entirety and the calculated flame profiles are compared with the experimental ones; and
2. The analytic approach, in which the data for each species are treated individually using the experimental profiles to derive fluxes and rates.

These approaches complement one another. However, the synthetic approach requires substantial computational facilities that are not at present available to every experimentalist, although the rapid increase of computational capabilities makes one optimistic about the future. There is not sufficient space in this book for a meaningful treatment of synthetic flame theory. Therefore, treatment in this chapter will concentrate on the analytic use of the flame equations for data interpretation.

The full, direct solution of the flame equations has been described in some detail by a number of previously mentioned authors. Despite powerful algorithms for sensitivity analysis that allow isolation of effects of individual reactions, the author is of the opinion that at the present time the most productive for synthetic flame theory calculations is the testing of the validity of overall reaction mechanisms rather than direct analysis. The reader should keep in mind that as computational capabilities increase this value judgment may well be reversed. The analytic application of the flame equations to experimental data is a more modest approach. It is simpler because, instead of having to solve this set of nonlinear differential equations simultaneously, each species may be treated separately using experimental profiles. The analytic approach yields experimental fluxes, net species production rates, and rates of heat release. With the aid of an assumed reaction mechanism, these data can yield rate coefficients that can be compared with those derived from other flames and by other experimental techniques. Since this could potentially involve reactions of every species with every other species with perhaps several channels, it can be seen that except for the simplest flame systems chemical kinetics is an art rather than a science. The key is to choose the minimum set of reactions that will quantitatively describe the data. The problem is often obscured by experimental errors and deficiencies in transport coefficients and occasionally even thermodynamic data. Further, some of the elementary chemical kinetic data on high-temperature rates are questionable in many cases. These deficiencies are being reduced, and the long-term prospects are excellent. However, at present, the realistic approach is one of cautious skepticism. *The only unambiguous information that can be derived from flame data are fluxes and net rates of species change.* To go beyond this requires the assumption of a reaction mechanism.

The Complete Three-Dimensional Equations

For present purposes, a laminar flame is assumed to fulfill the following conditions:

1. It is a steady-state system, so that all macroscopic variables at any point in the flame zone are independent of time;
2. It is so close to constant pressure that kinetic energy is negligible, allowing a separable treatment of conservation of momentum considerations;
3. Dissipative effects due to radiation, viscosity, and external forces may be neglected; and
4. Higher-order transport processes are quantitatively negligible.
5. Continuum flow with logarithmic gradients of the intensive variables are large compared with the mean free path.

Mass Conservation

$$\nabla \cdot (\rho \mathbf{v}) = 0 \tag{9-3}$$

This relation states that the divergence of total flux of matter (g cm^{-2} sec^{-1}) is zero at every point in the system. Since steady state is assumed, there is no local accumulation or depletion of mass with time.

Species Continuity

$$\nabla \cdot [N_i(\mathbf{v} + \mathbf{V}_i)] = K_i \tag{9-4}$$

This specifies that the divergence of flux of any species, i, must be its net rate of appearance (or disappearance). This is the consequence of chemical reaction. Note that the presence of the diffusion velocity accounts for the effect of the concentration gradient on total flux.

Energy Conservation

$$\nabla \cdot \left(\rho \mathbf{v} H_{sp} + \sum_{i=1}^{n} N_i H_i \mathbf{V}_i - \lambda \nabla T \right) = 0 \tag{9-5}$$

In this equation H_{sp} is the specific enthalpy of the mixture defined as: $H_{sp} = (1/\rho)\Sigma_i N_i H_i$. This states that the divergence of total energy flux, that is, the sum of the terms due to convection, diffusion, and conduction, must equal zero at every point.

Equation of State

$$P = NRT \tag{9-6}$$

The ideal gas law is quantitatively adequate for flames up to perhaps 3 atm. Beyond this point higher order equations are required. These extensions are discussed in the books of Hirschfelder, Curtiss, and Bird [1944] and Reid, Prausnitz, and Poling [1987].

Diffusion Velocities

$$\mathbf{V}_i = \frac{N^2}{N_i \rho} \sum_{j=1}^{n} M_j D_{ij}^{*} \nabla \frac{N_j}{N} - \frac{D_i^{T*}}{N_i M_i} \nabla \ln T \tag{9-7}$$

This is Fick's law generalized to three dimensions and multicomponent systems with an additional term (the last) to account for the contribution of thermal diffusion. The inverse formulation, called the Stephen–Maxwell equations, are not commonly used for experimental analysis and will be only briefly considered.

The Steady-State One-Dimensional Form

Although flames are three dimensional, there are two major reasons to expect that a one-dimensional model would provide a reasonable approximation to real flames. The first is that steady-state propagation must be in a single direction, otherwise the system would either dissipate or explode. This occurs in the ignition and extinction of flames. The second factor is that a unique burning velocity appears to be characteristic of premixed flames. These factors were mentioned in Chapter II and are explored in more detail in Chapter X.

This book is therefore restricted to flames that can be quantitatively described using a steady-state, quasi-one-dimensional model [Fig. IX-2(a,b)]. This eliminates time and two of the three spatial coordinates, which reduces the problem to a set of ordinary differential equations. Lateral expansion is introduced as an area ratio, $A(z)$, which requires auxiliary consideration.

$$A(z) = \frac{a(z)}{a(0)} \tag{9–8}$$

where $A(z)$ is the area ratio, $a(z)$ is the stream tube area at point z, and $a(0)$ is the initial stream tube area.

This is an adaptation of an aerodynamic concept of fluid flow in a duct of varying cross section. In this approach measured quantities, which in principle may vary across a stream tube, are replaced by a value averaged across the stream tube. It is valid as long as the gradients across the stream tube are small compared with those along the propagation direction, z. For flames this requires that the burner diameter (for flat flames) or flame height (for conical and other flame geometries) be at least ten times the flame front thickness. The mathematical justification of this approach has been given by Hirschfelder, Curtiss, and Bird [1964]. The experimental validation has been discussed by Fristrom [1965] and Dixon-Lewis and Islam [1982].

The one-dimensional model is obvious in the case of the "flat flames," but less so in other geometries. The applicability of this model to other flame geometries can be appreciated by viewing them in a coordinate system that is stationary with respect to lateral gas movement (Fig. IX-3). Discussions of flame geometry can be found in Chapters II, III, and VI.

Spherical and cylindrical flames offer simple geometries that are experimentally realizable. A spherical hydrogen–bromine flame has been studied and is discussed in Chapter XI. Burners are described in Chapter III. Area ratios are:

$$\begin{aligned} \text{spherical: } A(r) &= \frac{4\pi r^2}{4\pi r_0^2} = \left(\frac{r}{r_0}\right)^2 \\ \text{cylindrical: } A(r) &= \frac{\pi r L}{\pi r_0 L} = \frac{r}{r_0} \end{aligned} \tag{9–9}$$

Diffusion Velocities

The definition of diffusion velocity in one dimension is:

$$V_i = \frac{N^2}{N_i \rho} \sum_j M_j D^*_{ij} \frac{d}{dz} \frac{N_j}{N} - D^*_{ij} \frac{\alpha_{ij}}{N_i M_i} \frac{d \ln T}{dz} \tag{9–10}$$

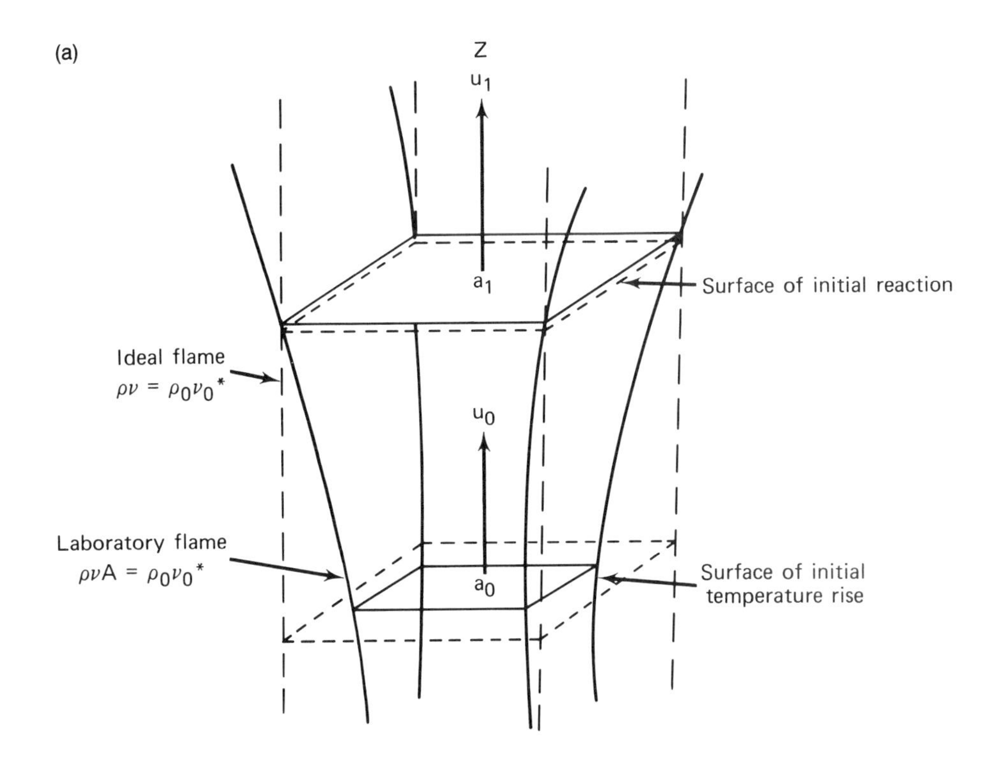

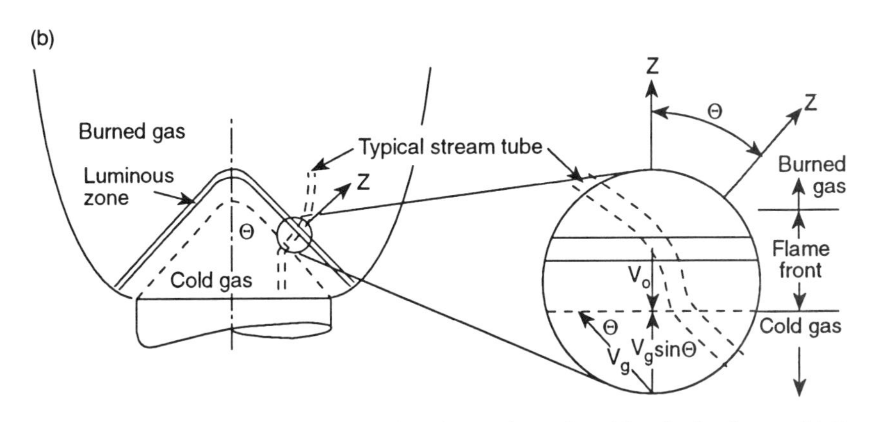

Fig. IX-2 Flame coordinate Systems. (a) Quasi-one-dimensional flow in flat flames; (b) Quasi-one-dimensional flow in conical flames.

In this equation the D^*_{ij}'s (cm^2 sec^{-1}) are the multicomponent diffusion coefficients of species i with the other components of the mixture. The α_{ij}'s (dimensionless) are the binary thermal diffusion factors of species i with the other components.

It should be observed that the product $(D^*_{ij}\,\alpha_{ij})$ has been used in place of the thermal diffusion coefficient $(D_i)^{T*}$. The author feels that this formulation of thermal diffusion is easier to comprehend and handle because in the high-temperature trace limit α_{ij}

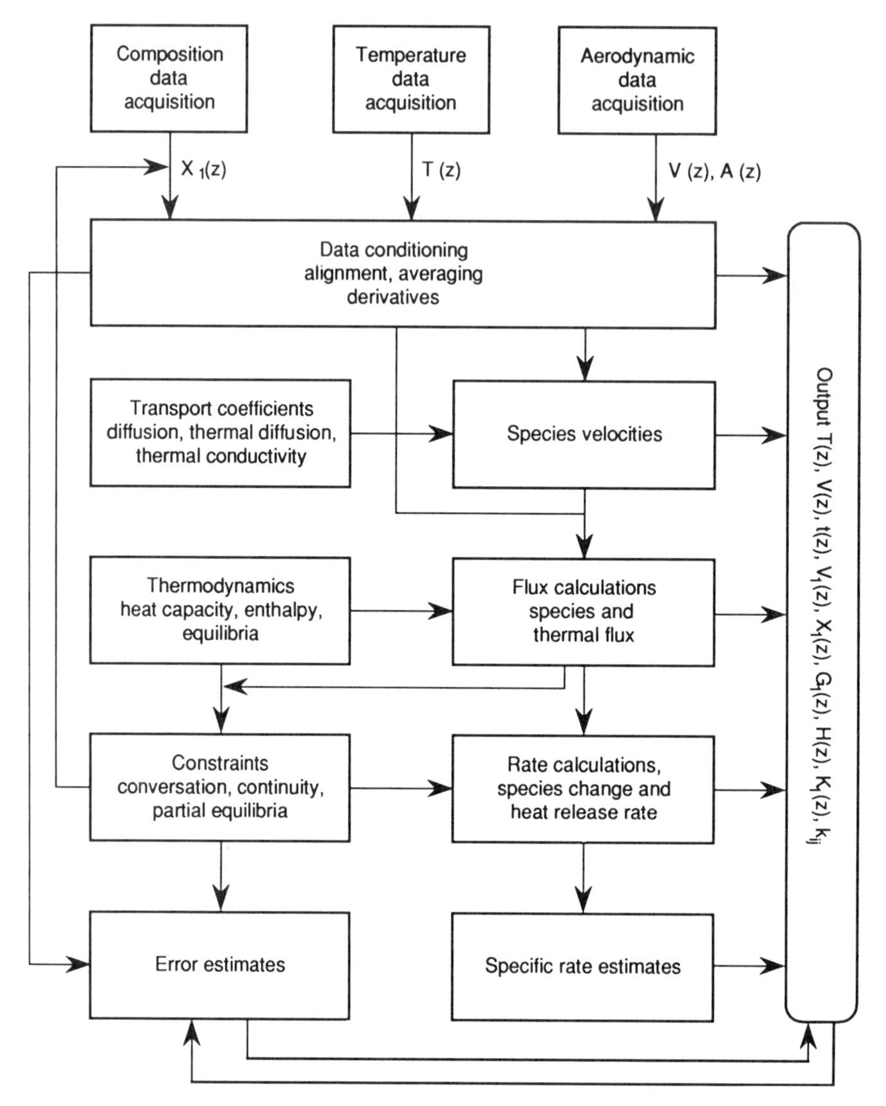

FIG. IX-3 Block diagram of scheme for flame structure data analysis.

becomes a pure number characteristic of the molecular pairs (Fristrom and Monchick [1989].

Fractional Mass Flux Definition

The species fraction mass G_i is defined as

$$G_i = \frac{N_i M_i (v + V_i)}{\rho v} \tag{9–11}$$

Flux includes contributions from diffusion as well as convection. It should be observed that the contribution of diffusion V_i can be either positive or negative. If the concentration of species i decreases downstream (i.e., the direction of flow), V_i will be positive, and to a stationary observer the species will contribute a greater fraction of the total mass flux than would be predicted on the basis of simple convection. If the concentration increases in the flow direction, the reverse is true.[1]

There can be no net transport of mass by diffusion, since by definition the sum of the diffusion velocities must equal zero.

$$\sum_i N_i M_i V_i = 0 \tag{9–12}$$

Combining this result with equation eleven yields

$$\sum_i G_i = 1 \tag{9–13}$$

since as required of all fractional variables:

$$\sum_i N_i M_i = \rho \tag{9–14}$$

The necessity for considering flux variables in flames rather than concentrations stems from the steep concentration and temperature gradients. This complication is characteristic of flames. In simple flow systems where molecular transport is negligible, the flux of a species at any point is simply the product of its concentration and the velocity at that point. In flames, by contrast, the flux depends not only on the concentration but also the concentration and temperature *gradients*. This makes the analysis of flame data more complex.

A Compact Formulation Using G_i

The flux variable G_i includes the contributions of diffusion velocities. This allows the one-dimensional equations to be presented in the compact form to be given in the following.

Mass Conservation

In the one-dimensional construct mass conservation takes on the familiar form of the following equation for overall flow generalized to include individual elemental con-

1. The difference between these two measures may be visualized by considering the differences that occur when one compares the estimates of the ratio of trucks to cars that a driver would make with those of an observer on the side of the road. If one is driving in a car at the (mass) average speed and counts the number of cars and trucks within sight, one obtains a local concentration. A quite different ratio is obtained by an observer on the side of the road who measures a flux. If the truckers drive faster than the average speed, a higher fraction will pass the sideline observer than that seen by the moving observer. This corresponds to a positive "diffusion velocity." The reverse will be true if the truckers drive slower than the average. The "diffusion velocity" will be negative, and the fraction will be smaller.

servation for these multicomponent systems. This extension comes about because nuclear reactions play no role in combustion.

$$\rho v A = \dot{m} = \sum_{e=1}^{n} \dot{m}_e = \rho_0 v_0 = const \tag{9–15}$$

In this equation $\dot{m}$ is the mass flow per stream tube, which is the eigensolution to the flame equations, while $\dot{m}_e$ are the individual elemental fluxes.

Species Continuity

In the one-dimensional form species continuity takes the form

$$\frac{\rho_0 v_0}{M_i} \frac{dG_i}{dz} = K_i A \tag{9–16}$$

where N_i is the number of moles of species i per unit volume (moles cm^{-3}), V_i is the species diffusion velocity (cm sec^{-1}), and K_i is the net reaction rate of species i (moles sec^{-1} cm^{-3}).

Note that although experimentally v and V_i are both still vectors, they have become pseudoscalars in the analysis since only the components along the direction of propagation (z) are involved. v is inherently positive, but V_i can be either positive or negative, since either upstream and downstream diffusion may occur. As will be seen later, the effects of diffusion are balanced so that there is no net transfer of mass.

Energy Conservation

After integration with respect to z, energy conservation in one dimension becomes:

$$\rho_0 v_0 H_{sp} + A \sum_{i=1}^{n} N_i H_i V_i - A\lambda \frac{dT}{dz} = const \tag{9–17}$$

For practical purposes it may be assumed that the transport terms (second and third terms on the left) vanish at the cold boundary and the integration constant may be set equal to $\rho_0 v_0 H_{sp0}$. This condition is not always fulfilled in practical burners, since a heat sink flame holder is often used. If this is done, the heat loss parameter becomes one of the inlet parameters and should be reported. A number of earlier studies have failed to do this, making them difficult or impossible to model.

In the ideal case the transport terms also vanish at the hot boundary, denoted by the subscript f, and the integration constant may be set equal to $\rho_0 v_0 H_{spf}$. In practical systems there may be losses due to transport to the walls and radiation. These are usually ignored, but they should be measured. If transport vanishes at both boundaries, then $H_0 = H_f$.

The Mechanics of Data Handling

The experimental results of a flame structure study are a family of profiles. Earlier chapters have outlined techniques for making such measurements. The usual variables are: temperature, gas velocity, gas density, and composition. As was indicated earlier, these are not all independent, and some may be derived rather than measured. Redundant profiles can be used to check the reliability of the measurements.

In addition to the experimental profiles, it is necessary to have reliable transport coefficients to allow the intensive measurements of temperature and composition to be interpreted in terms of fluxes and rates.

Obtaining Diffusion Velocities

In many cases thermal diffusion is unimportant and the total diffusion velocity is due to molecular diffusion. In such cases the superscript will be dropped and the total molecular diffusion velocity will be given as V_i. This will be obvious from context and should not cause confusion.

The first step in applying the equations is the computation of the diffusion velocities V_i for all species for which concentration profiles are available. The rigorous expression involves the multicomponent diffusion coefficients D^*_{ij}. As may be seen in examples in Chapter XI, the rigorous expression is rarely used in practice. There are two levels of approximation in the simplest case in which one component is in such excess that it can be considered the carrier and all other species are traces. This is often adequate for the description of air flames or when argon or oxygen are in excess. In this case the multicomponent diffusion coefficient is replaced by the binary diffusion coefficients of trace species with the carrier designated by the subscript c.

$$V_i^M = -D_{ic}\left(\frac{N}{N_i}\right)\frac{d}{dz}\left(\frac{N_i}{N}\right) = -\frac{D_{ic}}{X_i}\frac{dX_i}{dz} \tag{9–18}$$

where V^M is the molecular diffusion contribution to the diffusion velocity.

In the more general case where there is no dominant carrier, the next level of approximation for multicomponent diffusion coefficients is the Wilkes equation discussed by Reid, Prausnitz, and Poling [1987].

The diffusion velocity then becomes

$$V_i^M = \frac{-D_{i,mix}}{X_i}\frac{dX_i}{dz} \quad \text{where} \quad D_{i,mix} = \frac{1-f_i}{X_i \sum_{i \neq j} X_i/D_{ij}}\frac{dX_i}{dz} \tag{9–19}$$

Where full multicomponent diffusion coefficients are required, the reader is referred to Hirschfelder, Curtiss, and Bird [1964] for definitions and discussions. Programs have been written for the full calculations by Svehela and McBride [1973] and Kee, Miller, and Jefferson [1983]. These programs are available through NASA and Sandia National Laboratory (Livermore), respectively.

Evaluation of diffusion velocities require: compositions, their spatial derivatives, and an assignment of temperature. The appropriate transport coefficient, is calculated at that temperature. This coefficient, taken with the profile data, allow calculation of the local molecular diffusion velocity. It should be observed that the gradients are the most difficult factor because it is necessary to smooth most data to obtain reliable spatial derivatives. This can be done either graphically or numerically using computer fitting procedures. The practical problems have been discussed by several authors (Fristrom and Westenberg [1960, 1965]; Papp, Lazzara, and Biordi [1974, 1975]; Pauwels, Carlier, and Sochet [1990]). The bottom line is that precise data are required to yield meaningful second derivatives.

Thermal Diffusion

The effects of thermal diffusion are significant for light species. A feel for the relative importance (compared with molecular diffusion) can be obtained by assuming that temperature and concentration gradients are of the same order of magnitude. In this case the values of thermal diffusion factors provide a direct comparison. A table of these factors for some common molecular pairs are given in Table A–1 of the appendix at the end of the book. In many of the studies in the literature the factor has been ignored because of its complexity. This is no longer necessary since the first approximation (high-temperature trace limit) for the binary coefficients is a pure number, and Fristrom and Monchick [1989] have shown that this approximation is reasonable. For a more rigorous treatment the reader is again referred to Hirschfelder, Curtiss, and Bird [1964], the discussion of Dixon-Lewis [1968], and the documentation of the two transport programs (Svehala and McBride [1973]; Kee, Dixon-Lewis, Warnatz, Coltrin, and Miller (1986)].

Binary coefficients only apply directly to the carrier trace approximation. However, by analogy an inverse averaging of the product $\alpha_{ij}D_{ij}$, similar to the Wilkes equation used for diffusion, should give a similar approximation.

$$\alpha_{i,mix}D_{i,mix} \approx \frac{1 - f_i}{\displaystyle\sum_{j=1,\neq i}^{n} \frac{X_i}{\alpha_{ij}D_{ij}}} \tag{9–20}$$

Thus the thermal diffusion velocity v_i^T is given by:

$$V_i^T \approx \frac{\alpha_{i,c}D_{i,c}}{T}\frac{dT}{dz} \approx \frac{\alpha_{i,mix}D_{i,mix}}{T}\frac{dT}{dz} \tag{9–21}$$

To evaluate these velocities requires concentration profiles and temperature profiles including the local gradient. At a given position the product of the thermal diffusion factor and the diffusion coefficient is evaluated at the appropriate temperature. This is then multiplied by the local logarithmic temperature gradient. Again the main source of error is the derivatives, and as with molecular diffusion either graphical or numerical smoothing methods may be applied.

Total Diffusion Velocities

The total diffusion velocity is the sum of molecular diffusion and thermal diffusion. Since both factors can be important, it is best to calculate them simultaneously along with the local mass velocity. These three factors together with the local concentration allow direct calculation of the flux through Eq. (9–17). If the binary trace approximation is used, the carrier flux G_c may be taken to be its mass flux fraction $N_c M_c/\rho$. It is interesting to observe that often an even closer approximation is obtained by treating the carrier as a trace species using its self-diffusion coefficient. Computed G_is should be checked against the requirement that $\Sigma_i G_i = 1$.

Derivation of Net Reaction Rates

Perhaps the most important and useful information that can be obtained from flame structure analysis is the net reaction rates for the individual species in the flame front. These are the quantities K_i, the local net molar rate of appearance or disappearance of a species per unit volume. This is given directly by:

$$K_i = \frac{\rho_0 v_0}{M_i A} \frac{dG_i}{dz} \tag{9–22}$$

What is required are the spatial derivatives of the fluxes. These are obtained by either graphical or numerical integration of the flux profiles derived using the methods described in previous sections. This procedure yields more consistent results than if the raw data of concentration and temperature are used directly, since this approach would require both first and second derivatives of concentration. It has been the experience of many workers (Dixon-Lewis and collaborators [1967–1976]; Fristrom and Westenberg [1960, 1965]; Peeters and Mahnen [1973a, b]; Papp, Lazzara, and Biordi [1975]; and Van Tiggelen and Vandooren, [1980]) that taking derivatives presents the most difficulties. A secondary problem is the alignment of temperature, velocity, and composition data coming from different experimental techniques. These problems are examined in Chapter X. Experimental errors magnify as each successive derivative is taken. As a result, there can be large local inaccuracies, such as false extrema. Averaged quantities and global trends are less affected. The bottom line is that data of the highest precision are required to obtain reliable rate data, and even the best data require some smoothing. It is convenient to check the computed rates using the following equation, (9–23) which is an expression of conservation of elemental species

$$\sum_{i=1}^{n} \nu_{ei} K_i = 0 \tag{9–23}$$

This requires that for any given elemental species the weighted sum of the production rates must be zero.

Heat Release Rates

The determination of heat release rates in flames is of general interest. There are two independent ways of doing this from a given piece of data. Perhaps the simplest and most direct approach is to use the calculated net reaction rates K_i and the absolute molar enthalpies H_i in the relation

$$Q = \sum_{i=1}^{n} H_i K_i \tag{9–24}$$

where Q is the volumetric heat release (a negative number in our chemical convention). The alternative method is to combine the energy and species continuity equations.

In this case energy conservation may be written as:

$$\frac{d}{dz}\left(\rho_0 v_0 \sum_{i=1}^{n} \frac{H_i G_i}{M_i} - A\lambda \frac{dT}{dz}\right) = 0 \tag{9–25}$$

or

$$\rho_0 v_0 \left(\frac{H_i}{M_i}\frac{dG_i}{dz} + \sum_{i=1}^{n} \frac{G_i}{M_i}\frac{dH_i}{dz}\right) - \frac{d}{dz}\left(A\lambda \frac{dT}{dz}\right) = 0 \tag{9–26}$$

Since H_i is a function only of temperature, and by definition of the molar heat capacity at constant pressure,

$$\frac{dH_i}{dz} = \frac{dH_i}{dT}\frac{dT}{dz} = C_i \frac{dT}{dz} \tag{9–27}$$

Using this with Eq. (9–26) in the previous relation gives

$$A \sum_{i=1}^{n} H_i K_i + \rho_0 v_0 \frac{dT}{dz} \sum_{i=1}^{n} \frac{G_i C_i}{M_i} - \frac{d}{dz}\left(A\lambda \frac{dT}{dz}\right) = 0 \tag{9–28}$$

or in view of Eq. (9–24)

$$Q = -\frac{1}{A}\left[\rho_0 v_0 \frac{dT}{dz} \sum_{i=1}^{n} \frac{G_i C_i}{M_i} - \frac{d}{dz}\left(A\lambda \frac{dT}{dz}\right)\right] \tag{9–29}$$

This equation states that the rate of chemical heat release at any point in the flame front must be equal to the gradient of the energy flux due to mass transfer (both convection and diffusion) and conduction.

The first method [Eq. (9–24)] of computing Q involves only the net chemical rates and enthalpies; so it is a direct measure of the rate of enthalpy change per unit volume due to reaction.

In the second method [Eq. (9–29)] of computing, Q is regarded as a source term in the enthalpy flux conservation equation and involves directly the G_i and the second derivative of the temperature profile. Ideally, of course, both methods should give the same result, and the degree to which they do provides a useful check. An example is discussed in Chapter X.

Another calculation of importance is the integrated heat release, that is, the total heat released in a stream tube of unit initial area up to any point z in the flame. This is given by the integral

$$\int_0^z QA\, dz \tag{9–30}$$

which may be obtained by numerical or graphical integration of the Q profile.

As Friedman and Burke [1954] have pointed out, the value of the integrated release at the hot boundary should be given by

$$\int_0^\infty QA\, dz = \rho_0 v_0 (\Delta H_{spc})_{T_0} \tag{9–31}$$

$$\int_0^\infty QA\, dz = -\rho_0 v_0 \int_0^\infty \sum_i \frac{G_i C_i}{M_i}\, dT + \int_0^\infty d\left(A\lambda \frac{dT}{dz}\right) \tag{9–32}$$

where $(\Delta H_{\text{spc}})_{T_0}$ is the heat of combustion per unit mass of the unburned mixture. This must be so because the second integral vanishes because $dT/dz = 0$ at $z = 0$ and $z = \infty$ and the first integral is simply the heat of combustion at the cold gas temperature. As a result, Eq. (9–31) is only useful as a check on the internal consistency of the computed heat release profiles. It proves nothing about the accuracy of the experimental data. Similarly for reactants (a) and for products (b).

$$\text{(a)}\quad -\int_0^\infty K_i A\, dz = \frac{\rho_0 v_0 (G_i)_0}{M_i} \qquad \text{(b)}\quad \int_0^\infty K_i A\, dz = \frac{\rho_0 v_0 (G_i)_\infty}{M_i} \tag{9–33}$$

Tests for Data Reliability

A convenient check on the consistency of the data is elemental conservation, which requires that at every point in the flame the net rate for each elemental species must be zero. This check is obtained by calculating the product of each species net reaction rate, K_i, by $\nu_{e,i}$, the number of atoms of the element e the species i possesses. For example, for the species methane $\upsilon_{\text{H,CH}_4} = 4$ for the element hydrogen, while for carbon $\upsilon_{\text{C,CH}_4} = 1$. This provides an excellent check of the local consistency.

Element Conservation

In complex experiments such as laminar flame structure studies it is desirable to be able to check the data to assess the general reliability and internal consistency of the operation. Since no nuclear reactions are involved, it is clear that the elemental species flux should be conserved individually. Once the molecular fluxes G_i have been calculated, the data can be examined straightforwardly for element conservation. Since the sums of the elemental rates are zero, the spatial derivatives are also zero.

$$\frac{d}{dz} \sum_{i=1}^{n} \frac{\nu_i G_i}{M_i} = 0 \tag{9–34}$$

or

$$\sum_{i=1}^{n} \frac{\nu_i G_i}{M_i} = const \tag{9–35}$$

The degree to which the sum is constant for each element is a useful indicator of the reliability of the flame structure data and the diffusion contributions. It is important because all of the experimental data enter this conservation computation. Composition and temperature profiles enter the computation of fluxes both directly and by their derivatives and the temperature-dependent diffusion coefficients. This makes the computation particularly sensitive to the alignment of temperature and composition profiles as well as the reliability of the coefficients themselves. A convenient form of presenting the material is to plot the fractional deviation of the elemental fluxes as a function of position

$$\Delta_G = \frac{\sum_i \nu_i G_i/M_i - \sum_i \nu_i (G_i)_0/M_i}{\sum_i \nu_i (G_i)_0/M_i} \qquad (9\text{–}36)$$

Ideally this expression should be zero at every point. The conformance of real data is examined in Chapter X.

Energy Conservation

The energy relations expressing conservation have already been set forth. Experimental data can be examined in the light of these equations. The molar enthalpies H_i are readily obtainable from standard sources (Gardiner [1984]; Reid, Prausnitz, and Poling [1987]; Stull and Prophet [1986]; Kee, Rupley, and Miller [1987]). The specific enthalpy at any point is given by

$$H_{sp} = \sum_{i=1}^{n} \frac{N_i H_i}{\rho} \qquad (9\text{–}37)$$

If there is a dominant carrier molecule, thermal conductivities may be approximated simply using its coefficient. Otherwise one of the higher-level approximations for mixtures discussed in Appendix B may be used.

The thermal conduction term in the enthalpy balance is always negative since the temperature gradient in flames (dT/dz) is always positive. Similarly, since the system is exothermic, the diffusion contribution is always negative because high-enthalpy reactants diffuse downstream, while low-enthalpy products diffuse upstream. These terms are of the same order of magnitude, and in an ideal flame, if the molecular diffusivity of all species were equal to the average thermal diffusivity, there would be complete cancellation. It should be observed that complete cancellation requires consideration of the thermal diffusion contribution. The controlling parameter is the ratio of molecular to thermal diffusivity and is called the Lewis number ($\rho C_{av} D_{ij}/\lambda$), where C_{av} is the mean specific heat (at constant pressure) of the mixture. In the idealized flame of theory it is asumed that this number is unity for all species. For real gases it varies between 0.7 and 1.3, so the cancellation in experimental flames is only an approximation.

Comments on Data Handling

In these days of powerful microcomputers, manual data reduction is neither necessary nor desirable except for spot checking. Smoothing operations may be done graphically, but modern smoothing routines that utilize inverse Fourier transform filtering introduce less bias. Several such programs are available. The choice is clearly one of using the facilities at hand. If the experiment utilizes computerized data gathering, then clearly a data program is almost a necessity. In the not too distant future it will probably be possible to have real-time outputs of the analyzed flame profile data, but this would require a large computer capacity. Requirements depend on the experimental techniques used:

1. Whether temperatures are derived from thermocouples, pneumatic probes, spectroscopic measurements, or is inferred;
2. Whether velocity is measured directly or derived;
3. Whether all compositions are available from a single technique such as mass spectrometry or whether it must be assembled from several sources such as laser resonance fluorescence, laser Raman, etc.

Figure IX-3 offers a block diagram of required packages and a suggested ordering. A number of these can be assembled from standard sources such as Chemkin [Kee, Miller, and Jefferson (1980)]. Papp, Lazzara, and Biordi [1975] give a clear, detailed documentation of the program they used to interpret flame data. This is available either directly or through the government document (National Technical Information Service) NTIS service (this is separate government agency not to be confused with NIST). It should be remembered that this program is specific to mass spectral composition and thermocouple temperature sources and should be generalized to multicomponent diffusion and thermal diffusion using the simplifications outlined in this chapter.

Although in principle complete automation is possible, the author feels that some elements should be monitored to avoid clear inconsistencies. This is particularly true in taking derivatives and aligning profiles from different techniques. Figure IX-4 shows an example of an analysis of profile data from a low-pressure methane–oxygen flame showing the relations of experimental concentration and temperature to the derived fluxes due to ordinary and thermal diffusion and the rates derived from the total flux curve. The effects of various errors on derived rates are illustrated in Figure IX-5. They include errors in profile alignment, diffusion coefficient choice, temperature, and composition.

The Stephen–Maxwell Approach

In principle it is possible to avoid use of the cumbersome multicomponent diffusion coefficients in the analysis of flame data by using the Stephen–Maxwell relation. This is the inverse of the generalized Fick equation for diffusion. This formulation is used in Dixon-Lewis's [1974] flame theory program. Here the concentration gradient is expressed in terms of the ratios of fluxes to concentration and a group of matrices that

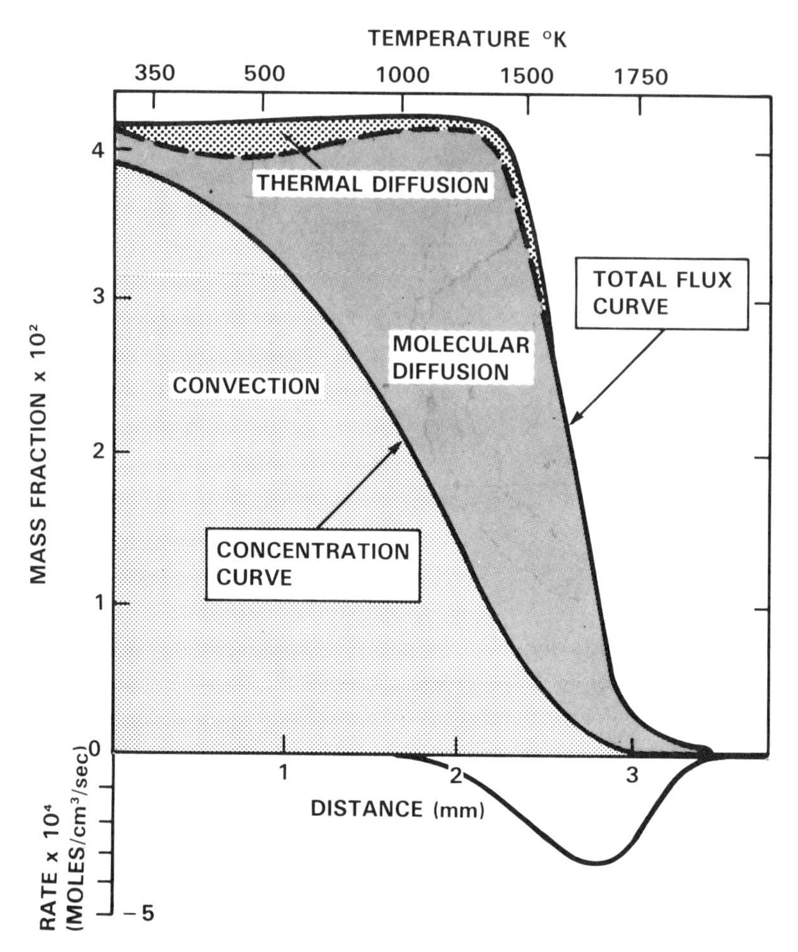

FIG. IX-4 Example of a profile showing relation of composition, flux and rates. CH_4 from a 0.1 atm. $0.078CH_4$ - O_2 flame. (Replotted data of Westenberg and Fristrom [1960]).

are related to and as complex as the multicomponent diffusion formulation. The implementation of this expression for species flux requires a knowledge of concentration gradient, the concentration, and in addition the ratios of flux to concentration for all other species, the temperature gradient, and the relevant binary diffusion coefficients and thermal diffusion factors. Such a solution requires simultaneous solution for all fluxes using matrix algebra or an iterative solution. Machine computation is required for practical implementation. It is not clear that this would provide an advantage over the conventional method of data analysis.

Testing Kinetic Mechanisms

Having determined the net reaction rates K_i, the crux of the problem of inferring chemical kinetic information from flame structure measurements lies in examining the

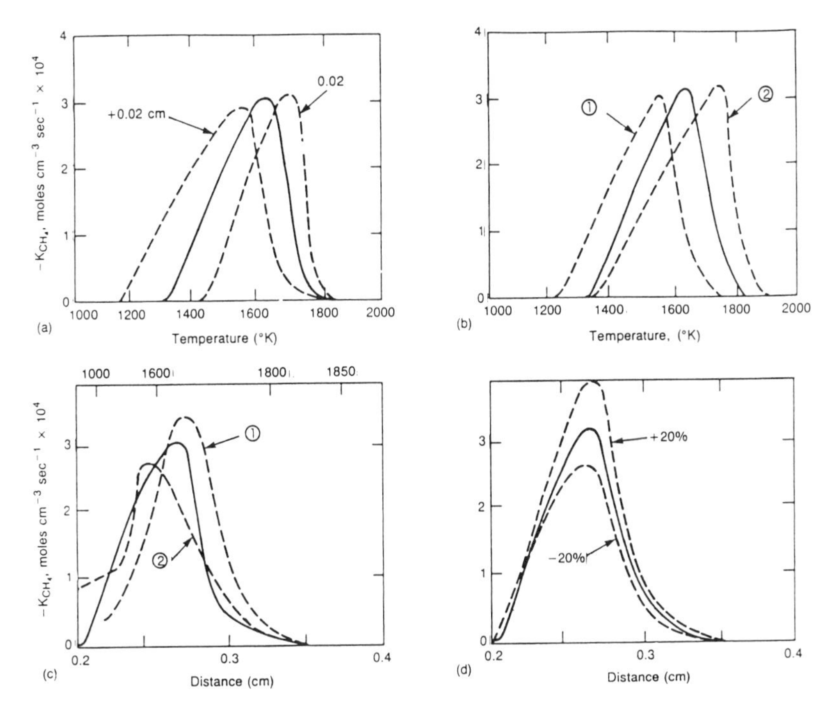

FIG. IX-5 Effects of errors in various experimental flame data. Calculated rate of disappearance of methane in the exemplar flame (see Fig. IX-4). (a) Effect of alignment ±0.1 mm of temperature with composition profiles; (b) effect of temperature distortion. Temperature profile (1) 5% too high at the low end and 5% too high on the high end and (2) conversely; (c) effect of concentration profile distortion in the same manner as temperature in (b); and (d) effects of errors in methane diffusion coefficient.

experimental rates in terms of reaction mechanisms or elementary steps. Since, as will be seen, most flame reactions involve free radicals or atoms, the data analysis is straightforward if the results include radical profiles. Often these data are missing. Then the problem becomes more involved, but it is often possible to derive relative rates by comparing the rate of another stable species that reacts with the same radical. In any case it is necessary to assume a functional form for rate processes, otherwise one is left with the impossible situation of not having enough data points for the fit (Fristrom and Westenberg [1965, p. 84]). The conventional method of resolving this problem is to assume the Arrhenius form for bimolecular reactions.

$$k(T) = AT^B \exp(-E/RT) \tag{9–38}$$

A similar functional form applies to termolecular recombinations, in which often the exponential factor is dropped and a numerical factor C_3 is added to account for the differing three-body efficiency factors of molecules. These questions are discussed in most texts on physical chemistry or chemical kinetics.

This reduces the problem to the determination of three coefficients for each reaction. It should be recognized that this is an approximation. However, this convention is generally accepted for chemical kinetic studies and most rate data are presented and tabulated in these terms. They will be used throughout this book.

Overall Flame Analysis

An alternative to individual species analysis, which is popular in laboratories with access to large computational facilities, is to assume a mechanism with standard coefficients and calculate the flame structure. The calculated profiles are then compared with the experimental ones. Coefficients are then adjusted for best fit. Individual reactions can be examined using sensitivity analysis. This complex computational approach is beyond the scope of this book. Interested readers are referred to discussions such as those in the books edited by Gardiner [1984] and Peters and Warnatz [1982] and reviews such as that of Oran and Boris [1981]. A neophyte would be wise to consult individuals at one of the many laboratories with experience in the area. Many of them are willing to share programs and experience with newcomers.

The author feels that this approach should be used with discrimination. Literature values should be used for parameters wherever possible since their number is so large. Care must be exercised so that modeling does not degenerate into a curve fitting exercise. A comment by the late G. B. Kistiakowski is perhaps appropriate: "Give me five adjustible parameters and I'll write my name."

In comparing full-scale modeling with experimental data, it is important to consider the system as a whole. Ideal experiments do not disturb a flame, but real experimental methods affect flames to varying degrees. Laser diagnostics offer the least disturbance while molecular beam inlet probing offers the most, with quartz microprobing falling in between. Each technique offers special problems and special advantages. The optimal approach depends on the requirements of the problem.

Appendix A Data Sources for Flame Structure Analysis

Quantitative analysis of flame structure data requires physical, thermodynamic, transport, and rate data. As a general rule the species involved in combustion are simple and well understood and with few exceptions have well-known physical and thermodynamic properties. The most compact source of such information is the authoritative text of Reid, Prausnetz, and Poling [1987].

The most reliable general source of data is the National Institutes of Standards and Technology (NIST) Reference Standard Data Program. NIST maintains a clearinghouse which critically evaluates and publishes data on physical properties. NIST Reviews include: transport (Boushehri, Bzowski, Kistin, and Mason [1987]; Marreno and Mason [1972]; and Fristrom and Westenberg [1968]), thermodynamics (Stull and Prophet [1986]), and reaction kinetics (Westley, Herron, Cvetanovic, Hampson, and Mallard [1990]). In addition to monographs, they publish a journal of *Physical and Chemical Data,* and make their information available in computer-compatible form both for personal computers and mainframes. Some researchers, such as Warnatz [1977], Dixon-Lewis [1977] and Westbrook and Chase [1982] have published their flame databases. Gardiner's book [1984] contains a number of useful compilations. For comparison of flame kinetic data and for modeling the NRDC chemical kinetic database Westley et al. [1990] is invaluable.

Appendix B Transport Coefficients

The method for evaluating transport coefficients using kinetic theory is given in detail by Hirschfelder, Curtiss, and Bird [1964]. Two programs are available for the estimation of transport coefficients of complex high temperature mixtures. The first (Svehela and McBride [1973]) is available through NASA, and the second (Kee et al. [1980]) through the Combustion Laboratory of the Sandia National Laboratory.

In many cases, however, the full treatment is an overkill, and approximate methods are quantitatively adequate. These are based on kinetic theory supplemented by empiricism. A critical discussion of the reliability of such methods is given by Reid et al. [1977]. The situation has been discussed by Westenberg and Fristrom [1965]. Some approximate techniques applicable to flame studies are outlined in the following sections.

Intermolecular Potentials

Transport coefficient estimation requires an intermolecular potential. The true potentials are complex and nonanalytic, but because the thermal and pressure regimes in combustion are moderate a simple potential such as the two-parameter Lennard–Jones and Devonshire 6–12 (LJ) is adequate.

$$\phi(r) = 4\epsilon\left[\left(\frac{\sigma}{r}\right)^{12} - \left(\frac{\sigma}{r}\right)^{6}\right] \tag{9–39}$$

in which σ is a characteristic diameter of the molecule (the "collision diameter") and ϵ is a characteristic energy of interaction between the molecules. Between different molecules, the effective values for bimolecular collisions are:

$$\begin{aligned} \sigma_{12} &= (\sigma_1 + \sigma_2)/2 \\ \epsilon_{12} &= \sqrt{\epsilon_1 \epsilon_2} \end{aligned} \tag{9–40}$$

Where values are not directly available approximate coefficients may be derived from critical parameters or, failing this, from boiling points and densities. Such data can usually be obtained from the compilation of Reid et al. [1977]. To relate this to transport coefficients requires both the parameters and a collision integral Ω, which is a function of the temperature and the ϵ/k parameter through a dimensionless function called the "reduced temperature" T^*, which is defined as:

$$T^* = \frac{T}{\epsilon/k} \tag{9–41}$$

In this equation T is the temperature (K), ϵ/k is the Lennard–Jones well depth parameter (Table A–1 in the appendix at the end of the book).

The "collision integrals" for the LJ potential can be approximated by the Nuffield–Jansen equation (Reid, Prausnitz, and Poling [1987]) to about 1%.

$$\Omega_D = \frac{A}{T^{*B}} + \frac{C}{\exp(DT^*)} + \frac{E}{\exp(FT^*)} + \frac{G}{\exp(HT^*)} \tag{9–42}$$

where: T^* is kT/ϵ_{ab}, $A = 1.06036$, $B = 0.15610$, $C = 0.19300$, $D = 0.47635$, $E = 1.03587$, $F = 1.52996$, $G = 1.76474$, and $H = 3.89411$.

For the high-temperature regime of interest in flames one can often use an approximation due to Westenberg (eqns. 46,47) good to about 2%, providing T^* exceeds 3. This is usually the case in flames.

$$\Omega^{11} = \frac{1.12}{(T^*)^{0.17}} \tag{9–43}$$

$$\Omega^{22} = \frac{1.23}{(T^*)^{0.17}} \tag{9–44}$$

where Ω^{11} applies to diffusion and Ω^{22} applies to viscosity and thermal conduction.

Diffusion Coefficients

The basic information resides in the binary pair diffusion coefficients. Fortunately they are almost composition independent so that $D_{ij} \approx D_{ji}$. They vary accurately with inverse pressure over the same range of validity as the ideal gas law. Table A–1 at the end of the book presents a set of experimental binary coefficients for species of flame interest over a range of temperatures. These were determined for a flame where oxygen was the dominant carrier (Walker and Westenberg [1958, 1960]). However, since oxygen and nitrogen are very close in diffusivity, they also represent approximate values for air flames. Reasonably reliable binary coefficients can be calculated using Lennard–Jones parameters (Table A–1) using the following equation.

$$D_{ij} = 1.86 \times 10^{-3} \frac{\sqrt{\frac{(M_i + M_j)}{M_i M_j} T^3}}{P\sigma_{ij}^2 \Omega_{ij}^D} \tag{9–45}$$

In this equation Ω^D is the collision integral appropriate for diffusion.

The diffusion velocity of a species depends on gradients and a set of multicomponent diffusion coefficients and thermal diffusion ratios. Each of these, in turn, depends in a complex way on the mixture composition and the binary diffusion coefficients of all the pairs of species in the mixture (Hirschfelder et al. [1964, p. 487]).

Even the simplest flame, ozone decomposition, involves three species, O, O_2, and O_3. Other flame systems are more complex, so that rigorously flames must be treated as multicomponent mixtures. For diffusion, this means that multicomponent diffusion coefficients or the Stephen–Maxwell relation must be used. This approach is too involved for inclusion here. A discussion is given by Dixon-Lewis [1974].

Complete rigor requires machine computations using programs such as those of Svehela and McBride [1973] or Kee, Warnatz, and Miller [1983]. Fortunately, many common systems such as air flames have one dominant component as a diluent or excess reactant. *In such cases, each species may be regarded as a trace in a binary mixture with the excess component.* If no single species is in excess, a reasonable approximation may be obtained by using the Wilkes equation discussed in the section on diffusion velocities. These binary approximation methods should be distinguished from engineering usage, where species are often divided into two classes, light and heavy, and each class is assigned an average diffusion coefficient.

Measurements have been made over a wide enough temperature range and compared with viscosity-derived information to feel reasonably confident that these standard LJ potentials reproduce the diffusion coefficients and other coefficients with sufficient precision for most flame studies. Thus it would appear that transport coefficients for most species can be calculated with a reliability of perhaps 20–30% over the temperature range of most flames.

Thermal Conductivity

Heat transport is more complex than mass transport because the latter only involves the translational motion of the molecules, while the former also involves internal degrees of freedom. Table A–1 at the end of the book presents experimental thermal conductivities for some common flame species over a range of temperatures. For a pure gas the one-dimensional flux of energy due to a temperature gradient is given simply by:

$$q = \lambda \frac{dT}{dz}$$

where λ is the thermal conductivity.

In a pure gas an approximate thermal conductivity can be estimated using kinetic theory and the modified Euchen correction due to Hirschfelder, Curtiss, and Bird [1964].

$$\lambda^0 = \frac{0.994 \times 10^{-4}\sqrt{T/M}}{\sigma^2 \Omega^T} \tag{9-46}$$

In this equation λ^0 is the translational thermal conductivity, where the molecule is considered as a monatomic gas. This can be related to the true thermal conductivity by adding the contribution due to internal degrees of freedom using the modified Euchen correction due to Hirschfelder.

$$\lambda = \lambda^0(0.115 + 0.178C_p) \tag{9-47}$$

In mixtures with gradients in composition there will be other terms due to diffusion. Expressions for the thermal conductivity of binary and multicomponent mixtures of monatomic gases are given (Hirschfelder et al. [1964, pp. 534–38]). For extensive calculations there are a NASA program (Svehala and McBride [1973]) and a more recent CHEMKIN related program (Kee, Dixon-Lewis, Warnatz, Coltrin, and Miller [1986]). They can be used in computations on any mixture of species available in the JANAF species list (Stull and Prophet [1986]); however, the reader should be warned that many of the parameters only represent educated guesses, since basic thermal conductivity data are not available for many species.

Assuming full equilibrium of internal and translational degrees of freedom, Hirschfelder proposed the following equation for mixtures.

$$\lambda_{mix} = \lambda^0_{mix} + \frac{\sum_{i=1}^{n} (\lambda_i - \lambda_i^0)(X_i/D_{ij})}{\sum_{j=1}^{n} (X_j/D_{ij})} \tag{9-48}$$

In this equation the superscript 0 indicates thermal conductivity treated as if the molecules had no internal degrees of freedom. The subscript m refers to the mixture.

Where low-molecular-weight species are involved, Brokaw [1964] has recommended averaging the sum of the linear mole fraction average with the inverse mole fraction average of the inverse. These are empirical procedures useful for estimates.

Mention should be made of the so-called "effective thermal conductivities" in a mixture of chemically reacting gases, a term that has enjoyed considerable popularity. Flames certainly fall into this category. Whenever chemical reactions occur in a flow system there are concentration gradients so that energy is transported by diffusion as well as conduction and convection. In this book the heat transfer in flames is considered from the fundamental point of view, that is, regarding the different mechanisms separately, so that at any point (temperature) in a chemically reacting system such as a flame the mixture has a certain thermal conductivity that is a true molecular property of the mixture. This is in contrast with the "effective" thermal conductivity concept, which defines a local parameter that, when multiplied by the local temperature gradient, gives the local heat transport. It should be understood that this concept is only valid if the reactions are rapid compared with diffusion so that the mixture can be regarded as being in chemical equilibrium with the local temperature. This is definitely not the case in flames. This is analogous to the Eucken correction of polyatomic gases (Hirschfelder, Curtiss, and Bird [1964]). The diffusive contributions to heat flux are incorporated in the "effective" thermal conductivity. This is not generally useful for flames, since local equilibrium is violated. The thermal conductivity of a reacting mixture is not fundamentally different.

Thermal Diffusion

There are two common methods of presenting thermal diffusion: the thermal diffusion ratio $(k^T)_{ij}$ and the thermal diffusion factor α_{ij}. The thermal diffusion factor is the most convenient form because it is independent of pressure, slightly dependent on composition, and for high temperatures close to a two-species dimensionless parameter. By contrast, the thermal diffusion ratio is a linear sum of thermal diffusion factors that shows a quadratic dependence on the composition and varies inversely with the pressure. The thermal diffusion ratio can be related to the thermal diffusion factor α_{ij} by

$$k^{T_i} = X_i \sum_{j=1, \neq i}^{n} X_j \alpha_{ij} \tag{9–49}$$

As was noted previously, α_{ij} is only weakly concentration dependent, and at temperatures above $T^* = 4$ it is also temperature independent. This makes it particularly useful in flame studies because in the binary trace approximation the thermal diffusion factor α_{ij} becomes a single-valued function of the molecular pair parameters independent of composition, temperature, and pressure.

Table A–1 in the appendix at the end of the book offers some thermal diffusion factors for molecular pairs of interest in combustion. These are useful for the estimation of thermal diffusion effects, and for many systems they are quantitatively adequate. Conjugate factors α_{ij} and α_{ji} are of opposite sign and almost equal in magnitude. These differences reflect a residual composition dependence that is regained when the binary trace assumption is removed. Fristrom and Monchick [1989] showed that the next level of approximation for α_{ij} can be calculated using binary diffusion coefficients, their temperature dependence, and the viscosities.

Appendix C A Compact Method for Presenting Flame Structure Data

The full presentation of flame structure information requires either extensive tabulation or graphing. For many purposes the more compact display of Table IX-3 is convenient and adequate. This approach makes use of the observation that for most flames there is an approximately linear relation between temperature rise and composition change.

Such a table is constructed by assembling the composition and temperature gradient data at certain key temperatures and recording the pressure, P_0, inlet enthalpy of the mixture H_0 and burning velocity v_0. In addition, if a heat extraction burner is used, either the heat extraction rate or temperature gradient at the burner surface should also be recorded to establish the flame system enthalpy. In free burning flames the enthalpy can be calculated from the inlet concentrations and temperature.

Using this approach it is possible to reconstruct a complete *approximate* profile by using the interpolation formulae given on pages 245–46.

The usual choice for defining temperatures are: (1) the inlet temperature, T_0; (2) the temperature of initial reaction, T_i; (3) the temperature where the total radical concentration passes through a maximum, T_r, and the final adiabatic flame temperature, T_f. These choices are adequate for simple systems such as many flames of hydrogen

Table IX–3 Compact Presentation of Flame Structure Data

Reactants and Products (Mole Fraction × 100)				
	T_0	T_i	T_r	T_f
	350	920	1780	2000
z(mm)	−2.0	0	1.8	+∞
A(z)	1.0	1.0229	1.044	—
dT/dz	75	700	29	0
M_{av}	30.9			
CH_4	7.8	5.03	0	0
O_2	91.5	87.6	75.36	75.4
H_2O	0.6	4.5	14.6	15.4
CO_2	0.22	0.75	7.6	8.02
V(m/S)	0.67			

Stable Intermediates (Mole Fraction × 10^4)							
	H_2	CO	H	O	OH	HO_2	HCO
T_a	1200	1300	1360	1360	1360	—	—
T_p	1490	1700	1790	1790	1790	950	1700
Peak conc	24	31	11.5	0170	150	—	—
T_f = 2000	1.4	3.3	0.5	15	58	$<10^{-5}$	$<10^{-6}$

Fuel Intermediates (Mole Fraction × 10^4)			
	CH_3	C_2H_5	OCH_2
T_a	1130	—	1150
T_p	1350	1350	1490
Peak conc	7	$<10^{-6}$	10
T_d	1570	—	1580

[a]temperatures are in K, temperature gradients in K/mm, time in s, concentrations in mole fractions, velocities in m/s, density in moles/cm^3, enthalpy in cal/mole

CH_4 = 0.078 O_2 = 0.915 CO_2 = 0.0022 H_2O = 0.0006 Ar = 0.0034. Fristrom, Grunfelder and Favin [1960].

Initial Conditions: ER = 0.17; ZS = 1.8 mm; ρ_0 = moles/cm^3; M_0 = 30.9; Z_0 = −2.0 mm; ZS = 1.8 mm; A = 1 + 0.0229 (Z + 2); enthalpy = −2000 cal/mole; burning velocity = 0.67 m/s; pressure = 0.1 at

with the halogens and oxygen and the carbon monoxide–oxygen flame. However, in the more general case the stable intermediates and radicals do not all peak at T_r and the individual peak compositions, X_i, and temperatures, T_p, should be recorded.

Hydrocarbons introduce the additional complexity of fuel intermediates. This information can be included by tabulating the peak concentration, X_{ip}, and temperature, T_p, of each intermediate and the temperatures at which half height is reached on the ascending (T_a) and descending (T_d) sides of each curve. In some systems these temperatures differ only slightly and only three averaged temperatures are required to describe these intermediates. It should be borne in mind that they are produced sequentially, so some differences are to be expected (see, for example, the ethylene flame of Fig. XII-28).

It should be emphasized that this is an empirical fitting method. It offers a qualitatively correct functional form in matching both the functional values and their first derivatives at the defining temperature stations[1]. However, it does constrain the func-

1. The linear interpolations for intermediates introduces a non-physical discontinuity in gradient at the peak of intermediate species. This is a minor problem which would be removed by using a suitable non-linear interpolation such as a low order Fourier expansion.

tion by forcing the half value of the function to occur at the half temperature change point. This, of course, could be modified, but this would double the number of zones considered.

The success of this approach is a consequence of the approximate balancing of diffusional and thermal conduction contributions to energy flux in flames (Zeldovitch [1949]). This makes many flames approximately constant enthalpy throughout. Klein [1957] used this to develop an integral, iterative approach to solving the flame equations. He pointed out that there would be a rigorous linear relation between temperature rise and composition change if the system were: (1) pointwise constant enthalpy, (2) the Lewis number of all species were the same, and (3) a linear relation existed between reaction progress and heat release. These requirements are reasonably met in many flame systems. Fristrom, Favin, Linevsky, Vandooren, and Van Tiggelen [1992] extended the concept in their approximate flame theory by observing that the unity Lewis number restriction would be significantly relaxed by dividing the flame into three zones: (1) a pre heat zone ($T_0>T>T_i$) where there is no reaction or heat release; (2) a zone dominated by biomolecular reaction ($T_i>T>T_r$), and (3) a thermolecular reaction zone ($T_r>T>T_f$) where partial equilibrium is valid. The degree to which these approximations reproduces flame data can be seen in the examples of Figures IX-1, X-5, and XI-9.

Interpolation Formulae[2]

Fractional Temperature Rise Variables

$$FTI = (T - T_i)/(T_i - T_0) \quad (T_0 \leq T \leq T_i)$$
$$FTII = (T - T_i)/(T_r - T_i) \quad (T_i \leq T \leq T_r)$$
$$FTIII = (T - T_r)/(T_F - T_r) \quad (T_i \leq T \leq T_F)$$
$$FTa = [T + T_p/-3T_a/2]/[T_p - T_1]\ (T < T_p)$$
$$FTd = [T + T_p/2 - 3T_d/2]/[T_p - T_d]\ (T > T_p)$$

Temperature Gradient

$$(dT/dz)_I = (dT/dz)_i\,(1 - FTI^2)$$
$$(dT/dz)_{II} = [(dT/dz)_i - (dT/dz)_r]\,(1 - FTII^2) + (dT/dz)_r$$
$$(dT/dz)_{III} = (dT/dz)_r\,(1 - FTIII^2)$$

where $(dT/dz)_r \approx P0/30(dT/dz)_i$

2. It is assumed in this table that the there is no significant stream tube expansion through the flame front. If this is not the case the stream tube area function must be added.

In these equations, subscripts: 0 = inlet conditions, i = position where heat release begins, r = position of maximum total reactive radical concentration, f = position of maximum fuel reaction, F = adiabatic flame conditions; a = the ascending branch of an intermediate, d = the descending branch of an intermediate.

Term T = temperature (K).

The following variables are functions of temperature: X_i = mole fraction of species i; M_i = molecular weight of species i, M(T) refers to the average at the T station; v = velocity (m/sec); t = time (s); ZS = distance between points T_i and T_r; z = distance (mm); and dT/dz = Temperature Gradient (K/mm).

Distance

$z(T)_I = \int_{Ti}^{T} dT/(dT/dz)_I$
$z(T)_{II} = \int_{Ti}^{T} dT/(dT/dz)_{II}$
$z(T)_{III} = \int_{Ti}^{T} dT/(dT/dz)_{III}$

Velocity

$v(T) = v_0(T/T_0)(M_0/M(T))$

Density

$\rho(\text{moles/cm}^3) = \rho_0 v_0/v(T)/A(T)$
$\rho(\text{g/cm}^3) = \rho(\text{moles/cm}^3)\ M(T)$

Molecular Weight

$M = \Sigma X_i M_i$

Reactant and Product Concentrations[2]

$X_i(T) = X_{i,0} + (X_{i,i} - X_{i,0})FTI\ (T_0 \leq T \leq T_i)$
$X_i(T) = X_{i,0} + (X_{i,r} - X_{i,i})FTII\ (T_i \leq T \leq T_r)$
$X_i(T) = X_{i,r} + (X_{i,r} - X_{i,F})FTII\ (T_r \leq T \leq T_F)$

Intermediate Concentrations

$X_i(T) = X_{i,p}FTa$ or $X_{i,p}$ (ascending branch)
$X_i(T) = X_{i,p}FTa$ or $X_{i,p}$ (descending branch)

X

FLAME PROCESSES

The preceding chapters have covered experimental techniques and the mathematical methods required to obtain and analyze flame structure data. This chapter will outline the information on flame process that can be derived from these studies. Wehner [1964], Wolfrum [1986], and Dixon-Lewis [1993] have reviewed various aspects of flame and combustion processes.

The flame that is used as an example is a lean low-pressure methane–oxygen flame with which the author is familiar. The complete set of profiles showing the zones and linearity of the data when plotted against temperature is presented in Figure X-1. A compact tabulation of the critical parameters is given in Table IX–3. The experimental profiles have been presented in figures in previous chapters (velocity, Fig. IV-9; temperature, Fig. V-9; and composition, Fig. VI-1).

Physical processes are emphasized in this chapter because they are common to all flames, but some general characteristics of the chemistry will also be presented. Individual chemistries will be discussed in more detail in subsequent chapters.

Flame Processes

As was previously indicated, flames are exothermic systems where reaction is strongly coupled with the molecular transport processes of diffusion, thermal conduction, and thermal diffusion. They possess the unique ability to propagate at a constant velocity through a suitable mixture after ignition. The propagation is driven by the fluxes of heat and reactive radical species that initiate reaction in the unburned gases. These fluxes result from the transport processes, which are driven by the temperature and concentration gradients produced by the reactions. This is a positive feedback loop where reaction induces temperature and concentration gradients; the transport processes convert them into the feedback of heat and radical species that speed the reaction. The feedback is opposed and limited by the propagation of the flame. The constraints of conservation and continuity combined with the differential equations of reaction and molecular transport make the system overdetermined. Only one stable propagation rate exists called the *burning velocity* (see Chapters II, IV, IX, and X). This is the rate of propagation of a flame front into a stationary gas and should be distinguished from *flame speed,* which is the speed in laboratory coordinates.

If minor effects of geometry are neglected, the burning velocity and microstructure

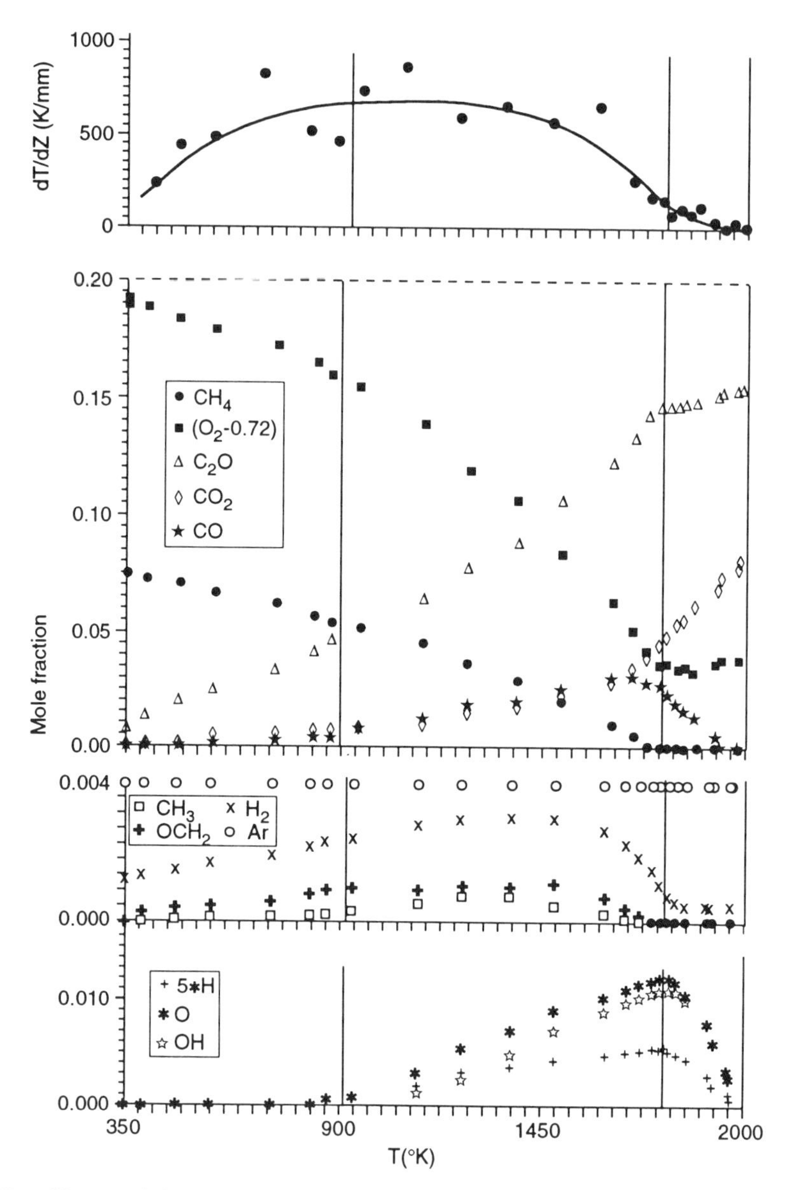

FIG. X-1 Characteristic profiles of the flame used as the example in this chapter: (CH_4 = 0.078, O_2 = 0.82, P = 0.1 atm, T_0 = 350 K). (Experimental data of Fristrom, Grunfelder, and Favin [1961]). Points are replotted against temperature to show the approximate linearity of profiles in this plane. See Figs. IV-9 (velocity), V-1 (temperature), and VI-1 (composition) for the experimental data plotted against distance.

of the flame are independent of whether the propagation is into a quiescent gas, as in a propagating flame, or into an opposing gas flow, as occurs in a Bunsen burner. This is simply a question of the coordinate system in which the observer views the flame.

Flame Reactions

The reaction mechanisms for flame systems are dominated by radical–molecule and radical–radical reactions. Flame chemistry appears to be the domain of the high-temperature reactions of a few reactive radical species (Table X–1).[1] Bi and ter molecular reactions are important and unimolecular reactions occur from excited states. However, unimolecular lifetimes are so short that they cannot be distinguished from direct transitions. Radicals are generated by rapid bimolecular reactions and destroyed by slow ter molecular recombination. This produces a nonequilibrium maximum in reactive radicals at normal pressures. This is discussed in the section on partial equilibrium. The mode of radical generation provides the key to flame chemistry. Two general mechanisms are found.

1. The dissociation of a stable molecule: $Br_2 + M^* \rightarrow Br\bullet + Br\bullet + M$. Here molecular bromine is broken into two atoms, each with an unpaired spin.
2. Exchange reactions in which an existing stable radical produces a reactive radical: $O_2{:} + H\bullet \rightarrow OH\bullet + O{:}$.[2] The net result is that one reactive spin and two unreactive spins become three reactive spins, with a net gain of two reactive spins. The spins are subsequently redistributed by bimolecular exchange reactions (Table X–1).

1. Radicals are species that possess one or more unpaired electron spins. They should not be confused with ions, which are charged species with an excess or deficiency of electrons. For purposes of flame chemistry a distinction should be made between reactive radicals and stable radicals. Reactive radicals are often called *free radicals* because they are not stable under normal laboratory conditions. They can, however, be stable under the high temperatures occurring in flames. Examples in common C/H/O flame systems are: H, O, OH, and the hydrocarbon radicals CH_3 and C_2H_5. Other radicals become important in the flames of other elements. Stable free radicals are radical species that are stable under normal laboratory conditions. Most such radicals involve elemental oxygen. The most important is molecular oxygen, O_2 itself, but oxides of nitrogen and the halogens such as NO, NO_2, and ClO_2 are also stable and form the basis for several important flame systems. For purposes of flame chemistry a distinction should be made between the reactive radicals, which react rapidly under most conditions, and the stable radicals, which are stable under normal laboratory conditions (STP) and form the initial reactants in flame systems. These stable radicals such as molecular oxygen can be premixed with fuels because they do not react directly with most fuels. In flames as a general rule the inlet fuel and oxidizer do not react directly but only indirectly though the agency of the reactive radicals. For example, in oxygen flames oxygen reacts dominantly with hydrogen atoms while the fuel reacts dominantly with the reactive radicals H, O, and OH, which are produced by the oxygen reaction. As a consequence such oxidizers may be mixed with fuels without reaction for indefinite periods. Their flames require finite ignition energies and show ignition temperatures. The detailed chemistry of flames is controlled by the mode of reactive radical generation. There are a few flames where this is not the case, and they are difficult to premix. They are called *hypergolic*. The flames of ClO_2 with hydrocarbons are examples.

2. In these equations we have added dots to indicate unpaired electrons or spins. These dots are often assumed and left off or occasionally replaced by dashes indicating ruptured bonds.

Table X–1 Radical Reactions in Flame Systems[a]

Flame Stoichiometry	Radicals	Generation Steps
$O_2 = 3/2O_2$	O	$O_3 + M^{*b} = O_2 + O$ $O + O_3 = 2\,O_2$
$H_2 + X_2 = 2HX(X = F, Cl, Br)^c$	H, X	$X_2 + M^* = X + X + M$ $H + X_2 = HX + X$ $X + H_2 = HX + H$
$2H_2 + O_2 = 2H_2O$ $(2CO + O_2(trH)^d = 2CO_2)$	H, O, OH $(HO_2)^e$	$H + O_2 = O + OH$ $O + H_2 = OH + H$ $OH + H_2 = H_2O + H$
$CH_4 + 2O_2 = 2H_2O + CO_2$	H, O, OH, (HO_2), CH_3, HCO	(See $H_2 + O_2$ flame) $CH_4 + R^f = CH_3 + RH$ $CH_3 + O = CH_2O + H$ $CH_2O + R = HCO + RH$
$2NH_3 + 2O_2 = N_2 = 3H_2O$	H, O, OH, (HO_2), NH_2, NH, (N)	(See $H_2 + O_2$ flame) $NH_3 + O_2 = NH_2 + OH$ $NH_2 + O_2 = NH + OH$
$2S_2 + O_2 = 2SO_2$	O, S, SO, $(S_2O)^g$	$S_2 + M^* = S + S + M$ $S + O_2 = SO + O$ $O + S_2 = SO + S$ $SO + S_2 = S_2O + S$ $SO + O_2 = SO_2 + O$
$H_2 + N_2O = N_2 + H_2$	H, O, OH, (HO_2)	$O + N_2O = O_2 + N_2$ (see $H_2 + O_2$ flame)
$H_2 + 2NO = H_2O + N_2$	H, O, OH, N, O_2	(See NO flame) (see $H_2 + O_2$ flame)

$2NO = N_2 + O_2$	O, N	$NO + M^* = NO + NO + MN + NO = N_2 + O$ $O + NO = O_2 + N$
$2H_2S + 3O_2 = 2SO_2 + 2H_2O$	H, O, OH, (HO_2), SH, S	(See $H_2 + O_2$ flame) (See $S_2 + O_2$ flame)
$2H_2 + NO_2 = 2H_2O + N_2$	NO_2, H, O, OH, (HO_2)	$(NO_2 + H2 = NO + H_2O)$[h] (see NO flame) (See $H_2 + O_2$ flame)
$ClO_2 = Cl + O_2$	ClO, Cl, O	$ClO_2 + M^* = Cl + O_2$ $O + Cl_2O = ClO + Cl$ $Cl + ClO = 2ClO$
$(2Na_2 + O_2 = 2Na_2O)$[i]	Na, O, NaO	$Na + O_2 = NaO + O$ $O + Na_2 = NaO + Na$ $NaO + Na_2 = Na_2O + Na$ $Na + NaO + M = NA_2O + M^*$

[a]References can be found in Chapters 11, 13.

[b]M^* represents any thermally excited molecule.

[c]Iodine does not form a flame (see Chapter 11). Chlorine reacts with fluorine. Here chlorine takes the role of H and fluorine is *X*.

[d]The carbon monoxide–oxygen flame does not propagate in the absence of traces of hydrogen or any hydrogen-containing compound such as water. This occurs because in the absence of H radicals must be produced by high activation dissocation of oxygen.

[e]HO_2 plays a major role in recombination, but only a minor role in direct propagation.

[f]*R* represents any radical (H, O, OH).

[g]There are two isomers of S_2O. One is the radical in this reaction and the other is a sulfur-substituted sulfur dioxide.

[h]NO_2 is a reactive radical that forms a hypergolic flame.

[i]Sodium flames with oxygen bear a formal similarity to hydrogen–oxygen flames with atomic sodium playing the role of H. The major difference is that the Na_2 molecules are dissociated to Na under normal conditions. This increases the importance of the recombination step $NaO + Na + M = Na_2O + M^*$. It should be observed that dry CO–oxygen flames are catalyzed by the addition of sodium.

A third possibility would be an internal rearrangement unpairing a spin pair, but the author is not aware of any examples. One possibility would be singlet delta oxygen, whose spins are paired (see Chapter XIII). Troe [1989] and Gray [1993] have written excellent reviews on combustion chemistry and Mulcahy [1982] has reviewed experimental methods.

Dissociation reactions have high activation energies and become important only at high temperatures. By contrast, radical exchange reactions have moderate to low activation energies, and radical generation and reaction become important at lower intermediate temperatures. In systems where more than one generation mechanism compete, the exchange mechanism usually dominates because it occurs earlier in the flame. However, there may be interference. This is the basis for the phenomena of inhibition and promotion (see Chapter XIII). For example, oxygen flames are inhibited by halogen compounds, while halogen flames are promoted by oxygenated compounds. If the reactant-stable radical reacts with the fuel at the inlet temperature, as is the case with ClO_2, the flame self-ignites. Such flames are called *hypergolic.*

Flames where radical generation is controlled by dissociation are discussed in Chapter XI. The most common radical exchange flames that involve molecular oxygen with C/H/O fuel molecules are discussed in Chapter XII. Chapter XIII is a catchall that includes: (1) flames of oxygen with fuels of other elements; (2) fuels with other stable free radical oxidizers such as the oxides of nitrogen and the halogens; (3) mixed chemistries involved in flame inhibition and extinction and air pollution.

Most flame radicals are derived from oxidizers, but a few systems (boranes, silanes, and phosphene) may be driven by fuel radicals. At present the evidence for these fuel-derived radicals is circumstantial.

Final Flame Conditions

The final flame composition and temperature can be calculated from thermodynamic considerations, providing losses due to conduction and radiation are negligible or can be quantified. Results from such calculations for a number of flame systems can be found in the appendix at the end of the book. This is commonly called the "adiabatic flame temperature (and/or composition)" because the underlying assumptions are adiabaticity and total equilibrium. Where these criteria are met thermodynamics provides a powerful, reliable tool. It can also often be applied if losses or departures from true equilibrium can be quantitatively characterized. Examples of such applications are flames with heat extraction and the partial equilibrium problem discussed later in this chapter. It can even be applied to cases where a final product is suppressed kinetically, as occurs in the ethylene oxide decomposition flame (Chapter XI), where carbon formation is suppressed.

Manual calculations are laborious since neither the composition nor the temperatures are known initially and an iterative search for matching conditions must be made. This is best done using one of the several programs available for the purpose. The most common is the NASA Program (Svehela and McBride [1973]), which includes both thermodynamic and transport programs. The thermodynamics is fitted to the JANAF tables (Stull and Prophet [1986]). This is a general program applicable to flames, detonations, rocket exhausts, and shocks. Adiabatic expansions can be calculated and they can be modified to account for frozen equilibria. Most common elements and compounds are included; so almost any flame system can be calculated. Since most flames are close to adiabatic, these calculations provide a convenient method for characterizing combustion systems. Although the results are generally reliable, problems can occur in high-temperature phase equilibria and with radical species where thermodynamic parameters are sometimes only poorly known. Substantial computing capacity is required, but this problem is now usually minor. Simpler, more rapid algorithms can be used in cases where computational capacity is pressed to the limit. This situation often occurs in practical engineering modeling problems.

Burning Velocity

The most common method for characterizing premixed laminar flames is by *burning velocity*. This is defined as the rate of propagation into a stationary gas or be balanced by an opposing flow. The difference between them is simply between stationary and moving coordinate systems of observation. The overall flame geometry adjusts itself so that everywhere along the front the local component of velocity normal to the flame surface is equal to the burning velocity. Measurement methods are discussed in Chapter IV. Examples for various flame systems can be found in Chapters XI through XIII. In the appendix at the end of the book there are tables giving references for systems that have been studied. Burning velocity is directly related to the eigensolution to the flame equations. This conserved quantity is the product of burning velocity by initial density ($m = \rho_0 v_0$). One should distinguish between the ideal one dimensional flame of theory and experimental flames. In burner flames stream tubes expand (see Fig. IX-2), while in spark ignited flame kernels the reverse is true. This modifies the velocity profile. Fristrom [1953] has suggested using the surface of initial reaction as a reference. This reduces the importance of stream tube expansion, but requires additional measurements for characterization. Dixon-Lewis [1991] points out that the only common simple experimental measurement which can be directly related to theory is the expansion velocity of a spherical flame kernel (Dowdy et al. [1991]). These questions are explored in the section on the one-dimensional approximation.

Rosen [1958] proposed a theorem that identifies the burning velocity as the most rapid (subsonic) processes connecting reactants with products. It seems intuitively reasonable since if a more rapid process existed it would eventually outrun the initial processes. This simple viewpoint is often obscured by the complexities of reaction and transport interactions, but it contains the essence of flame propagation.

There are, however, certain practical limits. On the low side propagation is limited to a few millimeters per second by gravity-induced buoyancy instabilities. The upper limit is controlled by the onset of compressibility, which can induce a transition to detonation. If a flame shows a positive change in burning velocity with pressure, as do most fast flames, then the velocity will be unstable because increasing the velocity increases the stagnation pressure, which in turn increases the velocity, leading finally to detonation.

Chapman [1899] and Jouget [1905] observed that momentum–energy conservation placed constraints on burning velocities. The conclusions from and arguments underlying these relations are too lengthy for inclusion here, but the reader can find critical discussions by several authors (Emmons [1959]; Williams [1985]).

Flame Thickness and Residence Time

Elementary considerations from the ignition approximation indicate a balance between the thermal wave velocity and the mass velocity at the point of initial reaction. This suggests that the product of burning velocity with flame thickness corrected for thermal diffusivity $[\lambda/(\rho C_p)]$ should be a dimensionless constant of the order of unity. To test this the author has recalculated the measurements of Morgan and Kane [1953] of luminous region thickness and burning velocity for a number of systems. The product was constant within an average deviation of 20%. Luminosity outlines the primary reaction region where radical concentration and reaction are highest. This is the case because the radiation is generated by molecules excited by bimolecular radical recombination.

Dependence of Flame Propagation on Composition

Burning velocity depends on composition. The most important single factor is the mode of radical generation. This depends on temperature and the details of the individual flame chemistries. In flames where radicals are generated by dissociation, temperature is the controlling factor, and the peak usually occurs close to the stoichiometric point. Chemical factors are more important in flames where radical generation is controlled by exchange reactions. This is the case in C/H/O flames where radical generation is controlled by the reaction $H + O_2 = OH + O$. In these systems the burning velocity maximum usually occurs just beyond stoichiometric on the fuel-rich side. This shift is particularly pronounced in the hydrogen system. Illustrations for specific flame chemistries can be found in the figures in Chapters XI through XIII.

These generalities can be strongly modified by the details of the chemistry, since some systems such as rich hydrocarbons, are self-inhibiting (Kaskan and Reuther [1977]). The extreme of this is the methyl bromide flame, which cannot be burned richer than stoichiometric. Questions of inhibition are discussed in Chapter XIII. There are usually moderately well-defined limits of flame propagation. These are important for safety considerations and have been measured for many systems (Coward and Jones [1954]; Lewis and von Elbe [1987]; Hibbard and Barnett [1959]). Some of the values for some common fuels with air are collected in Table A–1 in the appendix at the end of the book. Fenn and Calcote [1953] related the lean limits of combustion to the energy required to bring the mixture to a temperature where radical generation can occur. Stull [1971] discussed the correlation of thermodynamics with combustion limits. The adiabatic flame temperature must generally exceed 1000 K, where the $H + O_2 = OH + O$ reaction becomes significant. This limit is increased for fuel-rich systems such as methane, which is self-inhibiting.

Effect of Inlet Temperature on Flame Propagation

Preheating a combustible mixture generally increases the burning velocity. It is more meaningful to compare the mass burning rate $\rho_0 v_0$ rather than the volumetric burning rate because this eliminates the physical effect of density change. Effects are smaller than might be expected because temperature changes in the reaction region are less than changes in inlet temperature. This is the case because heat capacities are higher and there is a strong buffering of temperature by the dissociation equilibria. A small positive effect is usually found. Some examples can be found in Chapter XII.

Kaskan [1957] in an early study found that there was a linear relation between the log of the mass burning velocity and reciprocal temperature in C/H/O flames. He suggested that these flames were controlled by similar high-temperature reactions.

The dependence of burning velocity on inlet temperature has been used to deduce engineering "effective activation energies of flames." These purely empirical factors bear only indirect relation to the true activation energies in flames because it is usually assumed that fuel reacts directly with the oxidizer close to the adiabatic flame temperature. Flame structure studies (Chapters X through XIII) show that primary reaction occurs at much lower temperatures and that the inlet species react with radicals rather than with each other. For example, in the hydrocarbon–oxygen flames the highest activation energy of any propagating step is 16.8 kcal/mol yet the empirical value using these assumptions varies between 35 and 45 kcal/mol. Detailed modeling resolves these problems.

It should be emphasized that burning velocity depends on reaction temperature and only indirectly on the inlet and exit temperatures. This point was made by Weinberg [1975] in designing his "bootstrap" burner, in which the inlet gases are preheated by the exit gases. Using this technique he was able to burn methane–air flames with an overall temperature rise of 200 K. Despite this the reaction zone temperature remained around 1400 K. A discussion of this burner system can be found in Chapter II.

Effects of Pressure on Flame Propagation

Assuming that flame propagation is controlled by initial bimolecular reactions, it might be expected that pressure would have little effect on flame propagation. There are, however, second-order effects. Pressure affects the ratio of two-body reaction rates to three-body reaction rates. As a result the peak radical concentration drops with increasing pressure. This would suggest that burning velocity might increase with decreasing pressure. On the other hand, increasing pressure increases the adiabatic flame temperature and the temperature in the reaction zone, which suggests the opposite behavior. In low-velocity flames the first factor (peak radical concentration) dominates because radical concentrations are low and can respond to pressure changes. In high-velocity flames radical concentrations are high, and the relative change with pressure will be smaller and the thermal factor should dominate.

Lewis [1954] pointed out that for hydrocarbon-air flames the pressure coefficient of burning velocity, (dv/dP), was negative for slow flames ($v < 0.3$ m/s) and positive for fast flames ($v > 1$ m/s) with a short pressure independent transition region. This behavior appears to be confined to the pressure region around atmospheric in C/H/O flames. Warnatz's [1981] calculations on the hydrogen–oxygen–nitrogen system fit the correlation at atmospheric pressure but show more complex behavior over a wider range of pressures (see Fig. XII-5). This is a situation where only detailed experiments or modeling can provide a reliable answer.

The low-pressure limit in practical systems is controlled by quenching distance and burner diameter. Since quenching distance varies inversely with burning velocity, fast flames can be burned at lower pressures than slow flames. In the laboratory a very fast flame such as hydrogen–fluorine (V_0 = 50 m/sec) can be burned as low as a ten thousandth of an atmosphere.

Hirschfelder, Curtiss, and Bird [1964] pointed out that the flame equations predict a theoretical low-pressure limit for unimolecular decomposition. Another interesting system is the hydrogen–oxygen flame, where the bimolecular reactions are almost thermoneutral and three-body reactions are required to reach temperatures where the radical-generating reaction $H + O_2 = OH + O$ can take hold. As a result, in this system a low-pressure limit is set by thermal conditions. These questions have been discussed by Zeldovich [1949] and Dixon-Lewis [1974].

On the high-pressure side the limits of combustion depend on the details of the chemistry. For hydrogen–air flames both rich and lean limits are independent of pressure up to 125 atm (Jost [1946]; Zeldovich [1949]; Hibbard and Barnett [1959]). Carbon monoxide–air flammability limits were slightly narrowed by pressure, but as Zeldovitch pointed out, this may be due to the decreasing concentration of water impurity as pressure increased. He assumed the gas was in equilibrium with liquid water, as may well have been the case in these early studies. Water has a strong effect on CO flames. The lean limit of methane–air flames is also pressure independent, but the rich limit varies strongly with pressure. This may be connected with self-inhibition by bimolecular reactions in this system by reactions of CH_3 and CHO. These questions are discussed further in Chapter XII.

There appears to be no particular upper pressure limit to flame propagation, although one expects mechanisms to change from dominantly two body to dominantly three body as pressure increases. This transition in oxygen-based flames is probably controlled by competition between $H + O_2 = OH + O$ and $H + O_2 + M = HO_2 + M^*$, which cross over around 39 atm. Unfortunately there are no flame structure measurements of high-pressure flames. This problem remains a challenge for the next generation.

Experimental Flame Structure

Experimentally flame structure is usually measured as a function of distance, but theoreticians favor temperature as the independent variable. The relation between schemes is illustrated in Figure IX-1 and Table IX–1. Techniques for making the measurements are discussed in Chapters IV through VIII. An example of a typical set of profiles for a lean low-pressure methane–oxygen flame (Fristrom, Westenberg, Grunfelder, and Favin [1960, 1961, 1965]) is given in Figure X-1. This flame is used as the exemplar system in this chapter.

Aerodynamic Variables

The aerodynamic variables (see Chapter IV) are: stream tube geometry, velocity, and density. Pressure is considered constant in flames,[3] stream tube geometry is usually a minor variable, and density is often derived from continuity or a combination of temperature and composition measurements. As a consequence, the principal experimental aerodynamic variable is velocity. In the example (Figs. IV-9, X-1) stream tube and velocity were derived using the particle track technique.

Temperature Profiles

Temperature (see Chapters V and VIII) is the single most important and useful variable for characterizing flames. Many of the parameters used in analyzing flame data are temperature dependent. The temperature profile of the exemplar flame is presented in Figure V-1. Examples from other systems can be found in the figures of Chapters X through XIII.

All analyses of flame structure assume that major species are in local thermal equilibrium with this temperature so that it can be used to calculate local transport coefficients and collision rates, heat release, and reaction rates. The only documented exceptions are the hydrogen–fluorine and hydrogen–chlorine flames, where energy chains may play a role in radical generation. They are discussed in Chapter XI.

Although flame luminosity clearly indicates the presence of nonequilibrium processes, the excited species represent such a small fraction of the system that they can be neglected quantitatively for an overall analysis. The secondary processes such as ion production and chemiluminescence are interesting in their own right, but in this book their effects will generally be ignored.

Composition Profiles

The detailed set of composition profiles provides a complete chemical description of a flame (see Chapters VI through VIII). The primary experimental variable is local concentration. This is the amount of a species per unit volume. It is often convenient to report data in dimensionless units of mole fraction X_i, or mass fraction f_i, with local density being an implicit variable. The exemplar methane–oxygen flame (Figs. X-1, VI-1) was obtained using microprobe sampling with mass spectral analysis. Atom and radical profiles were added later using scavenger sampling measurements (Fristrom [1963a, b]). This is a typical complete composition profile. Examples from other flame systems can be found in Chapters XI through XIII.

Transport and Fluxes in Flames

To derive quantitative information on flame processes from experimental flame structure data it is necessary to calculate flux variables. Assuming the transport coefficients

3. This is justified for all but high-velocity flames such as that of hydrogen with fluorine. Vanpee, Cashin, Falabella, and Chintappili [1973] observed that the fastest of these flames approaches the upper limit for burning velocity set by the lower Chapman–Jouget point.

are known, this can be done using the experimental profiles and a knowledge of local temperature. In a system with negligable transport, the fluxes could be obtained by multiplying the local intensive variable by the local velocity. This is not possible in flames because the steep gradients produce fluxes of species and energy that are superimposed on the convective fluxes. Energy is transferred not only by convection but also by thermal conduction, which in turn is opposed by the diffusion of energetic species. Each species has its own species diffusion and thermal diffusion velocities superimposed on the local mass flow. Thus, when expressed in dimensionless fractional units, the local flux of a species can differ significantly from its local concentration. Reactants that are disappearing will have fluxes that are higher than their concentrations. The reverse is true for products. Only at maxima (or minima) of concentration are flux and concentration equal. The relations between concentration, flux, and rates are illustrated in Figure IX-5, where the mechanics of these calculations are developed.

Diffusion Velocities

This complex situation is best visualized by considering the individual species velocities. These are the differential velocities that diffusional transport adds. They can be calculated from a knowledge of local temperature and concentration gradient and the requisite species (multicomponent) diffusion coefficient. The derivation of these quantities is discussed in Chapter IX. An example of diffusion velocities for the exemplar flame is presented in Figure X-2. Notice that many species velocities greatly exceed the mass velocity.

The steep temperature gradients in flames produce a secondary diffusion processes that is superimposed on the common concentration or molecular diffusion. This results in an additional species velocity, which is called the *thermal diffusion velocity*. In flames this is generally smaller than diffusion velocities. This is a complex phenomenon, but in the high-temperature low-concentration limit binary thermal diffusion factors become pure numbers. A few of these for CHO flame species are given in Table A–1 of the book appendix. The derivation of these quantities is discussed in Chapter IX. An example of thermal diffusion velocities in the exemplar flame is given in Figure X-3.

Species Fluxes in Flames

The flux of a species through a flame is important because reaction rates are the spatial gradient of the fluxes (see the discussion in Chapter IX). Local flux fraction is the product of the local mass fraction multiplied by the sum of convective, molecular diffusional, and thermal diffusional velocities.

$$G_i = f_i(v + V_i + V_{it}) = \frac{N_{PT}X_iM_i}{M_{av}}(v + V_i + V_{it})$$

where G_i is the fractional mass flux of species i (g cm^{-2} sec^{-1}); f_i is the mass fraction of the species; v is the convective velocity (cm sec^{-1}); V_i is the molecular diffusion velocity of species i (cm sec^{-1}); V_{it} is the thermal diffusion velocity of species i (cm sec^{-1}; X_i is the mole fraction of species i; M_i is the molecular weight of species i; and M_{av} is the average molecular weight. N_{PT} is the molar density 0.012186 P/T (moles cm^{-3}).

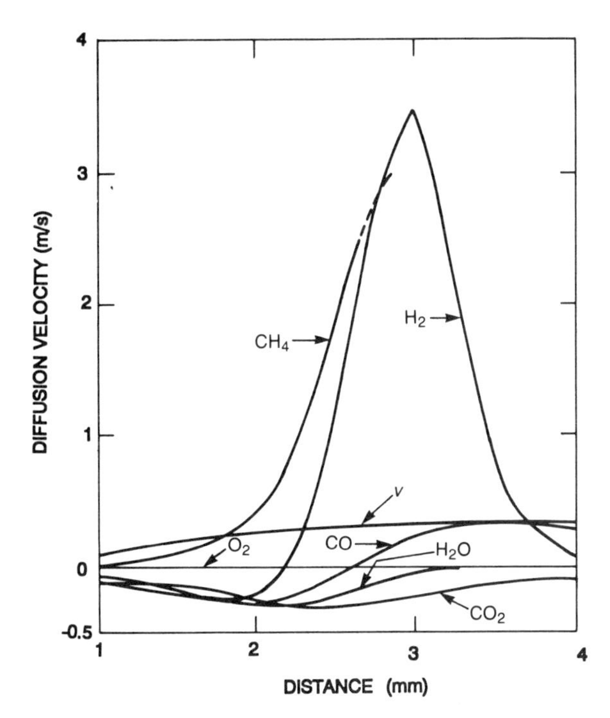

FIG. X-2 Diffusion Velocities in the exemplar low-pressure methane–oxygen flame (see Fig. X-1). (After Fristrom and Westenberg [1960].)

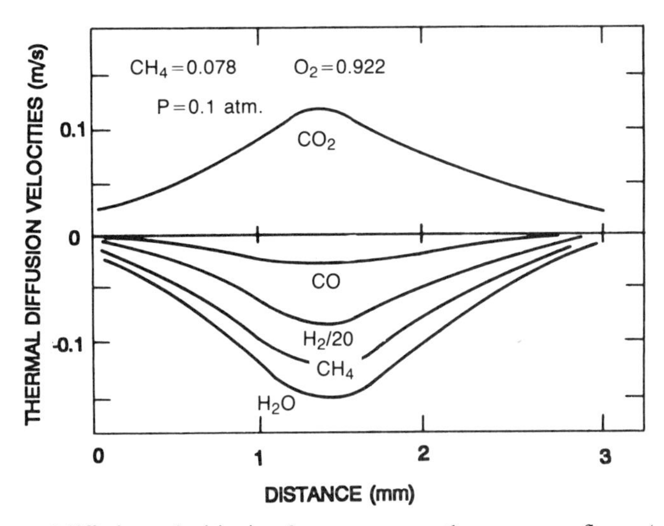

FIG. X-3 Thermal diffusion velocities in a low-pressure methane–oxygen flame. (see Fig. X-1). (After Fristrom and Monchick [1989].)

The effects of molecular diffusion on flux depend on the sign of the gradients. Reactants that are disappearing have negative gradients, and the concentration is lower than the flux. The reverse is true of products. Intermediates that pass through a maximum behave as products before the maximum and as reactants after the concentration maximum.

In flames the temperature gradients are always positive (unless there are heat losses), and therefore the effects of thermal diffusion depend on the molecular weight of the species. If it is lower than the local average, it will increase the species flux. If the molecular weight is higher than the local average, it will decrease the flux. In contexts other than combustion one should remember that this is the high-temperature-limit behavior. Thermal diffusion at low temperatures, which is more complex, is discussed in detail by Hirschfelder, Curtiss, and Bird [1964].

The contributions of convection, diffusion, and thermal diffusion to the total flux in the exemplar flame are illustrated for: a reactant (methane, Fig. IX-4), a product (water, Fig. X-4), and an intermediate (hydrogen, Fig. X-5). These are recalculations by Fristrom and Monchick [1989] taking thermal diffusion into account. The original data (Fristrom et al. [1960–1961]) have been corrected for probe recombination.

It should be remembered that for conservation of momentum the sum of the individual species momenta due to molecular diffusion, V_M, and thermal diffusion, V_{TD}, must vanish, that is, that $\Sigma N_{TP}(V_M + V_{TD})_i X_i M_i / M_{av} = 0$.

Thermal Conduction and Energy Flux

The energy transferred by thermal conduction in flames is always against the flow because temperature gradients are positive. Thermal conductivity is a more complex transport process because the contribution of internal molecular energy is important for all species except atoms. The derivation of multicomponent thermal conductivity of mixtures is complex. A short discussion of the problems can be found in Chapter IX. There are several programs for making such calculations (Svehela and McBride [1973]; Kee, Miller, and Jefferson [1980]).

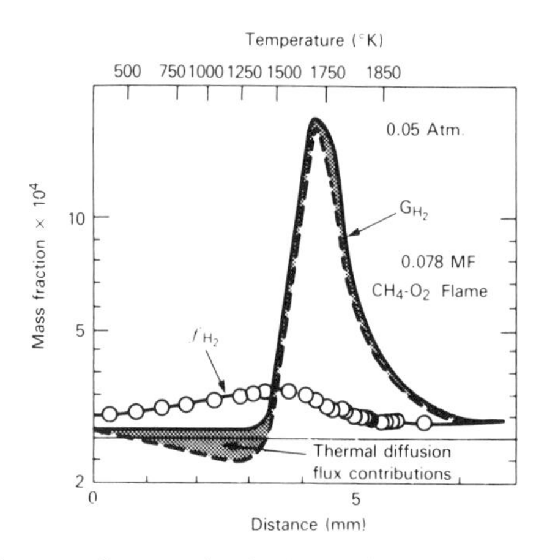

FIG. X-4 Fluxes of the intermediate species, hydrogen in the $P = 0.05$ atm, 0.078 CH_4, 0.92 O_2 flame. (After Fristrom and Monchick [1989].)

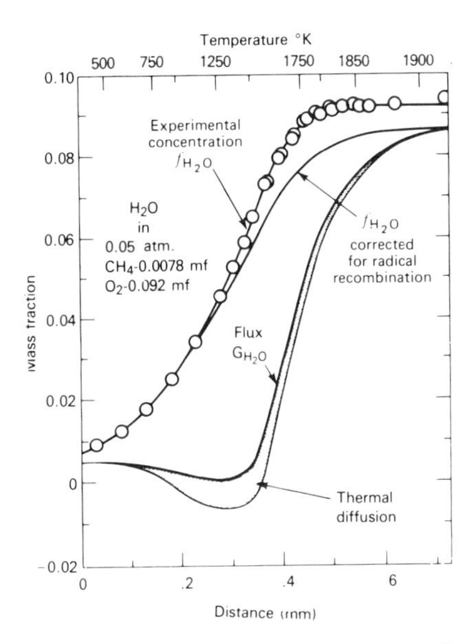

FIG. X-5 Fluxes of the product, water, in a low-pressure methane flame ($P = 0.05$ atm, 0.078 Methane–oxygen flame). (After Fristrom and Monchick [1989].)

Energy can be transferred in flames by four mechanisms: convection, conduction, diffusion, and thermal diffusion (see Fig. X-6). In the region of steep gradients the transfer of energy by the transport processes greatly exceeds the contribution of convection. Even the contribution of the second-order process of thermal diffusion is not negligible. It is interesting to observe that thermal conduction transfers energy against the flow because temperature gradients in a flame are always positive. In contrast, the contribution of diffusion is always in the direction of flow because the high-enthalpy products diffuse with the flow and the low-enthalpy products diffuse against the flow. It should be observed that the contributions of diffusion and thermal conduction are always opposed and of the same order of magnitude as is required by kinetic theory. They do not exactly balance because individual species diffusivities differ. The ideal balanced case where all species have the same diffusivity is discussed in the section on the Lewis number approximation. The thermal diffusion contribution could lie in either direction because of the complex interplay between molecular weight and energy flux contribution. The contribution of convection depends on the local enthalpy.

Energy Conservation

Local energy conservation provides a severe test of the quality of flame structure data since it depends on concentrations, temperature, and their gradients, as well as the local transport coefficients. There are deviations from local conservation in the experimental data of the exemplar flame (Fig. X-6). They arise from the neglect of radicals, which were not measured in the original study. This does not change the qualitative picture because they are minor species. An attempt was made to use the deviations to

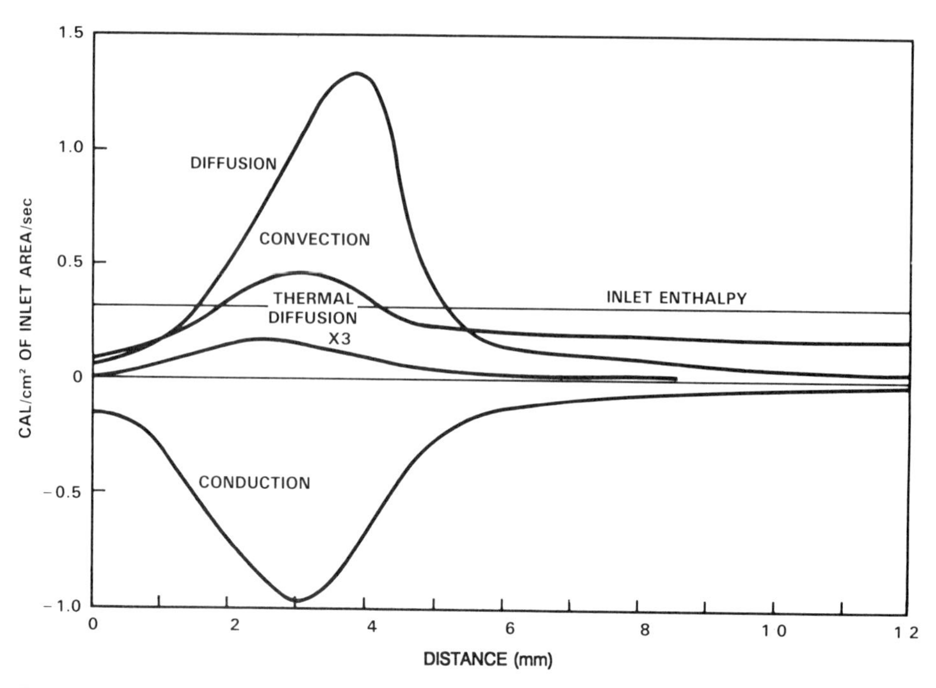

Fig. X-6 Energy fluxes in the exemplar hydrocarbon flame. (Data of Fristrom and Westenberg [1960] are replotted with thermal diffusion added.)

estimate radical contributions (Fristrom and Grunfelder [unpublished]). This gives the proper order of magnitude for concentrations, but does not allow identification of individual species contributions. Other sources of deviations include radiation and heat losses to the burner and walls.

Elemental Flux Conservation in Flames

In flames not only mass should be conserved, but also the flux of each elemental species. This provides a valuable check on the quality of flame structure data and the validity of the analysis. As was the case with energy conservation, elemental conservation involves all of the aspects of measurement and data manipulation. Figure X-7 gives an example of the precision to be expected with carefully handled data. As would be expected, deviations are worst in regions of high gradient. Absolute errors were comparable for all elements. The fractional deviation for oxygen appear better than those of carbon and hydrogen because it is the major species.

Radiation Losses in Flames

Although flames are luminous, the radiant flux is small compared with the total energy transferred in the flame front by convection, conduction, and diffusion. One should distinguish between radiation from the active flame front and the final burned gasses. Flame gases can transfer substantial amounts of energy, as anyone who has stood in

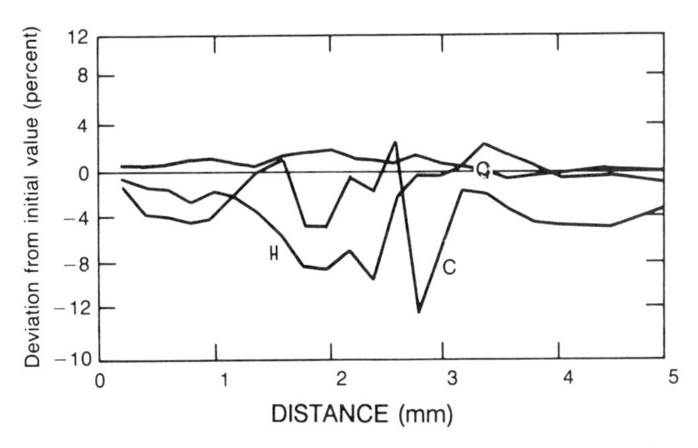

FIG. X-7 Conservation of elemental species fluxes in the exemplar methane flame (see Figs. X-1 through X-3). (After Fristrom and Westenberg [1960]).

front of a fireplace can testify. Radiation transfer and losses in the flame front proper, however, are small because of the short residence time and the transparency of the flame gases. Soot containing flames, of course, represent an exception to this generalization.

Net Production Rates

The final piece of information that can be derived from flame structure data is the species net rate of appearance (or disappearance) and the net rates of heat release.

The net production rate of each species can be obtained by differentiating the flux curves with respect to distance. To analyze the system beyond this point requires the assumption of a chemical mechanism and a functional form for the elementary rate processes. The mechanics of obtaining this information is outlined in Chapter IX. Questions of reaction mechanisms are treated in the discussions of specific flame systems (Chapters XI-XIII). Some typical rate data for the exemplar methane flame are given in Figure X-8.

The net rate of heat release is also of interest. It can be derived from the energy conservation equation or from the species production rates. This is not a popular mode of analysis, and the only available example (Fig. X-9) was done before radical concentration measurements were feasible. The inclusion of such information would improve the agreement.

Errors and Reliability Checks

Since flame structure measurements are subject to many sources of error, it is desirable to understand their effects and provide methods of assessing the reliability of data.

Experimental error sources are discussed in the respective experimental chapters. The most reliable method of detecting errors is to derive a measurement using more

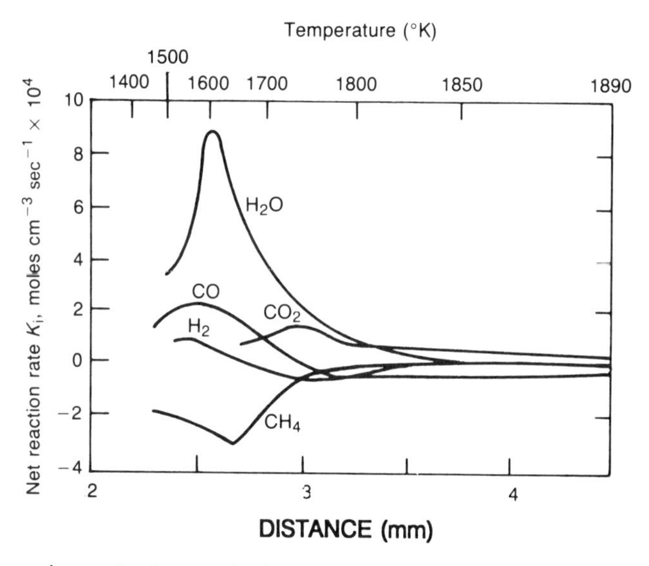

Fig. X-8 Net species production rate in the exemplar methane–oxygen flame of Fig. X-7. (After Fristrom and Westenberg [1960].)

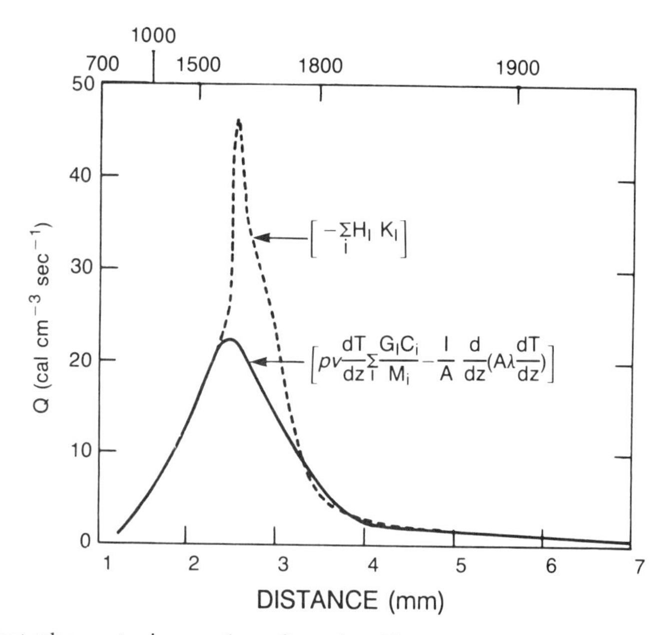

Fig. X-9 Heat release rates in a methane flame (see Fig. X-1). (After Fristrom and Westenberg [1965].)

than one technique. Unfortunately this is not commonly done except to validate a new experimental technique. Beyond direct errors the analysis can introduce errors if profiles are misaligned, if gradients are distorted by the smoothing processes, and if incorrect transport coefficients are assigned. It is these errors that will be discussed here. Experimental problems have been discussed by Fristrom and Westenberg [1965] and Biordi, Lazzara, and Papp [1976]. The transport coefficient problems have been discussed by these authors and theoreticians such as Dixon-Lewis [1974], Tsatsoronis [1978], and Warnatz [1977, 1981]) as well as in the books of Fristrom and Westenberg [1965], Hastie [1975], and Ksandropolo [1980].

Some of the effects of errors in misalignment, resolution and transport coefficients were discussed in Chapter III (Figs. III-1 and III-2). The most critical problem is the alignment of temperature with composition profiles when they are obtained by different methods. In the ideal situation all measurements are made using the same technique, but usually two or more methods are necessary. An example would be a thermocouple for temperature, LDV for velocity, molecular beam inlet mass spectrometry for compositions, microprobe sampling with gas chromatographic analysis, and laser fluorescence for radicals. This combines the strong points of each technique, but it is a nightmare for aligning the profiles with each other. Thermocouples tend to distort the flame and sample temperatures a few diameters downstream of their physical location. Sampling probes, by contrast, sample a region several throat diameters ahead of the sampling point. Optical probes do not distort and, except for small effects due to density gradients, sample the point of focus. The major problem is defining the sampling volume and providing a common reference point. The luminous zone, which is the obvious choice, is too diffuse in most flames to be satisfactory for quantitative analysis. The burner surface is a common choice, but great care must be taken in reproducing and keeping flows, pressure, and heat extraction constant, otherwise the stabilization height of the flame will vary. The best alignment method is to connect methods by measuring a common species or temperature.

Generalities and Approximations

There are a number of generalizations and approximations that are used in flame studies, which structure studies can validate or clarify. Some of these will be discussed. Questions relative to specific chemistries are discussed in Chapters XI through XIII.

Applicability of the One-Dimensional Model

The one-dimensional approximation is used in almost all analyses in the literature. It presumes that the gradients in flame variables in the direction of propagation are large compared with those along the orthogonal "indifferent" coordinates. This is usually accomplished by using a burner with diameter large compared with the flame thickness. Most experimentalist test their systems, but the results are rarely reported. An early published test was made by Fristrom [1965] comparing velocity profiles of various flame geometries and along a conical flame front. Some of these results are shown in Figures X-10 and X-11. Dixon-Lewis and Islam [1982] addressed the problem with

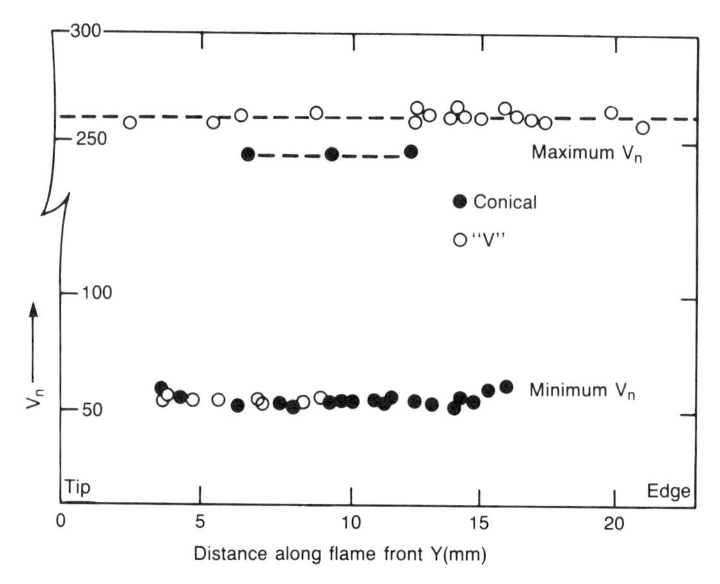

FIG. X-10 Constancy of velocity along the flame front comparing conical and inverted Vee flames testing the applicability of one-dimensional models. Quarter atmosphere stoichiometric propane–air flame. (From Fristrom [1956].)

respect to burning velocity and concluded that the concepts were viable. It should be recognized that these considerations do not apply to "cellular flames." These interesting systems are covered in Markstein's book [1964].

It should be observed that the congruence of flame profiles is confined to the primary bimolecular reaction zone. The transport region can be considered as a ducting region preceding the reaction. Similarly, in the termolecular recombination region the gradients are so low and the coupling is so weak that it is a region of constant enthalpy

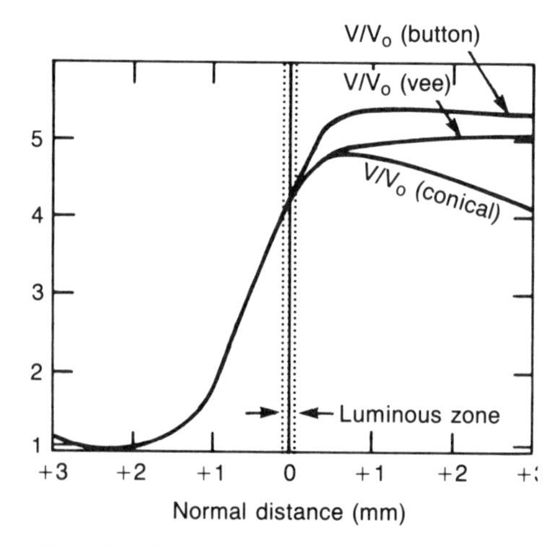

FIG. X-11 Velocity profiles of conical, button, and Vee flames testing one-dimensional models (Stoichoimetric, 0.25 atm propane–air flame).

where the invariant independent variable is time. This region can be ducted and in extreme cases be removed without strongly affecting the flame propagation. Flame separation was discussed in Chapter II.

Zones in Flames

Observers as early as Bacon in the 1500s and Faraday in 1860 noted that flames show structure and identifiable zones (see Fig. I-3). More recently microstructure studies have furnished more quantitative details. Three zones are distinguished (Fig. X-I): (I) a transport region dominated by diffusion, thermal conduction, and thermal diffusion; (II) a bimolecular reaction zone where reactive radicals are generated and the chain processes attack the initial reactants; (III) a termolecular reaction zone where the excess radicals in the second zone are recombined toward final equilibrium. This behavior is a consequence of the varied time scales shown by flame processes. *All integrated species reaction rates are constrained by the same eigenvalue or burning velocity*; as a consequence reactions with rapid rates will be confined to narrow regions, while slow reactions will occupy extended regions in space.

Another consequence of disparate time scales is decoupling and flame separation phenomena. No matter how slow a reaction is, it must potentially go to completion; however, if the gradients fall too low, a secondary region can be decoupled. This phenomena occurs in mixed fuels or oxidizer systems (Chapters II and XII). Examples are mixtures of boranes with hydrocarbons in air. It can also occur if a fast-reacting fuel (or oxidizer) produces a reacting intermediate. The most common example is the CO oxidation in rich hydrocarbon flames.

Pressure Scaling

The structure of flames changes with pressure. The most obvious difference is that in zones one and two distances scale inversely with pressure. This occurs because the transport processes and bimolecular reactions are direct functions of bimolecular collisions and therefore distances scale with mean free path. This is shown clearly in Figure X-12, which compares results of flame structure studies of the exemplar methane flame at two pressures.

By contrast, in zone III, where the time (and distance) scales depend on termolecular reactions, distance scales inversely with the second power of pressure. The consequence of this is that many low-pressure flames have such extended third zones that they do not go to completion in the restricted space available in most burner chambers. This has only a minor effect on flame propagation because of the loose coupling between zones II and III. The coupling drops in importance with inverse pressure. These questions have been discussed by Hirschfelder et al. [1964] and Fristrom and Westenberg [1960–1965]. In the low-pressure limit the effect of zone III on propagation approaches zero, but at intermediate pressures some effects may be noticed. For this reason care should be taken in interpreting experimental studies on the pressure effects on burning velocity. In the high-pressure limit termolecular reactions will dominate bimolecular reactions. This should result in changes in reaction mechanisms in high-pressure flames. In the case of oxygen flames with C/H/O molecules discussed in Chapter XII the critical reaction is the competition between $H + O_2 = O + OH$ and $H + O_2 + M = HO_2 + M^*$, whose rates cross over around 30 atm.

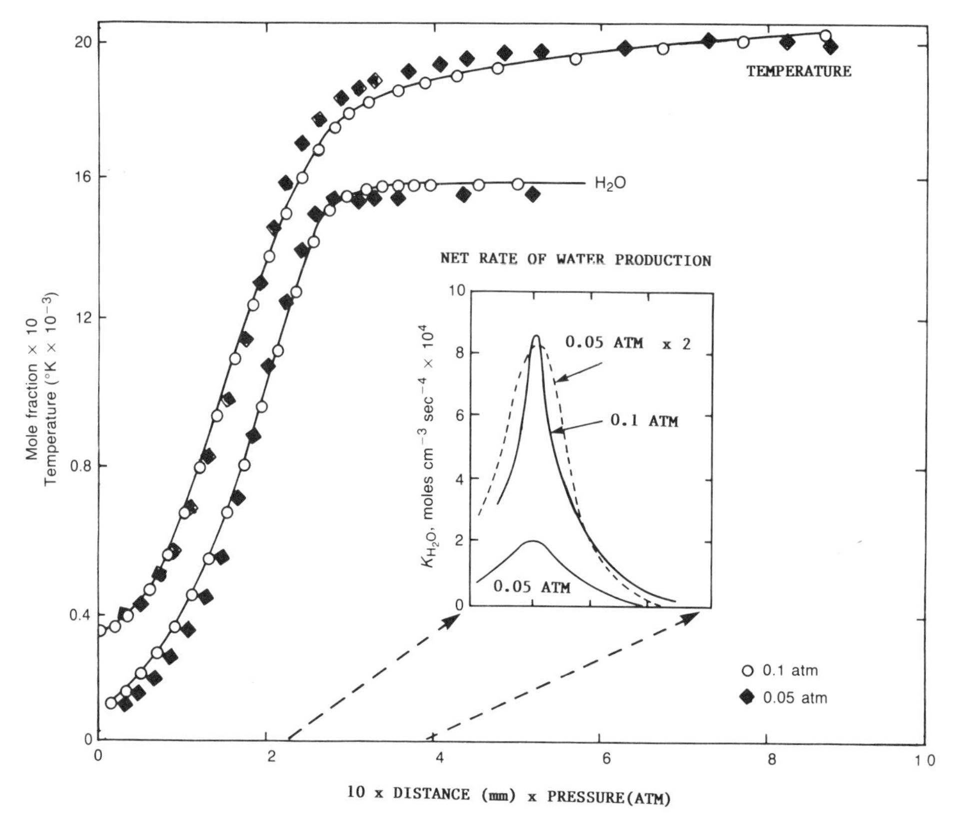

FIG. X-12 Pressure scaling of Temperature and water in the exemplar methane–oxygen flame (see Fig. X-1). (After Fristrom and Westenberg [1965].)

Partial Equilibrium

In flames radicals are generated by bimolecular reactions but are destroyed by termolecular reactions. Both types of flame reactions are rapid in the sense of being high-probability reactions per collision under flame conditions. However, their relative speeds are controlled by the ratio of two-body to three-body collisons, and this is a function of pressure. This means that at low and moderate pressures three-body reactions will be relatively slow and there will be an initial nonequilibrium overproduction of radicals peaking at the boundary between zones II and III. This is slowly brought into final equilibrium by the recombination in zone III. This nonequilibrium radical peak is characteristic of flames: The lower the pressure, the higher the peak. In oxygen and other spin-exchange-type flames this is limited to the availability of oxygen spins. For example, Eberius, Hoyermann, and Wagner [1971] obtained a 22 percent peak in H atoms in a 25 percent oxygen–hydrogen flame. Because the bimolecular reactions are rapid compared with recombination, a reasonable approximation for the region can be made assuming that the bimolecular reactions stay in equilibrium with one another while the system slowly recombines toward final equilibrium. This is called the *partial equilibrium assumption.* It has been used in many fields and was suggested for combustion by Bulewicz, James, and was Sugden [1956] and was formalized by Kaskan and Schott [1962]. By adding the assumption of constant enthalpy, which is reasonable for this zone of low gradients, it becomes possible to calculate the composition at any given temperature (Fristrom, Favin, Linevsky, Vandooren, and Van Tiggelen [1992]). This occurs because the combination of constant enthalpy, elemental conservation, and the bimolecular equilibrium completely constrain the system. A test of the results of such calculations for estimating the peak OH concentration in methane flames is shown in Table X–2. The approximation has been examined theoretically and experimentally by a number of authors (Kaskan [1958]; Schott [1960]; Biordi, Lazzara, and Papp [1975c]; Dixon-Lewis, Greenberg, and Goldsworthy [1975]). The general conclusions are that it is applicable to many, but not all, flames.

Unity Lewis Number Approximation

A correlation would be expected between temperature rise and composition change in flames. In regions where no reaction occurs the condition for a linear relation is that the Lewis number (Hirschfelder et al. [1964]) [$\lambda/(\rho C_p D)$] be unity. It is the dimensionless ratio of species diffusivity to the thermal diffusivity. The ideal kinetic theoretic relation is 25/12. The concept was extended by Klein [1957] to include reaction. The requirement for this is that heat release be linearly related to the rate of disappearance. These conditions are only approximately met in real flames because for polyatomic molecules in flames the Lewis number varies between 0.7 and 1.3 (Westenberg and Fristrom [1965]), and reaction rate is not a linear function. A modification of this concept is used in the zonal flame theory of Fristrom, Favin, Linevsky, Vandooren, and Van Tiggelen [1992]. This improves the applicability for deviant species such as hydrogen by dividing the problem into two regimes, one with pure transport (Zone I) and one dominated by reaction (zone II). Despite limitations the concept can provide a reasonable approximation for many species (Fig. X-1).

Table X–2 Comparisons of Absolute Partial Equilibrium Calculations of Peak OH Concentration with Experimental and Flame Theory

ER Ref[a]	MB	LA	SC	SM	TS	WE	PE
0.8	—	0.0053	—	0.0053	0.0065	0.0050	0.0072
1.0	0.0062	0.0060	0.0053	0.0072	0.0072	0.0065	0.008
1.2	—	0.0042	—	0.0057	0.0047	0.0042	0.0071

[a]MB, molecular beam sampling; LA, laser fluorescence studies by Cattolica, Yoon, and Knuth [1982]; SC, scavenger microprobe by Fristrom and McLean [1981]; SM, flame theory calculation by Smoot et al. [1976]; TS, Tsatsaronis [1978]; WE, Westbrook [1979]; PE, Unpublished absolute partial equilibrium calculations by Fristrom, Favin, and Linevski; ER = $[Fuel/O_2]/[Fuel/O_2]_{STOICH}$

The Steady-State Approximation

Flames are overall steady state, but steady state cannot be generally applied locally through the flame front because of the strong effects of molecular transport processes. The question was examined for the hydrogen–bromine flame by Campbell et al. [1967]. Despite this there are points in all flames and all species where it can be quantitatively applied. These points occur wherever a species flux passes through a minimum or maximum where the net production rate of that species is zero by definition and steady state, applies rigorously. The concept is useful in identifying the temperature of initial radical production in CHO flames. Here the onset of radical production is identified with the steady state, where the rate of radical production by $H + O_2 = OH + O$ is balanced by recombination by $H + O_2 + M = HO_2 + M^*$. This lies close to 1000 K for hydrogen and most CHO flames, as was indicated by Dixon-Lewis and Williams [1979] in radical generation rate calculations (see Fig. X-13). It also lies close to the minimum in H, O, and OH radical flux and the peak in HO_2 concentration (Fig. X-14).

The Ignition Approximation

Mallard and Le Chatelier [1883] observed that if a temperature exists below which reaction is negligible, that conservation of energy requires that flame propagation be balanced by transport at and below that point. The consequent relation between diffusivity, burning velocity, and the temperature gradient here can be related to the characteristic flame thickness of the primary reaction zone (Evans [1952]). Flames commonly show an initial region of pure transport preceding the primary reaction zone. This occurs in flames because reaction is delayed until radicals can be generated. This

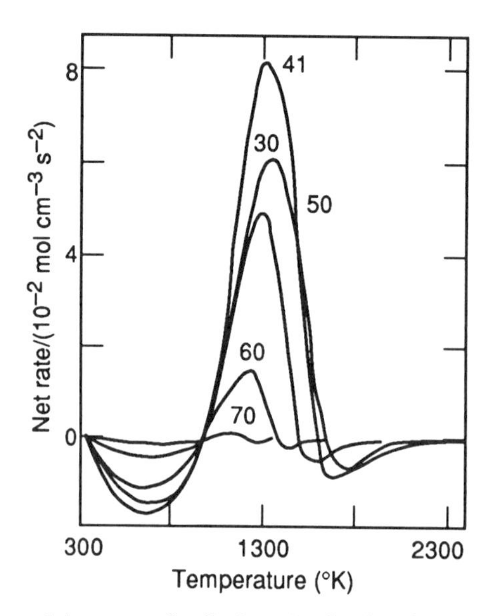

FIG. X-13 Steady state and the onset of radical production in a family of hydrogen–oxygen flames. (After Dixon-Lewis [1979a].)

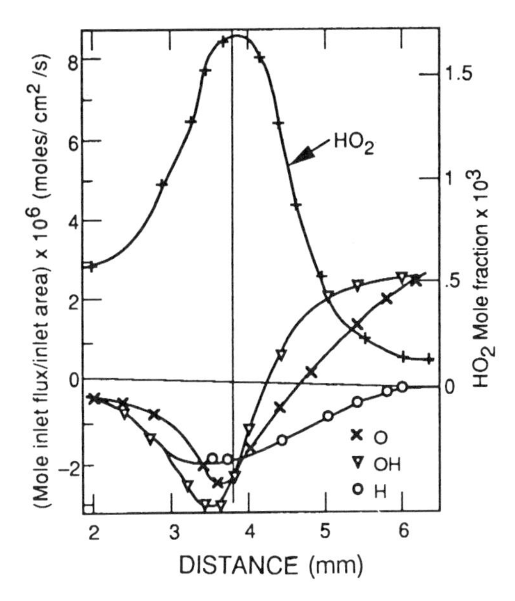

FIG. X-14 Congruence of peak of HO_2 with minima in fluxes of H, O, and OH in a low-pressure methane flame. (Replotted data of Peeters and Mahnen [1973].)

behavior, often called *the ignition approximation,* is a common but not universal behavior in flames. It has been validated for many hydrocarbon–oxygen flames (Fristrom et al. [1960–1967]) (see, for example, Fig. IX-3), but there are exceptions, such as fuel-rich hydrogen–oxygen flames (see Chapter XII), where hydrogen atoms can back diffuse. Hirschfelder [1959] observed that an important consequence of this approximation is decoupling of energy and species flux, removing the mathematical necessity for a flame holder. It is used in approximate flame theories (Evans [1952]).

It is interesting to observe that where the ignition approximation applies the "excess enthalpy" concept proposed by Lewis and von Elbe [1987] of approximate flame theories also applies, providing one identifies the relevant concentration, variable as flux rather than concentration as was done in the original work.

XI

DECOMPOSITION AND HALOGEN FLAMES—DISSOCIATION-INITIATED SYSTEMS

As was pointed out in the previous chapter, flames are driven by the reactions of radicals, and their mode of generation provides a key to the chemistry. There are two general mechanisms for producing radicals: (1) dissociation and (2) radical exchange. This chapter is devoted to the first class, where reactive radicals are produced by dissociation. Dissociation reactions require breaking of chemical bonds and characteristically have high activation energies lying between 35 and 60 kcal/mol (Table XI–1). The major members of this class are unimolecular decomposition flames and the halogen flames with various fuels.

Decomposition Flames

In the early days of flame theory, unimolecular decomposition flames were sought out because it was thought that they would provide simple systems with minimal transport, molecular weight, and kinetic complications. This simplified ideal flame was called the "$A \rightarrow B$" flame by Hirschfelder. The concept played an important role in the early development of flame theory by Hirschfelder, Curtiss, and Bird [1964] by providing simple theoretical illustrations of the effects of various physical processes in flames as documented by Hirschfelder and McCone [1959]. Unfortunately no such ideal flames have been found. Even the simplest experimentally realizable system, the ozone decomposition flame, involves three components undergoing three reversible reactions and requires consideration of multicomponent diffusion and thermal diffusion. Other decompositions have even more complex chemistries.

Unimolecular decomposition flames are a mixed lot. They can be fuels such as acetylene (C_2H_2) or hydrazine (N_2H_4), oxidizers such as ozone (O_3), nitric oxide (NO), or chlorine dioxide (ClO_2). Alternatively, they can be molecules containing both oxidizing and reducing substituents (fuel) such as methyl nitrite (CH_3NO_2), ammonium nitrate (NH_4NO_3), ammonium perchlorate (NH_4ClO_4), or perchloric acid ($HClO_4$). Some systems such as H_2O_2 have suitable thermodynamics but have not been observed. The adiabatic combustion conditions for some representative unimolecular decomposition systems are given in Table A–6, and a list of references on their burning velocities are given in Table B–1 in the appendix at the end of the book.

Table XI–1 Approximate Bond Strengths (kcal/mole) (Pauling [1980])

Bond	Strength	Bond	Strength	Bond	Strength
C–H	100	C–N	73	Cl–Cl	58
O–H (H_2O)	111	C=N	147	Br–Br	46
N–H (NH_3)	93	C≡N	213	I–I	36
Si–H (SiH_4)	76	C–O	86	N≡N (N_2)	170
S–H (H_2S)	83	C=O	176	N–H (N_2H_4)	39
C–C	83	C–F	116	B–C	89
C=C	146	C–Cl	81	S–F	68
C≡C	200	F–F	64	Si–Cl	91

The Ozone Flame

The ozone decomposition flame exhibits all of the elements of complex flames, while its chemical simplicity has made it a favorite proving ground for flame theories.

Species and Kinetics

The flame has three components: O, O_2, and O_3, which undergo three reactions ($O_3 + M^* = O + O_2 + M$, $O + O + M = O_2 + M^*$, and $O + O_3 = 2O_2$). The irreversible, bimolecular chain-breaking step makes this one of the few examples of a chain reaction with a chain length of unity. As a consequence, the reaction depends globally on the decomposition reaction. Ozone is also a strong oxidizer and supports combustion of a number of fuels. These systems are described in Chapter XIII. Ozone (O_3) is an allotrope of oxygen with a pungent odor that one associates with ultraviolet lamps and sparking electric motors. It is poisonous in high concentrations. Although at atmospheric pressure it appears colorless, it absorbs weakly in the red (Chapius bands), the liquid is a deep blue, and the solid is purple. Even diluted ozone detonates and is dangerous; so experimental studies are limited, and no structural studies have been reported. It is so unstable that it must be made in situ using an ozonizer, which is a silent electric discharge at atmospheric pressure. This produces 5–10% ozone in oxygen, which can be concentrated by low-temperature distillation. Information on ozone properties and production can be found in the bibliography of Thorpe [1955] and the *Encyclopedia of Chemical Technology* (Grayson [1988]).

The product, oxygen, is a colorless gas that is difficult to liquify. It is available in high purity in cylinders pressurized to about 150 atm. It is an unusual molecule in having a triplet ground state with two unpaired spins, making the molecule diamagnetic.

Oxygen atoms can be produced transiently in a low-pressure electric discharge.

Ignition and Related Studies

The Heidelberg combustion group of Maas, Rafel, Wolfrum, and Warnatz [1986, 1988] have studied laser ignition and modeled it and the subsequent flame propagation. The ozone decomposition has been studied in shock tubes by Jones and Davidson [1962]. The kinetics has been reviewed by Johnston [1968]. More recent information may be obtained from the NIST compilation of Westley et al. [1990]. Campbell [1978] reviewed the parameters required for the flame.

Burning Velocity and Detonations

Streng and Grosse [1957] observed detonation in ozone–oxygen mixtures containing as little as 20% ozone and measured the burning velocity of various atmospheric ozone–oxygen mixtures (Table XI–2). No structural studies were found. References on the burning velocity of unimolecular flame systems are collected in Table B–1 of the appendix at the end of the book.

Modeling

This proving ground for premixed laminar flame theory has been reviewed by Heimerl and Coffee [1980] and Oran and Boris [1981]. The pioneering study in 1934 by Lewis and von Elbe [1987] is now principally of historic interest because of the required simplifications and the poor state of transport and rate information at the time. However, this was the first computation of detailed flame structure. This was followed by efforts of Hirschfelder, Curtiss, and Campbell [1953] and Campbell [1955, 1978] at Wisconsin. Other efforts were made by Von Karmen [1957] in the U.S., Cramerossa and Dixon-Lewis [1971] in Leeds, and Bledjian [1973] in Russia. The next efforts were by Margolis [1978], Wilde [1972], and Warnatz [1978].

Approximate models have been offered by Rogg and Wichman [1985], using an asymptotic analysis, and Fristrom [1993], applying the zonal model. An example of the structure of the ozone flame is given in Figure IX-4. Some results that show the dependence of burning velocity, and structural parameters on dilution are given in Table A–15 of the appendix.

Talbe XI–2 Comparison of Burning Velocity Predictions of Various Models for the Ozone Flame with Experiment

O_3[a]	SG	CDL	Wi	Wa	HC	RW	Fr[b]	Fr
1	463	725	600	445	497	597	483	374
0.75	333			350	396	468		
0.5	193			225	248	281		
0.25	52	100	85	65	64	67		
0.2	24			37	33			

[a]Ozone concentration is in mole fraction. Diluent is molecular oxygen. Initial temperature is 300 K and pressure is 1 atm. References are: SG, Streng and Grosse [1957]; CDL, Cramerossa and Dixon-Lewis [1971]; Wi, Wilde [1972]; Wa, Warnatz [1978]; HC, Heimerl and Coffee [1980]; RW, Rogg and Wichman [1985]; and Fr, Fristrom [1992].

[b]The results from the zonal model Fristrom [1992] in column eight uses the kinetic constants of Warnatz [1978]; the results in column nine use the kinetic constants of Himerl and Coffee [1980].

Recent theories have successfully modeled the composition dependence of burning velocity with equivalence ratio (see Table XI–2), despite the fact that they used significantly different rate coefficients and transport coefficients. Hiemerl and Coffee [1980] compared their results with Warnatz [1978] and found that the two theories differ significantly in predicted oxygen atom profiles. The difference was removed, however, when the same coefficients were used. Unfortunately experimental results are not available for comparison.

The success of approximate theories and insensitivity of burning velocity predictions probably occurs because the limiting slow step is the initial decomposition. Such systems can be modeled using a single global reaction. Although the kinetic fits have differing numerical constants, they agree in the critical temperature region. The down side of this is that burning velocity becomes insensitive to chemical details. Experimental studies of structure over a range of pressures and concentrations are required. Theoretically, this is a well-documented system, but experimentally it is weak.

Hydrogen Peroxide Decomposition

Hydrogen peroxide (H_2O_2) is a pale blue viscous liquid boiling at 425 K (Schumb [1955]). Its decomposition into water and molecular oxygen yields a significant amount of heat. As a 3% aqueous solution it is a familiar household disinfectant for minor injuries. Flames with hydrogen peroxide as the oxidizer are mentioned in Chapter XIII. The decomposition reaction catalyzed by an iron oxide reactor was used for torpedo propulsion during World War II and as an oxidizer in the first jet fighters introduced by the Germans at the end of WW II.

Thermodynamics suggests that hydrogen peroxide could support a gas-phase decomposition flame. The chemistry would involve the standard hydrogen–oxygen mechanism (see Chapter XII) with the addition of reactions for hydrogen peroxide itself (Table XI–3).

No studies of the system were found in the literature. This may stem from a lack of interest, because low-pressure and/or high-temperature are required to initiate homogenous combustion or perhaps because of the violent reputation of the concentrated liquid. Hydrogen peroxide with over 90 percent purity is commercially available in plastic bottles containing an inhibitor. Even the inhibited liquid should be treated with respect since it is potentially explosive, particularly in contact with organic material. Monger et al. [1984a, b] discuss hydrogen peroxide explosions.

Table XI–3 Reactions of Hydrogen Peroxide

Reaction
$H_2O_2 + M^* = H_2O + O + M$
$H_2O + M^* = OH + OH + M$
$H_2O_2 + M^* = H_2 + O_2 + M$
$H_2O_2 + O = OH + HO_2$
$H_2O_2 + OH = H_2O + HO_2$
$H_2O_2 + H = H_2 + HO_2$

The Nitric Oxide Decomposition Flame

Nitric oxide (NO) forms a gas-phase unimolecular flame that was originally thought to be the simple four-center bimolecular reaction, $NO + NO = O_2 + N_2$. The true reaction mechanism is more complex and probably involves the mechanism suggested by Zeldovich [1946] (Table XI–4).

Using a single-step mechanism with a rate suggested by Farrington Daniels based on the thermochemistry, Henkel, Hummel, and Spaulding [1949] predicted the existence of a decomposition flame with a burning velocity of 2–3 cm/sec.

The flame was demonstrated experimentally by Parker and Wolfhard [1949], who found a substantially higher burning velocity. It is difficult to ignite and must be preheated to almost 1000 K to propagate. This can lead to prereaction, which inhibits propagation. They surmounted these problems by using a short preheat bed and igniting the system initially with mixed hydrogen and slowly decreasing the hydrogen until only a pure NO flame was left. Even this was not straightforward, since the authors also note that the ignition of H_2–NO mixtures requires initial boosting by added N_2O or NH_3. The NO decomposition flame is described as grey with a relatively sharp, inner luminous zone. The spectrum comprised essentially oxygen Schumann–Runge bands that extend from the ultraviolet into the visible.

The system would appear to be worthy of further study using the better kinetic and transport parameters available today. Nitric oxide supports combustion of a number of fuels. These systems are described in Chapter XIII.

The Hydrazine Decomposition Flame

Murray and Hall [1951] and Pannetier and Guedeney [1954] showed that hydrazine (N_2H_4) supports a decomposition flame even when diluted with water or inert gases (Fig. XI-1). The flame is described as yellow-brown with a persistent afterglow by Gray and Kay [1955]. NH emissions were identified by Hall and Wolfhard [1956]. Hydrazine is relatively involatile (BP = 387 K) so that it is necessary to study the flame at reduced pressure. It has been used as a rocket fuel and was one of the early systems tested by flame theory computation by Henkel, Hummel, and Spaulding [1949]. Hussain and Norrish [1963] studied the decomposition using flash photolysis techniques and concluded that there were two modes of decomposition: $2N_2H_4 = 2NH_3 + N_2 + H_2$ and $N_2H_4 = N_2 + 2H_2$. Partial decomposition gives maximum heat release (34 kcal/mole) because ammonia has a negative heat of formation. The mechanism presumably involves hydrogen atoms and NH, NH_2, and N_2H_3 radicals.

Table XI–4 Zeldovitch [1946] Nitric Oxide Mechanism

$NO + M^* = N + O + M$
$O_2 + M^* = O + O + M$
$O + NO = N + O_2$
$N + NO = N_2 + O.$

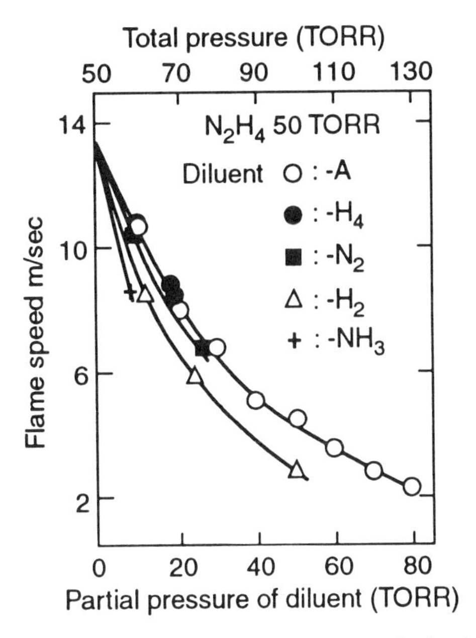

Fig. XI-1 Effect of dilution on the flame speed of low-pressure hydrazine decomposition flames. Burning velocity can be estimated by dividing by the expansion ratio estimated to be 10 ± 2. (Replotted data of Gray, Lee, Leach, and Taylor [1957].)

Antoine [1962] measured the decomposition velocity of the liquid at pressures ranging from 1 to 18 atm. He suggested the mechanism involved hydrogen atoms.

Gray et al. [1957] studied the burning velocity of pure hydrazine and the effect of various diluents (see Fig. XI-1). They studied the pure material over a range of pressures and found a slight increase with increasing pressure.

Maclean and Wagner [1967] studied the structure of hydrazine decomposition flame at two different pressures (Fig. XI-2). They found the flame stable and were able to measure reactants and products. However, they failed to detect N_2H_3, NH_2, and NH using absorption spectroscopy. Detection limits were 1% for N_2H_2 and NH_2 and 0.1%

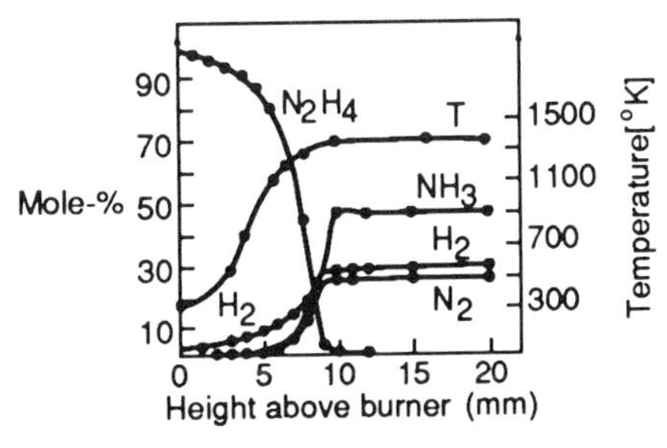

Fig. XI-2 Structure of a low pressure hydrazine decomposition flame (P = 14 Torr; T_0 = 353 K, v_0 = 152 cm/sec). (Data of MacLean and Wagner [1967].)

for NH. Emission from excited NH was found. A maximum was found in NH_3, indicating some decomposition of NH_3 formed in the flame, probably attack by H atoms. No measurements were made of atoms.

Early attempts by Adams and Stock [1953] to model this flame used the overall reaction scheme suggested by Hussain and Norrish [1963]. Spalding [1956] used the hydrazine flame as a proving ground for his nonsteady-state "marching" method of solving the flame equations. This method was utilized by later workers, such as Adams and Cook [1960] and Dixon-Lewis et al. [1967,1970]. A complete mechanism should involve reactions of H, H_2, N, NH, NH_2, NH_3, N_2H, N_2H_2, N_2H_3, and N_2H_4.

Hydrazoic Acid Decomposition Flame

Hydrazoic acid (HN_3) supports a decomposition flame and detonates. It is a colorless, low-boiling (310 K) liquid. Both burning and detonation were investigated by Hajal and Combourieu [1961, 1962] and by Laffitte et al. [1965]. They measured the burning velocity of some mixtures of hydrazoic acid with nitrogen (See Fig. XI-3) as well as quenching diameters and flashback gradients. The pressure dependence of burning velocity was small, and they concluded that the driving reaction was bimolecular. They found the principal products were elemental H_2 and N_2. The calculated flame temperature for the pure gas was 3024 K. The mechanism could involve the species H, NH, N_2, H_2. They found the burning velocity of DN_3 to be slower by a factor of 1.13. This is significantly lower than the 1.4 factor expected if H atom reactions dominated. No structural or modeling studies were found.

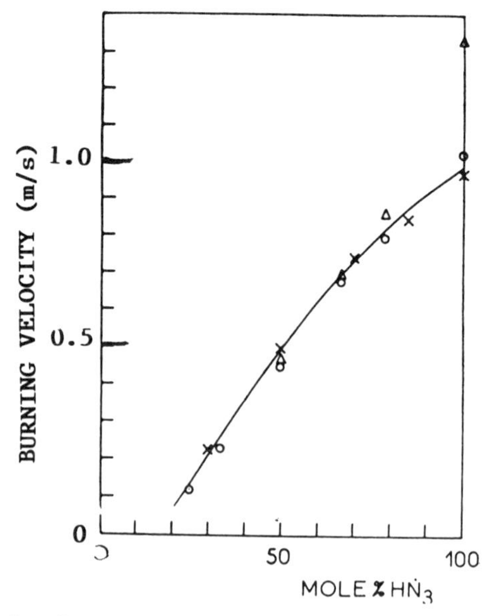

Fig. XI-3 Burning velocity of low-pressure (50 torr, T_0 = 296 K) hydrazoic acid as a function of dilution by nitrogen. (Data of Laffitte, Hajal, and Combourieu [1965].)

The Chlorine Dioxide Decomposition Flame

Chlorine dioxide (ClO_2) is a greenish-yellow gas. Schumacher and Stieger [1930] observed that it begins to decompose thermally at 313 K and explodes at 323 K. Despite this violent behavior, a decomposition flame can be supported at reduced pressure, as was investigated by Laffitte et al. [1967] with He, Ar, and O_2–N_2 as diluents. They measured burning velocity and quenching distance as a function of dilution (see Fig. XI-4) and deduced an overall activation energy of 8 kcal/mole.

This flame probably involves Cl atoms and the ClO radicals. It was suggested by McHale and von Elbe (comments on the paper of Laffitte et al. [1965]) that the adduct ClO–ClO_2 may play a role in the degenerate branching. ClO_2 is a strong oxidizer and supports flames with many fuels, which are discussed in Chapter XIII.

Perchloric Acid Decomposition Flame

As part of a program of study on perchloric acid ($HClO_4$) flames (see Chapter XIII), Cummings and Pearson [1965] investigated the decomposition of concentrated aqueous perchloric acid. It should be noted that because of the low vapor pressure of perchloric acid relative to water, the 72% acid concentration in the liquid phase becomes 32% in the gas phase. Even this relatively dilute flame propagates at 19 cm/sec. They were unable to obtain flame propagation at higher dilutions. Perchloric acid is dangerous to handle and explodes on contact with traces of organic material. It supports the combustion of a number of fuels, which will be discussed in Chapter XIII.

Acetylene and Methyl Acetylene Decomposition Flames

Both acetylene (C_2H_2) and methyl acetylene (C_3H_4) support decomposition flames. The major products are hydrogen and solid carbon with traces of hydrocarbons. Acet-

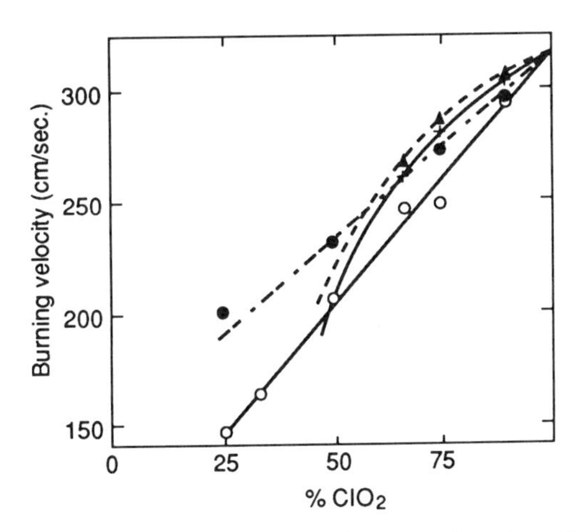

FIG. XI-4 Burning velocity of ClO_2 decomposition flame as a function of dilution by various inert gases, P = 200 torr; T_0 = 300 K). Diluents are designated as follows: Solid circles, helium; open circles, Argon; solid triangles, oxygen; and crosses, nitrogen. (Data by Laffitte, Combourieu, Hajal, Caid, and Moreau [1965].)

ylene detonates when compressed and is very flammable. The decomposition flame was first observed by Jones at the Bureau of Mines [1944].

The system is unusual in having a lower pressure limit. Duff, Knight, and Wright [1954] found that the limit for detonation lies below that for flame propagation. This behavior is discussed by Gaydon and Wolfhard [1979]. An extensive study was made by Cummings, Hall, and Straker [1962] of the burning velocities of acetylene and methyl acetylene as a function of pressure (see Table XI–5). They attributed this to radiation losses, which increase in importance as pressure drops.

Chase and Weinberg [1963] constructed an ingenious "peristaltic" pump that allowed safe handling of the material. They studied the limiting pressure for acetylene and acetylene–air mixtures and concluded that the mechanism involved H and C_2H.

Acetylene is believed to be a key intermediate in soot production in flames. As a result there have been a large number of studies of acetylene chemistry related to soot production. Soot production in flames has been reviewed by Homann [1985], Wagner [1979], and others. Gaydon and Wolfhard's book [1979] provides a good introduction.

It is not clear to the author that the acetylene flame propagation is truly homogeneous, since the soot particles migrate toward the cold gas, where they can induce heterogeneous decomposition reactions. This migration is driven by the temperature gradient in the flame front and is called the *thermomechanical effect*. It is the macroscopic analog of thermal diffusion. For small (compared with the mean free path) soot particles, this increases the residence time threefold.

Acetylene and methyl acetylene burn readily in air or oxygen. These flames are described in Chapter XII.

The Ethylene Oxide Decomposition Flame

Burdon and Bourgoyne [1949] first demonstrated that ethylene oxide (C_2H_4O) could support a decomposition flame. It is an exothermic molecule in which the oxygen forms a triangle with the two carbon atoms. It is available commercially. It begins to decompose at 600 K, and the kinetics has been studied by Mueller and Walters [1951] and Lossing, Ingold, and Tickner [1953]. The work was reviewed by Polanyi and Lossing [1953].

Table XI–5 Burning Velocities and Combustion Products of Acetylene and Methyl Acetylene Decomposition Flames as a Function of Pressure

	Acetylene			Methyl Acetylene		
Pressure (Atm)	2.02	4.06	6.1	10	20	30
V_0 (mm/sec)	28	56	65	20	21	22
H_2	0.60	0.8	0.85	0.67	0.69	0.71
Unreacted	0.3	0.11	0.05	0.05	0.04	0.01
CH_4	0.06	0.06	0.07	0.27	0.26	0.29
C_2H_4	0.04	0.03	0.03	0.02	0.01	0.005

After Cummings, Hall, and Straker [1962].

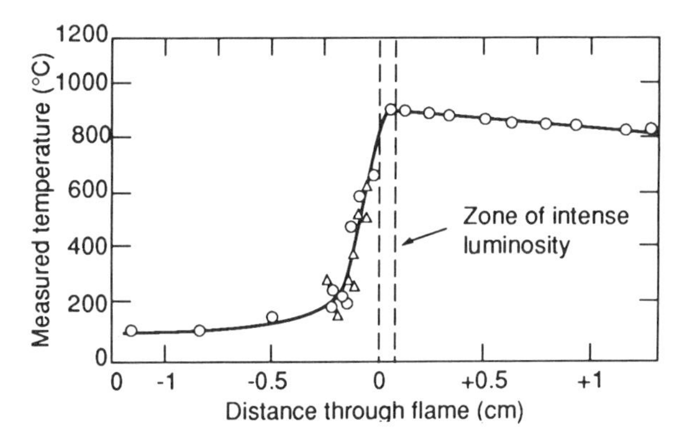

FIG. XI-5 Temperature profile in an ethylene oxide decomposition flame (T_o = 300 K, P = 0.9 atm). (Data of Friedman and Burke [1954].)

The burning velocity was measured by Gerstein, McDonald, and Schalla [1953]. Friedman and Burke [1954] stabilized the flame on a flat flame burner and measured burning velocity, the temperature profile on a 0.9 atm flame (Fig. XI-5), and the composition of the burned gasses. They found the burning velocity only changed from 30 mm/sec at 2 atm to 50 mm/sec at 0.2 atm. The product gases corresponded to 54 percent, yielding CO and CH_4, with 46 percent yielding CO + H_2 + 0.5 C_2H_4. Calculations by the present author show that this is close to the adiabatic condition with soot formation suppressed. This mechanism is supported by the lack of soot in this flame and the fact that the temperature profile peaks immediately beyond the luminous primary reaction zone (see Fig. XI-5). Ethylene oxide burns readily in air and oxygen. These flames are considered in Chapter XIII.

Decomposition Flames of Organic Nitrogen Compounds

Some of the common organic nitrogen compounds that can support decomposition flames are listed in Table XI-6. They differ in the ratio of oxygen to nitrogen and whether the carbon is bonded to oxygen or nitrogen. Nitroso compounds (R–NO) are

Table XI–6 Exoergic Organic Nitrogen Compounds[a]

Type	Formula	Character
(1) Diazo	R–N–N–R'	Nitrogen–nitrogen double bound
(2) Nitro	R–NO_2	Carbon nitrogen bond
(3) Nitrite	R–ONO	Oxygen–carbon bond
(4) Nitrate	R–ONO_2	Oxygen–carbon bond
(5) Nitramine[b]		H_2N–NO_2 Nitrogen–nitrogen bond
(6) RDX		(Cyclotrimethylenetrinitramine)
(7) HMX		(Cyclotetramethylenetrinitramine)
(8) TNT		(Trinitrotoluene)

[a] R and R' indicate aliphatic organic radicals (C_nH_{2n+1}), for example, methyl (CH_3) ethyl (C_2H_5), etc.

[b] Nonorganic included as prototype nitramine propellant.

not included because they isomerize so readily that they are usually not isolatable (Feiser and Feiser [1950]). The inorganic nitramine is included because it is the simplest of the nitramine propellant and explosive family that do contain carbon. They represent both high-energy molecules such as diazo methane and molecules containing both oxidizing and reducing constituent groups, such as the organic nitrite, nitrate, and nitro compounds. There have been more studies of the latter type because they provide prototypes for rocket propellant combustion.

The Azomethane Decomposition Flame

Azomethane (CH_3–N–N–CH_3) is a reactive gas. Studies by Rice and Sickman [1936] showed that it is readily decomposed thermally. Allen and Rice [1935] showed that it can decompose explosively. It is used as a reagent in organic chemistry. It should not be confused with diazomethane (CH_2–N_2), which also is an explosive organic reagent. The material was thought to undergo a unimolecular decomposition; so it was one of the early flame models tested by Hirschfelder's group at Wisconsin. The calculations by Henkel, Hummel, and Spalding [1949] were among the first to utilize machine computations for flame studies. Using low-temperature kinetics they concluded that the flame would be fast (0.8 m/sec) unless diluted. This work is of historic interest as an early example of numerical modeling.

Comparison of Methyl Nitrate, Methyl Nitrite, and Nitro Methane Flames

There are three methyl nitrogen–oxygen compounds (CH_3NO_3, CH_3–ONO, and CH_3–NO_2. All of them can be detonated, but they differ in thermal stability and in their ability to support decomposition flames (Gray and Pratt [1957]).

Methyl nitrate forms a decomposition flame that at low pressures shows two zones. The first is bright blue, the second an orange color. Hall and Wolfhard [1957] attributed the blue luminosity to excited formaldehyde. Methyl nitrite shows only one orange luminous zone and has a much slower flame. They were not able to produce a nitromethane decomposition flame, even though it is isomeric with methyl nitrite, has a similar heat of formation, burns when mixed with oxygen, and can be detonated.

Adams and Scrivener [1955] found that ethyl nitrate (C_2H_5–NO_3), nitrite (C_2H_5–ONO), and nitro ethane (C_2H_5–NO_2) show behavior similar to their methyl analogs. Steinberger [1955] studied ethyl nitrate, deuterated ethyl nitrate, and triethylene glycol dinitrate at pressures as high as 1000 psig. He concluded that radicals, probably hydrogen atoms, were involved in the mechanism. Hicks [1962] has measured the burning velocity as a function of pressure and found it to be 0.135 m/sec and pressure independent below 150 torr and dropping to half that value at 250 torr. He also measured the structure of a low-pressure ethyl nitrate decomposition flame (Fig. XI-6). The chemistry is complex involving the nitrite, formaldehyde and oxides of nitrogen. Unfortunately, radical profile measurements were not feasible at the time. The data were analyzed for heat release rate. Steinberger and Schaaf studied decomposition flames of ethylene glycol dinitrate [1958]. Melius and Binkley [1988] and Mitani and Williams [1989] studied the deflagration of nitramines.

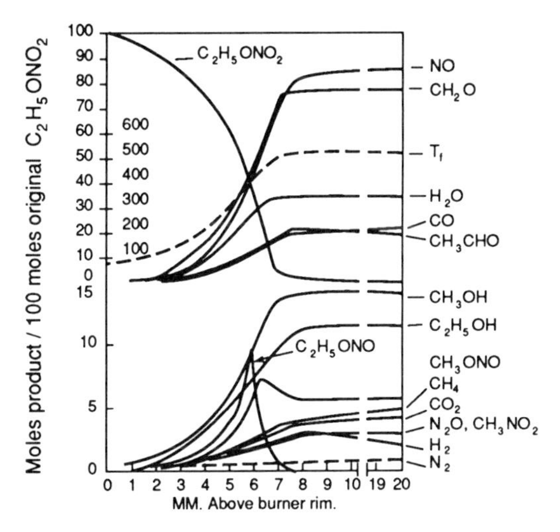

FIG. XI-6 Characteristic profiles of an ethyl nitrate decomposition flame (P = 35 torr; T_0 = 300 K). (a) Composition; (b) temperature and heat release profiles. (After Steinberger [1955].)

Halogen Flames

The second group of flames that are driven by dissociative production of reactive radicals are the halogens with various fuels. They form a family of four elements with related chemistries. There would be five, if the radioactive astatine were considered, but it is an even weaker oxidizer than iodine. In order of increasing atomic number and atomic weight and decreasing reactivity, they are: fluorine, (F), chlorine (Cl), bromine (Br), and iodine (I). The standard state (STP 298 K and 1 atm) is gaseous for fluorine and chlorine, liquid for bromine, and solid for iodine. In the gas phase they are diatomic at STP conditions. At flame temperatures the atomic form dominates. The exception is iodine which forms flame systems. Fluorine is a light yellow gas that is so reactive that it destroys most metals and glasses and must be handled in metal vessels that have been passivated by an adherent metal fluoride coating. Copper is particularly suitable for this purpose. The handling of fluorine is very dangerous and requires rigorous safety precautions. Chlorine is a greenish-yellow, poisonous, easily liquefied gas that is less corrosive than fluorine and can be handled using Monel vessels and valves. It is available in cylinders as a liquefied gas. Chlorine is a dangerous material, though less so than fluorine. Bromine is an easily volatilized red liquid. It is available as a laboratory chemical. The odor is suffocating, living up to its name, which is Greek for stench. It presents a health hazard. Iodine is a black solid, which is unusual in subliming at atmospheric pressure. The combustion behavior of the halogens ranges from the violently reactive fluorine through the mildly reactive iodine, which forms flames only with vigorous reducing agents and acts as a strong inhibitor in oxygen flames. In addition to the halogens, there are a number of interhalogen compounds (Table XI–7) that are also strong oxidizers. Their properties are dominated by the more reactive substituent. For example, chlorine trifluoride forms flames very similar to those of fluorine.

Table XI–7 Heats of Formation of Halogen Molecules at STP

X Formula	F	Cl	Br	I
X	18.86	28.92	28.18	25.54
XF	0	−12.14	−13.98	−22.19
XCl	—	0	3.5	−4.18
XBr	—	—	0	9.77
XI	—	—	—	0
XF_3	—	−37.97	−61.09	—
XF_5	—	−57	−102.5	−200.8
XF_7	—	—	—	−229.7
XCl_3	—	—	—	−23.4

The halogens have received less study than oxygen flames, but there is a significant body of work. Their kinetics has been critically reviewed by Baluch, Duxbury, Grant, and Mantague [1981]. Their physical and transport properties can be found in the treatise of Reid, Prausnitz, and Poling [1987] and the *International Critical Tables.* Their chemistry is discussed in inorganic chemistry texts such as: Partington [1939], Sidgwick [1950], and Cotton and Wilkinson [1988]. Flames with hydrogen are considered first, then interhalogen flames, and finally, flames with other fuels.

Hydrogen–Halogen Flames

The flames of hydrogen with the various halogens offer interesting contrasts. They are among the simplest of flames, each set having only five species with five reversible reactions (Table XI–8). Their burning velocities (Fig. XI-7) illustrate the influence of reactivity on burning velocity. The stoichiometric fluorine has the highest burning velocity observed, estimated at nearly one-third the speed of sound, while the least reactive member, iodine, is too slow to measure experimentally.

The original Christiansen mechanism is a radical chain (Table XI–8) that was developed for the hydrogen–bromine system. Fluorine and chlorine may require the additional consideration of vibrationally excited species and energy chains, and iodine is complicated by either the direct four-center reaction of hydrogen with iodine or, more likely, by the $I + I + H_2 = 2\,HI$ (see section on iodine). The original scheme

Table XI–8 The Basic Hydrogen–Halogen Mechanism (for Bromine[a])

$Br_2 + M \rightarrow 2Br + M$ (chain initiation)
$Br + H_2 \rightarrow HBr + H$ (chain propagation)
$H + Br_2 \rightarrow HBr + Br$ (chain propagation)
$H + HBr \rightarrow Br + H_2$ (chain retardation)
$2Br + M \rightarrow Br_2 + M$ (chain termination)
$H + H + M \rightarrow H_2 + M^*$ (chain termination)
$Br + H + M \rightarrow HBr + M$ (chain termination)

[a]For other halogen systems substitute the appropriate halogen (F, Cl, I) for Br. For the chlorine–fluorine flame substitute Cl for H and F for Br.

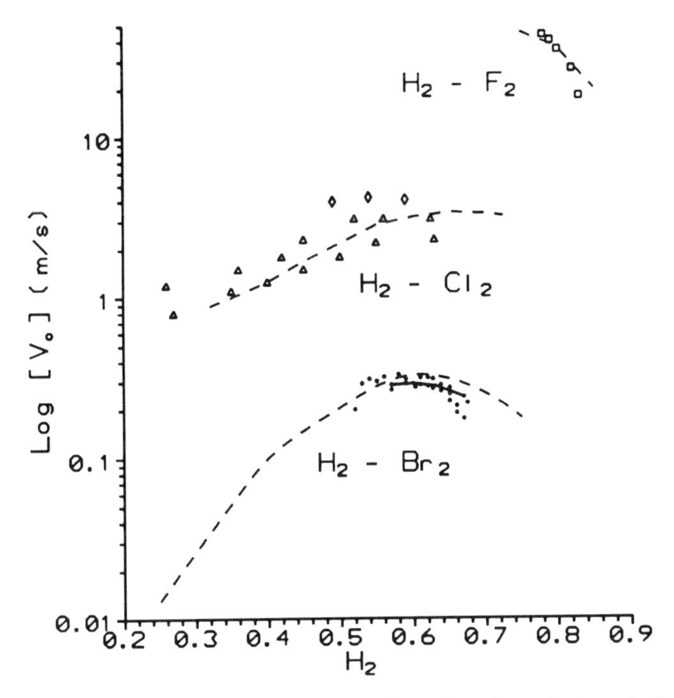

FIG. XI-7 Comparison of atmospheric hydrogen–halogen burning velocity calculations of Fristrom, Favin, Linevsky, Vandooren, and Van Tiggelen [1992] (dotted lines) and Spalding and Stephenson [1971] (solid line for H_2-Br_2) with experimental studies (F_2, 90 K, Gross and Kirshenbaum [1955] (squares); Cl_2 at 293 K Bartholome [1949] (diamonds); Rozlovsky [1956] (triangles); Br_2, Gooley, Lasater and Anderson [1952] (points).

neglected dissociation and recombination reactions. Quantitative structural and modeling studies are required for the complete understanding of these flame systems. A comparison study of hydrogen–halogen flames has been made by Fristrom, Favin, Linevsky, Vandooren, and Van Tiggelen [1992] using the approximate zonal flame theory.

The Hydrogen–Fluorine Flame

Fluorine–hydrogen flames produce among the highest flame temperature and burning velocity (Fig. XI-7). Fluorine is so reactive that it is difficult and dangerous to premix because of possible detonation. Containers must be made from metals that form adherent nonporous fluoride films. Copper is often used. Unfortunately, these films catalyze the reaction of hydrogen with fluorine, compounding the difficulty in studying premixed flames. This problem has been addressed in several ways. Grosse and Kirshenbaum [1955] studied burning velocities using strongly cooled systems. This was feasible because the reactants can be cooled to liquid nitrogen temperatures without condensation. Another approach was that of Slootmaekers and Van Tiggelen [1958], who studied stoichiometric flames diluted with argon. Vanpee, Cashin, Falabella, and Chintappili [1973] studied the flames at low pressure (2.3 torr) and found that near stoichiometric, the burning velocity approaches the lower Chapmann–Jouget velocity, which is the maximum possible for subsonic combustion. Experimental velocities on either side of stoichiometric fall below the calculated CJ maximum values. It is interesting to note that the Mach number of these flames is so high that there is a significant pressure change across the flame front, so that the usual constant-pressure assumption made in flame theory breaks down.

Homann and Maclean [1971a] and Maclean and Tregay [1973] studied low-pressure flame structure, using a multiple diffusion burner (see Chapter III). The structure of this flame has been modeled by Warnatz [1977] (see Fig. XI-8). These studies provide the most comprehensive treatment for the hydrogen–fluorine flame system. The agreement between the modeling and the structure studies was reasonable.

Maclean and Tregay [1973] made a detailed spectroscopic study of the structure of three low-pressure argon-diluted hydrogen fluorine flames and a stoichiometric undiluted flame. The undiluted flame showed a maximum temperature of 3900 K, which is 800 K higher than the calculated adiabatic value, suggesting that HF is not in local equilibrium with the atoms. They observed nonequilibrium excitation of HF vibrational levels up to $v = 8$. The inversion in hydrogen–fluorine combustion is so pronounced that it provides one of the more powerful infrared lasers. Vibrationally excited HF opens channels for atom formation not available normally. These are bimolecular energy transfer followed by dissociation: $HF^* + F_2 = 2F + HF$; $HF^* + F_2 = H + F + HF$. The reactions of hydrogen with fluorine atoms, of fluorine with hydrogen atoms, and the hydrogen atom–fluorine atom recombination, all produce

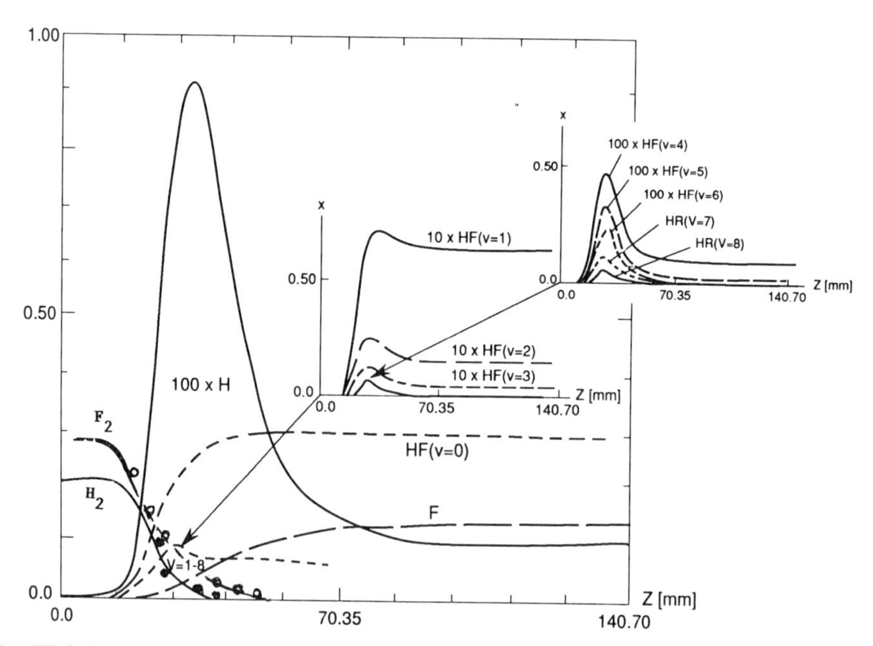

FIG. XI-8 Structure of a low-pressure (7.63 mbar) hydrogen–fluorine flame ($H_2 = 0.199$, $F_2 = 0.272$, $Ar = 0.529$, $T_0 = 298$ K, gas velocity 6 m/sec. Comparison of experimental points by Homann and MacLean [1971] with calculations by Warnatz [1977]. Radical and vibrationally excited species. Each insert is magnified by an order of magnitude from the previous one.

vibrationally excited molecules because of the high heat of formation of hydrogen–fluoride. This is such a stable molecule that it shows no significant dissociation at atmospheric pressure and temperatures below 3000 K. As a result, maximum hydrogen–fluorine flame temperatures approach 4000 K.

Warnatz [1977] employed a family of state-to-state rates in modeling this system and tested several relaxation schemes against the observed vibrational and rotational temperatures in the flame (see Fig. XI-8). Considering the uncertainty in the data, the test was a success.

The Hydrogen–Chlorine Flame

Hydrogen burns vigorously with chlorine. The heat of formation of HCl is 22 kcal/mol. The flame temperature of the stoichiometric mixture is 2494 K (Table A–7 in the appendix at the end of the book). Study of the system is complicated by the problems with handling the corrosive, poisonous chlorine and, in addition, the system is sensitive to actinic light. Corbeels and Scheller [1965] found a complex response of the flame to impurities. The problem of additives is addressed in Chapter XIII. Partington [1939] reports that the photochemical ignition was first observed by Gay-Lussac and Dalton separately. More refined studies showed that true ignition was only obtained in mixtures containing traces of oxygen or water, although significant reaction is observed even in exhaustively dried mixtures.

The system was considered by Kitawaga [1934] as a possible candidate for an energy chain in which excited HCl would induce dissociation. Indeed, vibrationally excited HCl has been observed by Cashion and Polanyi [1958, 1959] in low-pressure hydrogen–chlorine flames.

Burning velocity studies have been made by Bartholeme [1949a, b], Rozlovskiy [1956] (Fig. XI-7), and Slootmaekers and Van Tiggelen [1958]. Cellular behavior was observed by Corbeels and Scheller [1965] in fuel-lean mixtures. This effect compromises burning velocity measurements, even though vigorous combustion occurs in this region. The limits of combustion were reported by Simmons and Wolfhard [1955] as being 12% for lean and 95% for rich mixtures. They investigated the air–hydrogen–chlorine ternary system and reported that mixtures obeyed Le Chatelier's rule, which assumes adaptivity of burning velocities.

Istratov and Librovich [1962] used the theory of Zeldovich [1949]. Van Tiggelen [1952] used his molecular theory. These approximate flame theories were unable to account for the variation of burning velocity with dilution. More recently Fristrom, Linevsky, Favin, Vandooren, and Van Tiggelen [1992] used the zonal flame theory to model the dependence of burning velocity on composition, pressure, initial temperature, and dilution. They were able successfully to model the composition dependence of burning velocity (Fig. XI-7). However, this has been brought into question by a recent experimental study of the structure of a low-pressure flame reported by Vandooren, Fristrom, and Van Tiggelen [1992], who found that an extended reaction scheme was required to model the experimental structure.

The Hydrogen–Bromine Flame

The hydrogen–bromine flame has been a favorite testing ground for flame theory. The kinetics appear straightforward, without the energy chain complications of the fluorine and chlorine flames. The standard mechanism (see Table XI–8) suggested by Christiansen [1919] is based on Bodenstein's [1899, 1907] studies. The thermodynamics, transport, and kinetic constants are relatively well known. The survey of Campbell and Fristrom [1958] is now only of historic interest. Transport is complicated because both thermal diffusion and a multicomponent treatment is required. A substantial effort was mounted in the 1950s to use this system as a proving ground for rigorous flame theory with the hope that this would pave the way for meaningful engineering modeling applications of the theory. The effort was supported by the U.S. Navy and managed by W. H. Avery of The Johns Hopkins University Applied Physics Laboratory. The project's theoretical work at the University of Wisconsin was done under Hirschfelder and Curtiss, their students, and collaborators. They were among the first to recognize the potential of high-speed computers for combustion problems and pioneered modern flame computations using "Model T" computers. The H_2–Br_2 system was undertaken by Campbell [1957]. The results were primitive, but they laid a major foundation for modern modeling studies. Experimental flame studies were made in Anderson's laboratory at New Mexico (see section on burning velocity) and kinetic studies by Pease's [1942] group at Princeton.

The burning velocity of the hydrogen–bromine system has been reviewed by Frazier [1962]. Ohmann [1920], Sagulin [1928], and Kitagawa [1938] demonstrated that a flame could be stabilized in H_2–Br_2 mixtures, and Kitagawa investigated the emission spectrum of the flame, which he suggested was due primarily to bromine. The limits of flammability and the velocity of propagation were studied by Kokochashvili [1951].

Garrison, Lasater, and Anderson [1949] measured flame propagation in a glass tube. Burning velocities from 30 to 50 cm/sec were observed in the pressure range of 380–760 mm Hg. Diluents such as He, N_2, and Ar reduced the burning velocity. Cooley, Lasater, and Anderson [1952] studied burning velocities on a Bunsen-type burner (see Fig. XI-7). Flames were stabilized with up to 66 mole percent H_2, but mixtures near the stoichiometric value tended to flicker. Peacock and Weinberg [1961] noted that diluted bromine-rich flames became cellular.

Cooley and Anderson [1952, 1955] measured the effect on burning velocity of deuterium substitution for H_2 with a Bunsen-type flame. The observed values for the ratio of the burning velocity in hydrogen to that in deuterium were in the range 1.55–1.6. They report that raising the initial temperature of the gas mixture from 50 to 210°C increases the maximum burning velocity from about 32 to 80 cm/sec. During their flash-back and blow-off experiments, Phillips, Brotherton, and Anderson [1953] measured flame temperatures with a relatively large platinum–rhodium thermocouple. Measured values lay 300–400 K below calculated, as expected. Huffstutler, Rode, and Anderson [1955] extended the studies of diluent effect to the Bunsen-type flame. Nitrogen–hydrogen ratios and argon–hydrogen ratios of 0.33, reduced burning velocity considerably, except in mixtures rich in bromine. The effect of helium was to shift the maximum velocity from a composition of 40–44% bromine to approximately 36% bromine.

Experimental work on the structure of H_2–Br_2–N_2 flames has been reported by Peacock and Weinberg [1961]. Flames were stabilized on a flat Edgerton–Powling burner. They were the first to attempt to determine the structure of the hydrogen–bromine flame. The temperature profile was obtained using the "inclined slit" method described in Chapters IV and V. Molecular bromine concentrations were measured by light absorption. Although they were not able to provide a complete description of the flame, they did test published kinetic data by comparing the heat release using the measured temperature profile with that computed from an overall reaction rate mechanism assuming steady state of Br atoms. The experimental and calculated volumetric heat release profiles differed markedly in both position and magnitude. Peacock and Weinberg believe that these large differences were significant.

The most complete studies are those of Frazier, Fristrom, and Wehner [1963] and Frazier and Wendt [1969] on the structures of a spherical hydrogen–bromine flame at reduced pressure (Fig. XI-9). They include temperature and stable species.The only experimental evidence for bromine atom concentration is based on assigning sampled Br_2 to Br in regions where bromine is not visible.

Early attempts to predict the speed of propagation of hydrogen plus bromine flames used approximate equations. This work has been reviewed by Evans [1952]. Campbell [1957] studied simplified models of the stoichiometric hydrogen–bromine flame to determine which factors were important to the steady state flame and investigate the validity of the steady-state approximation for Br atoms. His results indicated that chemical steady-state with respect to Br atoms did not exist in this flame, in contrast to Gilbert and Altman [1956, 1957].

Lovachev and Kaganova [1969] employed the non-steady-state method for solving the simultaneous differential equations that are necessary for the hydrogen plus

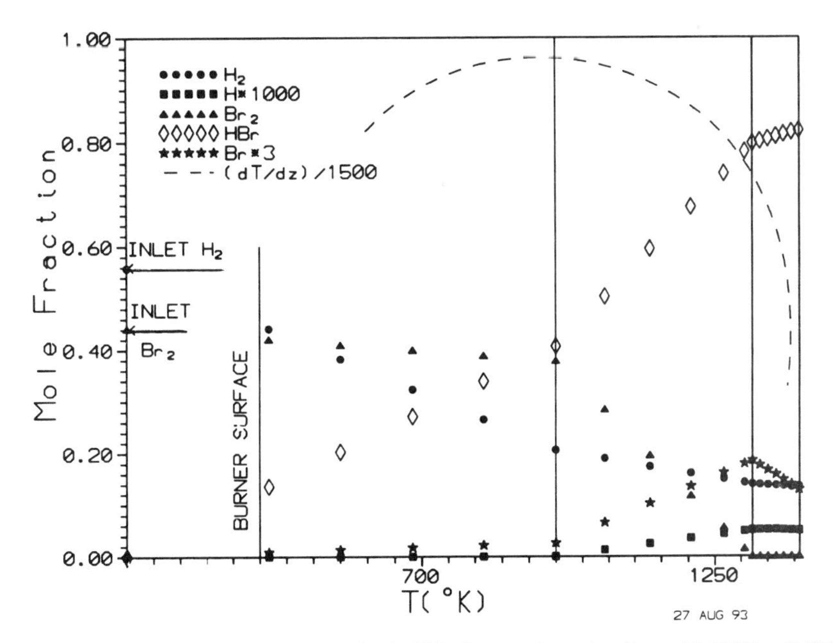

FIG. XI-9 Structure of a low pressure, spherical Hydrogen–bromine flame (0.422 Br_2, 0.588 H_2, T_0 = 594 K, R_0 = 1.25 cm, P = 0.12 atm). (Data of Frazier, Fristrom, and Wehner [1963] with calculations by Frazier and Wendt [1969] realigned 0.4 mm to account for shifts between temperature and composition measurements and replotted as a function of temperature to show linearity in this mode.)

bromine flame. They made simple assumptions concerning the variation of transport and thermodynamic properties through the flame. Halogen kinetics has been critically reviewed by the High Temperature Chemistry group at Leeds University, England, and more recent updates are available through the Reference Standard Data Center of NIST. A detailed comparison between theory and experiment was provided by Frazier and Wendt [1969], using a numerical model with two sets of kinetics. The burning velocity was reproduced within 6% using both sets of kinetics with modified transport coefficients. Some of the detailed structure was reproduced, but there were significant differences. The computation assumed steady state for hydrogen but not bromine and neglected thermal diffusion and the H + Br + M recombination step.

This is another case that shows that agreement with burning velocity does not guarantee agreement with detailed profiles. They showed that moderate changes in the kinetics and transport coefficients made only minor changes in the profiles and burning velocity, and suggested the most likely source of differences between experiment and model is the neglect of thermal diffusion. Spalding and Stephenson [1971] used the "marching method" on the system and obtained good agreement with the measured burning velocities of Cooley and Anderson [1955] (see Fig. XI-7).

Fristrom, Favin, Linevsky, Vandooren, and Van Tiggelen [1992] have applied the zonal model to this system. Agreement with burning velocities of Anderson was good (see Fig. XI-7), and reasonable agreement with structures predicted by Spalding and Stephanson's theory. It was also applied to the spherical flame study of Frazier, Fristrom, and Wehner [1963]. Key parameters were in reasonable agreement, providing account was taken of the radiation losses. The theory was used to derive the effects of composition, temperature, pressure, and dilution on structural parameters. These results are summarized in Table A-13 in the Appendix.

The Hydrogen-Iodine System

Iodine is the least reactive of the halogens, and no flames have been observed with hydrogen. This is related to the kinetics and reactivity of iodine, the low flame temperature (545 K), and consequent low H atom concentration.

In addition, the kinetic mechanism appears to be different. Since Bodenstein's [1899] original studies, the hydrogen–iodine reaction $H_2 + I_2 \rightarrow HI + HI$ had been considered the classic example of a bimolecular four-center reaction. Such reactions closely followed equilibrium. In a series of papers Sullivan [1959–1967] showed that under some circumstances the simple interpretation was incorrect. He found the rate of HI production to be consistent with the reaction $H_2 + 2I \rightarrow HI + HI$, which in thermal systems is kinetically indistinguishable from $I_2 + H_2 = 2HI$. This confirmed earlier calculations by Benson and Srinavasan [1955]. In crossed molecular beam studies, Anderson [1974] found no evidence of the direct bimolecular reaction.

The reactions important for the other halogens (Table XI–9) should occur, but the $I + H_2 \rightarrow HI + H$ is extremely slow. Comments on the H_2–I_2 reaction are given by Baluch, Duxbury, Grant, and Montague [1981] in their evaluation. From the standpoint of combustion the reaction $H_2 + I + I = 2HI$ is a three-body recombination that destroys spins.

Table XI-9 Energy Chain Reaction Scheme for H_2–Cl_2 Flame[a]

$HCl^* + Cl_2 = Cl + Cl + HCl$; $k = 115 \times 10^{13} \exp(14000T)$ $cm^3\ mol^{-1}\ s^{-1}$
$H + Cl_2 = HCl^* + Cl$
$Cl + H_2 = HCl^* + H$

[a]According to Vandooren et al. [1992].

Fluorine, Chlorine, and Interhalogen Flames

The Chlorine–Fluorine Flame

The only interhalogen flame that has been documented is the chlorine–fluorine system by Fletcher and Ambs [1968]. Burning velocities (Fig. XI-10) and a temperature profile of an argon-diluted flame have been measured (Fig. XI-11). Using this temperature profile Ambs and Fletcher [1971] calculated a composition profile to match this temperature profile assuming a Lewis number of unity, the measured burning velocity, and the low-temperature reaction rates.

The chemistry of the stoichiometric flames, where the formation and reactions of higher fluorides can be neglected, is straightforward, involving only the five reversible reactions of the Christiansen–Polanyi mechanism in which Cl plays the role of H and

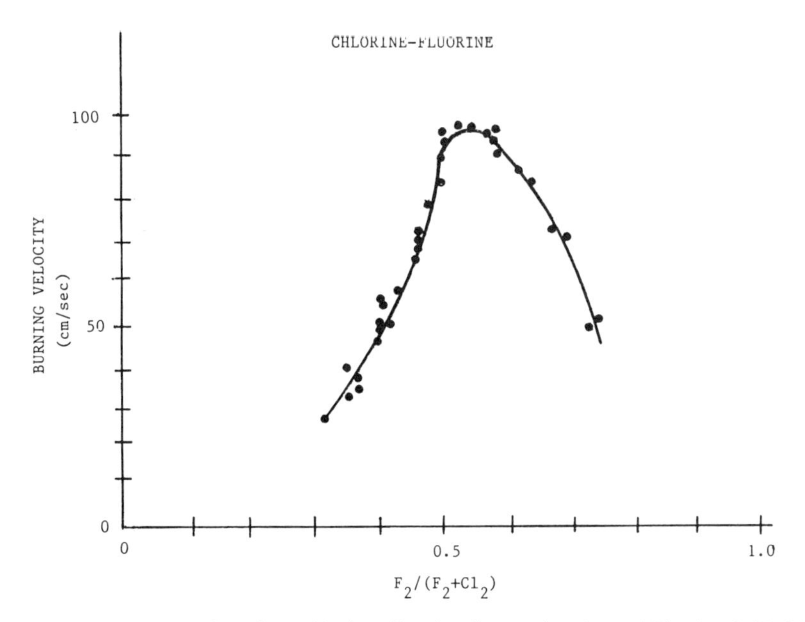

FIG. XI-10 Burning Velocity of the chlorine–fluorine flames. (Ambs and Fletcher [1971].) Approximate values were obtained by assuming burning velocity is proportional to measured flame speed measurements and scaling using the experimental value of 0.4 m/sec determined at 36% F_2.

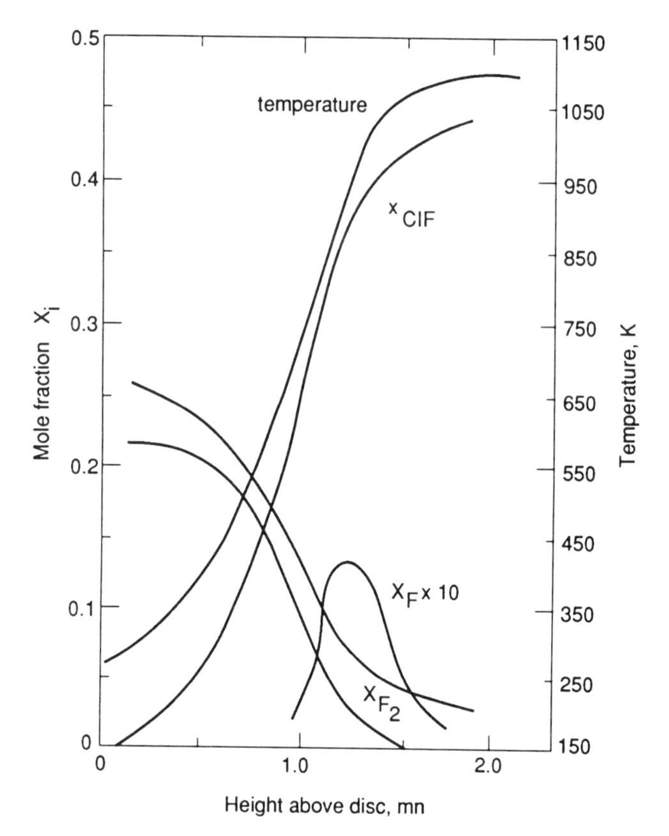

FIG. XI-11 Structure of a chlorine–fluorine flame (Cl_2 = 0.254, F_2 = 0.221, Ar = 0.525, P = 980 Torr, heat extracted to produce a final temperature of 1073 K; mass flow 0.0882 g/sec; 3 cm diameter burner. (Ambs and Fletcher [1971].) Temperature is experimental but compositions are calculated assuming unity Lewis number.

the halogen is F (see Table XI–8). Reaction appears initiated by halogen dissociation but dominated by the chain of radical exchange reactions.

Other Halogen-Supported Flames

Halogens form flames with many other molecules. Volatile elements such as phosphorous and sulfur form flames, and dusts of involatile metals can be burned by fluorine and chlorine. Hydrocarbons, as well as the hydrides of boron, silicon, phosphorous, and arsenic, also burn with fluorine and chlorine (Partington [1939]; Cotton and Wilkinson [1988]). Fluorine also burns many halocarbons. Only a few of these systems have been characterized by quantitative combustion studies. Fletcher and Kittelson [1968a 1969] at Minnesota have studied a number of flames of halocarbons with chlorine trifluoride and fluorine. They have characterized the final products, which do not always correspond to equilibrium adiabatic combustion. These systems undergo vigorous combustion and in some cases such as perfluorocyclobutane can even detonate with fluorine. The products are complex, being controlled by thermodynamics at low fluorine equivalence ratios and by kinetics for fluorine-rich systems (Fletcher [1983]).

Fluorine-Supported Flames

Fluorine burns an even wider variety of compounds than oxygen. Gaydon and Wolfhard [1979] comment that traces of oxygen have marked effects on fluorine flames. Where hydrocarbons or added hydrogen are involved, hydrogen atom reactions become possible and compete with the fluorine atom reactions that dominate the halogen–carbon flames. Hydrogen introduces the chain reactions that are discussed in the section on hydrogen–halogen flames. If oxygen is added to the system directly or as an oxygenated molecule or water, the rates appear significantly affected, probably as a result of radical generation by the $H + O_2 = OH + O$ reaction. The interaction between halogen and oxygen oxidizers is considered in Chapter XIII.

Fluorine diffusion flames were first investigated by Durie [1952], and Skirrow and Wolfhard [1955] investigated the structure of a diffusion flame using the Parker–Wolfhard flat burner (see Chapter V). Vanpee et al. [1974, 1978, 1979] have made extensive studies of the emission spectra of low-pressure premixed flames of fluorine with hydrogen, ammonia, cyanogen, carbon monoxide, methane, and the lower alkanes. Emission intensity structure in the recombination region was measured in these flames and measured peak rotational temperatures compared with calculated adiabatic flame temperatures. The bands of C_2, CH, C_3, C, F, HF, and OH were identified. OH appeared because of traces of oxygen in the fluorine. They concluded that C_2 resulted from CH reactions, while C_3 derived from evaporation from the solid carbon.

Homann and MacLean [1971a] have studied several low-pressure fluorine flames with hydrocarbons, halocarbons and ammonia (Figs. XI-12 and XI-13). These reactants ignite spontaneously upon mixing and have high burning velocities. The studies were made at reduced pressure using the multiple diffusion burner described in Chapter V. They also studied the inhibition of a hydrogen–fluorine flame by traces of ammonia (see Chapt. XIII). The inhibition was attributed to the formation of solid ammonium fluoride. The velocities given in the figures represent lower limits for free-burning flames since they were stabilized using significant heat extraction. In the acetylene–fluorine flame, the principal products are HF and the radical CF_2, which recombines to C_2F_4 as the burned gases cool. The intermediates include a number of unsaturated fluorine–carbon compounds and carbon fluorine radicals. There was soot production in the rich flame. The soot contained small but significant amounts of fluorine.

In the ethylene–fluorine flame, acetylene was a major intermediate. As a result, they behaved much like their acetylene relatives, producing similar intermediates and products. The ammonia–fluorine flame produced principally HF and N_2 as products with small amounts of equilibrium NF_2 radicals. There were a number of fluorinated ammonia intermediates found in low concentration. NH_4F was not observed, as it was in the argon-diluted hydrogen flames seeded with NH_3 (Section XII.4). This may be a result of the higher temperature of this flame. Homann and MacLean [1971b] also

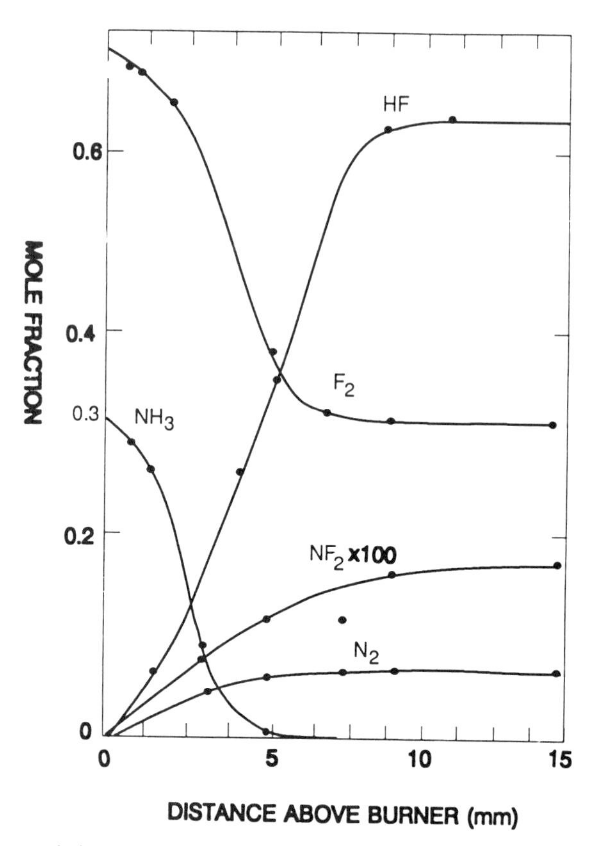

FIG. XI-12 Characteristic profiles of a low-pressure (3.8 torr) ammonia–fluorine flame, 0.25 NH_3, 0.74 F_2. Gas velocity 333 cm/sec (referred to 300 K). (Replotted data of Homann and MacLean [1971a].)

studied reactions of fluorine with halocarbons. These are vigorous flames that ignite spontaneously. A typical example is that of dichlorodifluoromethane with fluorine. This flame emits a grayish light that is weak compared with its hydrocarbon analog. They observed that if traces of hydrogen are added, the emission is reduced. With increased hydrogen addition, soot emission began to appear. The products of the flame are CF_4 and Cl_2, with traces of CF_3Cl and ClF. They propose a simple reaction scheme (Table XI–10).

All but the last reaction are considered irreversible. They note that its equilibrium constant, $K = (FxCl_2)/(ClxF_2)$, is not strongly temperature dependent, changing from 0.16 at 600 K to 0.48 at 1600 K. Attempts to measure temperature with Pt–Pt/Rh thermocouples failed because of attack by fluorine. The calculated adiabatic temperature was 2040 K, but the observed flame was far below equilibrium.

Parks and Fletcher [1969] have studied the combustion of halocarbons with fluorine. A particularly interesting case was that of perfluorocyclobutane, which can detonate. The product of this flame is principally CF_4.

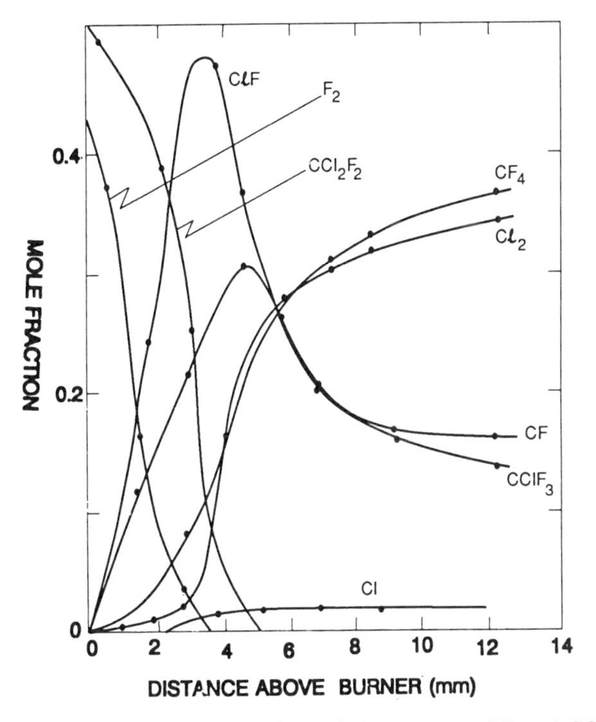

FIG. XI-13 Characteristic profiles of a stoichiometric low-pressure (77 torr) CCl_2F_2 + F_2 flame. Burner flow velocity was 13 cm/sec at 300 K. (Replotted data of Homann and MacLean [1971a].)

Chlorine-Supported Flames

Gaydon and Wolfhard [1979] observe that hydrogen and hydrocarbons burn well with chlorine both as premixed and diffusion flames. Soot and HCl are the principal products in hydrocarbon flames. Soot production appears to be quantitative. Because of radiative losses, these sooty flames show much lower experimental temperatures than adiabatic calculations. They quote the example of ethane–chlorine premixed flames with a measured temperature of 1000 K and a calculated value of 2000 K. There is another side to this. The soot irradiates the inlet gases, so that photosensitivity may

Table XI–10 Mechanism of F_2 + CCl_2F_2 Flame[a]

$F + CCl_2F_2 \rightarrow CClF_3 + Cl$
$F + CClF_3 \rightarrow CF_4 + Cl$
$Cl + F_2 \rightarrow ClF + F$
$Cl + ClF = Cl_2 + F$

[a]Homann and MacLean [1971a].

contribute to their stability. They demonstrated this by showing that the ethane–chlorine flame flashed back when illuminated with a flashlight.

Flames of the Interhalogen Compounds

The halogens react to form some nineteen covalent, interhalogen compounds, counting halogens and their atoms (see Table XI–7). They take the forms: X', XX', XX'_3, XX'_5, and XX'_7, where X is always the halogen of higher atomic number (weight). The XX'_5 and XX'_7 are only known as fluorides. The most common of these is chlorine trifluoride, which chemists consider as a convenient tool for handling fluorine. These compounds are strong oxidizers, and all, except perhaps IBr, should support combustion of hydrogenous fuels and carbon monoxide. As might be expected, the flame properties draw on both halogen substituents. The physical, transport, and thermodynamic properties of these halogens can be found in the previously mentioned sources. Their chemistry is discussed by Sidgwick [1950], Partington [1939], and Cotton and Wilkinson [1988]. They are usually formed by slow reaction of the elements and except for the chlorine–fluorine flame do not possess sufficient thermal stability to support combustion. There are few studies of interhalogen flames, but their chemistry should be that of the constituent halogens, with potential complication by the inhibitory effects a less reactive halogen may have on its more reactive partner. The formation of solid carbon and/or the formation of the stable lower fluorocarbons can occur at lower temperatures with the fluorinated interhalogens.

XII

OXYGEN FLAMES WITH C/H/O FUELS

This chapter is devoted to flames of molecular oxygen with fuel molecules containing carbon, hydrogen, and in some cases also oxygen. This includes most common flame systems. In these flames the source of reactive radicals is oxygen, which is a stable radical.[1] The driving reaction is $H\cdot + O_2: = OH\cdot + O:$, where the relatively unreactive, stable molecular oxygen radical is exchanged for the reactive O: and $OH\cdot$. The dominance of the reaction can be seen in Table XII–1, which compares radical generation processes in C/H/O flames. A number of other stable radicals exist, most of which are compounds of oxygen. These systems are discussed in Chapter XIII.

Differences among C/H/O flames lie in the fuel reactions. A rational hierarchy can be made passing from simpler to more complex fuels (see Fig. XII-1). The order is H_2, $CO(H_2)$, H_2CO, CH_4, C_2H_2, C_2H_4, C_2H_6, C_3H_8, C_4H_{10}. . . . The products are the same for all systems: (H, H_2, O, OH, H_2O, O_2) with CO and CO_2 added for carbon compounds. Oxygen-substituted and aromatic hydrocarbons form parallel systems with similar C/H/O chemistry. This ordering allows C/H/O chemistry to be built up step by step. The fuels are discussed in an order so that a given reaction mechanism involves only species and reactions from previously discussed fuels with the addition of the reactions peculiar to the fuel under discussion. This approach emphasizes the differences in individual systems and provides the basis for organizing the chapter. There are three sections: the first covers the generalities; the second is devoted to flames of the core fuels, hydrogen, carbon monoxide, and formaldehyde, whose chemistry is common to all C/H/O flames; and the third is devoted to flames of hydrocarbons and related molecules.

These simplifications are based on the premise that hydrocarbons are not stable under flame conditions,[2] and their reactions are irreversible. This is reasonable for fuel-lean and stoichiometric flames, but deteriorates as the equivalence ratio increases. In fuel-rich flames recombination of methyl and ethyl radicals produces hydrocarbons

1. The reader will recall that a radical is a molecule with one or more electrons with unpaired spins and as a consequence is diamagnetic. They should not be confused with ions that have an excess or deficiency of electrons and are charged. Most ions are also radicals, since they usually have an odd number of electrons. The reader is referred to texts on elementary or physical chemistry for more details.

2. This is true for normal flame temperatures and pressures; however, under high-pressure, fuel-rich conditions ($\approx$ 100 atm) acetylene and to a lesser extent methane become stable combustion products.

Table XII–1 Comparison of Radical-Producing Reactions in CHO Flames

Reaction	$k(1000)$	$k(2000)$
$H + O_2 = OH + O$	5.04×10^{10}	3.36×10^{12}
$H_2 + M^* = 2H + M$	2.32×10^{-7}	7.15×10^{3}
$H_2O + M^* = 20 + M$	2.67×10^{-10}	2.31×10^{3}
$H_2 + M^* = OH + OM$	2.33×10^{-7}	7.17×10^{4}
$H_2 + O_2 = 2OH$	$\sim 10^{0}$	10^{7}
$CO_2 + M^* = CO + O + M$	1.48×10^{-11}	5.44×10^{2}

through butane (Warnatz [1981]). As a consequence, the flames of all rich hydrocarbons are of comparable complexity once the initial fuel attack is accomplished. This makes it convenient to discuss fuel-rich flames as a class rather than individually.

It is important to observe with Warnatz [1981] that the chemistry of flames differs from that of ignition. This is illustrated in Figure XII-2, which compares the effect of changing rate coefficients for various reactions on calculated ignition times and burning velocities. Clearly the two processes are driven by different reactions. This occurs because in flames radicals can diffuse into regions where they are not otherwise available. This allows the rapid reactions of H, O, and OH to dominate the slower reactions

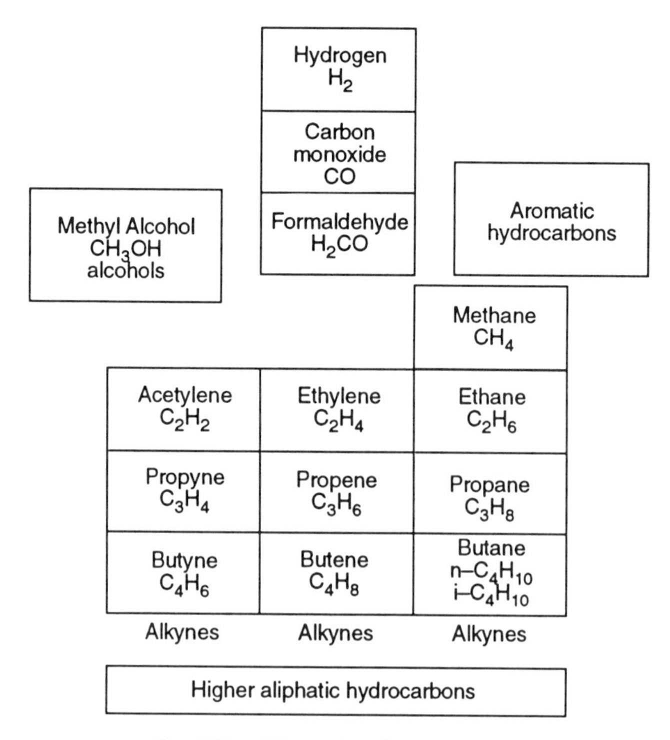

Fig. XII-1 Hierarchy of C/H/O Flames.

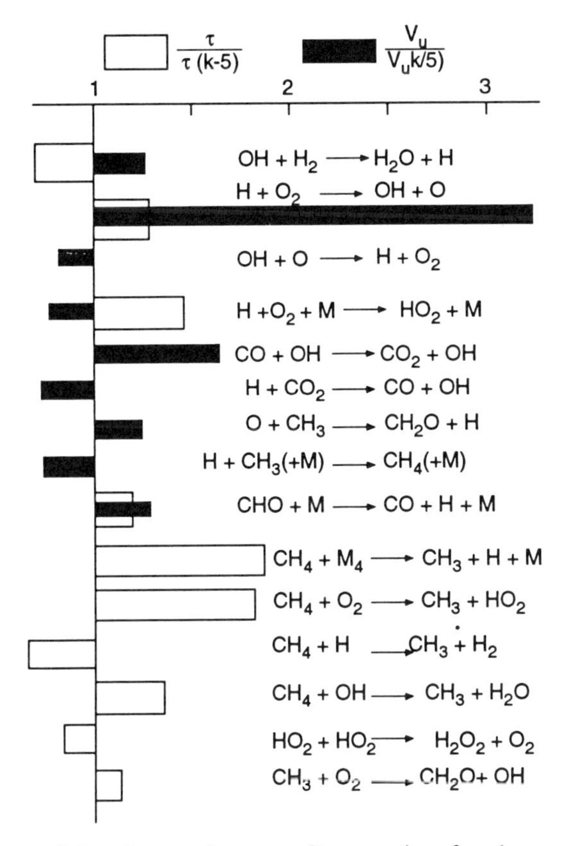

FIG. XII-2 Influence of changing reaction rates of key reactions for a lean methane–air mixture (1% CH_4 at 1400 K, $P = 2.87$ bar) by a factor of five on: (a) calculated ignition time (open bars) and (b) burning velocity (solid bars). (After Warnatz [1977].)

of ignition, which require dissociations to generate radicals. These questions are discussed in Chapter X.

These flames show the three zones discussed in Chapter X: (1) transport, (2) bimolecular reaction, radical generation, and (3) termolecular recombination. Fuel reactions are rapid compared with the subsequent CO/H_2 reactions so that they may be considered as a fast step creating a H_2/CO fueled flame. It often is convenient to present flame structure information in the compact form given in Appendix C of Chapter IX, where compositions are given at key temperatures. An approximate structure can be constructed from these data using the interpolation formulae.

The Core Systems (H_2, CO, and CH_2O)

The fuels whose reactions are basic to C/H/O chemistry are hydrogen, carbon monoxide, and formaldehyde. It may appear surprising that formaldehyde appears in the basic list. However, it must be remembered that formaldehyde and its radical fragment HCO appear as intermediates in all hydrocarbon flames.

Hydrogen–Oxygen Flames

Hydrogen is one of the most extensively studied flame systems. Although only the two elements H_2 and O_2 are involved, it is more complex than the hydrogen–halogen systems (see Chapter XI). This comes about because oxygen is divalent and OH and reactions of the radicals OH and HO_2 must be considered.

The heat of combustion of hydrogen is 57.8 kcal per mole in flames (68.3 kcal/mole in systems where liquid water is the product). Hydrogen–air mixtures between 4% and 74% burn, and with oxygen the upper limit is raised to 93.9%. Flame temperatures near the limits range from 1000 K (rich) to 1400 K (lean). A peak temperature of 3000 K occurs near stoichiometric. Air flames that are diluted by nitrogen show a peak temperature around 2300 K.

Species and Reactions

Hydrogen flames involve seven species: H_2, O_2, H, O, OH, HO_2, and H_2O. They can be adequately modeled with eleven reversible reactions (Table XII–2). Nevertheless, if low-temperature, high-pressure, ignition processes and surface reactions are considered, a formidable array of over fifty reactions requires consideration (Dixon-Lewis and Williams [1973]). Fortunately each regime has its own simplifications. At low temperature, radical concentrations are low so that radical–radical reactions are unimportant. In flames, ozone and hydrogen peroxide are unimportant trace species, and for some flames, bimolecular HO_2 reactions can be neglected. HO_2 is usually a trace intermediate because its formation is followed by destruction. This illustrates a point easily overlooked; that is, trace species may be important in a kinetic scheme. Low concentration only means that destruction reactions are more rapid than production.

Hydrogen (H_2) is colorless, odorless, and flammable. It is the lightest known gas, having a molecular weight of 2.016. The major commercial source is the steam reforming of natural gas, although significant amounts are produced from other hydrocarbons, by the partial oxidation of hydrocarbons, by coal gasification, and by the electrolysis of water (Grayson [1988]). It is commercially available in compressed gas cylinders, which vary in size from small steel laboratory bottles containing only a few liters (STP) to bottles containing many hundreds of cubic feet STP. The purity

Table XII–2 Minimal Mechanism for $H_2 + O_2$

$H + O_2 \rightleftharpoons OH + O$
$O + H_2 \rightleftharpoons OH + H$
$OH + H_2 \rightleftharpoons H_2O + H$
$H_2O + O \rightleftharpoons OH + OH$
$H\cdot + H\cdot + M \rightleftharpoons H_2 + M^*$
$H\cdot + OH\cdot + M \rightleftharpoons H_2O + M^*$
$O: + O: + M^* \rightleftharpoons O_2 + M^*$
$OH\cdot + OH\cdot + M \rightleftharpoons H_2O_2 + M^*$
$H + HO_2 \rightleftharpoons H_2 + O_2$
$O + HO_2 \rightleftharpoons OH + O_2$
$OH + HO_2 \rightleftharpoons H_2O + O_2$

ranges from 99% for commercial grades to the "five nines" grade (99.9995%) used in research and semiconductor production. In the laboratory it presents a hazard because of its flammability and detonatability and because of the high pressure of the storage bottles. A broken high-pressure cylinder can become a dangerous missile when propelled by the compressed gas. These cylinders should always be supported and placed in a well-ventilated area, preferably away from people. Hydrogen is an excellent fuel, but its most common use is as a component of synthesis gas mixtures with carbon monoxide. It is not widely used alone because of cost, difficulties in transport, and safety problems. Because of its low molecular weight, it has the highest specific impulse of any fuel, and liquified hydrogen is the fuel of choice for rockets, where weight is at a premium. It is not an outstanding fuel on a volume basis because of its low liquid density (0.07 g/cm^3). There have been recent proposals to utilize hydrogen as a general fuel because its combustion does not contribute to air pollution or the "greenhouse effect." Although a "hydrogen economy" is attractive, it requires the solution of infrastructure problems in production, transport, and safety.

Oxygen (O_2) is a colorless, odorless gas that is a strong oxidizer. Like hydrogen it is commercially available in high-pressure cylinders. Purity varies from commercial grades, which contain 1% argon, to the five nines grade for research purposes. Oxygen is a major basic chemical (Grayson [1988]). The ground state is a triplet with two unpaired electrons; this makes it diamagnetic. There is, however, also a low-lying electronic state singlet delta in which the spins are paired. Its possible flame chemistry is discussed in Chapter XIII.

Water (H_2O) is the major combustion product. At flame temperatures it is the vapor that we commonly call *steam*. At normal temperatures and pressures it is a colorless liquid (green along very long paths such as the ocean depths). Water is the cheapest chemical and finds many uses as a solvent and chemical intermediate. It is the key ingredient in agriculture and life in general. In the form of steam it is the major working fluid in most engines, both directly in steam engines and turbines and indirectly as a major species in internal combustion engines. We are quite literally "the water planet." It is the working fluid that drives the Earth's weather and climate. Water handling and purification are critical to all societies. In the laboratory it is usually purified by distillation or ion exchange columns. Commercially water is so cheap that it is sold for agricultural use by the acre-foot (an acre pool one foot deep). The major untapped source is seawater, but desalination by distillation has been uneconomical even in arid desert countries with abundant fuels. Recently a reverse osmosis processing in which water is purified by forcing it through a semipermeable membrane has been developed and used to desalinate the Colorado River and provide water for the island of Catalina. It is hoped that this technology will mature and eventually allow deserts to bloom.

Hydrogen atoms (H) have the lowest molecular weight of any species (1.007). At moderate pressures (<1 atm) measurable quantities of hydrogen atoms are in equilibrium with molecular hydrogen only at temperatures above 800 K. The heat of formation at standard temperature (298.15 K) is 51.63 kcal/mole. In the laboratory, they are best produced using an electric discharge (Bamford and Tipper [1977]; Rabek [1982]), preferably using the microwave region to avoid using metal electrodes, which catalyze radical recombination. Recombination can be inhibited by coatings such as phosphoric acid. Hydrogen atom concentrations over 90% have been reported at pressures as high as several torr. Excited molecular hydrogen and higher states of H atoms can be seen in the emission, but their lifetimes are so short that a few microseconds downstream the system is in translational thermal equilibrium with the walls and is a mixture of ground-state hydrogen atoms and molecular hydrogen. The lifetime of hydrogen atoms under these conditions can be as long as 0.1 sec.

Oxygen atoms (O) have a molecular weight of 16 and a heat of formation of 59.56 kcal/mole. This higher heat of formation means that at moderate pressures temperatures in excess of 2000 K are required to produce measurable amounts of atoms in oxygen mixtures. They can be produced in an electric discharge below a few torr pressure. The process is catalyzed by traces of moisture. Unfortunately, excited states of both atomic and molecular oxygen are also formed in electric discharges. Pure oxygen atoms can be produced by titrating nitrogen atoms from a discharge to produce oxygen atoms using the reaction $N + NO = N_2 + O$ (Kaufman [1985]).

Hydroxyl radicals (OH) has a molecular weight of 17.007 and the standard heat of formation from the elements of 9.43 kcal/mole. It is one of the dissociation products of water vapor and appears in water or hydrogen–oxygen mixtures above 1000 K. In the laboratory, it is commonly produced using a low-pressure discharge in water vapor. H and O are also present in such discharges, as well as excited molecular oxygen. It is produced in excited states by discharges, but these equilibrate with the apparatus walls with reasonable rapidity. OH can be produced without contaminations of H and O and excited O_2 by reacting hydrogen atoms with NO_2 ($H + NO_2 = OH + NO$) (Jeong and Kaufman [1982]). OH reacts at almost kinetic rates with itself or oxygen atoms ($OH + OH = H_2O + O$ and $O + OH = H + O_2$). As a result it shows a short half-life under laboratory conditions. In water discharges the apparent lifetime is increased by its continuous formation through $H + O_2 = OH + O$.

HO_2 radical is found in low concentrations as an intermediate in flames, but it is almost never present in significant concentration under final adiabatic conditions. It is present as a trace pollutant in the air formed from the reaction $H + O_2 + M \rightarrow HO_2 + M^*$. Hydrogen atoms are produced by photolysis of water by sunlight. It reacts with oxides of nitrogen and hydrocarbons in a complex cycle that produces the smog of modern cities (Sawyer [1981]). In the laboratory it is best formed by reacting hydrogen atoms with hydrogen peroxide ($H + H_2O_2 = HO_2 + H_2$). Concentrations of a few percent can be produced in this manner, but it is rapidly destroyed by bimolecular recombination with itself and other radicals. It is also formed in weak discharges in hydrogen peroxide itself, but the system is contaminated with H, O, and OH.

Kinetic Mechanisms

A minimal mechanism for hydrogen–oxygen flames is given in Table XII–2. Full coverage of high-pressure, heterogeneous reactions, ignition, and shock tubes in the manner of modelers such as Dixon-Lewis and Williams [1977], Warnatz [1981], and Westbrook and Dryer [1981] requires over fifty reactions. Vandooren and Bain [1991] have validated one mechanism experimentally. Some rate coefficients are collected in Table A–3 of the appendix at the end of the book.

Related Studies: Ignition Limits, Shock Tube, and Others

Hydrogen–oxygen chemistry can be investigated using relatively simple experiments. Reaction is negligible below 800 K. Above this point the system is self-catalytic and ignites spontaneously. When introduced quickly into an isothermal container, reaction may be slow for a prolonged period and then erupt suddenly in an explosion. The condition for these explosions and delays are reproducible and measurable. They depend on the temperature, pressure, size of vessel, and even the wall material. The temperature limit for explosion shows a complex dependence on pressure with three limits (Fig. XII-3). This intriguing behavior was the center of attention by a number of investigators in England, Russia, and this country during the first third of this century. The unraveling of this and other chain reactions resulted in the award of a Nobel Prize to Hinshelwood and Semenoff.

The first limit at low temperature is established because low pressure allows radicals to diffuse to the vessel wall, where they are destroyed by recombination. This limit depends strongly on the vessel wall material and conditions. The second limit reflects the competition between the bimolecular and termolecular reactions of hydrogen atoms with molecular oxygen, $H + O_2 = OH + O$; $H + O_2 + M = HO_2 + M^*$. The second limit for hydrogen has become so well understood that Baldwin and his collaborators [1965, 1977] use it as a tool for chemical kinetic studies.

The explanation of the region between the second and third limits requires consideration of reactions of H_2O_2. General accounts are given by Hinshelwood [1948] and Semenoff [1935] with combustion oriented discussions by Jost [1946] and Lewis and von Elbe [1987].

Most of the techniques of chemical kinetics have been used to study this system, and a few examples of importance to flame chemistry will be cited. The hydrogen sys-

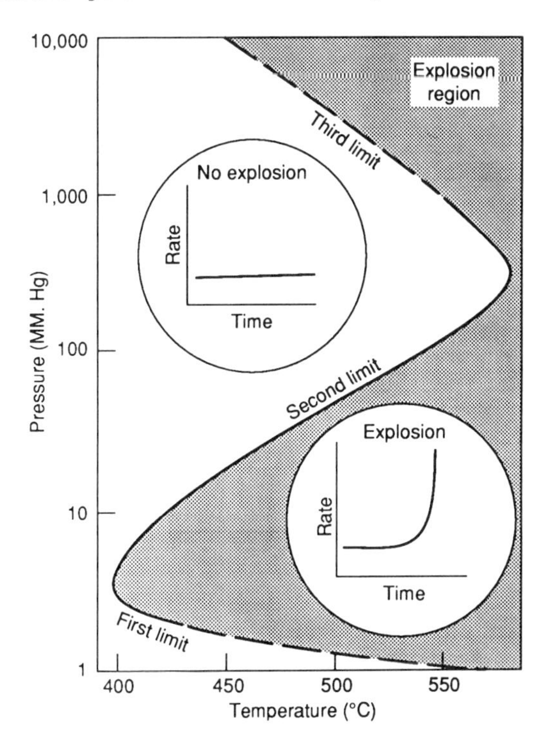

FIG. XII-3 Schematic of explosion limits in the hydrogen–oxygen system as a function of temperature and pressure. (After Lewis and von Elbe [1987] with additions.)

tem was reviewed by Dixon-Lewis and Williams [1977] and Warnatz [1983b]. Current information on specific reactions can be obtained from the NIST compilation (Westley et al. [1990]). One example of shock tube studies is the work of Hamilton and Schott [1967], who explored partial equilibrium considerations (see Chapter X). The study of Jenkins, Spalding, and Yumlu [1973] provides a good example of stirred reactor techniques. Many reactions of hydrogen atoms and oxygen atoms and hydroxyl radicals have been studied in low-pressure flow tubes using electric discharges as the source of radicals (Kaufman [1982]). Flash photolysis has been used to form atoms and radicals for combustion reactions following the reactions spectroscopically (Norrish [1965]). More recently atom–molecule reactions have been studied using flash photolysis/resonance fluorescence (Howard [1979]) (Chapter VIII). Temperatures in most of these studies lie well below those of flames, but they are important for establishing kinetic constants and activation energies for the reactions. The wider the temperature range considered in evaluating a reaction, the more accurate will be the resulting activation energy.

Burning Velocity and Detonations

Burning velocities of hydrogen–oxygen flames have been investigated over a wide range of compositions, pressure, inlet temperatures, and diluent species. References on burning velocity studies of hydrogen and other C/H/O flame systems are collected in Table B–2 in the appendix at the end of the book. Data from more recent studies usually agree within 20%, although some older studies show wider variations. Figure XII-4 shows comparisons between experimental studies and modeling by Warnatz [1981]. Recently Egolfopoulos and Law [1991] extended comparative studies into the very lean regime. Figure XII-5 shows the complex relation between burning velocity and pressure for a range of stoichiometric hydrogen–oxygen–nitrogen mixtures. Figure XII-6 shows the effect of inlet temperature (flame enthalpy) on propagation for stoichiometric hydrogen–oxygen and hydrogen–air mixtures. This is a comparison made by Warnatz [1981]. An equally successful set of comparisons of the effects of composition, pressure, and inlet temperature were made by Dixon-Lewis [1984]. The system appears to be well modeled.

Gray and Smith [1980] studied the burning velocity of hydrogen and deuterium mixtures with oxygen at 70 torr. The deuterium mixtures were slower; the ratio of speeds was 1.47, as might be expected if propagation were proportional to the square root of the peak hydrogen (deuterium) atom concentration. The effects of added steam has been investigated by Lui and MacFarlane [1983], Kumar, Tamm, and Harrison [1983], and Muller-Dethlefs and Schlader [1976].

Cellular flames occur near the lean limits of combustion because lateral diffusion in curved flame fronts produces locally enhanced fuel concentrations, which extends the lean limit locally. Discussions of these phenomena have been given by Markstein [1964], Lewis and von Elbe [1988], and Mitani and Williams [1980]. Hydrogen–oxygen mixtures produce detonations over almost as wide a range of composition as subsonic combustion. Propagation is limited by hydrodynamic considerations and can be calculated using the assumption that the detonation is at the Chapman–Jouget point, where the propagation velocity is equal to the sonic speed of the burned gases. This has been tested by Lewis and von Elbe [1987] and Berets, Green, and Kistiakowski

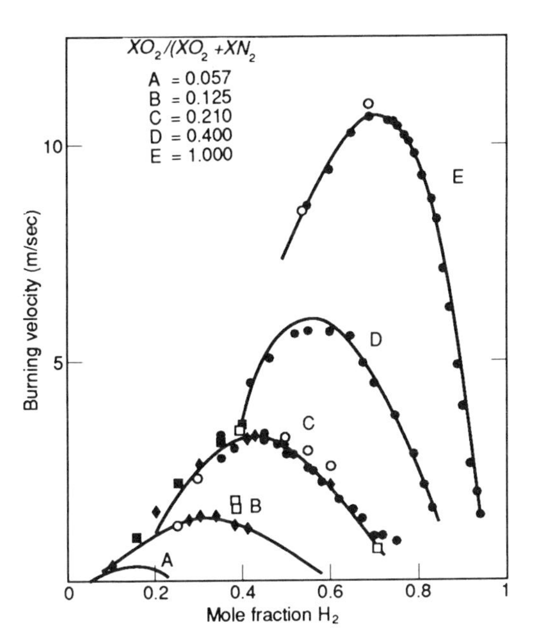

Fig. XII-4 Burning velocities in the hydrogen–oxygen–nitrogen system. Comparison of Warnatz computations with experimental measurements by: Jahn [1934], Senior [1961], Agnew and Graiff [1961], Dixon-Lewis et al. [1970], Edmonson and Heap [1971], Günther and Janisch [1959], and Andrews and Bradley [1973a]. Solid lines are calculations by Warnatz [1979]. (Replotted from figure by Warnatz [1977].)

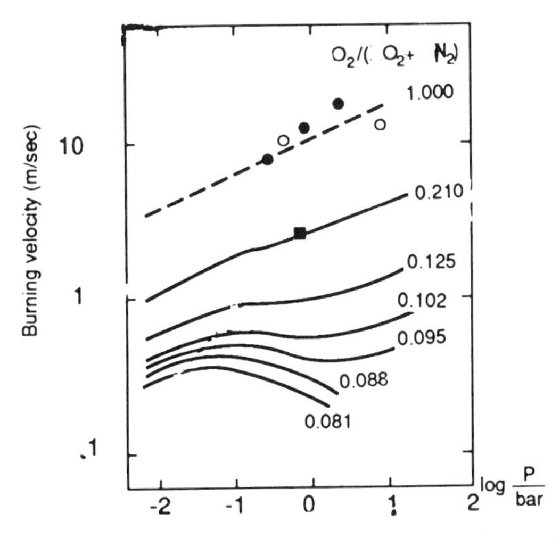

Fig. XII-5 Effect of pressure on the burning velocity of some stoichiometric hydrogen–oxygen–nitrogen flames. Comparison of computations by Warnatz [1981a] (solid lines) with measurements by Strauss and Edse [1959] (open circles) and Agnew and Graiff [1961] (solid circles). (After Warnatz [1981a].)

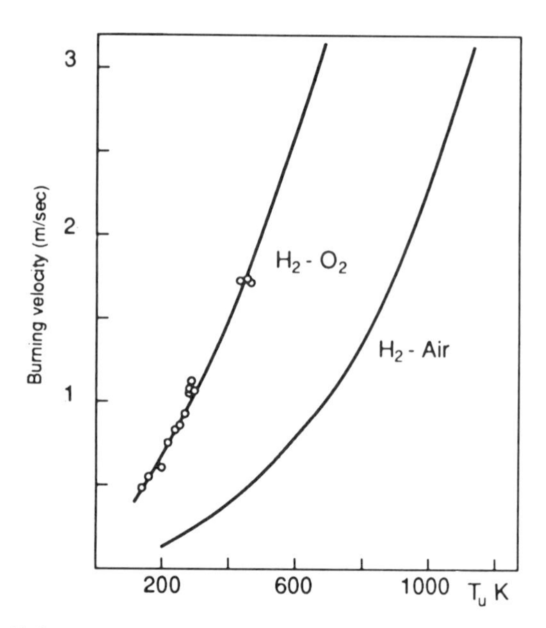

FIG. XII-6 Effect of inlet temperature on burning velocity in atmospheric stoichiometric hydrogen flames. Comparison of computations (solid line) by Warnatz [1981a] with measurements by Edse and Lawrence [1969] (open circles) scaled to agree with calculations at room temperature. (After Warnatz [1981a].)

[1950]. A concise description is given in the book of Lewis and von Elbe [1987]. The theory was borne out generally with deviations occurring in very rich or very lean mixtures near the detonation limits. Experimental structural measurements are not available, but Oran and Boris [1981] have modeled several systems. For further information see general combustion texts such as those of Lewis and von Elbe [1987], Kuo [1986], and Sterhlow [1986].

Flame Structure and Modeling Studies

The microstructure of a number of hydrogen–oxygen flames have been measured (Figs. XII-7 and XII-8) and compared with modeling. The work that established the kinetic scheme for this system and laid the basis for modern modeling studies in C/H/O chemistry were the experiments and modeling by Dixon-Lewis and his collaborators at Leeds University [1967–1982]. These influential, pioneering studies have been summarized in a review (Dixon-Lewis and Williams [1977]) and updated in an overall survey on laminar flame structure by Dixon-Lewis [1991]. Their results for a representative family of flames are summarized in Table A–17 in the appendix at the end of the book. They illustrate the effects of stoichiometry on flame structure. The system was also used by Dixon-Lewis [1968, 1974] to illustrate the interplay between transport processes in flames. They have been extended to spherically propagating flames by Dixon-Lewis [1983] and opposed jet diffusion flames by Dixon-Lewis and Missaghi [1989].

Warnatz [1977, 1981] has published an extensive study and comparison of this

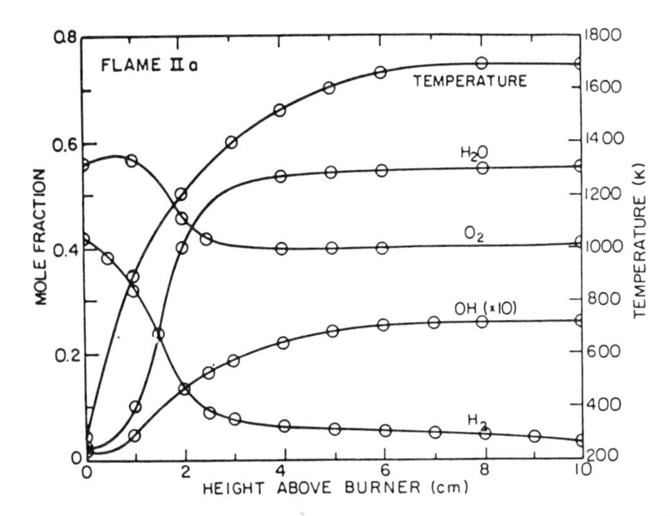

FIG. XII-7 Experimental structure of a fuel–lean, low-pressure hydrogen-oxygen flame (O_2 = 0.52; H_2 = 0.48; P = 7.2 Torr; T_0 = 298 K; υ_0 = 2.23 m/sec). (After Brown, Eberius, Fristrom, Hoyermann, and Wagner [1978].)

system covering the effects of composition, pressure, and initial temperature. Other groups who have modeled the hydrogen–oxygen flame system include: Stephanson and Taylor [1973] and Oran and Boris [1981, 1982]. All of the models produce equivalent results when using the same input parameters. Only minor changes are introduced in the profiles by using simplified mechanisms and neglect of thermal diffusion. This explains the success of approximate models such as those of the early theories reviewed by Evans [1952], the zonal theory of Brown, Fristrom, and Sawyer [1974], and the asymptotic theory of Peters and Williams [1987]. Simple models can be useful for engineering purposes. The calculations of Dixon-Lewis and Williams [1977] show a constancy in the onset temperature net radical generation that occurs because of the

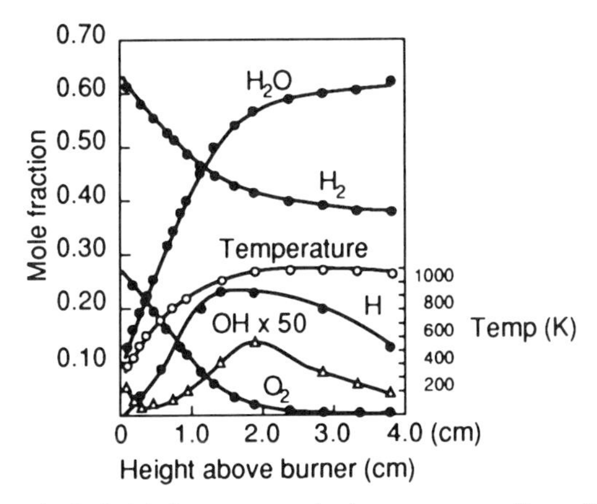

FIG. XII-8 Structure of a fuel-rich, low-pressure hydrogen–oxygen flame (H_2 = 0.81; O_2 = 019; P = 7.7 Torr; T_0 = 198 K; υ_0 = 1.56 m/sec). Comparison of experiments with modeling. (After Eberius, Hoyermann, and Wagner [1973].)

balance between the rates of $H + O_2$ and $H + O_2 + M$ reactions. This is discussed in connection with the ignition approximation in Chapter X (Fig. X-13).

The studies in the recombination zone made by the Cambridge University group under Sugden led to the partial equilibrium concept. This was outlined in a series of papers [Bulewicz, James, and Sugden [1956]; Padley and Sugden [1959]] relating radical and electron concentrations with chemiluminosity in flames. Their method of deducing radical concentrations from the nonequilibrium emission of trace metals is discussed in Chapter VI. Fenimore and Jones [1956] made a pioneering study of radical recombination in hydrogen flames, and Kaskan [1958] studied OH concentration in the equilibration zone. Low-pressure hydrogen flames have been studied at Göttingen (Eberius, Hoyermann, and Wagner [1971]; Brown, Eberius, Fristrom, Hoyermann, and Wagner [1978]) (see Fig. XII-8). Specialized studies include ESR measurements of H by Bregdon and Kardirgan [1981] and mass spectral studies of HO_2 by Hastie [1974].

Hydrogen–oxygen flames have been used as a high-temperature, radical bath for studying radical reactions. The technique was pioneered by Fenimore and discussed in his excellent book [1964].

Summary

Hydrogen flames show most of the common characteristics of C/H/O flames. However, because of the low molecular weight of hydrogen, the effects of thermal diffusion are enhanced. Another peculiarity is that reaction with hydrogen atoms produces no net result ($H + H_2 = H_2 + H$), so the fuel does not inhibit its diffusion. This contrasts with hydrocarbon flames, where the fuel reaction delays and inhibits regeneration of H atoms.

Modelers feel that computations for the hydrogen system approach or may be more reliable than experiments, but some experimentalists might take exception to this. These studies provide a firm basis for the understanding of more complex flames. Therefore, we may proceed to carbon-containing flames with a feeling of confidence.

Carbon Monoxide (Hydrogen Trace)–Oxygen Flames

Excluding the heterogeneous combustion of carbon, the simplest carbon-containing flame is carbon monoxide–oxygen. This system is important because its reactions occur in the combustion of all carbon-containing compounds.

One of the most remarkable facts about carbon monoxide flames is that despite strong exothermicity, completely hydrogen-free mixtures of carbon monoxide and oxygen show no significant flame propagation. For example, the hydrogen-free stoichiometric carbon monoxide–oxygen mixture has a negligible burning velocity, but the addition of a 1% trace of hydrogen, water, or other hydrogen-containing compound raises the burning velocity to over a meter per second (Fig. XII-9). This behavior results from its interaction with the hydrogen–oxygen chemistry, in particular, the radical generation step $H + O_2 = OH + O$. It also suggests that the reaction of CO with both atomic and molecular oxygen must also be slow compared with the dominant flame oxidation reaction, which is $CO + OH = CO_2 + H$. These studies are considered in more detail in the section on burning velocity.

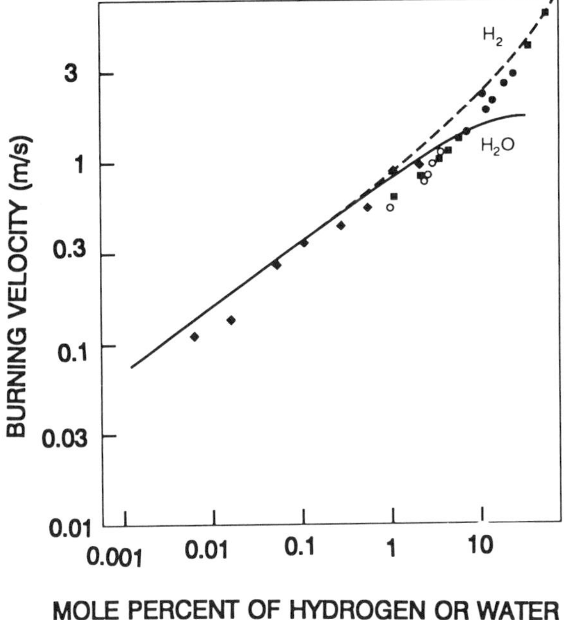

FIG. XII-9 Sensitivity of Carbon Monoxide flame burning velocity to added hydrogen (stoichiometric, atmospheric carbon monoxide–oxygen flames (T = 298 K) with added water). (Calculations, Warnatz [1979]. Experiments, Price and Potter (circles); Franz/Wagner (squares), Wires et al. (crosses). After Warnatz [1979].)

Moist carbon monoxide is flammable in air between 12.4% and 74%, and oxygen extends the upper limit to 93%. Peak flame temperatures are over 2300 K with air, and 3000 K with oxygen.

New Species and Reactions

The carbon monoxide flame introduces two new major species: carbon monoxide (CO) and carbon dioxide (CO_2), and one minor species (HCO). In addition, one must consider the species of the hydrogen flame (H, OH, HO_2, H_2, H_2O, O, and O_2). Neither solid nor gaseous free carbon appears in these flames because of the thermal stability of CO and the fact that its oxygen automatically prevents its flames from falling into the carbon-rich field. The stoichiometric regions in C/H/O flames are discussed in the next section.

Carbon monoxide (CO) is a colorless, odorless, poisonous, and flammable gas that is difficult to liquefy. Cylinders of the gas pressurized to 200 atm are the usual laboratory source for this gas. Although it is easily oxidized, it is one of the most thermally stable molecules, only showing significant dissociation (at atmospheric pressure) above 4000 K.

Carbon dioxide (CO_2) is a colorless gas with an acidic odor and taste. It is a suffocant that has the physiological effect of increasing respiration rate. It is easily liquefied and frozen. The critical point lies just above room temperature, which makes it a convenient material for classroom demonstrations. In the solid form CO_2 is a convenient refrigerant commonly called "dry ice" because it sublimes at atmospheric pressure. It also finds commercial use in fire extinguishers. The HCO radical occurs in high pressure, hydrogen-rich flames, but is unimportant otherwise.

Flame Kinetics

The new reactions introduced by carbon monoxide are collected in Table XII–3. Atomic oxygen reacts only slowly with CO. As a result, competing reactions involving trace hydrogenic contaminants usually dominate. The bimolecular reaction is responsible for some of the blue luminosity of C/H/O flames. The key fuel reaction in this flame is the reversible reaction, $OH + CO = CO_2 + H$. *This requires the presence of at least traces of some hydrogen-containing compound.* Under high-temperature, radical-poor conditions such as those found in shock tubes, molecular oxygen can react directly with CO, but in flames this reaction is preempted by the competing radical reactions. Reactions involving HCO are usually unimportant. These questions ares discussed by Dixon-Lewis and Williams [1977].

Related Studies

The carbon monoxide–oxygen system displays an ignition pattern similar to the hydrogen–oxygen system. At moderate temperatures there is a slow reaction that is strongly catalyzed by traces of moisture. Explosion limits exist, but the first and second limits are poorly separated, and there appears to be no third explosion limit. Cherian, Rhodes, Simpson, and Dixon-Lewis [1981a,b] provide a good discussion of the limits.

Early combustion studies are summarized by Lewis and von Elbe [1988] and Jost [1946]. The most comprehensive survey is by Dixon-Lewis and Williams [1977]. It has an excellent bibliography to 1977 that can be supplemented by the review in the Eighteenth Symposium by Cherian, Rhodes, Simpson, and Dixon-Lewis [1981a]. The system has been studied by a number of other investigators such as: Dryer and Glassman [1973]; Kozlof [1959]; and Hottel, Nerheim, Williams, and Schneider [1965] using high-temperature flow reactors. Jost, Schacke, and Wagner [1965] investigated

Table XII–3 Mechanism of the Oxidation of CO Reaction

$CO + H + H_2 \Rightarrow CHO + H_2$
$CO + O + CO \Rightarrow CO_2 + CO$
$CO + OH \rightleftharpoons CO_2 + H$
$CO + HO_2 \rightleftharpoons CO_2 + O$
CHO Reactions
$CHO + H \Rightarrow CO + H_2$
$CHO + O \Rightarrow CO + OH$
$CHO + O \Rightarrow CO_2 + H$
$CHO + OH \Rightarrow CO + H_2O$
$CHO + O_2 \Rightarrow CO + HO_2$
$CHO + M' \Rightarrow CO + H + M'$

the establishment of the water–gas equilibrium ($CO + H_2O = CO_2 + H_2$) in a flow reactor.

Shock tubes have been used to study the direct reaction of CO with molecular oxygen. The work is reviewed by Dixon-Lewis and Williams [1977]. One typical example is the study by Rawlings and Gardiner [1974].

One unusual phenomenon in this system is the existence of a luminous reaction below the explosion limits, which appears related to the afterglow found following explosions of carbon monoxide–oxygen mixtures in closed vessels. Linnett, Reuben, and Wheatley [1968] found that in some cases the glow persisted for as long as 20 sec. The emission consists of diffuse bands overlaid by a diffuse continuum. Gaydon [1979] has discussed these questions, as have Dixon-Lewis and Williams [1977]. A number of models have been proposed and reviewed by Gray [1975]. He reported that McCafferty and Berlad had observed as many as two hundred oscillations in one experiment.

Burning Velocity and Detonations

Burning velocity studies of the carbon monoxide system have provided a major clue for understanding H_2–CO flames. References to these studies are collected in Table B–2 of the appendix at the end of the book. This system shows a remarkable sensitivity to traces of hydrogen-containing compound. The effect is roughly proportional to the hydrogen content of the trace. For example, hydrogen and water traces are equally effective. This behavior was discovered by early workers in the field, as mentioned in the pioneering book of Bone and Townend [1927]. The effect was carefully studied by Jahn [1934] and others. Wires, Watermeier, and Strehlow [1959] showed that stoichiometric carbon monoxide–oxygen mixtures with no trace of hydrogen show a burning velocity of a few millimeters per second (Fig. XII-9), but with a 1% trace of hydrogen, water, or other hydrogen-containing compound in the carbon monoxide, the burning velocity jumps to over a meter per second. Related behavior is found in the explosion limits.

Yumlu [1967] has proposed a mixing rule for the burning velocity of CO–H_2, H_2O flames, which allows rough predictions.

In the carbon monoxide system nitrogen acts as a diluent, while carbon dioxide shows a chemical influence. Jahn [1934] found that it reduced the burning velocity and shifted the peak toward leaner mixtures (Fig. XII-10). This can be attributed to enhancement of the back reaction and reduction of the flame enthalpy due to the higher heat capacity of carbon dioxide (relative to nitrogen) and dissociation.

Carbon monoxide–oxygen mixtures detonate. The detonation velocity shows some sensitivity to added water vapor (Jost [1946]; Lewis and von Elbe [1987]). The effect shows a maximum, no doubt due to dilution overpowering the promotion of the water.

Vandooren, Peeters, and Van Tiggelen [1975] have studied CO flames, and the effect of halogenated additives on CO detonations was studied by Libouton, Dormal and Van Tiggelen [1975]. They visualized the detonation cell structure from patterns on presooted walls, which mark the path of Mach stem intersection. This region is so hot that the graphite is volatilized locally. Velocity varies periodically along the detonation front due to interaction between reaction delay time and the lateral driving

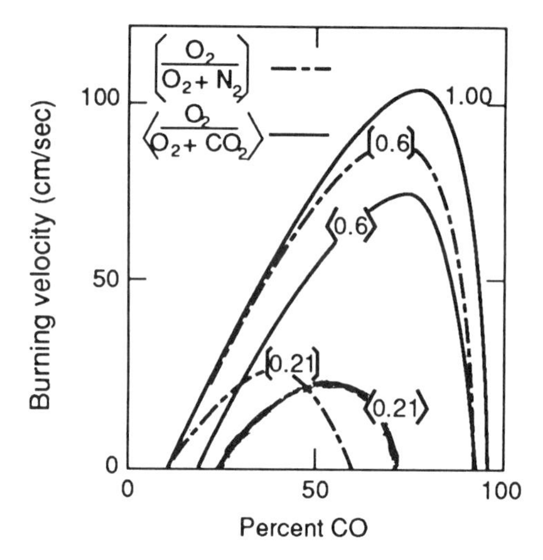

FIG. XII-10 Dependence of burning velocity of carbon monoxide flames on stoichoimetry and dilution showing the difference in effect of a pure diluent nitrogen, and a product carbon dioxide, which enters into the back reactions. (Replotted from data by Jahn [1934].)

pressure, which exchange chemical and kinetic energy. These considerations lie beyond our present discussion, and the interested reader is referred to the literature. An elementary discussion of detonation theory is given in Chapter VIII of the treatise of Lewis and von Elbe [1987], and a good discussion of detonation structure can be found in Strehlow's [1986] informative book.

Flame Structure and Modeling Studies

There have been fewer studies of the microstructure of carbon monoxide flames than might be expected considering the importance of the system. Work began with the pioneering studies of Friedman and Nugent [1959]. Studies of carbon monoxide in the postflame gases of hydrocarbon flames were initiated by Friedman and Cyphers [1955] and Fenimore and Jones [1958b, 1961a] and were continued by Singh and Sawyer [1971], Rogg and Williams [1989], and Howard, Williams, and Fine [1973]. Westenberg and Fristrom [1965] give a discussion of the overall kinetics of CO conversion in hydrocarbon flames, correlating flame structure with other kinetic studies. Postflame gas studies of carbon monoxide were used by Schoenung and Hanson [1981] and others to investigate quenching efficiency in microprobes. Probe-related questions are discussed in Chapters VI and VII. Since carbon monoxide is a major combustion pollutant, there have been many studies related to the interaction of this reaction and the formation of oxides of nitrogen. A cursory discussion of such questions is given in Chapter VIII. Dixon-Lewis [1972] and Dixon-Lewis, Sutton, and Williams [1965b] studied the effect of adding traces of CO and CO_2 to hydrogen–oxygen flames. Cherian, Rhodes, Simpson, and Dixon-Lewis [1981a] modeled a number of carbon monoxide–hydrogen flames. Table A–18 of the appendix summarizes their results for a range of stoichiometries. Temperatures and compositions are given at points of radical maxima and the initial and final adiabatic conditions. The Leeds work on the hydrogen–carbon monoxide combustion system was reviewed by Dixon-Lewis and Williams [1977]. Warnatz [1981] modeled the carbon monoxide system and made major contributions to the understanding of the complexities of burning velocity.

The most comprehensive experimental study of carbon monoxide flame structure is that of Vandooren, Peeters, and Van Tiggelen at Louvain [1975]. They investigated a low-pressure lean flame to study the flame reactions of CO (Figs. XII-11 and XII-12). This work was one of the first to document the strong non-Arrhenius behavior of the CO + OH reaction. This study was modeled by Cherian, Rhodes, Simpson, and Dixon-Lewis [1981]. Safieh, Vandooren, and Van Tiggelen [1982] made a flame structure study on $CO–H_2–O_2–Ar$ flames inhibited by CF_3Br to compare with the previously mentioned detonation studies. O atom reactions in CO flames were studied using ESR by Balakhnin, Egorov, Van Tiggelen, Azatyan, Gershenzov, and Korndratiev [1969]. An experimental study of this flame inhibited by CF_3Br was made by Safieh, Vandooren, and Van Tiggelen [1982], and a carbon monoxide oxygen flame seeded by sulfur dioxide was used by O. I. Smith, Wang, Tseregounis and Westbrook [1983] to elucidate the sulfur catalyzation of the recombination of atomic oxygen. These studies are also considered in the inhibition discussion of Chapter XIII.

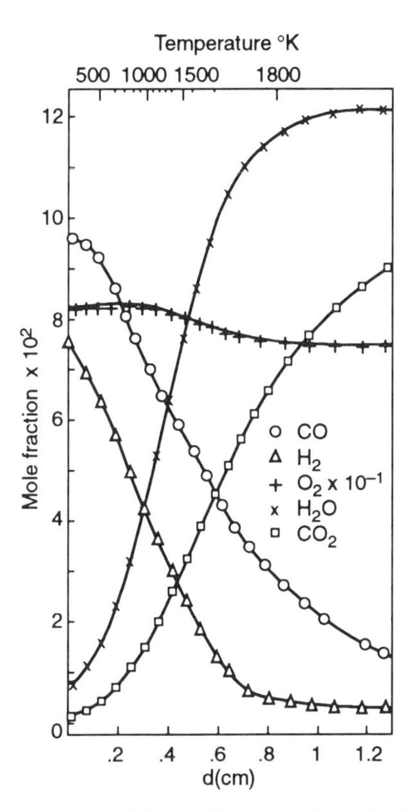

FIG. XII-11 Major species in a carbon Monoxide flame (CO = 0.094; H_2 = 0.114; O_2 = 0.792; P = 40 Torr; T_0 = 300 K; v_0 = 0.64 m/sec). (Data of Vandooren, Peeters, and Van Tiggelen [1975], replotted.)

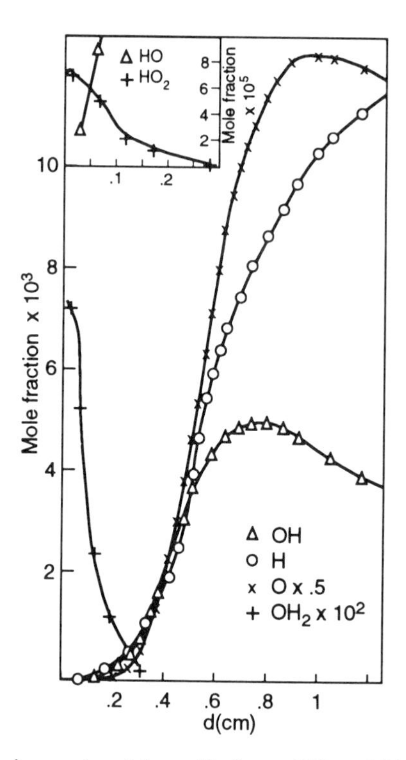

FIG. XII-12 Minor species in a carbon Monoxide flame ($CO = 0.094$; $H_2 = 0.114$; $O_2 = 0.792$; $P = 40$ Torr; $T_0 = 300$ K; $\upsilon_0 = 0.64$ m/sec). (Data of Vandooren, Peeters, and Van Tiggelen [1975], replotted.)

Summary

The carbon monoxide flame reactions are intimately related with and dependent on the hydrogen–oxygen chemistry. The only new reaction required beyond the hydrogen–oxygen chemistry is the reversible attack of carbon monoxide ($CO + OH = CO_2 + H$). Near the limits $O + CO + M = CO_2 + M^*$ and $H + CO + M = CHO + M^*$ play roles and $CO_2 + O_2$ becomes important in high-temperature shock tube reactions. Taken together, these two flame chemistries control the equilibration zone behavior of all C/H/O flames. Good-quality structural studies have been made and quantitatively interpreted by modeling. This is one of the best understood flame systems.

Formaldehyde Flames

It may seem odd that the third member of the C/H/O flame hierarchy is as complex a molecule as formaldehyde, OCH_2. It may be naively viewed as a complex of hydrogen and carbon monoxide with their combined flame chemistry. It is an intermediate in hydrocarbon flames and in the decomposition flames of many organic nitrate compounds (see Chapter XIII). Its major source in flames is the reaction $CH_3 + O \Rightarrow OCH_2 + H$.

Formaldehyde reacts rapidly with the flame radicals, H, O, and OH, yielding HCO and may undergo thermal decomposition to hydrogen and CO. Peeters and Mahnen [1973] make the point that in methane flames a decomposition reaction is necessary. Peeters and Mahnen [1973a] are of the opinion that in methane flames, the formaldehyde decomposes while Dixon-Lewis and Williams [1967] believe that HCO decomposition dominates in hydrocarbon flames. In flame studies it is difficult to distinguish between the two processes. The author favors the HCO mechanism, but both may contribute since in methane flames the formaldehyde may be formed in an excited state. In hydrocarbon flames most of the intermediate formaldehyde is formed by the reaction $O + CH_3 = CH_2O + H$, rather than the reaction with molecular oxygen $CH_3 + O_2 = (CH_3OO-)^* = CH_2O + OH$, as suggested by Fristrom and Westenberg [1965] in their study of an oxygen-rich methane flame.

New Species

In addition to the species of the hydrogen and carbon monoxide flames, only formaldehyde (CH_2O) requires consideration.

Formaldehyde (OCH_2) is a gas that is easily polymerized. In the most common form the carbons are linked through oxygen (paraformaldelhyde). It is a linear polymer having from 6 to 100 units that can be converted to the monomer by heating to 120°C. Sadequi and Branch [1988] give the design of a convenient apparatus for generating the monomer from paraformaldehyde using a castor oil carrier.

Reactions

The reactions of formaldehyde are collected in Table XII–4. It will be observed that these are all irreversible reactions.

Related Studies

Hall, McCoubrey, and Wolfhard [1952] studied the emission spectra of some low-pressure formaldehyde flames with oxygen and the oxides of nitrogen (NO, NO_2, N_2O) and compared their OH emission with that of related hydrocarbon and alcohol flames.

Burning Velocity

De Wilde and Van Tiggelen [1968] made measurements of the burning velocity of formaldehyde with O_2.

Table XII–4 Formaldehyde CH_2O Consumption

Reaction
$CH_2O + H \Rightarrow CHO + H_2$
$CH_2O + O \Rightarrow CHO + OH$
$CH_2O + OH \Rightarrow CHO + H_2O$
$CH_2O + HO_2 \Rightarrow CHO + H_2O_2$

Flame Structure and Modeling Studies

Olson and Smooke [1979] modeled formaldehyde flames. Oldenhove de Guertechin, Vandooren, and Van Tiggelen [1983a,b, 1986] studied a family of formaldehyde flames using molecular beam inlet sampling. The data (Fig. XII-13) has been analyzed and rate coefficients deduced for the reaction of formaldehyde with H, O, and OH, and compared with the literature. Sadequi and Branch [1988] measured the structure of a stoichiometric formaldehyde flame with NO_2 using microprobe sampling techniques.

Hydrocarbons

Before considering individual hydrocarbon flames, it is useful to examine the common behavior of these systems. Considering their diversity C/H/O fueled flames show remarkable similarities. This can be best appreciated through Figure XII-14, which shows burning velocities of C_1–C_8 hydrocarbon–air flames as a function of equivalence ratio. There are systematic differences in burning velocity and especially in rich flammability limits, but the general pattern is remarkably similar, suggesting that they possess a common basic flame chemistry. In addition, the final adiabatic conditions of temperature and composition and even the detailed flame structure profiles as can be seen in Tables A–19 and A–20, which provide a comparison among several stoichiometric air flames.

Early workers on flame structure (Fenimore [1964]; Kaskan [1959]; Friedman [1953]; Fristrom [1966]) observed that hydrocarbon flames may be considered as a family of fuel destruction reactions by H, O, and OH at the lower flame boundary feeding a H_2–CO–O_2 flame, which maintains a radical pool that diffuses into and reacting with the fuel. This simplistic view requires elaboration, but it has key elements.

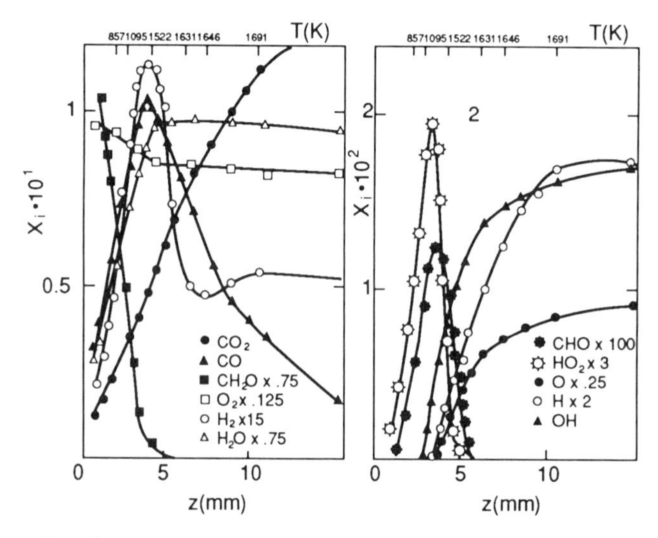

FIG. XII-13 Profiles of a lean formaldehyde–oxygen flame (0.179 CH_2O; P = 22.5 Torr; v_0 = 0.8 m/sec). (a) Major species; (b) radical species. (Replotted data of Oldenhove de Guertechin, Vandooren, and van Tiggelen [1983a].)

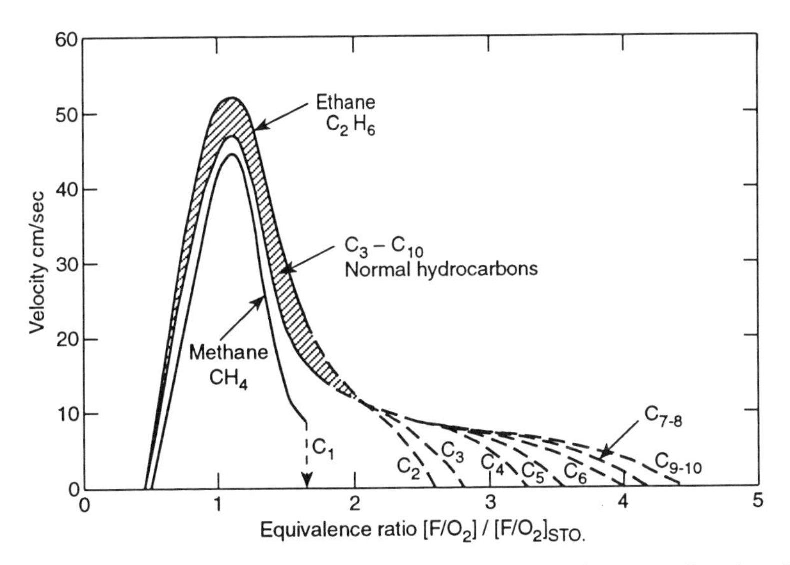

FIG. XII-14 Burning velocities for the saturated hydrocarbon–air flames as a function of equivalence ratio, showing similarities.

Another major factor is the leveling force of diffusion. In flames of comparable burning velocity profiles are very similar when plotted on a normalized time basis (z/v_0). These similarities are illustrated by the work of Fristrom, Grunfelder, and Avery [1959] (Fig. XII-15), which compares experimental major species profiles of C_2 hydrocarbon normalized and plotted on a time basis (distance/burning velocity).

These flames show a systematic dependence of burning velocity on the structure of the fuel molecule. These relations were summarized by Dugger (pp. 39–40 in Barnett and Hibbard [1959]). As a general rule for equivalent conditions of equivalence ratio (ER); initial temperature (T_0), and pressure (P), the burning velocities of hydrocarbons rise in the order: alkanes (–C–C–) < alkenes (–C=C–) < alkynes (–C≡C–). The effects of molecular structure may be summarized as: increased straight chain length, chain branching, added side-chain substitutions, or increased ring size all tend to decrease flame velocity. The exceptions are cyclopropane and cyclobutane, which have strained rings and hence higher heats of formation. Oxygenated compounds show the order: esters (R–COO–R') < ethers (R–O–R') < alcohols (R–OH) < aldehydes (RC=OH) < ketones (R–C=O–R') < alkyl oxides (R=OR'). They show molecular structure effects similar to hydrocarbons. For all substitutions and unsaturations the higher the molecular weight the closer the burning velocity approaches that of normal alkanes.

As would be expected the structural differences between flames of these fuel systems and the alkanes lie principally in the region of initial fuel breakdown.

Flame Stoichiometry

The conventional stoichiometry assumes that C/H/O chemistry goes to completion forming carbon dioxide and water with hydrogen and or carbon monoxide as residual traces. This would be the case if the reactions occurred at low temperature, but flame

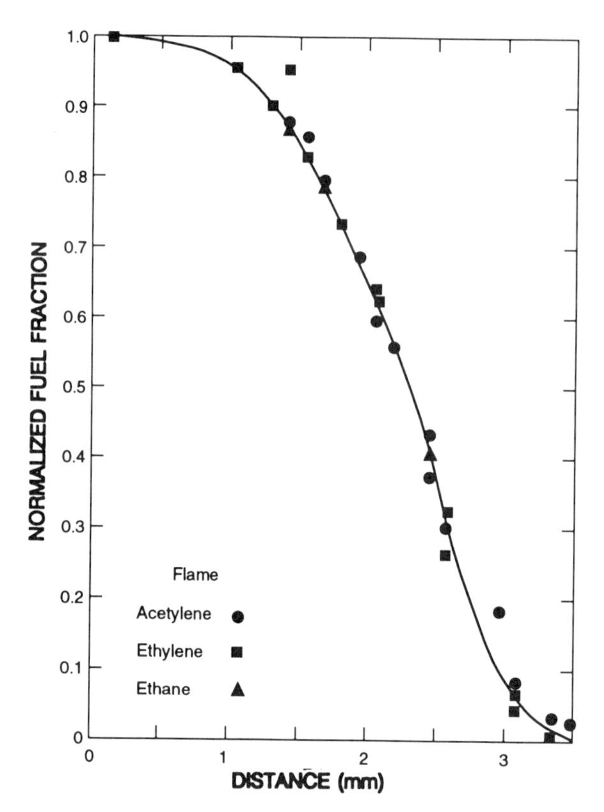

FIG. XII-15 Normalized fuel concentration profiles for three lean acetylene, ethylene, and ethane flames with oxygen with equal burning velocity, showing the importance of transport. (Data of Fristrom, Grunfelder, and Avery [1959].)

temperatures are high and hydrocarbon combustion yields a number of products (H, H_2, O, O_2, OH, H_2O, CO, CO_2, and C_{solid}), whose importance depends on the stoichiometry. If radicals were not taken into account, calculated flame temperatures would be thousands of degrees too high. They are discussed in Chapter IX. Figure XII-16 outlines the regions of importance for various species in this system.

Flame Kinetics

Hydrocarbons are not stable at flame temperatures, and therefore their reactions can be considered irreversible in lean flames. These fuels are attacked rapidly by H, OH, and O, forming radicals most of which are thermally unstable so that the only hydrocarbon radicals which appear in significant concentrations are methyl and ethyl. The reactions rapidly reduce the system to a mixture of simple molecules (H, O, O_2, H_2, OH, H_2O, CO, CO_2) with the radicals in superequilibrium concentrations. This mixture reacts under partial equilibrium conditions until final adiabatic flame conditions are attained.

In rich flames hydrocarbon radical recombination complicates the problem. Since

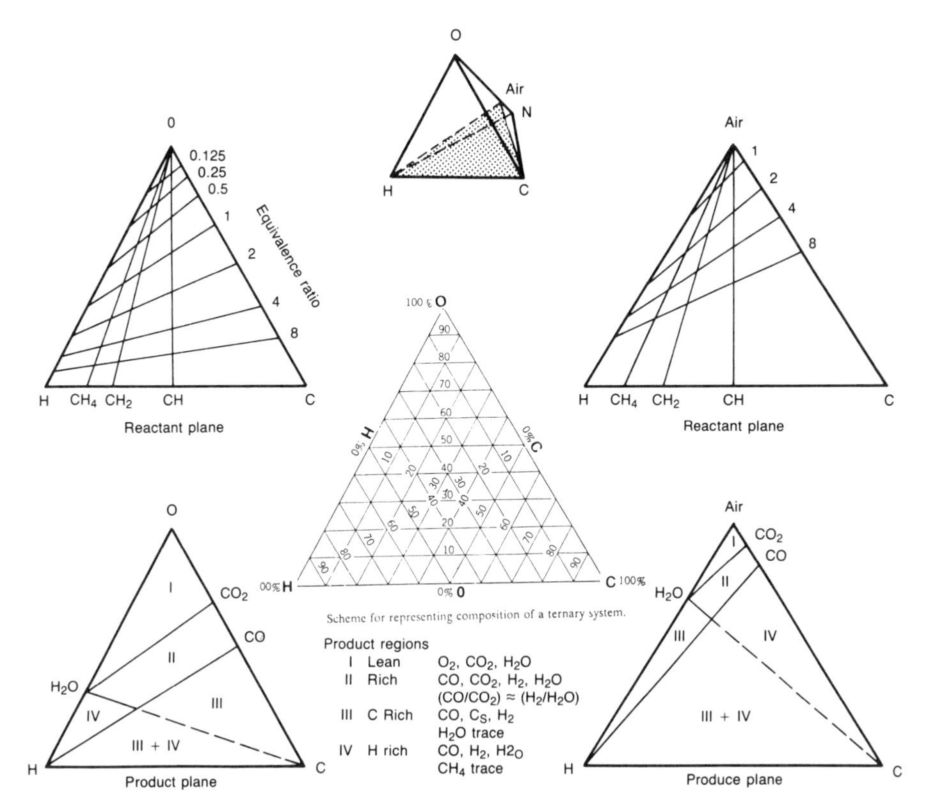

Fig. XII-16 Tetrahedral representation[1] of C–H–O–N mixtures, showing reactant and product regions in the oxygen (nitrogen-free) and air planes.

only the two lowest hydrocarbon radicals are stable at flame temperatures, recombination produces species through C_4 (Warnatz [1981]). This pattern is general for all rich hydrocarbon flames. Methyl introduces two *bimolecular radical recombination reactions* $CH_3\bullet + O: = CH_2O + H\bullet$ and $CH_3\bullet + CH_3\bullet = C_2H_6 + M^*$. A general reaction scheme based on Warnatz's [1981] observations is given in Figure XII-17.

The similarities and differences in profiles for the lower hydrocarbons form Table A–19 and A–20 in the appendix at the end of the book, presenting the results of calculations on eight hydrocarbon flames. They are summarized in the compact form described in Appendix B of Chapter VI. An approximate profile can be constructed using these tabulations. The author appreciates the courtesy of Professor Warnatz in making these results available.

[1] Tetrahedral representation of four components is the three dimensional extension of the planar triangular representation for three components. In this presention, due to Willard Gibbs, all possible combination of the elements is represented by a point in the volume. The scheme is shown for one of the faces. A more complete description of the representation can be found in most books on physical chemistry. The pure elements lie at the corners of the tetrahedron. Mixtures of three elements are the faces. Mixtures of two elements are the edges. Non-elemental mixtures such as those with air are also triangles which are isosceles rather than equilateral.

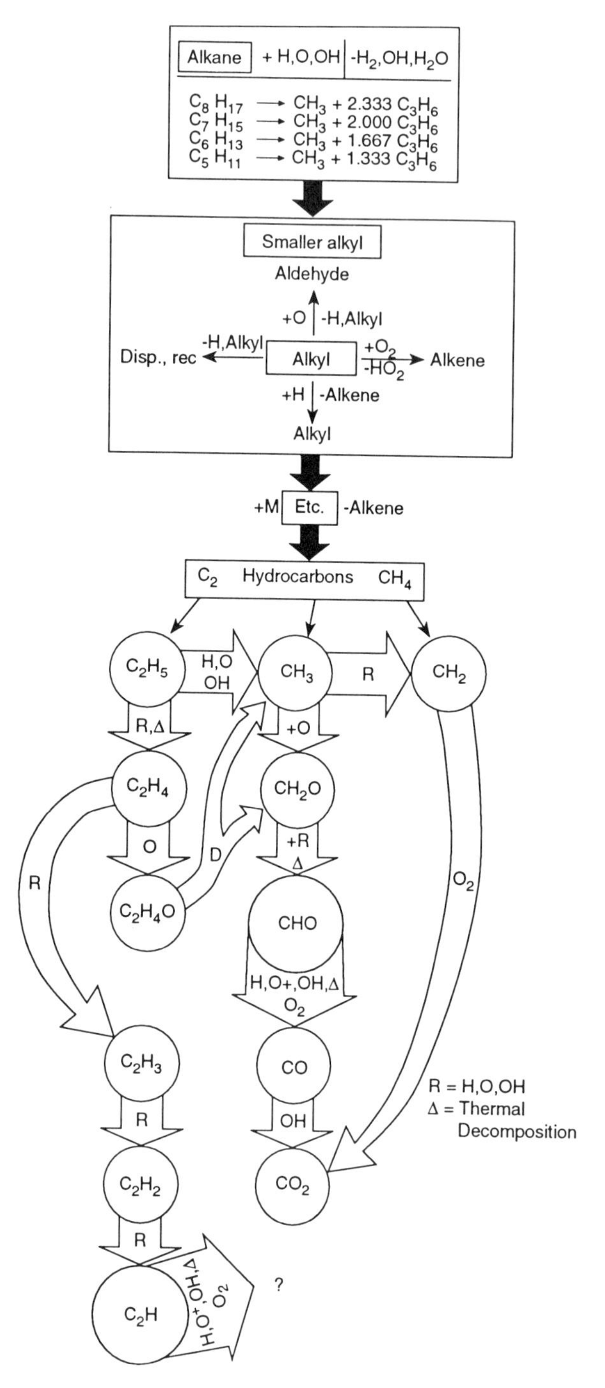

Fig. XII-17 Schematic of major paths in hydrocarbon. (After Warnatz [1981], with modifications.)

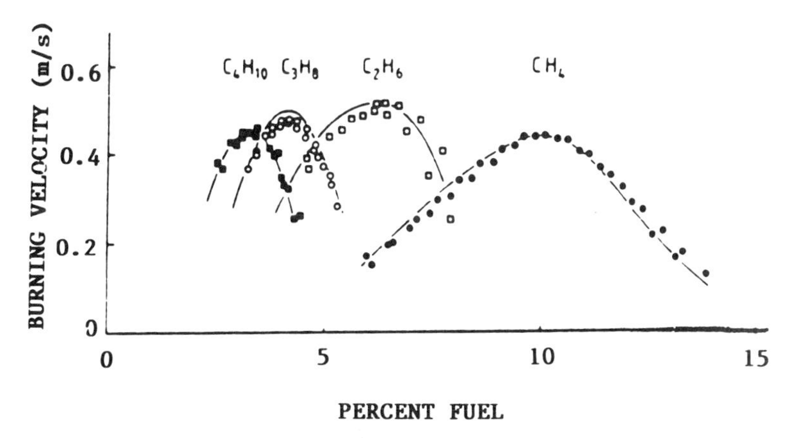

Fig. XII-18 Burning velocities for flames of the lower saturated hydrocarbon–air flames (C_1–C_4). Experiments by various authors (see Table XII-4). Values corrected as recommended by Gunther and Janish [1959]. (Lines are calculations by Warnatz [1981]. Figure after Warnatz.)

The chemistry of these flames can be divided into: (1) the basic H_2–O_2–CO–CH_2O subset; (2) the C_1/C_2 subset; and (3) the C_3/C_4 subset (see Fig. XII-17). Higher members and oxygen-substituted hydrocarbons are treated as fuels attacked by radicals, rapidly yielding a distribution of the species of the three reaction subsets. The general combustion properties of C/H/O fuels with air are summarized in Table A–4 of the appendix at the end of the book. The burning velocities of the lower alkanes are shown as a function of stoichiometry in Figure XII-18. The burning velocities of lower alkenes are collected in Figure XII-19.

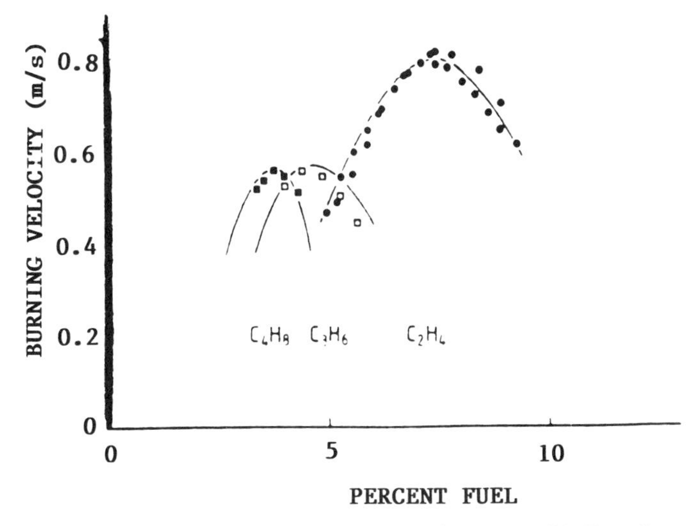

Fig. XII-19 Burning velocities for the lower alkene–air flames (C_2–C_4). Experiments by various authors (see Table XII-4) corrected as recommended by Gunther and Janish [1959]. (Lines are calculations by Warnatz [1981]. Figure after Warnatz.)

Rich and Sooting Flames

Rich flames present a complex chemical problem that is only partially understood and can only be discussed superficially here. A few general remarks are, however, appropriate. The literature in this evolving field is so voluminous that a meaningful summary is impractical. Interested readers are referred to the periodic sessions on soot and carbon formation held at the biannual Combustion Symposia as well as reports from other specialists meetings. The Combustion Symposia reports can be readily located by use of the decennial indices in the Tenth and Twentieth Symposium volumes.

Flames that are carbon rich are defined as those with a C/O ratio greater than one. They produce soot (see Fig. XII-16) and are important because many combustion systems are operated fuel rich to produce carbon black. Soot and smoke abatement is an important pollution objective. Pollution has been reviewed by Sawyer [1981].

Rich flames are discussed as a class because of the similarities induced by radical recombinations. Warnatz [1981] pointed out that they produce significant concentrations of hydrocarbons higher than the inlet fuel, making rich flames very similar in character after the initial fuel attack. The major exception is methane, whose behavior differs from the higher members. This is illustrated by the very low extinction limit shown by methane flames whose rich limit is at an ER of 1.6, with an adiabatic flame temperature of 1800 K, as compared with higher hydrocarbons, which approach a limit around an ER of 3, with an adiabatic flame temperature just over 1000 K. The author feels this occurs because the production of methyl radical is a maximum with this fuel. Methyl is an inhibitor in hydrocarbon flames because of the bimolecular recombination reaction $CH_3\cdot + O{:}\rightarrow CH_2O + H\cdot$ and $HCO + (H, O, OH) = CO + (H_2, OH, H_2O)$. These reactions reduce the peak radical concentration and completely remove the radicals produced by the oxygen reaction, leading to extinction.

Species in Rich Flames

The equilibrium situation is best appreciated by considering Figure XII-16, which shows the regions and dominant species in C/H/O flames. Regions where the ratio of C/O is higher than unity show soot, but soot can be detected in leaner systems. Oxygen goes preferentially to form CO, while the water–gas equilibrium ($CO + H_2O = CO_2 + H_2$) is maintained. In regions where solid carbon is in equilibrium the Boudart equilibrium ($C_{solid} + H_2O = CO + H_2$) is maintained.

As mentioned previously, the fuel attack region of rich hydrocarbon flames contains hydrocarbons through C_4 and radicals through C_2 and some of the peroxy and hydroperoxy radicals typical of cool flames.

Burning Velocity

The burning velocities of the individual hydrocarbons have already been considered. The similarities are clearly brought out in Figure XII-14, which compares the hydrocarbons at equal equivalence ratios and shows the systematic change in rich limits. Müller-Dethlefs and Schlader [1976] investigated the effects of steam addition on flame temperature, burning velocity, and soot formation in several flame systems. They found that water addition inhibits carbon formation and increases heat release.

Flame Structure Studies

Flame structure studies have contributed to the understanding of rich systems. Unfortunately the complexity of these flames makes quantitative interpretation difficult. As a consequence simpler techniques are more favored for developing mechanisms.

The groups at Göttingen and Darmstadt have systematically studied rich flame chemistry since the beginning of flame structure studies (Homann [1985] and Wagner [1979]). They have investigated the buildup of large molecules in the transition to soot and the formation of ions and suggested the key role of C_4H_2 and polyacetylene chemistry in sooting acetylene flames. The work of Weinberg's group at Imperial College on the electrical properties of soot and nucleation has been collected in the book by Lawton and Weinberg [1969].

In the United States Kaskan and Reuther [1977] studied self-inhibition of rich hydrocarbon flames near the limit. The MIT group under Howard made extensive studies of near-sooting flames and the buildup of particles. Their study of a benzene flame discussed in the section on aromatic flames at the end of this chapter, is a model of a comprehensive flame structure study. Studies of near-sooting flames were extended to aliphatic flames by Cole, Bittner, Longwell, and Howard [1984].

In the case of rich acetylene flames, several groups have successfully modeled these systems. Details can be found in the section on acetylene. The author does not feel competent to critique these complex mechanisms other than to observe that they are so complex that it is doubtful that present reaction schemes will be the last word.

In Alma Atta, Kazakhstan studies were made on the structure of very rich flames. Evidence of cool flame reactions in higher hydrocarbons was found. The work has been reviewed in the book by Ksandropolo [1980].

An interesting recent suggestion for the mechanism of soot formation involving aromatic compound buildup has been the discovery of the stability of spherical C_{60} called whimsically Buckminsterfullerene, fullerenes or "Buckey Balls" (Huffman [1992]; Smalley [1991]) after the architect of the geodome. It is suggested that this is formed in soot beginning with a cycle of five condensed benzene rings surrounding a phantom cyclopentagonal "defect," which curves the planar structure. This nonplanar molecule is postulated to grow by addition of acetylene and polyacetylene, sometimes forming the "perfect soccer ball," which has twelve symmetrically located pentagonal "defects" and sometimes growing in an indefinite spiral. Fullerenes have been recovered from flame soot by Howard, McKinnon, and Johnson [1991], and this has become an active area of research that would be fruitless to document at this early stage. Buckminsterfullerene (C_{60}) has higher members C_{70}, C_{240}, C_{540} and related "tubular analogs." Glumac and Goodwin [1992] reported that rich acetylene flames have also been used to prepare diamonds.

Reviews on soot production in flames have been written by Wagner [1979], Haynes and Wagner [1981], Homann [1985], and Howard [1991]. Smalley [1991] has prepared an extensive, continuing bibliography on Fullerines.

Methane

Methane (CH_4) is the simplest hydrocarbon. It is the lowest member of the alkane series whose general formula is C_nH_{2n+2}. Alkanes are stable molecules and are inert at normal temperatures though they are rapidly oxidized at high temperatures. In fact, the original chemical name of the series was paraffinic hydrocarbon, that is, without affinity. The heat of combustion of methane is 192 kcal/mole. Mixtures between 5% and 15.5% with air are flammable. Flame temperatures range from around 1000° K to a maximum around 2300° K near stoichiometric. With pure oxygen the rich limit increases to 59.2%. Normally, methane–air mixtures do not detonate, but there is speculation that this might occur in large systems. In any case, the explosion hazard is great and the widespread use of natural gas makes this fuel a common fire hazard. Oxygen and oxygen-enriched air mixtures with methane do detonate (Lewis and von Elbe [1987], p. 579 ff).

New Species in Methane Flames

Methane flame species include those of hydrogen, carbon monoxide, and formaldehyde flames (H, H_2, O, O_2, OH, HO_2, H_2O, CO, CO_2, CHO, CH_2, CH, and CH_2O). Additional species include the fuel itself, CH_4, and the methyl radical (CH_3). The lower radicals methylene, CH_2, and methyne, CH, play minor roles in lean flames. A minimal scheme for the new reactions which methane introduces is given in Table XII–5.

Methane (CH_4) is a colorless, odorless gas that boils at 90 K. Its principal source is natural gas wells, and it is piped throughout the country as a pressurized gas. It is also shipped and stored as a cryogenic liquid. For laboratory and commercial use it is available in pressurized cylinders with purity ranging from 99% for the commercial grade to 99.999% for research and electronic grades. These cylinders are hazardous both because of the flammability of methane and because of the high pressure involved. Methane is noncorrosive and compatible with all metals. It is easily handled using standard regulators, valves, and plumbing.

The methyl radical (CH_3) is a transient intermediate found in many C/H/O flames, particularly methane, where it is the first major intermediate. It can be produced in low concentrations from labile molecules containing methyl groups by: (1) an electric discharge; (2) pyrolysis; (3) reaction with another radical; or (4) photolysis. Precursor

Table XII–5 Mechanism of the Oxidation of Methane

$CH_4 + M^* \rightleftharpoons CH_3 + H + M$
$CH_4 + H \rightleftharpoons CH_3 + H_2$
$CH_3 + H_2 \rightleftharpoons CH_4 + H$
$CH_4 + O \rightleftharpoons CH_3 + OH$
$CH_3 + OH \rightleftharpoons CH_4 + O$
$CH_4 + OH \rightleftharpoons CH_3 + H_2O$
$CH_3 + H_2O \rightleftharpoons CH_4 + OH$
$CH_3 + O \rightleftharpoons CH_2O + H$
$CH_2 + H \rightleftharpoons CH + H_2$
$CH_2 + O \rightleftharpoons CO + H + H$
$CH_2 + O_2 \rightleftharpoons CO_2 + H + H$
$CH_2 + CH_3 \rightleftharpoons C_2H_4 + H$
$CH + O \rightleftharpoons CO + H$
$CH + O_2 \rightleftharpoons CO + OH$
$CH_3 + CH_3 + M \rightleftharpoons C_2H_6 + M^*$
$CH_3 + CH_3 \rightleftharpoons C_2H_4 + H_2$

molecules include tetramethyl lead [$Pb(CH_3)_4$], diazo methane (CH_3N_2), iodomethane (CH_3I), and acetone (CH_3–CO–CH_3). Electric discharges have such high energy that many bonds are broken and many different radicals are formed. Energies are lower in thermal dissociation, but unwanted species are still produced. Specific reactions for methyl include stripping of methyl halides by hydrogen atoms or vaporized alkali metals, or the reaction of methane with F atoms. Laser photolysis (Rabek [1982]) provides the cleanest production method since a chosen bond can be broken without secondary bond breaking. In addition to reacting rapidly with oxygen and other radicals, methyl reacts rapidly with itself to form ethane. This limits the lifetime and concentration that can be produced in steady-state flow systems.

Methylene radical (CH_2) is well known spectroscopically and can be found in low concentrations in many hydrocarbon flames. It can be produced by techniques similar to those used to produce methyl. Attainable concentrations are even lower, because of its greater reactivity. Precursor molecules include ketene (CH_2CO) and diiodomethane (CH_2I_2).

The methyne radical (CH) is found in even lower concentrations in flames. Again it is well known spectroscopically (Gaydon [1948]; Hertzberg [1971]) but difficult to produce cleanly. Similar production methods are applicable, with laser photolysis favored. Precursor molecules include acetylene and isocyanic acid (NCH). The concentration of gaseous monatomic carbon is very low in flames and other thermal systems because of reactivity and thermodynamic considerations. Higher polymers C_2, C_3, C_4, C_5, etc., are found during soot formation.

Related Studies

Study of the reactions between methane and oxygen in the nineteenth and early twentieth century concentrated on the slow, low-temperature reactions that involve peroxy radicals, formaldehyde, and its chemistry. These reactions are preempted in flames by the faster reactions of H, O, OH. This dichotomy between the slow reactions of ignition and cool flames and the rapid reactions of flames is typical of combustion systems. It results from the radical-poor conditions occurring during ignition, when radicals must be produced in situ with the radical-rich conditions occurring in flames where high temperature radicals are transported to cooler regions by diffusion. This is illustrated by Figure XII-2, which shows the relative sensitivity and importance of reactions in methane ignition and combustion. Ignition and cool flame chemistry can become factors in very rich flames and ignition. This chemistry can only be treated superficially here. The reader interested in the area is referred to the book of Minkoff and Tipper [1962] and Pollard's review in Volume 17 of Bamford and Tipper's series on kinetics [1977]. Methane flame chemistry consists of the reaction of simple radicals, H, O, OH, HO_2, CHO, CH_2, CH_3, and C_2H_5 with each other and nonradical molecular species H_2, O_2, CO, OCH_2, CH_4, H_2O, and CO_2. The relevant flame reaction chemistry has been reviewed by the modelers of the methane flame: Smoot et al. [1976], Tsatsaronis [1978], Westbrook and Dryer [1981], and Warnatz [1983b]. Rate coefficients for simple C/H/O flames are summarized in Table A–1 of the appendix. The complex hydrocarbon radicals important in rich flames are not included. An excellent survey of the field is given by Warnatz [1981b]. There is an ongoing series on the chemical kinetics of combustion by the Reference Data Group at NIST (Tsang and

Hampson [1985]). This is periodically updated, and the data are available in computerized form (Westley et al. [1990]).

Shock tube and high-temperature flow reactor studies have provided significant information on important methane reactions. Examples are the flow reactor studies of Glassman, Dryer, and Cohen [1976] described in the section on carbon monoxide and the shock tube ignition studies of Bowman [1975a]. These studies have been reviewed by Warnatz [1981, 1983b].

Burning Velocity and Detonation Studies

The propagation velocity of methane flames has been exhaustively studied (see Table B–2 of the appendix). Figure XII-18 shows the correspondence between experimental and modeling prediction of burning velocity in the methane–air system as a function of equivalence ratio. Modeling studies by Tsatsaronis [1978] outlined the general characteristics of these flames. Figure XII-20 shows the effect of stoichiometry on burning velocity and some flame structure parameters. Figure XII-21 shows the pressure dependence of burning velocity and flame parameters on the stoichiometric methane–air flame. Figure XII-22 shows the temperature dependence of burning velocity on a stoichiometric atmospheric methane–air flame. The atmospheric flames have been used as a proving ground for burning velocity techniques, and there is now a reasonable consensus on the burning velocity's dependence on composition. The dependence of burning velocity on pressure and inlet temperature are less certain. Methane systems have also been used to investigate the inhibitory effects of halogen additives and the promotion of added hydrogen. Promotion and inhibition are reviewed in Chapter XIII. The limits of flammability have been established, and Yamaoka and Tsuji [1984] have even measured the structure of a near-limit flame using the opposed double flame burner (Chapter III-2). The lean limit of methane flames is nearly pressure independent, but the rich limit increases with pressure (see discussion in Chapter X). Dilution reduces the detonation velocity and narrows the limits, and air mixtures do not detonate.

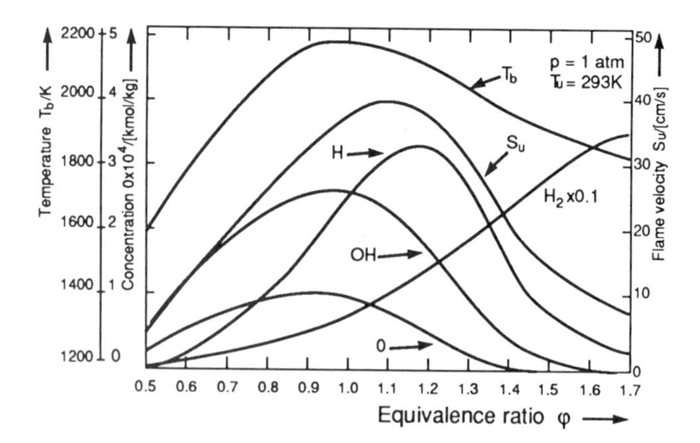

FIG. XII-20 Predicted dependence of Burning Velocity, hot boundary temperature, flame thickness, burning velocity, and peak concentrations of H, O, and OH as a function of equivalence ratio for STP methane–air flames. (Replotted from calculations by Tsatsoranis [1978].)

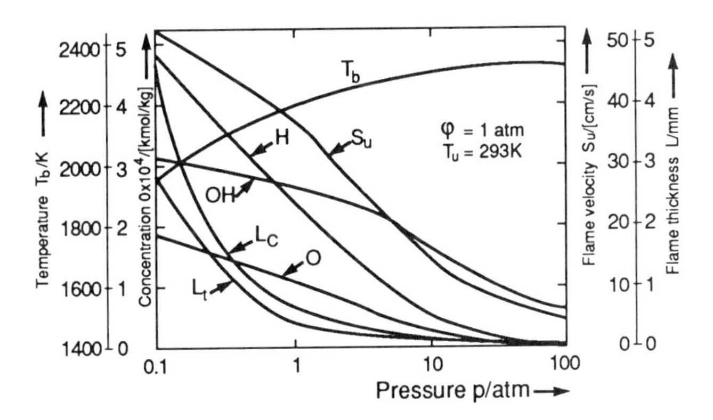

FIG. XII-21 Predicted burning velocity as a function of pressure for stoichiometric methane–air mixtures $T_0 = 300$ K. (Replotted from calculations by Tsatsoranis [1978].)

Flame Structure and Modeling Studies

As mentioned in Chapter VI, the methane flame has been used as a proving ground for both experimental structure techniques and for modeling. Microprobe techniques were applied by Fristrom, Grunfelder, and Favin [1960] to an oxygen-diluted flame. Measurements of radical concentration were added later using the scavenger probe technique of Fristrom [1963a] and the microprobe ESR method of Westenberg and Fristrom [1965]. The data were analyzed by Westenberg and Fristrom [1961] to derive species fluxes, net species production rate, and rate of heat release. This resulted in a

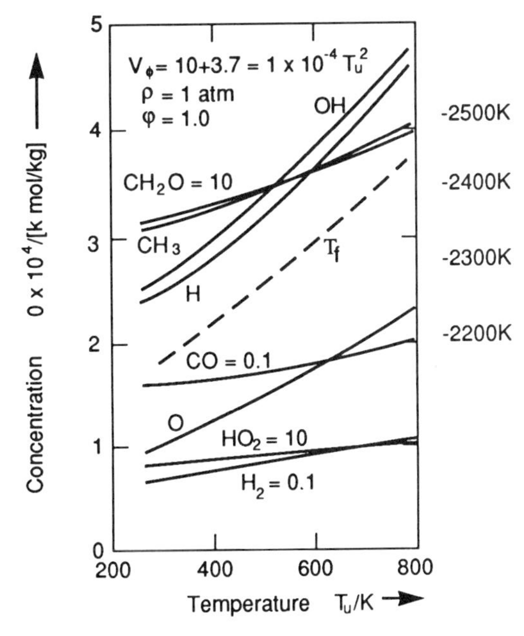

FIG. XII-22 Calculated dependence of peak intermediate concentrations for a stoichiometric atmospheric methane–air flame as a function of inlet temperature. (After Tsatsoranis [1978].)

definitive set of papers (Fristrom et al. [1960–1963]; Westenberg et al. [1960, 1961]) that clarified the relative importance of flame processes and allowed the derivation of significant kinetic information. The studies were used in Chapter IX to illustrate flame processes.

Other pioneering studies on the methane system clarified the mode of attack of methane in flames (Fenimore and Jones [1961b]; Dixon-Lewis and Williams [1967]). Soot formation was studied in rich flames of D'Alessio and collaborators [1973]. Studies of inhibition and NO_x pollutant formation are discussed in Chapter XIII.

The methane system was used to develop a molecular beam inlet with mass spectrometric detection at Göttingen (Bonne, Grewer, and Wagner [1960]) and the Midwest Research Institute (Milne and Green [1965b, 1969]). At the National Institute of Standards and Technology (NIST) Hastie [1973a] studied the inhibition of methane. This extensive work is summarized in his book on high-temperature vapors (Hastie [1975]). At UCLA Knuth pioneered in the interpretation of supersonic molecular beam sampling and applied the technique to the methane system (Gay, Young, and Knuth [1975]). Much of the work of these groups was aimed at the study of inhibition, which is discussed in Chapter XIII.

The first comprehensive studies using molecular beam inlet sampling on methane was reported in two papers at the Fourteenth Symposium on Combustion by Peeters and Mahnen of Louvain [1973b] and Biordi, Lazzara, and Papp of the U.S. Bureau of Mines [1973]. These landmark papers provided complete profiles of several low-pressure methane flames. They were similar to the earlier microprobe studies, but added the profiles of the radical species. The Louvain studies provided complete temperature and composition information, which, together with a derived velocity, allowed quantitative analysis of the flux and species appearance rates. Figure XII-23 shows a comparison between the experimental studies of Peeters and Mahnen [1973a] with modeling studies of Tsatsaronis [1978]. These data were used to investigate mechanisms and rates in methane flames. They found methane was attacked both by H and OH and were able to establish that the dominant reaction of methyl even in this oxygen-rich system was with oxygen atoms. They suggested, using "pool" arguments, that a significant fraction of the formaldehyde formed in this flame dissociated ($CH_2O + M^* = CO + H_2$). This question was discussed in the section on formaldehyde. The papers of Biordi, Lazzara and Papp [1973–1978]), which gave a critical evaluation of molecular beam inlet technique, are discussed in Chapters VII and VIII. The major thrust of these studies was flame inhibition, which is discussed in Chapter XIII. Their results on methane chemistry independently confirm and complement the Louvain studies and the earlier studies at APL Johns Hopkins.

The laser diagnostic methods discussed in Chapter VIII have contributed to many aspects of methane flames, particularly the radical distributions. The most complete study is that of Bechtel, Blint, Dasch, and Weinberger [1981] on a series of atmospheric flames ranging from lean to rich. Figure XII-24 gives an example of their results comparing experiments with modeling of the stoichiometric flame. Dreier, Lange, Wolfrum, Zahn, Behrendt, and Warnatz [1986] compare CARS (coherent anti-stokes resonance spectroscopy) measurements on a methane–air diffusion flame with detailed modeling.

The methane system has received more attention from modelers than any other hydrocarbon system. The development of modeling is discussed in Chapter VI. The

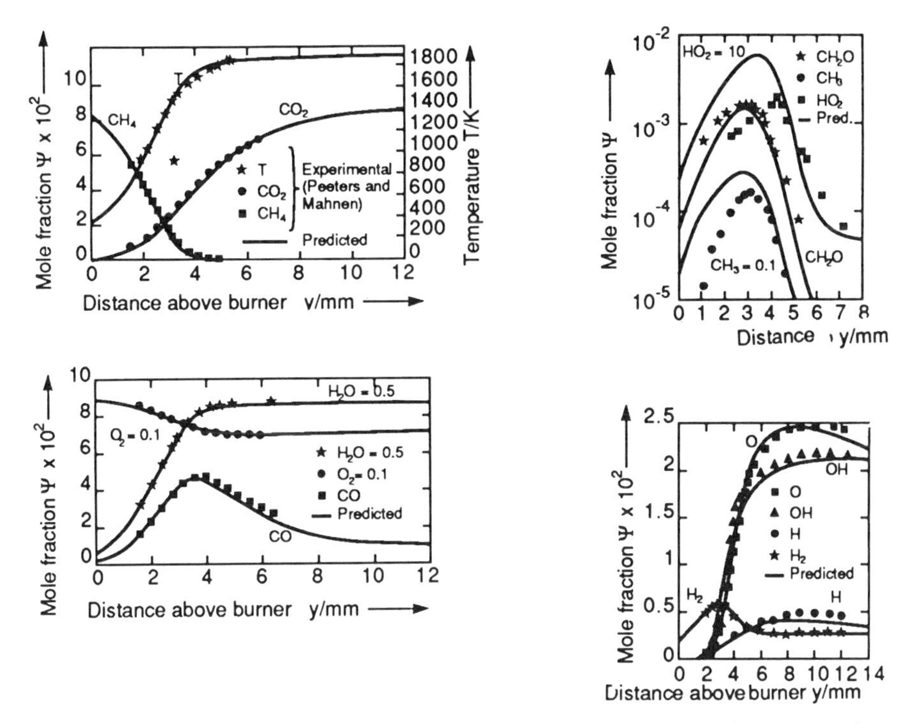

FIG. XII-23 Structure of a low-pressure (0.053 atm) CH_4 = 0.095–oxygen flame. (Experiments by Peeters and Mahnen [1973b] with calculations by Tsatsoranis [1978].) (a) Major species and temperature; (b) Radicals; (c) Fuel intermediates.

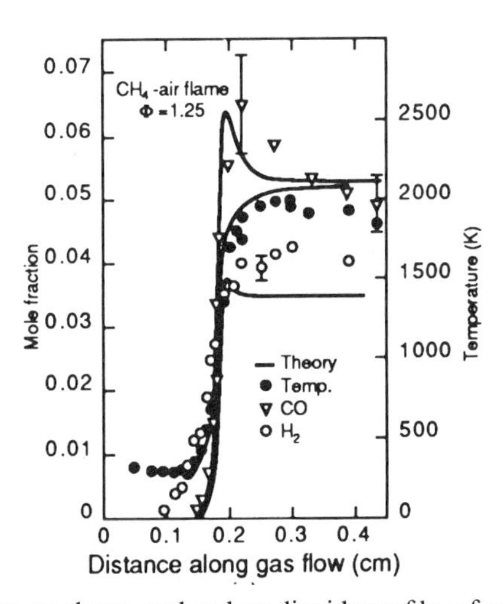

FIG. XII-24 Temperature, methane, and carbon dioxide profiles of an atmospheric fuel-rich methane–air flame (ER = 1.25) using laser-induced Raman scattering. (Replot of experiments and modeling results by Bechtel, Blint, Dasch, and Weinberger [1981].)

first synthetic attempt for methane was by the Utah group (Hecker [1975]; Smoot, Hecker, and Williams [1976]), who modeled the Peeters and Mahnen flame. They developed a full computer simulation and tested a number of kinetic mechanisms against burning velocity, flame thickness, and flame structure measurements. Their technique was based on the method of Spalding and Stephenson [1971]. They employed a 28-reaction mechanism, 5 of which they felt could be neglected. The model also was limited by the assumption of unity Lewis number.

Tsatsaronis [1978] used a 29-reaction model based on a literature survey with some modifications of rates (within limits of uncertainty) to improve the fit. The predictions were tested against: (1) burning velocity dependence on composition, pressure and initial temperature (Figs. XII-20–22); and (2) flame structure studies of Fristrom et al. [1960], Peeters and Mahnen [1973a], and Dixon-Lewis and Williams [1967]. The agreement of modeling with experiments was very satisfactory, although, as the author comments, the kinetic constants may not represent a unique set. Dixon-Lewis [1981] extended his flame model for hydrogen and carbon monoxide to include the methane system using the rigorous transport formulation of Monchick and Mason [1961a,b] with their polar, polyatomic multicomponent treatment including thermal diffusion. Warnatz [1981] included the methane system in his comprehensive modeling study of the lower hydrocarbons. Smooke [1982] also modeled the methane system, and Smooke, Crump, Seshadri, and Giovangigli [1991] extended the treatment to methane diffusion flames. Westbrook and Dryer [1981] reviewed their own work on methane and the general status of modeling of combustion systems. Other reviews that cover the methane system include those of Oran and Boris [1981], Warnatz [1981], and the book edited by Gardiner [1984].

Simplified treatments have been applied to methane. One by Olson and Anderson [1985] utilized experimental temperature profiles. This separates the energy equation from species production, which greatly simplifies the treatment. Peters and Williams [1987] have extended the steady-state and partial equilibrium treatments of Dixon-Lewis using asymptotic theory. They are able to simplify the mechanism to: $CH_4 + O_2 = CO + H_2 + H_2O$, $CO + H_2O = CO_2 + H_2$, and $O_2 + 2H_2 = 2H_2O$, while still retaining the kinetic information of the constituent elementary reactions.

Summary

The major aspects of the methane system appear reasonably understood. Its burning velocity and microstructure have been extensively studied experimentally and quantitatively analyzed and modeled. There is a reasonable consensus on the overall kinetic mechanisms, though there are many disagreements on minor points. The several groups active in modeling studies have been able to reproduce most of the experimental studies and generally agree with each other in this system, although there has been no systematic comparison of modeling on this system.

The fuel attack is principally by hydrogen extraction by H, OH, and O. In lean flames OH abstraction is dominant; in rich flames H atom extraction is dominant. Thermal decomposition of CH_4 is minimal in flames, though in shock tube this reaction can become important. The methyl radical formed in lean flames reacts dominantly with O atoms to produce formaldehyde, with minor contributions from other radical reactions. Direct reaction with molecular oxygen is apparently too slow to be important in flames. HO_2 appears to be important principally as a radical recombination path. In rich flames, methyl recombination to form ethane becomes important as well as stripping to form CH_2, which can react directly with molecular oxygen. On the lean side most of the methane follows the path methane to methyl to formaldehyde to HCO to CO to CO_2. Once formaldehyde is formed, the chemistry becomes identical with the formaldehyde–carbon monoxide–hydrogen systems discussed previously. On the rich side (beyond ER 1.2) the situation is more complex since methyl radical recombination produces ethane, which in turn produces ethyl radicals, which can recombine to form propane and butane. This means that the discussion of rich methane flames must include mechanisms of propane and butane systems. This is discussed superficially as one aspect of the section on rich flames.

Acetylene

The flames of acetylene are more luminous and have higher temperatures, faster burning velocities, and wider flammability limits than comparable hydrocarbon systems (Table A–6 in the appendix). Acetylene is also an important flame species in rich flames and it is believed to play a role in soot formation.

New Species

The new species that acetylene brings to flame chemistry are the C_2 molecules: C_2H_2, C_2H, C_2, CH_2CO, and HC_2O. In addition the lower members of the H/C/O hierarchy require consideration. Its rich flame chemistry is similar to other hydrocarbons.

Acetylene (HC≡CH) is the lowest member of the acetylene or alkyne family of hydrocarbons and is characterized by a single triple bond and the general formula C_nH_{2n-2}. It is a linear molecule consisting of two CH groups linked by a triple bond between the carbons. Because of symmetry it has no permanent dipole moment. Acetylene has a high heat of formation (54.2 kcal/mole). Despite this it is one of the most thermally stable hydrocarbons.

Acetylene is a colorless gas with a garliclike odor. It is hazardous because the compressed gas or liquid can detonate. This problem is avoided commercially by handling it in the form of a solution in acetone under modest pressure ($\backsim$ 140 psig) held on a porous medium such as asbestos fibers. Cylinders of various sizes are available from commercial sources with purity ranging from 99% for industrial grades to 99.999% for research grades. For research studies, it is necessary to remove the traces of acetone. This can be done using a cold trap or better by a chemical adsorption train consisting of aqueous sodium bisulfite ($NaHSO_3$) followed by aqueous sodium hydroxide and, finally, a silica gel column. The gas can be manipulated in the laboratory at or below atmospheric pressure using ordinary gas-handling procedures. However, because of the detonation hazard, it should always be handled with caution and never pressurized or liquefied.

Ethynyl radical (C_2H) is an intermediate in acetylene and other hydrocarbon flames. It is a significant equilibrium product under high-temperature and -pressure (>10 atm) conditions, but it is not isolatable under normal laboratory conditions. It can be produced for study by the standard methods of electric discharge, pyrolysis, photolysis, or radical reactions. The methods yielding the most satisfactory results are laser photolysis or the fluorine atom reaction, $F + C_2H_2 = HF + C_2H$ (Hoyermann [1979]; Blumenberg, Hoyermann, and Sievert [1977]; Bartels et al. [1982]).

C_2 radical is a transient intermediate in hydrocarbon flames. Its spectrum is prominent in rich flames (Gaydon [1974]; Hertzberg [1971]). It is not present at equilibrium under high pressures because of its instability relative to solid carbon. It can be produced using standard methods of electric discharge, pyrolysis, photolysis, or specific reaction.

Ketene ($O{=}C{=}CH_2$) is a very reactive gas with a strong tendency to polymerize. It has a sharp odor and is a lacharimeter. It is prepared by the pyrolysis of acetone or acetic acid.

Ketyl radical ($O{=}C{=}CH$) is a transient intermediate in acetylene and other hydrocarbon flames. It reacts rapidly with H_2 and O_2.

Reaction Mechanisms

Acetylene chemistry is complicated by the existence of addition as well as stripping reactions so that some oxygenated species assume importance. The minimal reaction scheme (Table XII–6) even for lean flames exceeds fifty reactions, and for rich systems the number exceeds 100 with a comparable number of intermediates. This chemistry is too complex for a simple summary, but a few important factors are worth bringing out. The radical acetylene reactions are complicated by multiple paths due to addition as well as stripping reactions, and in some cases more than one product path is significant in flames. Warnatz [1983a] suggests that a major source for carbon monoxide formation is the oxygen atom attack, $O + C_2H_2 = CO + CH_2$. Carbon dioxide is formed both from the attack of carbon monoxide by OH radicals as occurs in methane flames and also by destruction of the CH_2 radical by reaction with oxygen, $CH_2 + O_2 = CO_2 + 2H$. In rich flames water appears to be dominantly formed by $OH + H_2 = H_2O + H$ and $OH + OH = H_2O + O$, while in lean flames the OH stripping reaction $C_2H_2 + OH = H_2O + C_2H$ dominates. Radical addition and exchange reactions such as $C_2H_2 + H = C_2H_3$ and $OH + C_2H_2 = CH_2\text{-}CO + H$ play important roles. C_4H_2 and its reactions may play a major role in rich acetylene flames and soot formation. Thermal diffusion plays a significant role in these flames, tending to concentrate the

Table XII–6 Acetylene Reactions

$C_2H_2 + H \rightleftharpoons C_2H_3$
$C_2H_2 + O \rightleftharpoons CH_2 + CO$
$C_2H_2 + OH \rightleftharpoons CH_2CO + H$
$C_2H_2 + H \rightleftharpoons C_2H + H_2$
$C_2H_2 + OH \rightleftharpoons C_2H + H_2O$
$C_2H + O \rightleftharpoons CO + CH$
$C_2H + H_2 \rightleftharpoons C_2H_2 + H$
$C_2H + O_2 \rightleftharpoons CO + CHO$
$C_2H_2 + O_2 \rightleftharpoons CHCO + H$
$CHCO + H \rightleftharpoons CH_2 + CO$
$CHCO + O \rightleftharpoons CO + CO + H$

hydrogen in the cooler portions of the flame. In sooting flames the thermomechanical effect, which is the analog of thermal diffusion for particulates, may play a role by forcing migration toward colder gas and increasing early heterogeneous processes.

Related Studies

Acetylene and its radicals have been extensively studied using a number of techniques. This work has been summarized by Minkoff and Tipper [1962, p. 175], Lewis and von Elbe [1987], Jost [1946, p. 401], Pollard in Volume 17 of Bamford and Tipper's [1977] series on chemical kinetics, and Warnatz [1983a] in his survey of combustion chemistry. At moderate temperatures, <1000 K, the radical reactions have been studied using both electrically generated radicals and radicals generated by laser photolysis. At higher temperatures shock tube techniques have been used. Slow reaction and ignition studies have been made on acetylene, but its complex chemistry has made the interpretation of such studies in terms of elementary reactions difficult.

The shock tube studies of acetylene oxidation and pyrolysis are too numerous for a meaningful review in this book. Therefore, only a few typical examples will be given. Reviews include those of: Miller, Mitchell, Smooke, and Kee [1982], Warnatz [1981], Westbrook and Dryer [1981], and White and Gardiner [1979].

Homer and Kistiakowsky [1967] studied ignition delay in acetylene–oxygen mixtures by following the reaction with time-of-flight mass spectrometry while Gardiner [1961] used x-ray densitometry. One good shock tube study of an acetylene reaction is that of Frank, Bhaskaran, and Just [1988] on the $O+C_2H_2$ reaction. Pyrolysis studies include that of Colket [1988], which offers a good recent example of a shock tube study aimed at the elucidation of soot formation.

Burning Velocity and Detonations

The burning velocities of acetylene–air mixtures have been measured by a number of investigators. It has the highest burning velocity of any hydrocarbon. Acetylene shows no well-defined rich limit because it supports a decomposition flame whose propagation overlaps the oxidation system. The decomposition flame is discussed in Chapter XI. Its lean limit is the lowest of all hydrocarbons (Brown et al. [1969]). A comparison of experimental studies with modeling by Warnatz [1981b] is shown in Figure XII-25. references are collected in Table B–2 in the appendix. Similar agreement has been obtained by all the groups modeling this system (see following section). Despite this, the modelers agree that the kinetics are incomplete. The system also produces detonations over a wide range of compositions (Breton [1936]). Detonation structure is discussed by Strehlow [1983, p. 287].

Flame Structure and Modeling Studies

The first structural study of acetylene was a comparison by Fristrom, Avery, and Grunfelder [1959] of the C_2 hydrocarbons (acetylene, ethylene, and ethane) under oxygen-rich conditions with compositions adjusted to equal burning velocities. The initial microprobe studies were supplemented by temperature profiles and ESR determination of H and O atom profiles (Westenberg and Fristrom [1965]). As would be

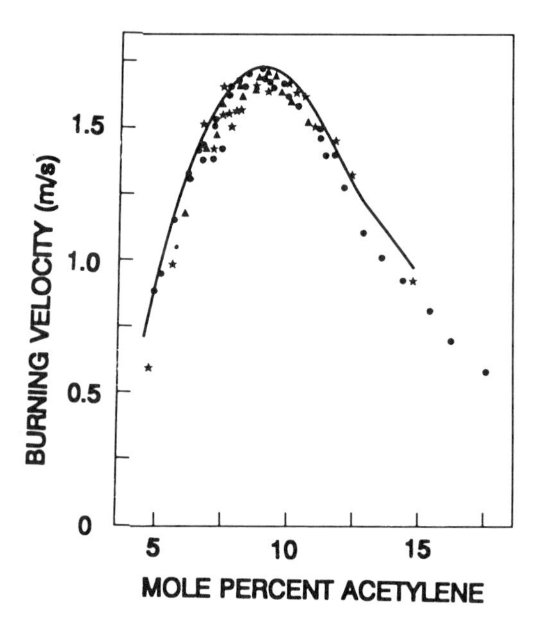

FIG. XII-25 Acetylene burning velocity as a function of compositon. (Calculations of Warnatz [1981–1983] compared with experiments (see Table XII-5). After Warnatz.)

expected with matched burning velocity and reactants of almost the same molecular weight, the major reactants and product profiles were almost identical when plotted on a normalized basis (see Fig. XII-15). This simply indicates a dominance of transport in determining these structures. By contrast, the intermediates differed, suggesting that under these strong oxidizing conditions stripping reactions were dominant and that the hierarchy was: ethane yields ethylene, which yields acetylene. The data have not been analyzed further. Other microprobe studies included those of Fenimore and Jones [1963, 1964], Porter, Clark, Kaskan, and Browne [1957], and Brown, Porter, Verlin, and Clark [1969]. Zeegers and Alkemade [1965] studied radical recombination in the postflame gases of acetylene flames.

The pioneer molecular beam inlet study was by the group at Göttingen on a near-sooting acetylene flame by Bonne, Homann and Wagner [1965] and a study of a low-pressure lean acetylene flame by Eberius, Hoyermann, and Wagner [1973]. At Louvain Vandooren and Van Tiggelen [1977] studied a lean acetylene flame, which they analyzed for rate data. The first comprehensive synthetic modeling of acetylene combustion was made by Warnatz [1981] in his landmark modeling of the lower hydrocarbon flames. He compared his model with burning velocities (Fig. XII-25) and the structure of the Eberius et al. acetylene flame. This required fewer than 60 reactions, but it is a lean system with minimal hydrocarbon intermediates. A major conclusion of the study was that the early formation of carbon dioxide in acetylene flames required a second path for its formation. He suggested the reaction $CH_2 + O_2 = CO_2 + 2H$. This is a kinetically awkward process that may not represent an elementary reaction. However, Warnatz maintains that this reaction or its equivalent is necessary. It provides an alternate path for forming reactive radicals from oxygen, but in the spin sense it is a recombination process reducing the spin pool by two.

Warnatz, Bockhorn, Moser, and Wenz [1982] studied the structures of acetylene flames between stoichiometric and sooting conditions and applied a 93-reaction model. This mechanism reproduces experimental results quite satisfactorily. The principal conclusions of this study were the mechanism for C_4H_2 formation, which the authors considered to be the prime precursor of soot.

Volponi, MacLean, Fristrom, and Munir [1986] used a stoichiometric low-pressure acetylene flame as a proving ground for scavenger probe studies of radical concentrations. These results were compared with the Sandia acetylene flame model of Miller, Mitchell, Smooke, and Kee [1982] (Fig. XII-26).

Joklik, Daily, and Pitz [1988] used laser fluorescence to measure CH concentration in an acetylene flame. The measurements were compared with model calculations.

The rich flame of Eberius et al. was restudied and modeled at MIT by Westmoreland, Howard, and Longwell [1988]. Experimental profiles of 25 intermediates were measured (Fig. XII-27). They used these experiments to compare mechanisms published by Miller et al. [1982], Warnatz [1981], Westbrook [1983], and Westbrook and Dryer [1981]. In dissecting the results using reaction path analysis for eight of the reactants and products, they identified the sources of differences among the models. They concluded that all of the models described the general acetylene flame features well, but the most satisfactory model appeared to be that of Warnatz with the reactions reversible.

Ions are more abundant in acetylene flames than in those of other hydrocarbons, and a number of studies have been made (Knewstubb and Sugden [1959]; Deckers and Van Tiggelen [1959]; Poncelet, Berendsen, and Van Tiggelen [1959]; Michaud, Delfau, and Barassin [1981]; Olsen and Calcote [1981]). Ion flame chemistry lies outside the scope of this book.

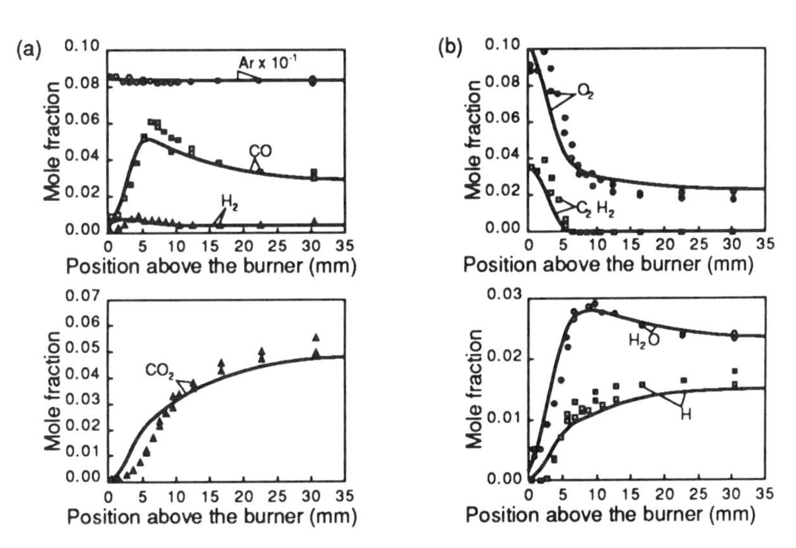

FIG. XII-26 Comparison of experiment (points) with computer modeling (solid lines) on a low-pressure (30 Torr) argon-diluted stoichiometric acetylene–oxygen flame. Radical concentrations measured using deuterium scavenger probe sampling. (Figure after Volponi et al. [1986].) (a) CO, H_2, Arb. O_2, C_2H_2; (b) H and H_2O.

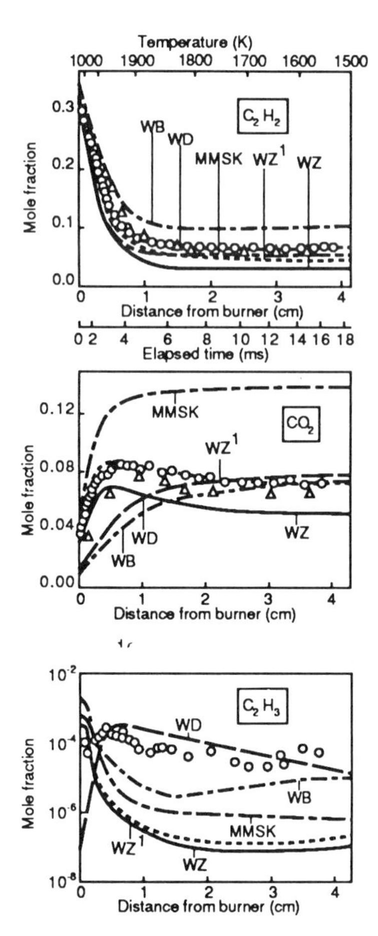

Fig. XII-27 Comparisons of several acetylene models with experiment on a rich acetylene flame. Profiles of C_2H_2 CO_2 and C_2H_3 are presented. Experiments by Westmoreland, Howard, and Longwell [1988] and Bonne et al. [1965] with published mechanisms by: WZ, WZ' (reversible)—Warnatz [1983a]; Wb—Westbrook and Dryer [1981]; MMSK—Miller, Mitchell, Smooke, and Kee [1982]; WD—Westbrook [1983]. (After Westmoreland et al. [1988].)

Ethylene

Ethylene forms one of the better known systems. The flames are typical of hydrocarbons but are somewhat hotter, faster, and more luminous than those of the alkanes. It has wider limits of combustion than any hydrocarbon except acetylene. This results from its higher heat of formation (0.52 kcal/g), and the consequent higher heat of combustion (12 kcal/g). In the early days it was widely studied because high-purity ethylene could be conveniently produced by the dehydration of ethyl alcohol.

New Species in Ethylene Flames

The ethylene flame introduces two new species, ethylene (C_2H_4) and the vinyl radical (C_2H_3), to be added to those of the lower C/H/O flames.

Ethylene (C_2H_4) is a planar symmetric molecule with no permanent dipole moment. It is the lowest member of the olefin family, also called *alkenes,* which are characterized by a single carbon–carbon double bond and the general formula C_nH_{2n}. Ethylene is a colorless gas with a slightly sweet odor that is not poisonous, although one should not breathe it. For laboratory use it is available as a compressed gas at pressures around 100 atm in the common cylinder sizes. Purities range from 99% for technical grades to 99.999% for research grades. The gas is noncorrosive and can be handled using standard equipment.

The vinyl radical (C_2H_3) is a transient found in many flames and combustion processes. It is not present in significant quantities under adiabatic flame conditions. It can be prepared by the usual radical-producing techniques of electric discharge, pyrolysis, and photolysis as well as the addition of acetylene to hydrogen atoms $H + C_2H_2 = C_2H_3$.

Reaction Mechanisms

To model ethylene combustion it is necessary to introduce at least six new reactions to the previously discussed schemes for hydrocarbon combustion. These are listed in Table XII–7. It should be observed that the reactions involving the oxygen radicals are essentially irreversible.

Related Studies

The pyrolysis and oxidation of ethylene have been studied by a number of techniques. The classic slow combustion studies of ethylene have been reviewed along with other hydrocarbons by Minkoff and Tipper [1962, p. 167] and Pollard's review in Volume 17 of Bamford and Tipper's [1977] series on kinetics. This chemistry is complex, involving oxygenated species and the formation of higher olefins. The time scale of these processes is too slow to be of importance in flames. Reactions of ethylene with H, O, and OH have been studied using discharge flow, laser photolysis, and crossed-beam techniques such as those of Hoyermann [1979] and his colleagues at Göttingen.

Ethylene combustion chemistry has been reviewed by Warnatz [1981b, 1983b]. Schug, Santoro, Dryer, and Glassman [1978] studied ethylene oxidation in a high-temperature flow reactor. The shock tube work has been reviewed by Tanzawa and Gardiner [1979]. Much of the ground work for flame modeling originated in mechanisms used to interpret shock tube ignition times. Westbrook, Dryer, and Shug [1983] used a 93-reaction model for comparison with the high-temperature flow reactor study

Table XII–7 Ethylene Reactions

$C_2H_3 + M^* \rightleftharpoons C_2H_2 + H + M$
$C_2H_4 + H \rightleftharpoons C_2H_5$
$C_2H_4 + H \rightleftharpoons C_2H_3 + H_2$
$C_2H_3 + H \Rightarrow C_2H_2 + H_2$
$C_2H_4 + O \rightarrow CHO + CH_3$
$C_2H_4 + OH \rightleftharpoons C_2H_3 + H_2O$
$C_2H_3 + O_2 \rightleftharpoons C_2H_2 + HO_2$

of Schug, Santoro, Dryer, and Glassman [1978]. They provide a good survey of work in the field. The modeling of these systems is less complex than flames because transport can be ignored. The authors made the point that mechanisms suitable for flames do not necessarily fit these systems and that more extensive models will be necessary to encompass data from flames, shock tubes, and flow reactors.

Burning Velocity and Detonations

Ethylene was an early favorite for burning velocity studies, and there is generally good agreement among investigators, probably because of the availability of relatively pure gas. References to these studies are collected in Table B–2 in the appendix. A comparison between measurements by Linnett and Hoare [1948] and Gibbs and Calcote [1959] with modeling by Warnatz [1981] is shown in Figure XII-19. Note that the experimental values are adjusted by 5–10%, in accord with the recommendations of Andrews and Bradley. The effect of dilution by nitrogen was investigated by Linnett and Hoare [1948] and the effect of pressure by Cullshaw and Garside [1949]. The limits of combustion were determined by Coward and Jones [1954] of the U.S. Bureau of Mines (see Table A–1 in the appendix at the end of the book). Some ethylene–air mixtures detonate, and the limits with oxygen are even wider.

Flame Structure and Modeling Studies

The ethylene flame microstructure was first studied by Fristrom, Grunfelder, and Avery [1959] in their comparative study of oxygen-rich C_2 flames. Atom profiles were measured by Westenberg and Fristrom [1965]. The temperature profiles have also been measured (Fristrom and Grunfelder [unpublished]). This work was followed by two perceptive studies by Levy and Weinberg [1958, 1959]. Temperature profiles were measured using optical deflection techniques. They demonstrated quantitatively that global models could be misleading for some flames, and first directly evaluated the radiation loss energy flux. Fenimore and Jones [1963b] studied the mode of fuel attack in ethane and ethylene flames.

The most comprehensive study of an ethylene flame is a molecular beam inlet study presented in the doctoral thesis of Mahnen at Louvain [1973]. Part of this work was published by Peeters and Mahnen [1973b] in a Combustion Institute meeting at Sheffield. This was a lean (0.0655 C_2H_4) oxygen-diluted flame (duplicating the previous Fenimore–Jones study) and a stoichiometric "argon–air" flame. Temperature and composition profiles were measured, including radicals, using molecular beam inlet mass spectrometry. Results from the lean flame are shown in Figure XII-28. The conclusions from this study were that the principal attack of ethylene was by oxygen atoms to form a complex that broke down by two paths, forming CH_2 on one hand and formaldehyde on the other. The flame has been analyzed by Basevich, Kogarko, and Posvyanskii [1977] using a specified temperature profile with an 86-step reaction scheme. They compared peak concentrations of intermediates and found reasonable agreement. In thesis work at the University of New South Wales, Australia, Haynes [1973] measured CH_4, C_2H_2, and H in the radical decay zone of a rich (ER = 1.66) flame. This work was extended by Levy, Taylor, Longwell, and Sarofim [1983], who mea-

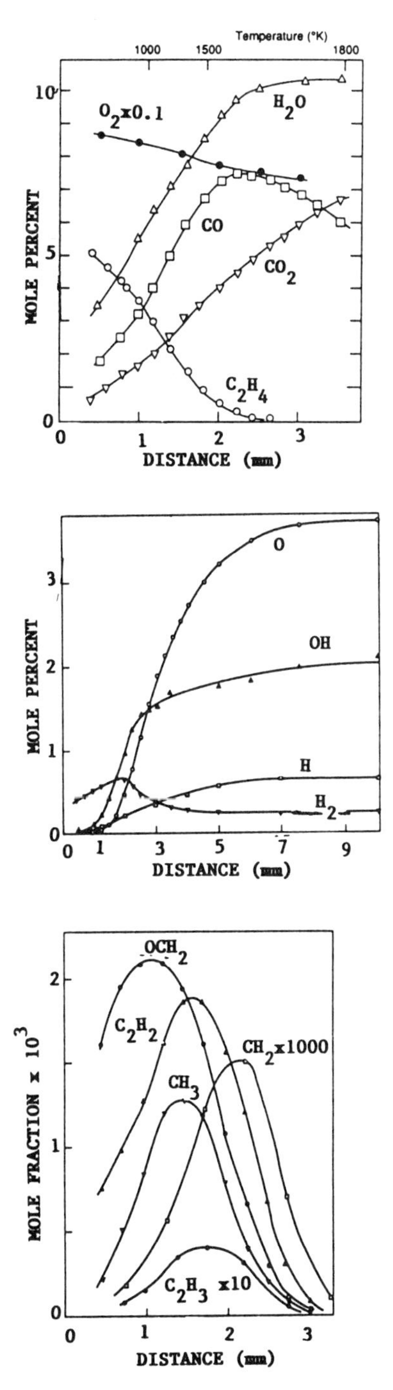

FIG. XII-28 Structure of a lean low-pressure ethylene–oxygen flame (0.0655 ethylene; P = 40 Torr. (a) Major Species and temperature; (b) Radical species; (c) Fuel intermediate species. (Reploted data from the Thesis of Mahnen [1973].)

sured and modeled profiles of methane, acetylene, and hydrogen atoms in the decay region of two fuel-rich ethylene–air flames with equivalence ratios of 1.71–1.83.

Levy, Taylor, Longwell, and Sarofim [1983] compared the measurements of Haynes of CH_4, C_2H_2, and H decay in a rich ethylene flame with predictions using previously published ethylene combustion mechanisms by Warnatz [1981] for flames, White and Gardiner for shock tubes [1979], and Westbrook for flow reactors. Results were disappointing; so they modeled their own decay profiles in a rich atmospheric ethylene–air flame (ER = 1.78) employing a 76-reaction model. The fit (Fig. XII-29), as might be expected, was good. This emphasizes the point that more detailed models are required to fit all types of combustion systems.

Ethane

Ethane is the least studied of the lower hydrocarbon flame systems. The flames are typical of hydrocarbons, and this might be considered to be the lowest typical hydrocarbon flame, since it is the first member of the series to have a C–C single bond and allow formation of the ethyl radical (C_2H_5) by stripping. It forms cool flames in rich systems and can be catalytically oxidized under low-temperature conditions (Minkoff and Tipper [1962]).

New Species in Ethane Flames

Ethane combustion introduces three new species: ethane (C_2H_6), ethyl radical (C_2H_5), and acetaldehyde (CH_3CHO). Lower members of the H/C/O hierarchy also require consideration.

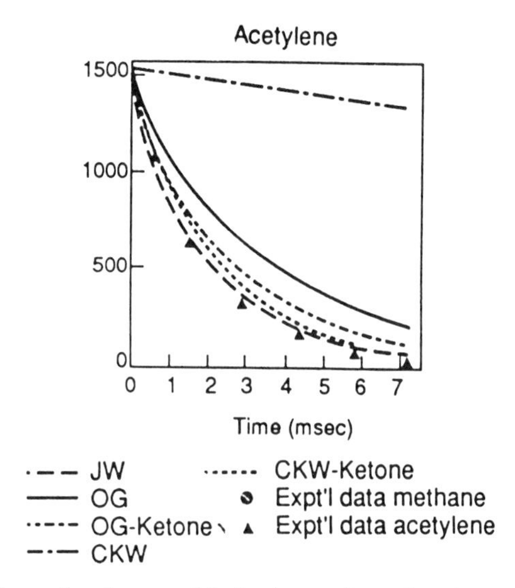

Fig. XII-29 Comparison of various models for decay of acetylene in post-flame zone of a rich ethylene flame. (Calculations by Levy, Taylor, Longwell, and Sarofim [1983] with experiments by Haynes [1973]. Published mechanisms were due to Warnatz (W); Olson and Gardiner (OG); and Westbrook (CKW).

Ethane (C_2H_6)

This is the second member of the alkane series (C_nH_{2n+2}). It consists of two methyl groups joined through a C–C bond. Because of symmetry it has no dipole moment. Ethane is a colorless, odorless gas that is difficult to liquefy. It shows the typical chemistry of aliphatic hydrocarbons (Minkoff and Tipper [1962]).

Ethane has a heat of combustion of 12.4 kcal/g. It is available in standard cylinders pressurized to 200 atm (3000 psig). Purities range from 99% to 99.999%. It is a non-corrosive, flammable gas that can be handled using standard techniques described in Chapter V.

Ethyl (C_2H_5)

This is a transient radical found as an intermediate in most rich hydrocarbon flames. This is the highest hydrocarbon radical that is thermally stable under flame conditions, but its equilibrium concentrations are negligible except in high-pressure fuel-rich systems. It is not isolatable under normal conditions, but can be prepared from precursor species containing an ethyl substituent such as tetraethyl lead [$Pb(C_2H_5)_4$], ethane, ethyl iodide (C_2H_5–I), etc. The techniques are similar to those used to produce methyl radicals, electric discharge, pyrolysis and radical reaction, and laser photolysis. A reaction unique to producing ethyl is the addition of H atoms with ethylene. This bimolecular reaction occurs because the CH bonds can redistribute the recombination energy so that no single bond is ruptured.

Acetaldehyde (CH_3–CHO)

This is a water-white liquid boiling at room temperature. Its odor is so acrid it brings tears to the eyes. It is the second member of the aldehyde family [formaldehyde (OCH_2) is the first]. Aldehydes are characterized by an oxygen atom double bonded to a terminal carbon. Like formaldehyde it polymerizes. The trimere, paraldehyde, which boils at 125°C, is available commercially. The polymerization is reversible, and acetaldehyde is often prepared from the solid by heating using techniques similar to those in preparing monomeric formaldehyde. The monomer can be handled as a gas using heated lines. The molecule is quite polar and adsorbs strongly in sampling lines. Its commercial source is the catalytic addition of water to ethylene under high-pressure conditions, $C_2H_4 + H_2O = CH_3CHO$; $C_2H_5OH + O_2 = CH_3CHO + H_2O$. It is an intermediate in the production of many organic chemicals.

Reaction Kinetics

The new reaction paths introduced by ethane flames are summarized in Table XII–8. As before the reactions of all of the lower members of the hierarchy are involved.

The oxidation of ethane under low-temperature conditions and in cool flames involves catalysis and aldehyde chemistry. These reactions are complex and unimportant in all save very rich flames. They will not be discussed here. Reviews of low-temperature kinetic studies of ethane oxidation can be found in Lewis and von Elbe [1987], Jost [1965], Minkoff and Tipper [1962], and Bamford and Tipper [1977].

Table XII–8 Ethane Reactions

$C_2H_6 + M^* \rightleftharpoons C_2H_5 + H + M$
$C_2H_6 + H \rightleftharpoons C_2H_5 + H_2$
$C_2H_6 + O \rightleftharpoons C_2H_5 + OH$
$C_2H_6 + OH \rightleftharpoons C_2H_5 + H_2O$
$C_2H_5 + M^* \rightleftharpoons C_2H_4 + H$
$C_2H_5 + H \rightleftharpoons CH_3 + CH_3$
$C_2H_5 + O \rightleftharpoons C_2H_4 + HO_2$
$C_2H_5 + O_2 \rightleftharpoons C_2H_4 + HO_2$
$CH_3CHO + H \rightarrow CH_3 + CO + H$
$CH_3CHO + O \rightarrow CH_3 + CO + OH$
$CH_3CHO + OH \rightarrow CH_3 + CO + H_2O$
$CH_2CO + H \rightarrow CH_3 + CO$
$CH_2CO + O \rightarrow CHO + CHO$
$CH_2CO + OH \rightarrow CH_2 + CHO$
$CH_2CO + M \rightarrow CH_2 + CO + M$

The attack in lean flames is dominantly by OH radicals with contributions from O atoms. In rich flames the dominant attack is by H atoms with the O atoms consumed by CH_3 radicals and hydrogen. Water formation is from the hydrogen formed by $H + C_2H_6 = H_2 + C_2H_5$.

The initial attack of ethane produces ethyl radical, which is thermally stable, in contrast with higher hydrocarbon radicals. In lean flames the ethyl radical can be stripped by reaction with H, O, or OH with the formation of ethylene, and as a result ethylene chemistry plays an important role in ethane flames. Minor paths for ethylene result in the formation of species such as acetaldehyde, ketene, etc., but the concentrations of oxygenated molecules are generally low. Since ethyl can recombine with H, CH_3, and itself to form CH_4, C_3H_8, and C_4H_{10}, this requires complex chemistry discussed in the section on rich flames.

Related Reaction Studies

Some of the ethane reactions with atoms and radicals have been followed using low-pressure discharge flow and laser photolysis. This work was summarized by Warnatz [1983b p. 261 ff] in his review of the kinetics of hydrocarbon combustion. He comments that these are all low-activation-energy reactions that show significant non-Arrhenius behavior. The most extensively studied system is the reaction with hydrogen atom by Clark and Dove [1973].

Shock tubes have been used by White and Gardiner [1979] to study ignition, by Olson, Tanzawa, and Gardiner [1979] to study pyrolysis, and by Cook and Williams [1971] to study oxidation. This work has been modeled by White and Gardiner [1979]. Baldwin et al. [1965, 1977)] used their hydrogen–oxygen bath plus trace technique to establish rate constants for ethane attack by radicals. The high-temperature flow reactor technique have been used by Glassman, Dryer, and Cohen [1976] to study the oxidation. Sorenson, Myers, and Uyehara [1971] studied ethane kinetics in engines.

Burning Velocity and Detonation Studies

The burning velocity of ethane–air flames has been measured by several groups (see Table B–2 in the appendix). Ethane has the highest burning velocity curve of any alkane (see Fig. XII-14), but it has a low rich limit of combustion, being second to methane. Some ethane–oxygen mixtures detonate. The composition dependence of burning velocity of atmospheric ethane–air mixtures have been successfully modeled by Warnatz [1981] (see Fig. XII-18).

Flame Structure and Modeling Studies

It is an indication of the limited interest in this system that all of the flame structure studies have been made to compare ethane with another flame system. The first microprobe studies on ethane were made by Fristrom, Grunfelder, and Avery [1958], with temperature profiles by Fristrom and Grunfelder [unpublished] and radicals by Westenberg and Fristrom [1965]. This was a comparison of lean, low-pressure oxygen (C_2H_2, C_2H_4, C_2H_6) flames at matched burning velocities (Fig. XII-15). The next was a study by Fenimore and Jones [1963b] comparing the attack of ethylene and ethane in flames. The next study was by Singh and Sawyer [1971] of the decay of carbon monoxide in comparable ethane and ethylene flames. A study by Hennessy, Robinson, and Smith [1988] compares methane and ethane flame chemistry. They used molecular beam sampling coupled with high-resolution mass spectrometry. This allowed resolution of formaldehyde from ethane, and ethylene from CO. Their results are shown in Figure XII-30. They used a 67-reaction scheme to model their experimental comparison of methane and ethane flames. This reproduced the behavior of stable species quite well but showed significant deviations with radicals. They blame experimental errors in calibrations for radical species.

Propane and Butane

The flames of propane and butane are discussed together for three reasons: (1) their chemistries are interrelated; (2) at the present time they represent the highest hydrocarbons for which it is reasonable to consider reaction schemes in detail; and (3) their chemistry completes the reactions required for rich hydrocarbon flame chemistry. Taken together they typefy the lower hydrocarbons, one having an even number of carbons, the other with an odd number of carbons. There is a striking contrast in the information available for the two systems. Propane is one of the most studied hydrocarbon flames, while butane has until recently been the least studied of the lower hydrocarbons. This results from the general perception of propane as the simplest typical hydrocarbon while butane is dismissed as the next in the series. As will be seen in the discussion of mechanisms, this is not quite accurate, since hydrocarbons with even numbers of carbon atoms show somewhat different decomposition patterns from those with an odd number of carbons.

New Species in C_3/C_4 Flames

The species in propane and butane flames include the lower members of the C/H/O hierarchy. It should be observed that butane is the first hydrocarbon to have isomers,

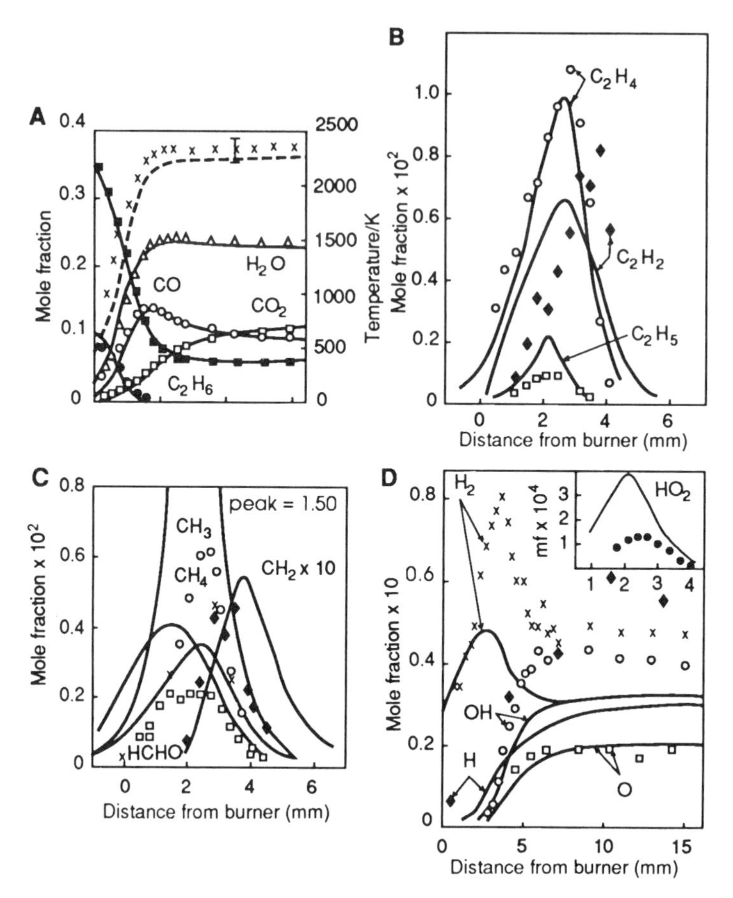

FIG. XII-30 Comparison of experimental structure with detailed modeling of a 20 Torr ethane 0.118, oxygen 0.415, argon 0.467 flame. (Experiments and modeling by Hennessy, Robinson, and Smith [1988].)

that is, two molecules with the same formula but different structure. The permutations are discussed in the following.

The new species introduced by these fuels are: C_3H_8, $(n,i)C_3H_7$, C_3H_6, C_3H_5, CH_3CHO, n-C_4H_{10}, i-C_4H_{10}, C_4H_9.

Propane (C_3H_8) and butane (C_4H_{10}) are normal saturated hydrocarbons. Only one type of propane exists, but there are two isomers of butane with slightly differing physical and chemical properties. In normal butane (n-C_4H_{10}), the carbons are bonded linearly (C–C–C–C), while in isobutane (i-C_4H_{10}) three of the carbons are bonded to the fourth but not each other ($C\equiv C_3$). These are easily liquefied gases that are colorless and have the typical hydrocarbon odor. They are widely used as a fuel called LPG (liquified petroleum gas). This mixture of the two species is obtained in fractionating petroleum. They are available in pressurized cylinders of liquefied gases with purity varying from a nominal 99% through the research grades of 99.999%. They are noncorrosive and at normal temperatures and pressures can be handled using standard gas handling hardware. Since the boiling point of butane lies just above room temperature, this may require heated lines.

Propylene ($H_3C{-}CH{=}CH_2$), propyne ($H_3C{-}C{\equiv}CH$), the Butylenes (C_4H_8), and the butynes (C_4H_6) are the second and third members of the alkene and alkyne series, respectively. Propylene and butylene have one double bond. Propyne and butyne are members of the acetylene series with one triple bond. They range from the easily condensible C_3 molecules to the C_4s, which are liquids boiling around room temperature. They are all available in liquefied form in standard cylinders with purity ranging from 99% to 99.999%. There are two isomers of normal butylene and butyne, but because of symmetry there is only one of the branched chain isomers. They are available as specialty chemicals produced by dehydration of the corresponding alcohol.

Propyl (C_3H_7) and butyl (C_4H_9) radicals have several isomers. In principle they can be produced from the parent hydrocarbons or from molecules containing propyl or butyl radical substituents. Of the standard methods laser photolysis yields the most well-defined radical product. There has been little direct study of these radicals because of their complexity. In flames they are probably formed in excited states and quickly fragment into more thermally stable species (propylene, ethylene, and ethyl, methyl, and methylene radicals). As a result, their steady-state concentrations are low, and reactions other than their decomposition play a minor role in combustion.

Reactions in C_3/C_4 Flames

It should be observed that there is only one propane molecule, but there are two isomers of butane and their combustion behavior differs. In all mechanisms, it is assumed that the stability of butyl radicals is low and that the system can be described in terms of C_3 and C_2 species. This instability is typical of hydrocarbon radicals higher than C_2. As a result, to first-order higher hydrocarbons can be treated as fuels that are attacked to yield C_3 and lower species. This view will be further explored in the section on higher hydrocarbons.

Synthetic modeling of these flames of course involves the chemistry of the lower flames in the hierarchy and the reactions of their species. The new reactions required beyond the lower C/H/O reaction systems discussed previously are collected in Table XII–9. This is a minimal set. It should be observed that most of the reactions of propane and butane are irreversible. The main problems introduced by the higher fuels are their initial reaction mode and the subsequent fate of the propyl or butyl radical. The radicals are generally formed in an excited state and decay or react rapidly so that their steady-state concentration is low in flames. These paths are limited by the availability of other radical species and stoichiometric considerations. Reaction can be with H, O, OH, O_2, or by thermal decomposition. The products in all cases will be C_1 and C_2 radical and stable molecules. Figure XII-17 shows paths available for propyl, butyl, and higher radicals. It is noteworthy that, although there are four distinct butyl radicals, they all lead to the same products, as does the propyl radical. Under conditions where higher hydrocarbon radical lifetimes are short, the system can be treated as if the fuel reaction led directly to C_1 and C_2 molecules. This reduces the propane and butane fuel attack problem to the rate of destruction and distribution among products without requiring detailed treatment of the paths.

Table XII–9 Propane Reactions

Reactions of C_3H_8
$C_3H_8 + H \rightarrow C_3H_7 + H_2$
$C_3H_8 + O \rightarrow C_3H_7 + OH$
$C_3H_8 + OH \rightarrow C_3H_7 + H_2O$
$C_3H_8 + HO_2 \rightarrow C_3H_7 + H_2O_2$
$C_3H_8 + CH_3 \rightarrow C_3H_7 + CH_4$
$C_3H_8 \rightarrow C_2H_5 + CH_3$
Reactions of C_2H_7
$i\text{-}C_3H_7 + H \rightarrow C_3H_8$
$C_3H_7 + O \rightarrow$ products
$i\text{-}C_3H_7 + O_2 \rightarrow C_3H_6 + HO_2$
$n\text{-}C_3H_7 + O_2 \rightarrow C_3H_6 + HO_2$
$i\text{-}C_3H_7 + i\text{-}C_3H_7 \rightarrow C_6H_{14}$
$n\text{-}C_3H_7 + n\text{-}C_3H_7 \rightarrow C_6H_{14}$
$i\text{-}C_3H_7 + i\text{-}C_3H_7 \rightarrow C_3H_6 + C_3H_8$
$n\text{-}C_3H_7 + n\text{-}C_3H_7 \rightarrow C_3H_6 + C_3H_8$
$i\text{-}C_3H_7 \rightarrow CH_3H_6 + H$
$n\text{-}C_3H_7 \rightarrow CH_3 + C_2H_4$
$n\text{-}C_3H_7 \rightarrow C_3H_6 + H$
Reactions of C_3H_6
$C_3H_6 + H \rightarrow i\text{-}C_3H_7$
$C_3H_6 + H \rightarrow n\text{-}C_3H_7$
$C_3H_6 + O \rightarrow$ products
$C_3H_6 + OH \rightarrow$ products
$C_3H_6 + CH_3 \rightarrow C_3H_5 + CH_4$
Reactions of C_3H_4
$CH_3CCH + H \rightarrow C_3H_5$
$CH_2CCH_2 + H \rightarrow C_3H_5$
$CH_3CCH + O \rightarrow$ products
$CH_3CCH + OH \rightarrow$ products
$CH_2CCH_2 + OH \rightarrow$ products

The kinetics has been reviewed by Warnatz [1983b], Dryer and Glassman [1979], and Illes [1971]. Mechanisms for low-temperature pyrolysis have been summarized by Minkoff and Tipper [1962]. Mechanisms for propane pyrolysis and oxidation in shock tubes were proposed by: Koike and Gardiner [1980]; Layokun and Slater [1979]; Lifshitz, Scheller, and Burcat [1973]; Lifshitz and Franklach [1975]; and Laidler, Sagert, and Wojiechowski [1960].

Related Studies

The study of the slow oxidation of hydrocarbons using classic chemical kinetic methods have included propane, propyne, and butane. However, the complexity of the products, which include aldehydes, peroxy compounds, and radicals, has not allowed a complete analysis of the information in terms of elementary reactions. The integration of low-temperature reactions with flame kinetics is a long-term goal that appears feasible using modern computer modeling capabilities. However, this will require much more and better quantitative information on elementary reactions studied individually. Ignition delay times and ignition temperatures and rich cool flames have all been studied, but the systems are too complex for any more than an empirical treatment. The information on propane far exceeds that on butane.

High-temperature pyrolysis and oxidation of propane and butane have been studied extensively using shock tubes (Burcat [1975]). This work has been reviewed for propane by Westbrook and Pitz [1984], for butane by Pitz, Westbrook, Proscia, and Dryer [1983, 1985], and for propane and butane by Warnatz [1981, 1983b, 1985]. Interpretation in terms of elementary reactions is feasible in these studies since these are dilute isothermal systems and computer modeling methods are feasible. Glassman, Dryer, and Cohen [1976] and Dryer and Glassman [1979] studied and modeled propane and other hydrocarbons in their turbulent flow reactor. These studies provided valuable information on the reaction sequence in these complex oxidations. Cathonnet, Boettner, and James [1981] also studied propane and butane in a flow reactor.

Baldwin, Bennett, and Walker [1977] applied their H_2–O_2 ignition inhibition studies to obtain rate constants for propane and other hydrocarbons.

Burning Velocity and Detonation

There have been a large number of studies of the burning velocity of propane using various techniques. These are collected in Table B–2 of the appendix. Propane has been considered a standard test system. Butane has received less attention. Warnatz [1981, 1983b] modeled the burning velocities of propane and butane as a function of equivalence ratio. Figure XII-18 shows the dependence of burning velocity on equivalence ratio. Westbrook and Pitz [1983, 1984] modeled propylene and propane pyrolysis and oxidation for interpreting shock tube and high-temperature flow reactor data. The model was extended to *n*-butane by Pitz, Westbrook, Proscia, and Dryer [1985] using a 238-reaction model. They also applied the model to butane flames using a one-dimensional flame code. The predicted burning velocity of the atmospheric, stoichiometric butane flame was 41 cm/sec, in reasonably good agreement with experiment. Metghalchi and Keck [1980] and Kuehl [1961] measured temperature and pressure dependence on both air–propane mixtures and vitiated air–propane using the combustion bomb technique. Vitiated air is air whose oxygen has been partially consumed. It is equivalent to dilution by CO_2 and H_2O with a corresponding temperature rise. In comparing the two studies it must be noted that Metghalchi and Keck normalized their approach velocities to a standard 298 K, while Keuhl reported the inlet velocity at the preheat temperature. When this factor is taken into account, the results agree reasonably well. Both propane and butane form detonatable mixtures with oxygen, but not with air. However, the examination of explosion damage from massive leaks of LPG in confined areas have suggested that large-scale mixtures may be detonatable. Adler et al. [1993] have suggested that the mechanism for such explosions may be radiant transfer from the flame to airborne dust particles ahead of the flame front, forming multiple ignition points, resulting in a volume explosion producing a shock front. In any case, these gases present the usual flammability hazard associated with volatile hydrocarbons.

Flame Structure and Modeling Studies

There have been a number of studies of propane flame microstructure beginning with a series of studies on the stoichiometric quarter atmosphere propane–air flame. Velocity profiles were measured by Fristrom, Avery, Prescott, and Mattuck [1954]; an initial composition profile by Prescott, Hudson, Foner, and Avery [1954]; and combined temperature and composition profiles by Fristrom, Prescott, and Grunfelder [1957]. The data were analyzed for fluxes and rates by Fristrom and Westenberg [1957], and kinetic considerations were analyzed by Fristrom, Westenberg, and Avery [1958]. This pioneering effort provided the first quantitative analysis of microprobe flame structure data. Its weakness was the lack of radical species and high-temperature experimental diffusion coefficients. These measurements were modeled by Warnatz [1981] using a mechanism that employed 123 reactions (see Fig. XII-31). Another pioneering study of this system was made at the Westinghouse research laboratory. A temperature profile was measured by Friedman [1953], followed by composition profiles by Fried-

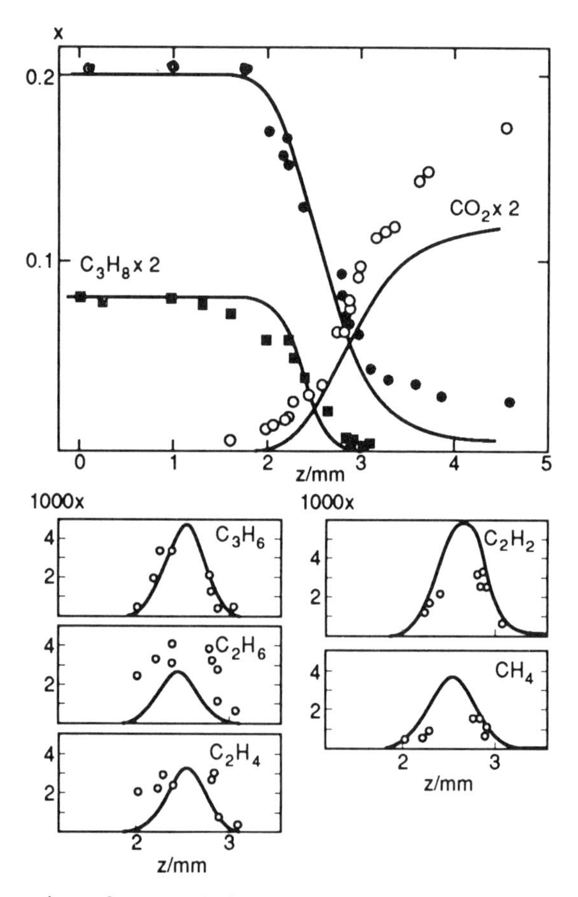

FIG. XII-31 Comparison of measured with calculated profiles for 0.25 atm. stoichoimetric propane-air flame. Experiments by Prescott, Hudson, Foner, and Avery [1956] and Fristrom, Prescott, and Grunfelder [1957]. Model calculations by Warnatz [1981].

man and Cyphers [1955]. At Manchester a series of studies on propane flames was made by Pownall and Simmons [1971] and Cook and Simmons [1979]. These studies of the inhibition of propane flames will be considered in Chapter XIII.

The group at Alma Atta in Kazakhstan have made extensive studies of propane and higher hydrocarbon flames. Ksandopolo, Kolesnikov, and Odnoroq [1974, 1975] made a complete experimental study and analysis of a rich atmospheric propane flame. Ksandopolo, Sagindykof, Kudaibergenov, and Mansurov [1975] measured hydrogen atoms and peroxyl radicals using ESR detection. Gukasyan, Mantashyan, and Sayadyan [1976] studied radicals in cold flames of propane using ESR detection. They were interested in rich chemistry and early reaction and peroxy compounds, and they found significant evidence for this complex chemistry. The original work is summarized in Ksandopolo's [1980] book. A more recent expanded version by Ksandopolo and Dubinin [1987] has been published in Russian. English versions of many of these papers are available in the translated journal *Combustion, Flames and Shock Wave* (Fizika Goreniya i Vzryva). They studied atmospheric propane and hexane flames using microprobe techniques supplemented by ESR and hydrogen atoms using the Westenberg–Fristrom method. They also froze samples and found long-lived peroxy radicals in the fuel disappearance zone. The signals were complex, suggesting the presence of several different peroxy species, which they were not able to identify positively. It was speculated that it might be a mix of oxy and peroxy radicals of methyl, ethyl, and propyl, together with HO_2. They made the point that wall adsorption of oxygenated compounds such as alcohols, ketones, and the peroxy compounds destroyed the H and O ESR signals. This suggests that H and O react with oxygenated wall species, which might explain why no H or O is detected by the method unless the fuel has completely disappeared. In their study of stoichiometric and rich atmospheric flames, they detected diffusion of H atoms into the preflame region as well as low levels of peroxy radicals, suggesting the existence of a region of complex chemistry preceding the primary reaction zone. This important finding does not invalidate the view that ascribes dominant flame propagation to high-temperature reactions of H, O, and OH. It simply means that a description of very rich systems must include these oxygenated species' competing reactions. As might be expected, the importance of these side reactions increases with richness of mixture and complexity of inlet fuel.

Studies by Kaiser, Rothschild, and Lavoie [1983] have been made on atmospheric flames at three compositions using laser Raman diagnostics to establish the level and identity of the intermediates, but the spatial resolution is not adequate for a quantitative interpretation. Their conclusions were that significant intermediate hydrocarbons were produced under all conditions, but that penetration of hydrocarbons beyond the luminous region only occurred in rich flames and that methane and acetylene are the dominant surviving species (Fig. XII-32).

Corre, Minetti, Pauwels, and Sochet [1991] at Lille, France, have studied the structure of a two-stage rich flame of butane as part of a program to study the complexities of engine ignition (Fig. XII-33). This has been supplemented by a burner and rapid compression study of ignition in the same system by Carlier, Corre, Minetti, Pauwels, Ribaucour, and Sochet [1991]. In addition, Minetti, Corre, Pauwels, Devolder, and Sochet [1991] studied the hydroperoxy radicals and the formation of hydrogen peroxide in the butane system.

Warnatz [1983a] applied his model to the butane flame, but at the time no struc-

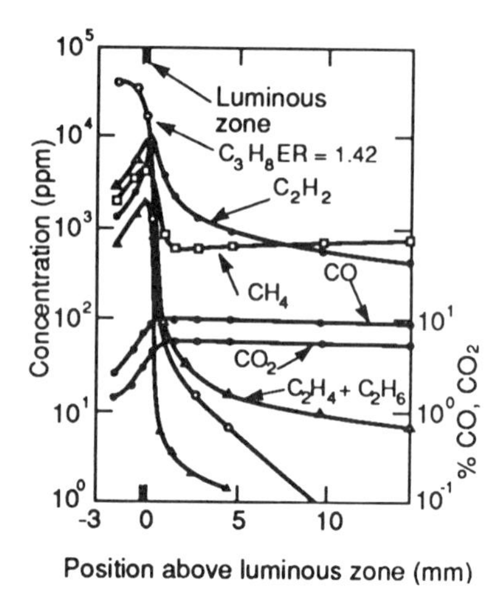

FIG. XII-32 Structure of a rich atmospheric propane–air flame. (After Kaiser, Rothschild, and Lavoie [1983].)

tural studies were available for comparison. Wilk, Pitz, Westbrook, Addagarlia, Miller, Czransky, and Green [1991] modeled normal and isobutane combustion and compared it with sampling studies on an internal combustion engine.

Higher Hydrocarbons

The combustion behavior of the higher alkanes, alkenes (hydrocarbons containing a single double carbon–carbon bond), and alkynes (hydrocarbons with a single triple carbon–carbon bond) all tend to converge when presented on an equivalence ratio basis. This occurs because the fractional contribution of the unsaturated bond becomes less important in the enthalpy and kinetics. As was seen in the discussion of propane and butane flames, hydrocarbon radicals higher than C_2 are so excited by their formation that they rapidly fission into lower products, leaving only the thermally stable single and double carbon radicals. Warnatz [1985] suggested that in lean and stoichiometric flames hydrocarbon breakdown can be simulated by a mixture of propylene and methyl radicals as (see Fig. XII-17). This, in turn, yields ethylene and one carbon and two carbon radicals, which finally produce mixtures of carbon monoxide, hydrogen, and varying amounts of water depending on the stoichiometry. The leaner the mixture, the more of the hydrogen is converted to water.

Species in Higher Hydrocarbon Flames

In addition to the inlet fuels, there are a myriad of lower hydrocarbons, both saturated and unsaturated, and their radicals. There are also oxygenated species such as the lower alcohols and aldehydes. All of the species considered for the lower members of the C/H/O flame hierarchy are represented in these flames. Beginning with butane hydrocarbons can have more than one form with slightly differing properties but the same elemental formula. This comes about because the carbon chain can be branched. These varieties are called *isomers.* The larger the number of carbon atoms in the mol-

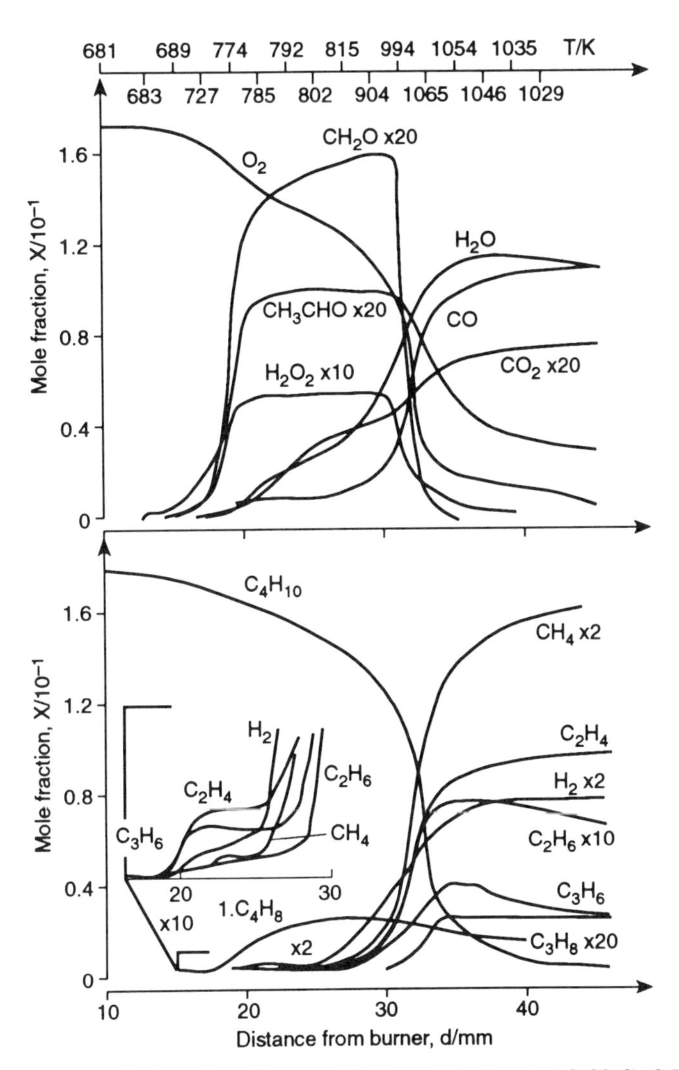

Fig. XII-33 Structure of a Two-Stage Butane–Air Flame (Carlier et al. [1991]). $C_4H_{10} = 0.18$; $O_2 = 0.18$; $N_2 = 0.64$; $P = 1.8$ bar; $T_0 = 670$ K.

ecule, the larger the number of possible isomers. For example, there is only one methane, ethane, or propane, but there are two butanes, five hexanes, seven heptanes, and seventy-five decanes. These are distinguished by the IUPAC (International Union of Pure and Applied Chemistry) systematic nomenclature. A summary of these rules can be found in most organic chemistry texts. Each of the higher hydrocarbons can yield a multiplicity of radicals, depending on which hydrogen is removed so that the number of species that would be required to describe these systems in detail is large and the radicals are beyond the scope of this work. Fortunately, as indicated, these complex radicals are not required to give a reasonable picture of flame propagation. If, however, one is interested in intermediate production, the problem must be treated at an appropriate level of detail.

Aliphatic hydrocarbons are molecules with the generic formula C_nH_m, where $m \leq 2n + 2$. They are divided into three families according to the C–C linkage: (1) the alkanes (C_nH_{2n+2}), in which the carbons are linked with each other by single bonds; (2) the alkenes (C_nH_{2n}), where two or more of the carbons are linked by a double bond; and (3) the alkynes (C_nH_{2n-2}) where two or more carbons are linked by a triple bond. Hydrocarbons with multiple double and triple bonds exist and the chains can be branched or cyclic. In addition, there is the class of aromatic hydrocarbons that possess cyclic, alternating double bonds that resonate and produce molecules of unusual stability. They will be discussed in the last section. The properties of hydrocarbons depend regularly on the chain length. Beginning with pentane, the hydrocarbons are clear liquids with boiling points increasing with each added carbon atom. Hydrocarbons higher than C_{12} are white, waxy solids often called "paraffins." Some of the common individual hydrocarbons are available as high-purity liquid chemicals. However, there are so many isomers and individual species that a specific hydrocarbon may only be obtainable as a mixture of isomers. Pure isomers require careful separation or synthesis. Their source is fractional distillation of petroleum. The lowest cut, consisting of propane and butane, is called *LPG*, the lower cuts (C_5–C_8) are called *gasolines*, the next higher is called *kerosene*. Liquids of even lower volatility are designated diesel fuel and heating oil. The lowest cuts are low-melting-point solids used to fuel ships and power plants. The U.S. Navy calls this cut "bunker C." They can be handled as liquids by using heated lines. These solids can be burned as diffusion flames as, for example, are candles, but premixed flames are rare since this would require strong preheating of the fuel.

Reaction Kinetics

Warnatz [1985] proposed modeling the flames of higher hydrocarbons using the existing mechanisms for propane–butane and lower hydrocarbons with the assumption that the higher alkane radicals rapidly are converted to propylene and methyl radical (Fig.XII-17). The approach requires only the initial attack rate of the initial hydrocarbon which can be obtained either from direct experiments or by using the approximate additive bond-specific rates of Baldwin et al. [1965, 1977] or those of Herron and Huie [1973]. This reduces the problem to determining the attack rate on the hydrocarbon and the distribution of secondary products. Once this is done, the problem is the same as the previously discussed mechanism for propane and butane flames.

Another approach suggested by Axelsson et al. [1988] is to assume the initial attack rates suggested by additivity of bond rates for hydrocarbons and for radical decomposition, assuming ring formation followed by further decomposition. In addition, the "beta scission" rule of Glassman and Dryer is used. This states that in the decomposition of a complex hydrocarbon radical the bond most likely to break is one carbon removed from the site of radical attack and that when there is a choice between C–C and C–H bonds, the C–C bond is usually the one ruptured. These assumptions allowed them to construct a detailed mechanism for attack of the two isomeric octanes in combustion despite the absence of direct experimental rates for many of the species involved. The details are too involved for the present discussion.

Related Studies

Because of their importance as fuels and sources of other organic chemicals, there have been many engineering studies of the combustion and pyrolysis of higher hydrocarbons. Summaries of some of the earlier work can be found in the books of Jost [1965], Lewis and von Elbe [1987], and Minkoff and Tipper [1962]. Experiments include ignition temperatures and delay times, which have been collected by Mullins [1957]. Work relative to aircraft fuels was reviewed in an early NACA report by Barnett and Hibbard [1959].

Brezinsky and Dryer [1986] and Dryer and Brezinsky [1985] studied normal octane and iso-octane combustion in a high-temperature turbulent flow reactor. They observed significant differences in combustion decay routes for the two isomers. This was compared with a study of the structure of similar flames.

Burning Velocity and Detonation Studies

The burning velocities of several higher hydrocarbons (pentane, hexane, heptane, and octane) mixed with air have been measured (see Table B–2 in the appendix at the end of the book) and modeled by Warnatz [1985] (see Figs. XII-34 and XII-35). The effects of the molecular structure of hydrocarbons on the peak burning velocity were discussed in the earlier section on general hydrocarbon combustion.

Flame Structure and Kinetic Modeling Studies

There have been several microstructure studies of hexane-air flames by Ksandopolo and his group at Alma Ata. Figure XII-36 shows results from a study of rich hexane; note the intermediates and peroxides. Results were similar to the studies on propane which were discussed in the previous section. Axelsson and Rosengren [1986] studied the structure of two premixed flames with that of normal octane-air and iso-octane air similar to the turbulent flow reactor study of Axelsson et al. discussed previously. The two isomer showed significant differences (see Fig. XII-37).

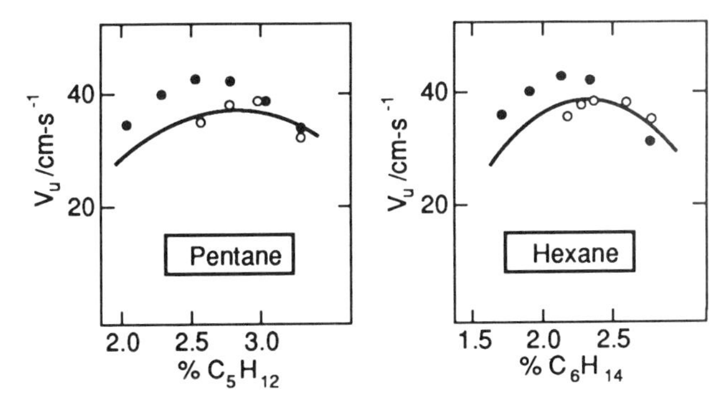

FIG. XII-34 Comparison of measured with predicted burning velocity for STP: *n*-pentane (C_5H_2)–Air flames and *n*-hexane (C_6H_{14})–air flames. (Experiments by various authors (see Table XII-5). Modeling by Warnatz [1985].)

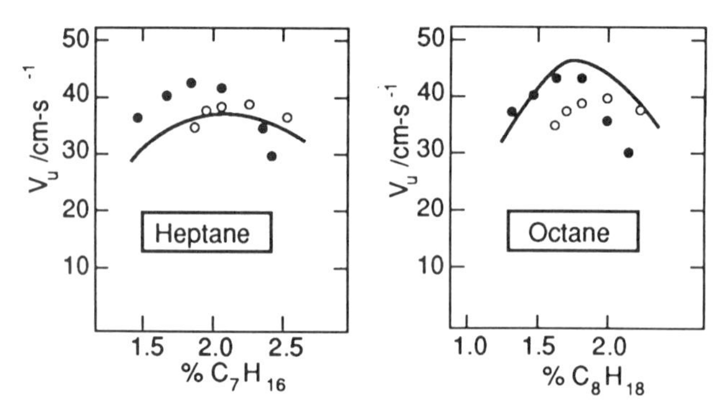

FIG. XII-35 Comparison of measured with predicted burning velocity for STP: *n*-heptane (C_7H_{16})–air flames and *n*-octane (C_8H_{18})–air flames. (Experiments by various authors (see Table XII-5). Modeling by Warnatz [1985].)

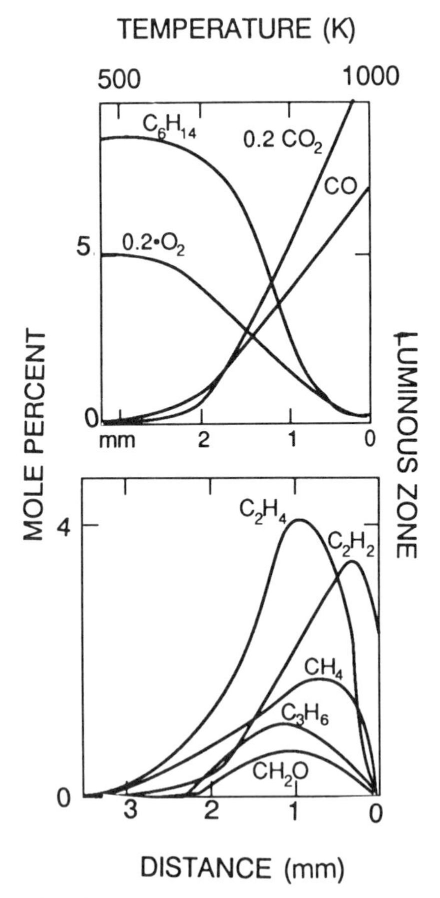

FIG. XII-36 Structure of a fuel-rich hexane–air flame. (Replotted data by Ksandopolo, Kolesnikov, and Dubinin [1977].) (a) Major species; (b) Fuel intermediates.

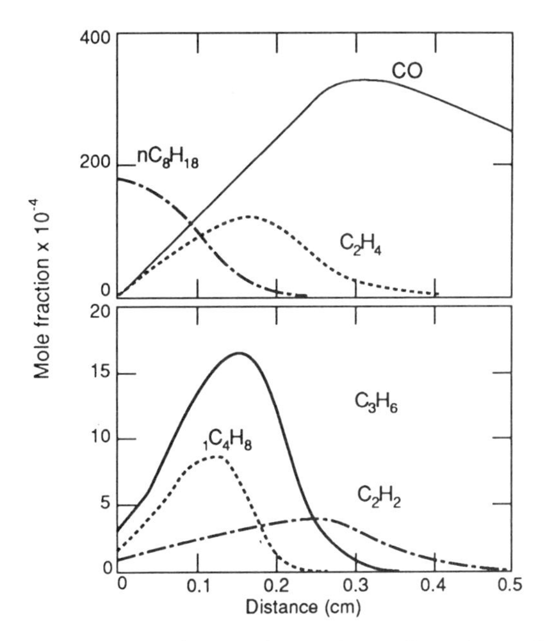

FIG. XII-37 Structure of octane–air flames. (Comparisons of experiments by Axelsson and Rosengren [1986] with modeling by Axelsson et al. [1988].) (a) Normal octane; (b) iso-octane.

Higher hydrocarbon oxidation modeling has been undertaken by Warnatz [1985] and Axelsson, Brezinsky, Dryer, Pitz and Westbrook [1988]. This requires a breakup scheme and consideration of more than 90 reactions through C_4 compounds. Both of the approaches which were outlined in the preceding section on reaction kinetics successfully model burning velocities. However, they leave the detailed pattern of hydrocarbon intermediates in doubt. Each group feels that their own approach is more realistic given present knowledge. They both agree that the beta scission rule offers a valuable guide, but differ on the mode of application.

These empirical approaches have been reasonably successful in modeling two octanes (normal and iso) oxidations in the turbulent flow reaction and in a premixed flame (Fig. XII-36). The approaches are approximate, but they represent a step forward. The ideal would be to have directly measured rates for all species and paths involved, but the number of isomers is large, for example, there are 89 C_8H_{17} radicals.

This is such a massive amount of information that it will require either standardized, automated experimental studies or massive quantum-mechanical computations on the reactive scattering of complex molecules. These approaches are feasible in the foreseeable though somewhat distant future, and it is likely that they will be used.

Related Fuels

There are a number of fuel molecules that have combustion chemistry similar to that of the hydrocarbons. The two most common classes are the alcohols and other oxygenated hydrocarbons and aromatic hydrocarbons. The fuel attack of alcohols and other oxygenated compounds is similar to that of the aliphatic hydrocarbons. By contrast the attack of aromatic compounds in flames appears to introduce a new fuel fragmentation pattern.

Alcohols and Oxygenated Fuels

The alcohols and other oxygenated C/H species produce flames with C/H/O chemistry. The same flame radicals (H,O, OH, HO_2) are involved in fuel attack. The principal difference is that the initial attack produces oxygenated radicals, which only appear as traces in normal hydrocarbon combustion. The intermediates are similar. The final adiabatic products are the same because the same elements are involved. The thermodynamics depends only on the C/H/O ratio and the enthalpy of the inlet mixture, not on the molecular details of the fuel. Alcohols are as rapidly attacked by flame radicals as were the hydrocarbons; so they also can be considered as carbon monoxide–hydrogen flames fed by fuel decomposition products. Alcohols and other oxygenated molecules are trace species in hydrocarbon flames, but with the exception of formaldehyde, which has been discussed, these are secondary paths in combustion. Therefore, it is appropriate to discuss these flame systems as an extension of C/H/O flame chemistry. Only the methanol system has received extensive study; so we will discuss it as a prototype, appending a few remarks on the other flame systems.

Alcohol and most oxygenated compounds are too valuable as chemicals to receive extensive use as fuels. The major exceptions are methanol (CH_3OH) and ethanol (C_2H_5OH), which are used as automobile fuel additives to raise the octane rating of fuels, replacing the environmentally undesirable lead. In the United States, the use is minor though significant. In Brazil, however, ethanol has replaced gasoline. The economics depend on the local, relative cost of fermentable agricultural products versus oil.

Species from Oxygenated Fuels

There are four types of oxygenated hydrocarbons that might be considered fuels and other oxygenated compounds that are potential intermediates in combustion processes. They may be visualized as substituted waters (or hydrogen peroxide in the case of peroxy compounds) (see Fig. XII-38), in which one or both of the hydrogens bonded to the oxygen are substituted by a hydrocarbon radical. In alcohols one of the hydrogens is substituted. The first member is methanol (CH_3OH), commonly called *wood alcohol,* with methyl as the substituting group; the next is ethanol (C_2H_5–OH), the common alcohol with ethyl substituting for the hydrogen. The general formula for this family is $O–C_nH_{2n+2}$.

Ethers are formed if both of the hydrogens of water are substituted by hydrocarbon radicals. They are named according to the hydrocarbon radical substitute. The lowest member is dimethyl ether ($CH_3–O–CH_3$). The generic ether is diethyl ether ($C_2H_5–O–C_2H_5$). There are also mixed ethers such as methyl ethyl ether ($CH_3–O–C_2H_5$). These are low-boiling-point liquids. They can be partially oxidized by air forming *explosive* peroxides. For this reason opened bottles of ethers are considered hazardous. Many lower ethers are available as laboratory chemicals of high purity.

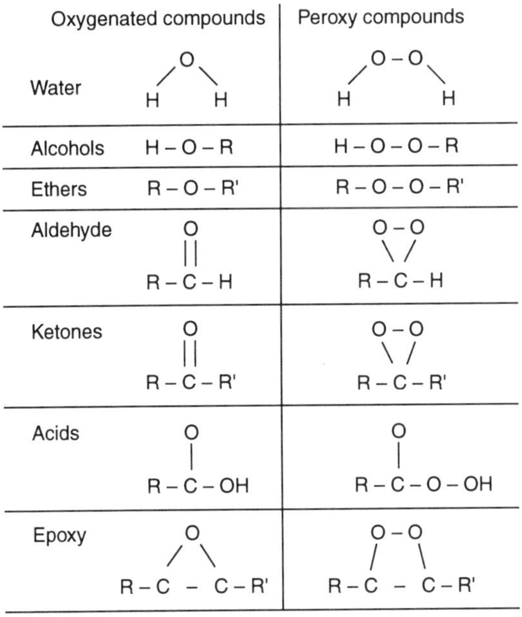

	Oxygenated compounds	Peroxy compounds
Water	H–O–H	H–O–O–H
Alcohols	H – O – R	H – O – O – R
Ethers	R – O – R'	R – O – O – R'
Aldehyde	R – C(=O) – H	R – C(–O–O–) – H
Ketones	R – C(=O) – R'	R – C(–O–O–) – R'
Acids	R – C(O) – OH	R – C(O) – O – OH
Epoxy	R – C – C – R' (O bridging the two C)	R – C – C – R' (O – O bridging the two C)

FIG. XII-38 Structural relations of various oxygenated hydrocarbons considered as substituted waters or substituted hydrogen peroxide.

If both of the valences of oxygen are bonded to carbon, the molecule is an aldehyde or a ketone. If the carbon atom is on the end of the chain, the compound is an aldehyde. We have discussed the two lowest aldehydes, formaldehyde (H_2CO) and acetaldehyde (CH_3CHO). Higher members are liquids with piercing odors. They are quite reactive, and aldehyde chemistry is intimately involved in low-temperature oxidation of hydrocarbons.

Ketones have a double-bonded oxygen on an interior carbon. The most common member of the group is acetone (CH_3–CO–CH_3). It is a low-boiling-point liquid that is used as a solvent. It is very flammable. Higher members are named for their radical substituents, as, for example, methyl ethyl ketone, which is a commercial paint solvent called MEK. They are less reactive than aldehydes, but appear as intermediates in low-temperature oxidations.

Ketene (CH_2CO), which was discussed in the section on acetylene, can also be considered as an unsaturated member of this family. It is very reactive.

There are several other families of oxygenated hydrocarbons that are usually unimportant in combustion, but may play roles in cool flames, slow oxidations, and in the low-temperature regimes of slow fuel-rich flames. These are:

1. Acids in which a terminal carbon has both a double-bonded oxygen and an OH group. The general formula is C_nH_{2n-2}–O_2. The two lowest members are formic acid, the essence of ant bites, and acetic acid, which is the active ingredient of vinegar.

2. Peroxy compounds, which can be considered as hydrogen peroxide with hydrocarbon radical substitutions for the hydrogens. These are very reactive compounds and are found in cool flames.

Oxy radicals are formed in many oxygenated hydrocarbon flames. In the methanol flame the radical CH_3O and CH_2OH play a role as initial products. They are found in trace amounts in hydrocarbon flames.

If a radical contains an O–O bond, it is called *peroxy*. The chemistry of these radicals is important in cool flames, slow oxidations, and ignitions. Some discussions of these chemistries can be found in Minkoff and Tipper [1962] and Jost's book on low-temperature oxidations.

Methanol Flames

The alcohols require the C/H/O reactions through methane and its own reactions for methanol and its two radicals CH_3O and CH_2OH. These are collected in Table XII–10.

Methanol (CH_3OH) is a clear liquid boiling at 337.7 K with a characteristic odor. It has a heat of combustion somewhat lower than the comparable hydrocarbon methane. However, it has the desirable property of burning without soot.

Related Studies on Methanol

Pyrolysis and oxidation reactions of these species have been studied. Again, methanol studies have been the favorite because of its relative simplicity. Cooke, Dodson, and Williams [1971] studied methanol oxidation in shock tubes; this was extended by Bowmann [1975c], who proposed a detailed mechanism for interpreting the results. Cribb, Dove, and Yamazaki [1985] studied the pyrolysis of methanol and were able to deduce rates for the two dissociation channels (CH_3 + OH and CH_2OH + OH). The slow oxidation of methanol has been studied by Bell and Tipper [1957] and Vanpee [1953]. Aronowitz, Santoro, and Glassman [1978] studied methanol oxidation in a high-temperature flow reactor and proposed a scheme to interpret the results. Their results were modeled by Westbrook and Dryer [1979a,b].

Table XII–10 Methanol Reactions

Reactions of CH_3OH
$CH_3OH + H \rightarrow CH_3O + H_2$
$CH_3OH + O \rightarrow CH_3O + OH$
$CH_3OH + OH \rightarrow CH_3O + H_2O$
$CH_3OH + Ar \rightarrow CH_3 + OH + Ar$
$CH_3OH \rightarrow CH_3 + OH$
Reactions of CH_3O/CH_2OH
$CH_3O + H \rightarrow CH_2O + H_2$
$CH_2OH + H \rightarrow CH_2O + H_2$
$CH_3O + O_2 \rightarrow CH_2O + HO_2$
$CH_3O + M \rightarrow CH_2O + H + M$
$CH_2OH + M^* \rightarrow CH_2O + H + M$

Burning Velocity and Detonation Studies

The burning velocity of several mixtures of air with the oxygenated compounds acetone, ethyl ether, ethanol, and methanol have been measured by Hartman [1931] (Fig. XII-39). A number of methanol mixtures have been also studied: (1) with ethanol and iso-octane by Gulder [1982–1984] and (2) with water by Hirano, Ode, Hirano, and Akita [1982]. De Wilde and Van Tiggelen [1968] measured burning velocities of mixtures of oxygen with: methanol, formaldehyde, and formic acid. Metghalchi and Keck [1982] measured burning velocities of mixtures of air with mixtures of methanol with iso-octane and indolene over a range of temperatures and pressures. The dependence of burning velocity on composition and equivalence ratio was calculated by Westbrook and Dryer [1979b, 1980]

Flame Structure and Modeling Studies

The pioneering flame microstructure studies on alcohol flames were made by S. R. Smith and Gordon [1956, 1958]. They studied a series of wicked diffusion flames of methanol through butanol. They identified many of the intermediates in these flames. Aldred and Williams [1969] studied a spherical diffusion flame of methanol. Akrich, Vovelle, and Delbourgo [1978] made a microprobe study of methanol–air flames at three equivalence ratios (0.77, 1.0, and 1.73) at 80 torr. They analyzed their data using estimated radical concentrations and Bowman's mechanism for shock tube studies [1975c]. The agreement was satisfactory and is a good example of the utility of microprobe studies.

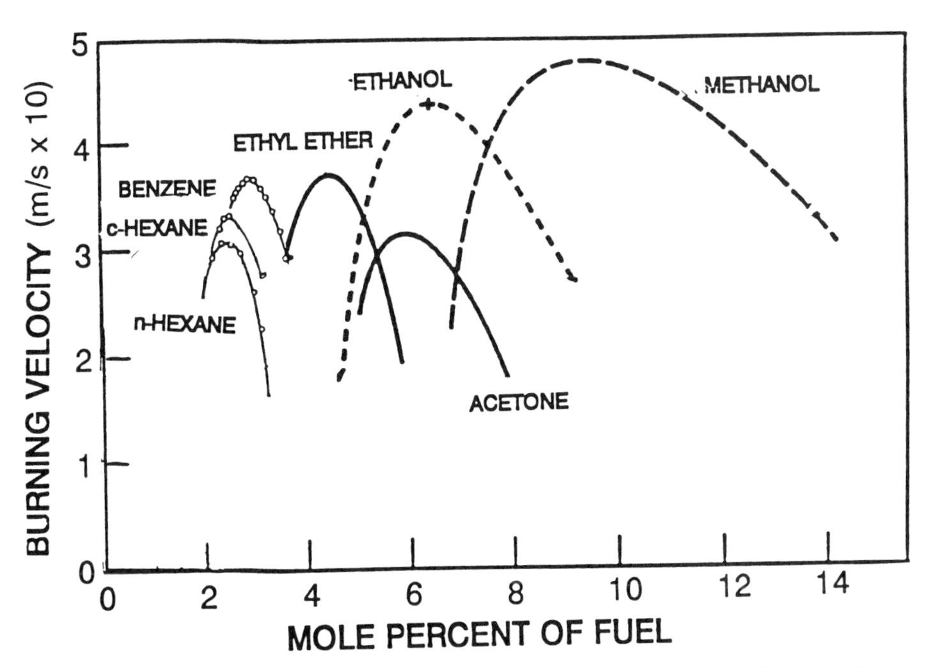

FIG. XII-39 Burning velocity of various STP C_6 hydrocarbons and oxygenated hydrocarbons with air. (Replotted from data reported by Hartman [1931].)

The first complete study of a methanol flame was by Vandooren and Van Tiggelen [1981] using molecular beam inlet techniques. Two fuel-lean methanol–oxygen flames diluted with argon were studied (ER = 0.36 and 0.88). They used a 24 step mechanism to analyze their results. The studies provided rate data for the reaction of methanol with H and OH (Fig. XII-40). Pauwels, Carlier, Devolder, and Sochet [1991] measured flame profiles of methanol doped with hydrogen sulfide. They also measured and modeled a stoichiometric, low-pressure methanol–air flame [1990]. Measurements included stable species as well as atoms using the ESR detection.

Bradley, Jones, Skirrow, and Tipper [1966] made a microprobe study of cool flames of acetaldehyde and propionaldehyde. They identified and followed many intermediates.

Methanol pyrolysis and combustion have been successfully modeled by several groups beginning with the shock tube experiments, especially those of Bowman [1975c]. A minimal mechanism is given in Table XII–10. The structural studies of Vandooren and Van Tiggelen [1981] provided mechanistic insights, as did the turbulent flow reactor studies of Aronowitz, Nageli, and Glassman [1978]. From these inputs and their earlier computer simulations of methane combustion, Westbrook and Dryer [1979a,1980] developed an 84-step mechanism, which has been successfully applied to the modeling of shock tube data, flow reactor data, burning velocity, and flame structure information. Figure XII-41 shows the predicted peak intermediate concentrations for methanol–air mixtures as a function of equivalence ratio. Westbrook, Adamczyk, and Lavoie [1981] used the model to investigate wall quenching, and Dove and Warnatz [1983] calculated the dependence of burning velocity on composition and pressure. The studies were motivated by the automotive unburned fuel problem.

Simplified models utilizing partial equilibrium and steady-state considerations have been successfully applied to the methanol system by Pacsko, Lefdal, and Peters [1988]. This approach compares reasonably well with more detailed models.

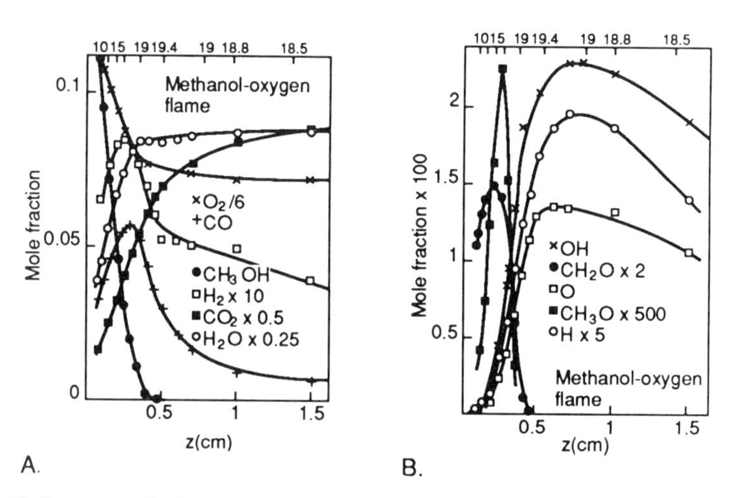

FIG. XII-40 Structure of a low-pressure (0.1 atm) methanol (0.194), V_0 = 44 cm/sec. (Experiments by Vandooren and Van Tiggelen [1981].)

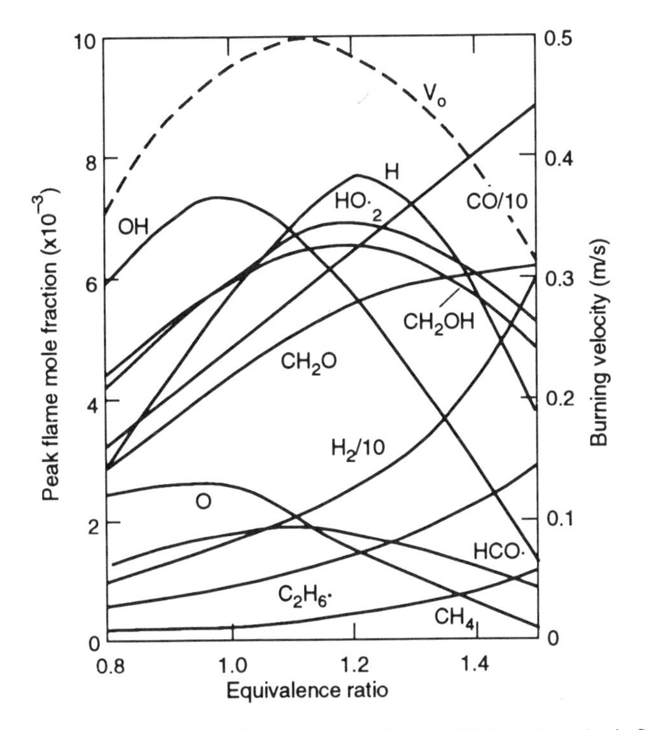

FIG. **XII-41** Calculated peak intermediate concentrations in STP methanol–air flames. (Calculations by Westbrook and Dryer [1979a].)

Benzene and Other Aromatic Hydrocarbons

Aromatic compounds represent the second major class of organic molecules, the other being aliphatic. This division is made because the chemistries differ sharply. Aliphatic hydrocarbon combustion has been the subject of previous sections. Aromatic compounds, which are discussed in this section, are characterized by cyclic alternating double bonds, which cannot be localized because of molecular symmetry. In terms of molecular bonding aromaticity requires $4n + 2$ delocalized π bonds. This is called the Hückel rule. A discussion of aromaticity and aromatic molecules can be found in most texts on organic chemistry. These two (and other) structures are interchangeable by electron movement with no nuclear shifts. This is often indicated by writing the benzene formula as a hexagon for the six carbons with a circle inside. This structure makes benzene and its relatives much more stable than corresponding aliphatic compounds; so as a result the character of fuel attack in combustion is changed. It should be observed that, despite significant differences in fuel attack, there is an overall similarity between the combustion of the aliphatic and aromatic families. This occurs because even with aromatic fuels the fuel attack is still rapid compared with the subsequent reactions and flame propagation. They can still be approximately modeled as hydrogen–carbon monoxide flames fed by fuel decomposition. The differences lie in the intermediates and the low hydrogen-to-carbon ratios found in these flames. As might be expected the ignition chemistries of the two families are significantly different. The combustion of aromatic compounds have received less attention than have the aliphatic families, but recurring oil crises have forced the use of more aromatic-based crude oils as fuels. These questions have been discussed by Hottel [1973] and Longwell [1977] and in a Project Squid workshop edited by Bowman and Birkeland [1978].

The hydrogen atoms of benzene can be substituted by aliphatic or aromatic radicals. Some simple examples are: methyl benzene, which is called toluene, dimethyl benzene, called xylene, which has three isomeric forms: ortho, meta, and para. An enormous number of substitutions and ring systems are possible. More details can be obtained from texts on organic chemistry.

Aromatic molecules are more resistant to radical attack than are corresponding aliphatic molecules. In combustion this shows up in the form of higher ignition temperatures and different fuel intermediates. Because of their higher carbon-to-hydrogen ratios, flames of aromatic fuels tend to be sootier than corresponding aliphatic flames. Benzene is the only aromatic fuel whose combustion has been studied in detail; so most of this section will be devoted to it. Other aromatic compounds follow similar patterns within physical limitations such as those set by their volatility. A primary problem with the combustion of the higher aromatic compounds is their low volatility. Another problem is the pollution caused by the formation of small amounts of PNA (polynuclear aromatic) compounds and PAH (polycyclic aromatic hydrocarbon) compounds. These are found at levels of a few parts per million, but this can be hazardous with exposures over a period of years. These problems are the subject of a considerable body of ongoing combustion research (see Longwell [1983]).

New Species in Benzene and Related Aromatic Flames

In addition to benzene itself (C_6H_6), there is a family of substituted benzenes in which one or more of the hydrogens on the benzene ring is substituted by an alkyl radical such as methane, ethane, etc. These fuels yield many of related radicals and ring fragments found in the early parts of the flames. The properties of some of a few common aromatic fuels are given in Table A–4 in the appendix at the end of the book. The physical properties of other aromatic compounds can be found in Reid, Prausnitz, and Poling [1987]. The identity of the radical species is in some doubt; so their reactions have not been extensively studied.

Benzene is a water-white, low-boiling-point liquid with a characteristic aromatic aroma. It is a commonly used solvent which must be handled with care because it is flammable, carcinogenic, and toxic.

Kinetic Mechanisms for Aromatic Compounds

Dissociation channel mechanisms were proposed by Brouwer, Muller-Markgraf, and Troe [1985] based on shock tube studies. The early shock tube mechanism proposed by Fujii and Asaba [1985] extended one proposed by Bowman for benzene oxidation in shock tubes. These were overall models that assumed that benzene was attacked by the flame radicals, H, O, OH, forming the radical benzin (C_6H_5), which they assumed reacted with molecular oxygen.

$$C_6H_6 + (H, O, OH) \Rightarrow C_6H_5\cdot + (H_2, H_2O, OH)$$

$$C_6H_5\cdot + O_2 \Rightarrow 2\,CO + H + C_4H_4$$

Santoro and Glassman [1979] proposed a more detailed scheme (Fig. XII–42) to interpret their flow reactor studies of benzene and substituted benzenes that adequately simulated the results.

More attention should be paid to the lean systems, where chemistry is less complicated. They could make significant contributions to the understanding of these systems. Clearly systematic studies of all aromatic families are necessary.

Related Studies on Aromatic Compounds

Benzene and the other aromatic compounds are stabilized by the resonance phenomena mentioned earlier. This makes them less reactive than their aliphatic analogues and more stable toward thermal decomposition. A number of studies of engine combustion of benzene have been made because of the engine knock problem. Its pyrolysis and ignition characteristics have been studied using shock tubes. An excellent bibli-

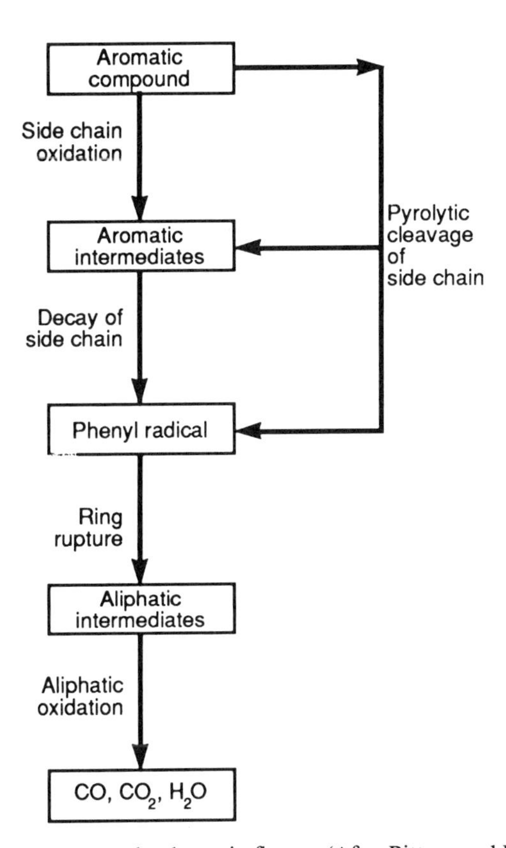

Fig. XII-42 Benzene attack scheme in flames. (After Bittner and Howard [1981].)

ography of studies is given by Vaughn, Lester, and Merklin [1982]. Hautman [1980] studied benzene oxidation in the Princeton high-temperature flow reactor. Results were reviewed by Santoro and Glassman [1979]. Smith and Johnson [1979] studied benzene pyrolysis using their high-temperature Knutson effusive reactor with mass spectrometric detection. They identified acetylene and diacetylene as major products.

Some substituted benzenes have been studied. Astholz, Durant, and Troe [1985] studied the pyrolysis of toluene using shock tube techniques, as did Kern, Singh, Esslinger, and Winekler [1983]. This has also been studied by Smith [1979] using a

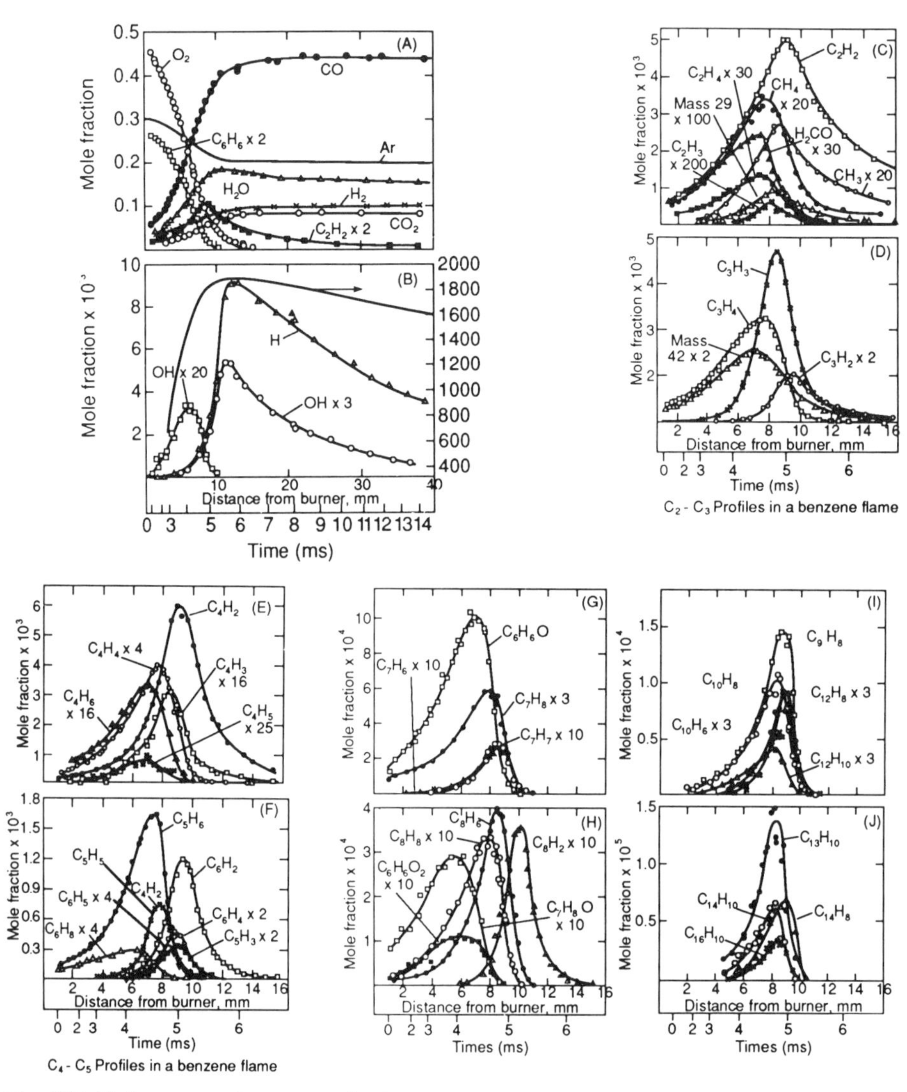

FIG. XII-43 Structure of a near-sooting benzene (ER = 1.8) [benzene (0.135), oxygen (0.565), argon (0.3)] flame at 2.67 kPa. Cold gas velocity 50 cm/sec. (Replotted experiments by Bittner and Howard [1981].)

high-temperature Knudson cell with mass spectrometric detection. They concluded that the C–H bond of the methyl side chain was usually broken, forming the radical benzyl (C_6H_5–CH_2). They derived rate constants for toluene and benzyl decomposition. Verkat, Brezinsky, and Glassman [1979, 1983] compared the oxidation of benzene (C_6H_6), toluene (C_6H_5–CH_3) and ethyl benzene (C_6H_5–C_2H_5) using their high-temperature flow reactor. They found the reaction products similar and related, suggesting similar reaction paths. They proposed a general combustion path for substituted benzenes and for the combustion of benzene and its phenyl (C_6H_5) radical (Fig. XII-42).

Spontaneous ignition temperatures have been discussed and collected in a book by Mullins [1957]. A number of aromatic compounds in air are included, and some of these are collected in Table A–4 in the appendix at the end of the book.

Burning Velocity Studies

Peak burning velocities and limits of combustion of benzene and a number of substituted benzenes have been measured by the NACA group (Hibbard and Barnett [1959]). A few values for some common aromatics are collected in Table A–4 in the appendix at the end of the book. The dependence of burning velocity on equivalence ratio has been measured by Hautmann [1931] for some aromatics (see Fig. XII-38).

Flame Structure Studies of Benzene

A few flame structure studies have been made on benzene, and no other aromatic flames appear to have been studied.

The earliest study was by Homann, Mochizuki, and Wagner [1963]. This was extended by Bonne, Homann, and Wagner [1965], who showed that polyacetylenes were equilibrated. Wagner's group at IPC Göttingen and later Homannn's group at Darmstadt have made extensive studies of sooting in flames and the role of aromatic compounds in this process. These are mentioned in the section on rich flames.

The most extensive work is Bittner and Howard's [1981, 1982] study of the structure of a near-sooting benzene. This was part of the MIT program of study on soot in flames that Howard [1991] reviewed recently. Their benzene study is a model of care-

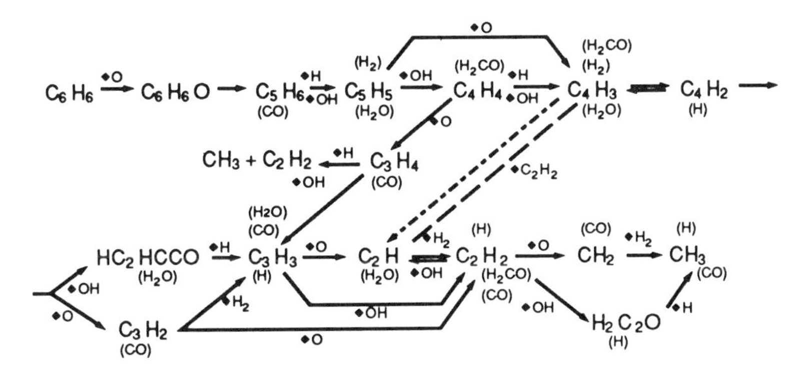

FIG. XII-44 Features of aromatic oxidation. (Suggested by Dryer and Glassman [1979].)

ful, detailed experimental work. Some of their results are shown in Figure XII-43 (a–j). They were able to show that after an initial attack region the flame had a slowly evolving pool of equilibrated hydrocarbons, with the rate controlled by reactions of C_2H_2 and C_4H_2, with O and OH slowly approaching the final equilibrium of CO, H_2, H_2O, CO_2. Their reaction scheme is given in Figure XII-44.

XIII

OTHER FLAME SYSTEMS

Most of the elements and many of their compounds with hydrogen and carbon are volatile and burn with a wide variety of oxidizers. As a consequence the coverage of this chapter is complicated. The situation can be appreciated by considering the relative electronegativity of the elements. This concept is due to Pauling [1939, 1988]. It is a measure of the power of an atom in a molecule to attract electrons. Relative values are derived from bond energies expressed in electron volts. To form a convenient scale the square root of this value is used and the value of H atoms is chosen to be 2.05. In early papers the value for H was set at zero, but this gave some negative values. On the new scale values range from 0.7 for Cs to 4.0 for F. The distribution can be seen when electronegativity is plotted against atomic number. This forms a somewhat skewed periodic table (compare Fig. II-1 with Fig. XIII-1). The significance of the concept for combustion is that whenever the difference in electronegativity exceeds 0.4, combustion is possible, and the larger the difference the more vigorous will be the combustion. This should be viewed only as an overall guide because it refers to the individual elements and only indirectly to their compounds. In addition, physical factors such as subdivision, temperature, and pressure must be considered. Thermodynamics provides a more reliable guide to the possibility of combustion. This approach is discussed by Stull [1971]. Complex equilibrium programs such as that of Svehela and McBride [1973], cover most of the elements and common compounds. If a significant temperature increase is indicated, then combustion is possible. The minimum required rise depends on the chemistry of the system. For CHO compounds it lies around 1000 K, while in the B–H–O system it can be as low as 300 K. Further, there are systems that are favorable thermodynamically but fail kinetically. The best example is carbon monoxide and oxygen, which burns vigorously as long as some trace of a hydrogen-containing compound is available, but otherwise is not reactive. The reasons for this lies in the carbon monoxide reaction mechanism, which was discussed in some detail in the previous chapter.

There are few systematic (combustion) studies on many compounds because their combustion is of little practical interest except as a fire safety problem. What little information is available has been included to make the coverage systematically complete. When not otherwise attributed, the information was obtained from the inor-

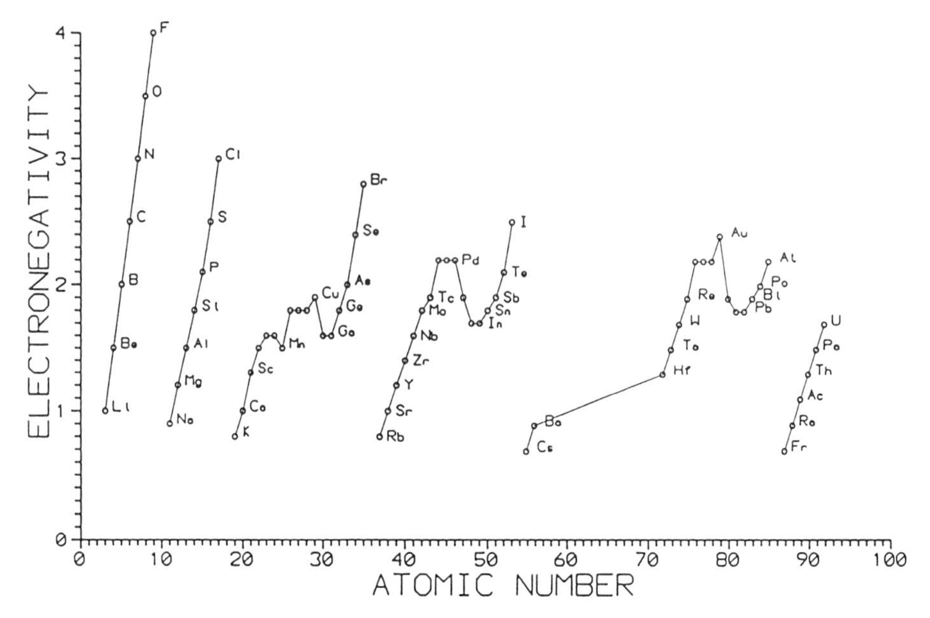

Fig. XIII-1 Electronegativity of the elements according to Pauling [1988] showing the periodicity. Elements differing by more than 0.5 may undergo combustion. The larger the difference, the stronger is the tendencey to burn.

ganic chemistry texts of Partington [1939], Sidgwick [1950], and Cotton and Wilkinson [1988] and the dust fire hazard compilation of Palmer [1973].

To systematize the coverage, the chapter is divided into three sections: (1) non-C/H/O fuels with oxygen; (2) oxidizers other than the halogens and oxygen; (3) mixed fuels, oxidizers, and additives.

Fuel elements and their volatile compounds are discussed in order of their (covalent) valence toward hydrogen. This groups elements and compounds with similar chemistry and hence similar combustion behavior. Within each group the elements are ordered according to atomic number or weight. This is not completely straightforward since many elements occur in several valence states. Elements are discussed with respect to their most common valence. The metals that show ionic valences are grouped in a separate section. Those unfamiliar with the intricacies of chemistry will find that Pauling [1988] gives a clear discussion of the concept of valence.

The oxidizers of this chapter are oxygen related or compounds of oxygen. They are discussed in order of atomic number of the element combining with the oxygen. Halogen flames were discussed in Chapter XI, oxygen flames with CHO compounds in Chapter XII, and oxygen flames with non-CHO compounds in this chapter. It should be noted that in addition to the common oxidizers considered in this book there are many (more than fifty) others that could support combustion, but will not be discussed here. These relatively exotic species were mostly compounds of oxygen and the halogens with other elements. They were studied for potential rocket usage, and this work was reviewed by Lawless and Smith [1975].

Non-C/H/O Fuels with Molecular Oxygen

A majority of volatile fuels are hydrides. However, there are some volatile nonhydrogenic compounds of carbon, nitrogen, and sulfur as well as a number of volatile metallic halides and carbonyls. Beyond the simple fuel molecules discussed here, there are many complex combustible compounds. The carbon–hydrogen–oxygen compounds of organic chemistry offer the largest single field, but several elements mimic carbon in forming homonuclear chain hydrides. They include: boranes, silanes, germanes, and phosphenes. These elements can also form compounds where one or more of the hydrogens is substituted by: hydrocarbon radicals such as methyl, ethyl, phenyl, etc. or borane, phosphene, or silane radicals. Another class of mixed element compounds are the heterocyclic organics, in which one of the carbons in an aromatic ring is substituted by another element. Commonly substituted elements include: O, S, Se, Te, N, P, As, Sb, Bi, Si, Ge, Sn, Pb, and Hg. Most of these compounds burn, so it can be seen that the range of combustible fuel molecules is very large. The permutations run into the millions and obviously cannot be discussed in detail here. However, systematic discussions of their chemistries can be found in texts on organic chemistry.

The combustion properties of some common non-CHO fuels with air are collected in Tables A–8–11 of the appendix. The combustion of organic molecules is covered in Minkoff and Tipper's book [1962]. The combustion of organic polymers is treated in the book of Cullis and Hirschler [1981]. Ignition and flammability of liquids is discussed by Mullins [1957], who gives tables of ignition temperatures with air and oxygen. A similar coverage for dusts has been given by Palmer [1973]. Related material can be found in the books by Jost [1965] on low-temperature oxidation, by Madorgsky [1964] on polymer degradation, and by Lyons [1970b] on fire retardants. The author is not aware of any systematic treatment of the premixed combustion of non-C/H/O fuels and/or nonoxygen oxidizers.

Fuels of Divalent Elements

Four nonmetallic elements in this family of the periodic table form volatile fuels (see Fig. II-1). They are: sulfur (S), selenium (Se), tellurium (Te), and polonium (Po). Their behavior ranges from oxygen and sulfur, which are typical nonmetals, through selenium, which shows some metallic properties and is photoconductive through tellurium and finally to polonium, which is a metal. Their combustion properties are related.

Sulfur

Elemental sulfur burns in air, and there are four volatile inorganic sulfur compounds that can form premixed flames with oxygen. These are: hydrogen sulfide (H_2S), carbon disulfide (CS_2), and carbonyl sulfide (OCS). The last two will be discussed as carbon compounds. Many other sulfur compounds can be burned as diffusion flames or as dusts. An unusual example is iron pyrite (FeS_2), which is a common contaminant of coal. It has been known to ignite piles of coal through spontaneous combustion.

The adiabatic flame temperature and composition of a number of sulfur compounds with oxygen are collected in Table A–8 in the appendix at the end of the book.In addition, in many combustion systems the diffusion of oxygen atoms from the flame gases into cooler regions containing SO_2 results in SO_3 formation. This can become a significant pollutant in effluent gases. In the presence of moisture SO_3 is usually converted into a fine mist of the relatively involatile sulfuric acid (H_2SO_4). At flame temperatures sulfuric acid and the other oxy acids of sulfur are unimportant. However, these molecules may play roles as intermediates in flames. The radical SO is an important chain carrier in the combustion of sulfur. Zacariah and Smith [1987] believe that S_2O, a sulfur-substituted analog of SO_3, may play a key role in sulfur–hydrogen–oxygen chemistry.

The kinetics of sulfur combustion has been reviewed by Cullis and Mulcahy [1972], Palmer and Seery [1973], and Muller, Schofield, Steinberg, and Broida [1979].

Elemental sulfur burns readily in air with a luminous blue diffusion flame and the characteristic odor of SO_2, which the major product of sulfur combustion. The Bible associates burning brimstone (crude sulfur) with Hell. The high boiling point of sulfur (718 K) makes it inconvenient for premixed atmospheric flames, but there is no bar to producing such flames at reduced pressures. A minimal reaction scheme is given in Table XIII–1.

Azatyan, Gershenson, Sarkissyan, Sachyan, and Nalbandian [1969] studied low-pressure flames of sulfur using ESR detection. They found that the O atom concentration was of the order of one-third of the initial oxygen concentration. They also identified SO radicals, in much lower concentration.

Linevsky and Carabetta [1974] found a low-pressure sulfur–oxygen flame to be a good source of oxygen atoms. They reported that one oxygen atom was produced for every three sulfur atoms consumed.

Many sulfur compound reactions with oxygen have been studied with shock tubes. This work was reviewed by Palmer and Seery [1973].

Hydrogen sulfide was studied by Homann, Krome, and Wagner [1969] through the reactions of O atoms with H_2S using ESR detection. The burning velocity of hydrogen sulfide with oxygen was measured by Van Wonterghem, Slootmaekers, and Van Tiggelen [1956]. The values are illustrated in Figure XIII-2. The structure of hydrogen sulfide–oxygen flames has been studied by Merriman and Levy [1971], and Levy and Merriman [1965] measured the structure of several low-pressure hydrogen sulfide–oxygen flames using microprobe sampling techniques. Some of their results are shown in Figure XIII-3.

Table XIII–1 Reactions for Sulfur with Oxygen

Reaction
$S_2 + M^* \rightleftharpoons S + S + M$
$S_2 + O \rightleftharpoons SO + S$
$S + O_2 \rightleftharpoons SO + O$
$SO + O_2 \rightleftharpoons SO_2 + O$
$SO + S_2 \rightleftharpoons S_2O + S$
$SO_2 + O + M \rightleftharpoons SO_3 + M^*$
$S_2O + O_2 \rightleftharpoons SO_2 + SO$

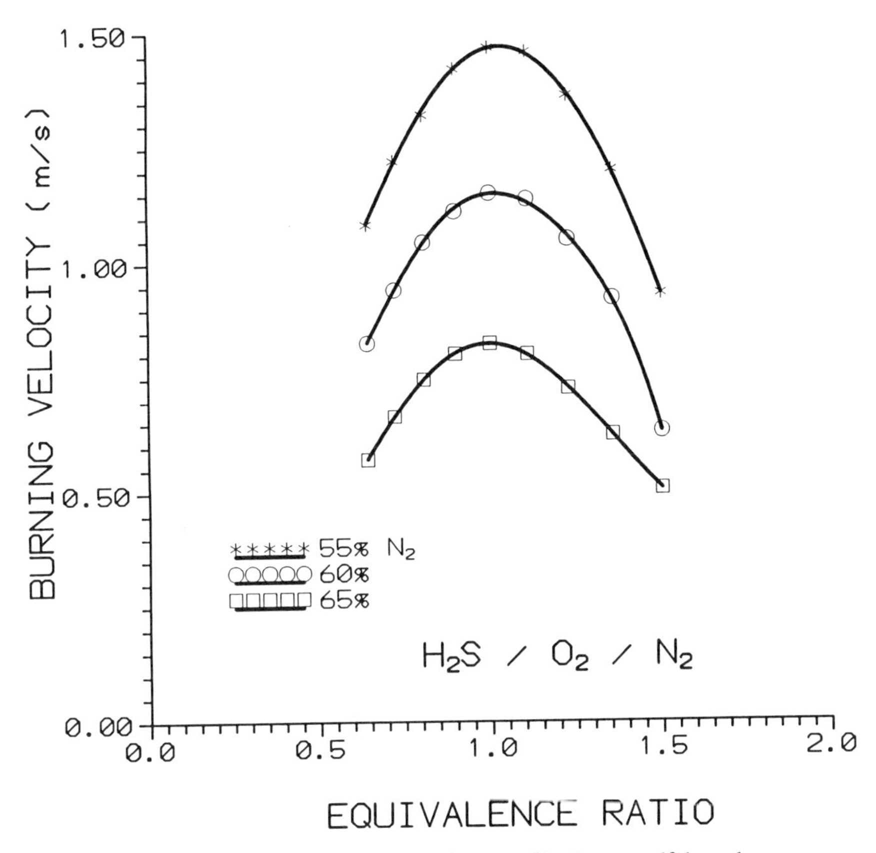

FIG. XIII-2 Burning velocity of atmospheric mixtures of hydrogen sulfide and oxygen. (Replotted from data of Van Wontergehm, Slootmaekers, and Van Tiggelen [1956].)

Webster and Walsh [1964] studied the effect of sulfur dioxide on the second explosion limit of the hydrogen–oxygen system. Spontaneous ignition temperatures have been measured for H_2S and CS_2 (Hibbard and Barnett [1959]) (see Table A–4).

O. I. Smith's group at UCLA (Smith, Wang, Tseregounis, and Westbrook [1983]; Tseregounis and Smith [1983]; and Zacariah and Smith [1987]) has made systematic studies of sulfur addition to hydrogen flames. The work is discussed briefly in the section on pollution.

Other Divalent Elements

The other divalent elements, Se, Te, and Po, form volatile hydrides that have even fouler odors than does H_2S with its aroma of rotten eggs. Both selene (SeH_2) and tellurine (TeH_2) are known to burn, and they are decomposed by moist oxygen. Polonium hydride has been prepared and studied using radioactive tracer methods, but no combustion studies appear to have been made.

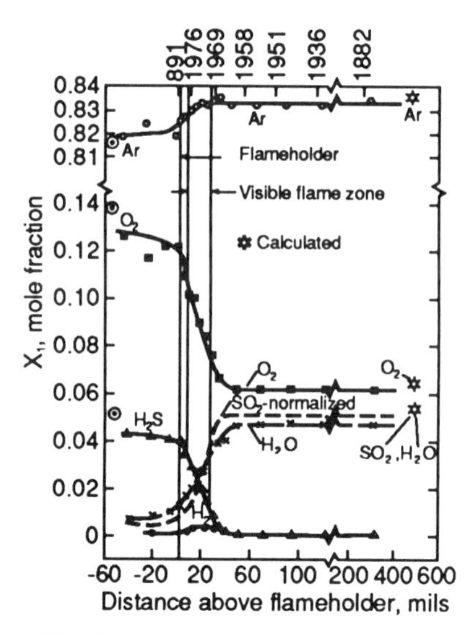

FIG. XIII-3 Compositon profile of a hydrogen sulfide–oxygen flame (0.051 H_2S–0.139 O_2–0.81 Ar). (Levy and Merriman [1965].)

Fuels of Trivalent Elements

Two classes of elements are trivalent. The first group in period three (of the periodic table; see Fig. II-1) begins with boron. The second group in period five beginning with nitrogen also show a valency of five. The groups are discussed together because of similar chemistries. Six of them are non- (or semi-) metallic elements whose volatile hydrides allow the formation of premixed flames. They are: boron (B), nitrogen (N), phosphorous (P), antimony (Sb), and bismuth (Bi). Their volatile compounds with carbon (but without hydrogen) will be discussed in the tetravalent section. As with the divalent elements properties range from the nonmetallic boron (B) and nitrogen (N) through phosphorous (P), to antimony (Sb), arsenic (As), and bismuth (Bi), which are typical metals. Adiabatic flame temperatures and compositions for a number of compounds of trivalent elements with oxygen are given in Tables A–9 through A–11 in the appendix at the end of the book.

Boron

Boron is a combustible element that forms a number of combustible compounds, notably hydrides. In addition to hydrides, there are a number of volatile, combustible alkyl-substituted boron hydrides that were investigated as possible "high-energy fuels" for aircraft and missiles. Boron carbonyl (BH_3–CO) is a volatile, combustible molecule. Boron also forms aromatic ring compounds with nitrogen (e.g., borazole, $B_3N_3H_6$). Discussions of the chemistry of boron and its compounds can be found in Cotton and Wilkinson [1988] and the Callery [1954] summary of work on boranes and alkyl boranes. High-energy fuel combustion was reviewed by Olson and Setze [1958]. These fuels were abandoned because of their toxicity and the discovery that the stability of lower oxides of boron and boric acids at flame temperatures made their heats of combustion lower than originally anticipated.

Boron flame chemistry includes some strange systems such as the diborane–hydrazine system, which produces boron nitride. This may offer a case where the flame radicals are derived from the fuel molecule, since both species are considered fuels. Hydrazine is the more likely candidate since it supports a decomposition flame (See Chapter XI), but diborane dissociation cannot be ruled out. The flame has been studied by Vanpee, Clark, and Wolfhard [1963] and Berl and Wilson [1961].

Elemental boron is a refractory solid melting at 2450 $\pm$ 20 K and boiling at 3930 $\pm$ 20 K. Like carbon, it is too involatile for premixed combustion but can be burned as a powder. Because of its very high specific heat of combustion, boron dust has been studied as a potential high-energy fuel by Palmer [1973], Macek [1973], Takahashi, Dryer, and Williams [1983], and others. Hydrocarbon slurries of boron have been suggested as potential jet fuels.

There are a number of boron hydrides since boron mimics carbon in forming chain and cyclic compound. They are called *boranes*. They show an unusual behavior called "electron-deficient bonding" in which certain boron-boron pairs appear to have two protons imbedded in the electron bond between them. The lower members are gases, but higher members are liquids and solids. They are all flammable. Two protons appear to be located in the electron cloud between certain pairs of boron atoms. As a result their molecular geometry is more complex than their hydrocarbon analogs. A short discussion of these molecules is given by Cotton and Wilkinson [1988, p. 272ff] and Pauling [1988]. The effect of electron-deficient bonding is so pronounced that the lowest expected member of the series BH_3 has not been detected. The boranes have properties resembling the hydrocarbons and silanes. The lowest member of the borohydrides, B_2H_6, is a gas with properties similar to ethane.

The chemistry of oxygen–borohydride flames is only poorly understood. The radicals BH_3, BH_2, BO, and HBO are probably involved. This could be a flame system where the radicals are produced by the fuel molecule rather than the oxidizer since the temperatures of lean borane flames is not high enough to allow the oxygen system radical formation reaction $H + O_2 \Rightarrow OH + O$ to function. Berl et al. [1966] discuss some of these questions. Metaboric acid is an early product in these flames. They suggest dissociation of B_2H_6 to BH_3 as the radical-generation step followed by reaction with oxygen forming a radical chain involving HBO, HOBO, and BOOH and various boric acids and HBO radicals. The rapidity of water reaction suggests that BH_3 also reacts with moisture with the formation of hydrogen. Measurements in boron–oxygen flames are difficult because of the condensation of the product B_2O_3, which coats thermocouples and probes.

If moisture or traces of higher borohydrides are present, spontaneous ignition occurs, but dry mixtures are stable. Schalla [1957] measured the flammability limits with air. Earlier contradictory studies by Price [1951] and Whatley and Pease [1954] were resolved by Roth and Bauer [1955]. They attributed previous difficulties to the presence of decomposition products. Parker and Wolfhard [1956] studied the ternary mixtures of diborane–ethane and air and found a wide area of flammability.

The burning velocity of diborane–air mixtures was measured by Berl, Gayhart, Maier, Olsen, and Renich [1957] (Fig. XIII-4).

The structure of a low pressure diborane–air flame and some diborane–propane–

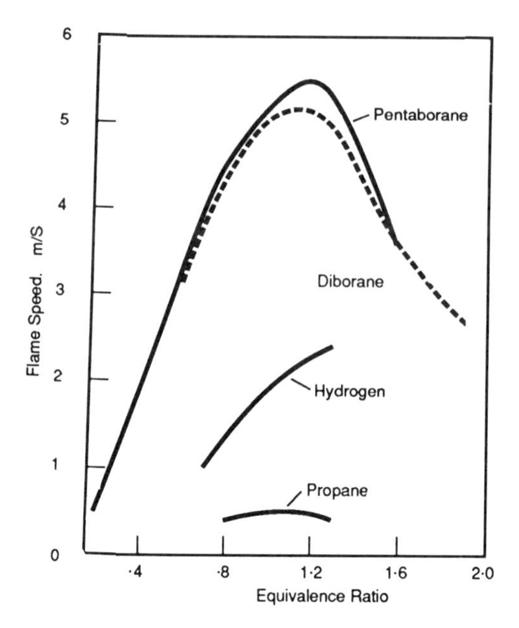

FIG. XIII-4 Burning Velocity of diborane– and pentaborane–air mixtures compared with hydrogen and propane. Replotted from data by Berl, Gayhart, Maier, Olsen, and Renich [1957].)

air flames have been studied by Breisacher, Dembrow, and Berl [1958]. Flame separation phenomena were found by Berl and Dembrow in the mixed flames with hydrocarbons [1952]. This subject will be revisited in the section on mixed chemistries.

Pentaborane

Pentaborane is a low-boiling-point liquid (321 K) that takes fire spontaneously on contact with moist air. Its ignition limits in the vapor phase were studied by Schalla [1957], and burning velocity has been measured (Fig. XIII-4) by Berl, Gayhart, Maier, Olsen, and Renich [1957]. Some mixtures of pentaborane with hydrocarbons and the structure of a flame of propyl–pentaborane–helium–oxygen mixture were studied by Berl, Breisacher, Dembrow, Falk, O'Donovan, Rice, and Sigillito [1966]. Figure XIII-5 shows their results for propyl pentaborane. They concluded that the structure of the mixed compound were well simulated by mixtures of the two pure compounds.

Nitrogen

Nitrogen compounds with hydrogen (NH_3, N_2H_4), carbon (C_2N_2), and sulfur (S_4N_4) are fuels, while its oxides (N_2O, NO, NO_2) and halides (NF_3) act as oxidizers.

Elemental nitrogen (N_2) is diatomic and relatively inert. However, it is attacked by fluorine with the formation of NF_3 and can attack metals such as lithium, titanium, and zirconium forming solids. Nitrogen does not react with oxygen directly, but in flames some nitrogen is converted into nitric oxide through the radical-driven Zeldov-

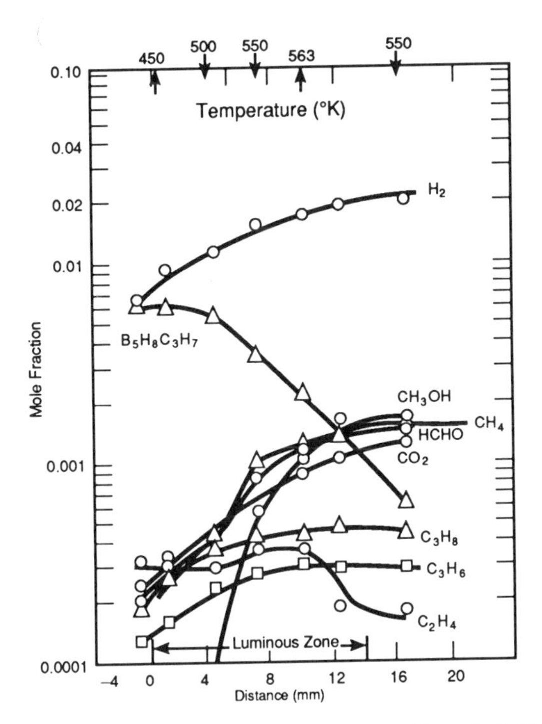

FIG. XIII-5 Structure of a 29 torr 0.0062 propyl pentaborane flame in 0.21 O_2 + 0.784 He showing initial fuel attack. (Berl, Breisacher, Dembrow, Falk, O'Donovan, Sigillito, and Rice [1958].)

itch mechanism (see Chapter XII). This is one source of the pollution problems in combustion, which are outlined in a later section.

There is an interesting report that a film of nitrogen atoms condensed at liquid helium temperatures ignited centrally by a laser pulse developed a luminous ring propagating outward at a constant velocity with a final temperature of 15 K. A theory of Reed and Herzfeld [1960] suggested that this was a flame. If this is considered a flame, it holds the record for low temperature and minimal temperature rise.

Ammonia

NH_3 is the simplest nitrogen fuel. It is familiar to most of us as the active ingredient of the household cleanser called *ammonia*. This is a strongly basic, aqueous solution of ammonia often containing added soap. The pure compound is a colorless, flammable gas that is easily condensible (BP, 240 K). It has a sharp characteristic odor. High concentrations induce eye watering and suffocation. High-purity ammonia is available as cylinders of liquified gas.

It has a lower heat of combustion than comparable hydrocarbons (NH_3, 91.5 kcal/mole versus CH_4, 212 kcal/mole) and is not considered flammable for purposes of transportation.

Ammonia flames have not been as extensively studied as hydrocarbons. Verwimp and Van Tiggelen [1953] measured burning velocities of ammonia-oxygen mixtures.

Murray and Hall [1951] studied ammonia-hydrazine-oxygen mixtures. Andrews and Gray [1963] and Ausloos and Van Tiggelen [1951] compared burning velocities of ammonia with oxygen and nitrous oxide. Some results are shown in Figure XIII-6. Starkman and Samuelsen [1967] studied flame propagation rates of partially dissociated ammonia–air mixtures. They also tested it as a fuel for internal combustion engines. Effective use required about 5% hydrogen (by weight) added for ignition. This was produced by decomposing part of the ammonia using a heated catalytic nickel predissociator. There have been a number of shock tube studies of ammonia decomposition and oxidation. The studies have been reviewed by Hanson and Salimian [1982]. Examples of such studies include that of Holzrichter and Wagner [1982] on dissociation and that of Takeyama and Miyama [1967] on ammonia oxidation.

Studies on ammonia flame structure began with the pioneering work on the oxygen diffusion flame by Wolfhard and Parker [1949]. They studied both absorption and emission and were able to measure oxygen and hydroxyl radical concentrations and identified NH and NH_2 radicals. Their burner is described in Chapter III. Fenimore and Jones [1961c] and Kaskan and Hughes [1973] made early microprobe studies on the decay of ammonia added to hydrocarbon flames.

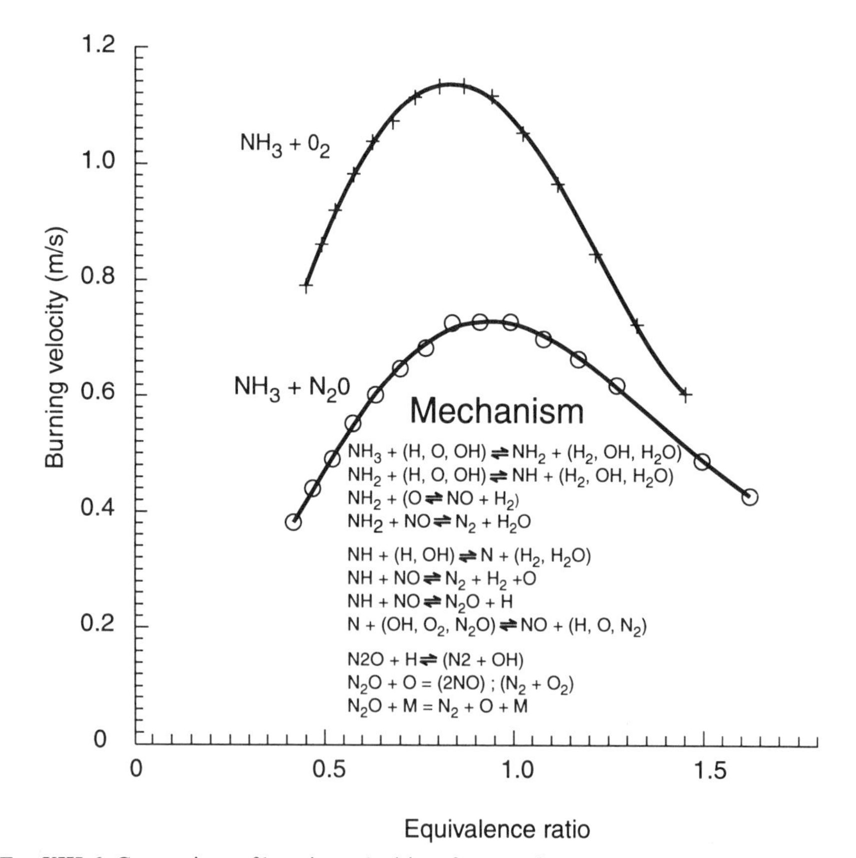

FIG. XIII-6 Comparison of burning velocities of ammonia oxygen and ammonia nitrous oxide flames. Replotted data of Ausloos and Van Tiggelen [1951]. Insert is the ammonia oxygen flame mechanism suggested by Bain, Vandooren, and Van Tiggelen [1988].

The first comprehensive structural measurements were by MacLean and Wagner [1967] in their comparison of ammonia and hydrazine flames. They studied three low-pressure (20 Torr) flames: one oxygen rich, one with balanced ammonia and oxygen, and one with excess ammonia. In the fuel-rich mixtures, nitrogen and hydrogen appeared in the flame gasses. The modeling by Miller, Smooke, Green, and Kee [1983] employing ninety reactions was successful for the lean and stoichiometric flames but poor for the rich flame. The results are shown in Figure XIII-7.

Fisher [1977] studied the structure of ammonia-rich flames diluted with nitrogen, reporting temperature measurements and the concentrations of NH_3, NH_2, NH and OH. Dean, Hardy, and Lyon [1982] studied ammonia oxidation in a flow reactor and modeled the processes.

There have also been laser fluorescence studies of radical profiles in ammonia flames by Dean, Chou, and Stern [1984] Chou, Dean, and Stern [1982] Lyon and Benn [1979] and Dean, Hardy, and Lyon [1982].

The most comprehensive study was by Bain, Vandooren, and Van Tiggelen [1988]. This is a low-pressure (35 Torr) near-stoichiometric ammonia–oxygen flame. They measured temperature and both stable and radical species using molecular beam inlet with mass spectrometric detection. The data were analyzed, and species production rates were derived. The authors concluded that the major paths of ammonia destruction are radical stripping forming NH_2 and NH and that these then react with N forming N_2. NO formation was thought to result from reaction of O atoms with NH.

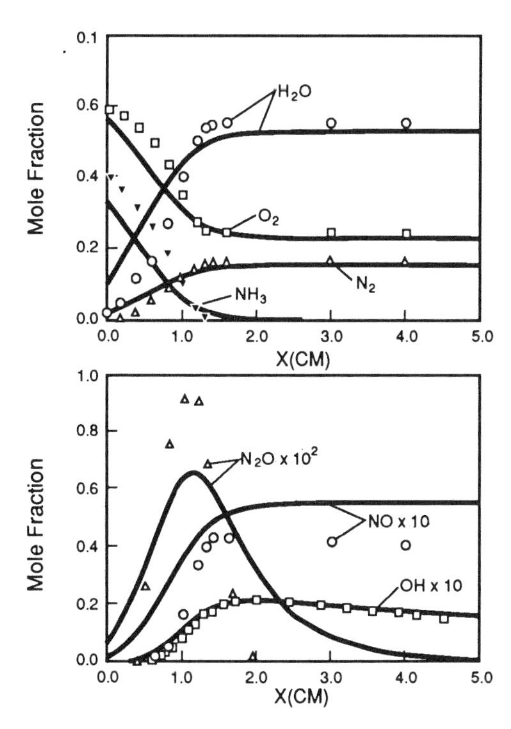

FIG. XIII-7 Structure of a low-pressure ammonia–oxygen flame ($NH_3 \Rightarrow 0.4$, $O_2 = 0.6$; P = 20 torr; flame speed 600 mm/sec). (Comparison of experiments (points) by MacLean and Wagner [1967] with model by Miller, Smooke, Green, and Kee [1983].)

Hydrazine

N_2H_4 is a reactive liquid. The burning velocity of hydrazine with oxygen, N_2O, and NO were compared by Gray, McKinven, and Smith [1967] (Fig. XIII-8). The only structural study of hydrazine found was of the decomposition flame, which was discussed in Chapter XI. Sawyer and Glassman [1967] studied the combustion of hydrazine with oxygen, nitrogen dioxide, and nitric oxide using an adiabatic turbulent flow reactor. Their reaction scheme used the ammonia chemistry, adding hydrazine attack by H, O, and OH to form N_2H_3. This radical is also rapidly attacked with the ultimate formation of NH and NH_2. A reaction scheme is given in Table XIII–2.

Hydrazoic Acid

Hydrazole acid (HN_3) forms a low-pressure decomposition flame that produces hydrogen and nitrogen, which is discussed in Chapter XI. The products of decomposition are flammable; so it should be flammable in air or oxygen. *The material is extremely dangerous.* A simple scheme for its combustion is outlined in Table XIII–2.

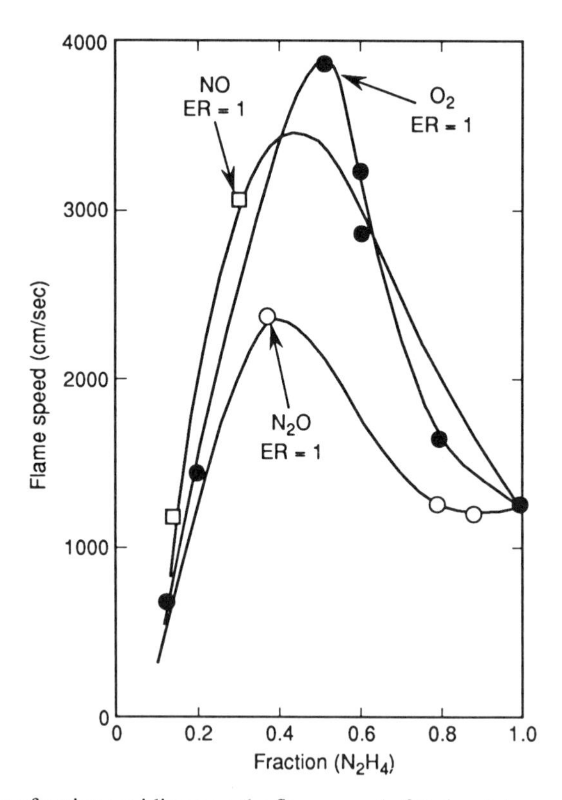

Fig. XIII-8 Effect of various oxidizers on the flame speed of their mixtures with hydrazine. Approximate burning velocities can be estimated assuming a nominal expansion ratio of eight. (Redrawn from data of Gray, McKinven, and Smith [1967].)

Table XIII–2 Nitrogen Fuel Reactions
Hydrazine Reactions[a]

$N_2H_4 + M^* \Rightarrow 2NH_2$
$N_2H_4 + (H, O, OH) \Rightarrow H_2H_3 + (H_2, OH, H_2O)$
$N_2H_4 + NH_2 \Rightarrow NH_3 + N_2H_3$
$N_2H_3 + M^* \Rightarrow NH_2 + NH$
$N_2H_3 + (H, O, OH) \Rightarrow N_2H_2 + (H, OH, H_2O)$
$N_2H_3 + O_2 \Rightarrow NH_2NO^* + OH \Rightarrow OH + H_2O + N_2$
$N_2H_2 + M^* = N_2H + H; (2NH) \Rightarrow$
$N_2H + (H, O, OH) \Rightarrow N_2 + (H_2, OH, H_2O)$

[a]For the complete scheme add the reactions of ammonia in Figure XIII-6.

Hydrazoic Acid Reactions[b]
$HN_3 + M^* \Rightarrow NH + N_2$
$HN_3 + (H, O, OH) \Rightarrow$ products

[b]Add ammonia reactions from Figure XIII-6.

Hydrocyanic Acid Reactions[c]
$HCN + (H, O, OH, N, NH) \Rightarrow CN + (H_2, OH, H_2O\ NH, NH_2)$
$CN + O \Rightarrow CO + N$

[c]Add reactions of C_2N_2 and NH_3.

Hydrogen Cyanide

HCN is an extremely poisonous, low-boiling-point (299 K) liquid with the characteristic odor of oil of bitter almond. It burns well in air and vigorously with oxygen-forming flames with three visible luminous regions: an inner violet cone, surrounded by a bright yellow mantel, which in turn is enclosed by a pale blue outer mantel. Cohen and Simpson [1957] have measured the burning velocity dependence on composition for both oxygen and air flames (see Fig. XIII-9). Hydrogen cyanide is believed to be an intermediate in the formation of NO_x pollutants.

Phosphorous

Phosphorous chemistry is described in texts on inorganic chemistry. The most authoritative treatment is that of Yost and Russell [1944]. Elemental phosphorous exists in five allotropic forms: white, yellow (alpha and beta), violet, and black. Common red phosphorous is a mixture of yellow and violet forms. The white form takes fire spontaneously, whereas the red and black forms are stable in air, and the black form can only be ignited with difficulty. At low pressures it undergoes a luminous reaction with oxygen that apparently also requires moisture. This behavior was the origin of the term *phosphorescence*. The element is manufactured by reducing phosphate rock with coke in an electric furnace from which it is vaporized and condensed. Its major use is in the manufacture of safety matches.

Phosphorous forms two volatile, combustible hydrides: phosphene, PH_3, and diphosphene, P_2H_4. Phosphene is a colorless, odorless, poisonous gas boiling at 259 K and igniting spontaneously in air at 373 K. The second member, diphosphene, P_2H_4, is a low-boiling-point liquid (BP, 330 K) that ignites in air at room temperature.

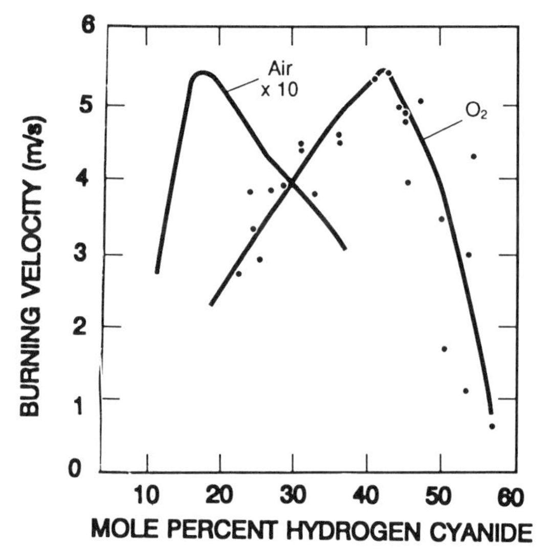

FIG. XIII-9 Burning velocities of various mixtures of HCN with oxygen and air. (Replotted from data by Cohen and Simpson [1957].)

Arsenic, Antimony and Bismuth

The remaining elements of this family, arsenic (As), antimony (Sb), and bismuth (Bi), are metals that are not volatile enough to form premixed flames but do burn as dusts. They all form volatile hydrides with the formula MH_3. A standard analysis for arsenic is the "Marsh test," in which arsine is decomposed on a hot surface forming a mirror. All of these hydrides are combustible, but no systematic combustion studies were found. The products of combustion are metal oxides and oxy acids.

Fuels of Tetravalent Elements

Carbon is the lowest member of this series. CHO flames were discussed in Chapter XII. In this section we will discuss volatile fuels of carbon with elements other than hydrogen. This will be followed by a discussion of the higher elements of the tetravalent family: silicon (Si), germanium (Ge), tin (Sn), and lead (Pb). These are involatile elements that form adherent, involatile oxide coatings, which contrasts sharply with carbon, whose oxides are gases. The higher elements do not normally burn in the bulk form; however, tin and lead dusts (<10 μmicron dia.) do burn. This is not the case with silicon and germanium, probably because their oxide coats are so adherent and involatile that reaction is inhibited. Silicon is a semiconductor, as is germanium. Tin and lead are metals.

These elements form hydrides similar to the hydrocarbons, but as the atomic number increases, the compounds become less stable. They are all combustible. Silicon forms hydrides through Si_8, mimicking hydrocarbon chemistry. Germanium forms at least three hydrides (GeH_4, Ge_2H_6, and Ge_3H_8). Tin forms only one stable hydride (SnH_4), and this decomposes at 420 K (Sidgwick [1950]). Lead hydride has been detected using radioactive tracer techniques, but it is too unstable to isolate.

Volatile–Nonhydrogenic Compounds of Carbon

The carbides of most elements are involatile solids that burn only with difficulty. The simplest volatile carbon compound is carbon monoxide, which was covered in Chapter XII. The other volatile carbon compounds are sulfur and nitrogen compounds.

Carbon Disulfide

Burning velocities have been measured by Vetter and Culick [1977] for carbon disulfide with both air and oxygen ($P = 0.01$ atm) (see Fig. XIII-10). Merriman and Levy [1971] studied structures of flames of carbon disulfide (CS_2) and carbonyl sulfide (OCS) flames using microprobe sampling. Bystrova and Librovich [1977] also studied the structure of a carbon disulfide–oxygen flame.

Homann, Krome, and Wagner [1969] used low-pressure discharge flow tube techniques with ESR detection to measure reactions of CS_2, H_2S, and OCS with O atoms (Fig. XIII-11). A group at the Institute of Physics in Moscow (Sarkissyan, Azatyan, and Nalbandian [1966]; Azatyan, Gershenson, Sarkissyan, Sachyan, and Nalbandian [1969]) studied low-pressure flames of sulfur and oxygen using a flow tube with ESR detection. They identified O atoms and SO radicals. Graham and Gutman [1977] used laser fluorescence techniques to identify and study sulfur radicals in flames.

Linevsky and Carabetta [1973, 1974] demonstrated a flame laser using carbon disulfide with oxygen or N_2O. The lasing lines were CO vibrations excited by the reac-

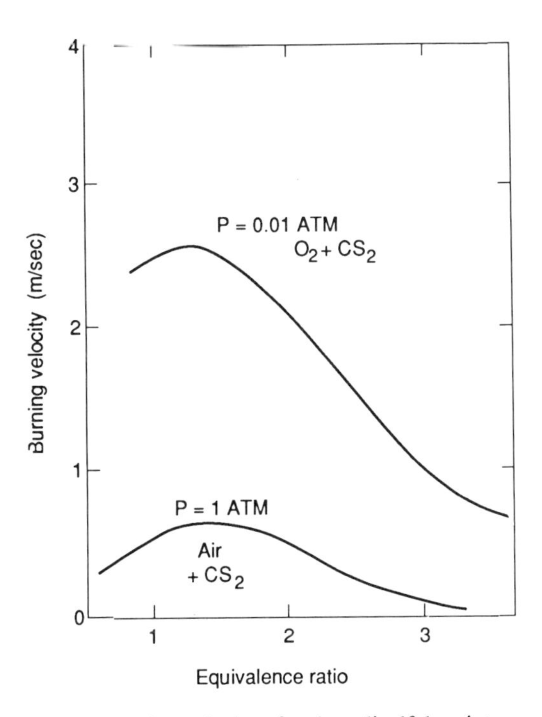

FIG. XIII-10 Dependence of burning velocity of carbon disulfide mixtures with air at atmospheric pressure and oxygen at 0.01 atm. (Replotted from data by Vetter and Culick [1977].)

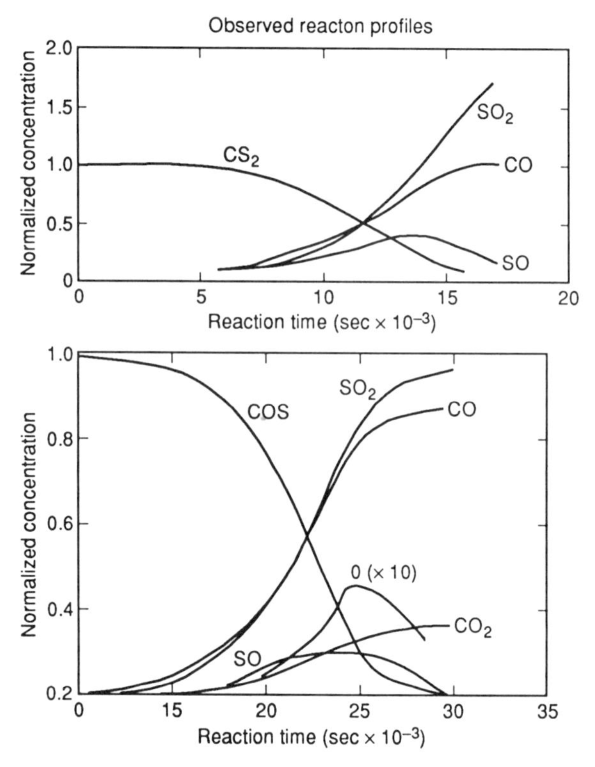

FIG. XIII-11 Reaction time profiles in low-pressure flame reactor. (a) Sulfur; (b) carbon disulfide; and (c) carbonyl sulfide. (Homann, Krome, and Wagner [1969].

tion $O + CS = CO^* + S$. The system is described in the book of Gross and Bott [1976] on chemical lasers. The burner is described in Chapter III.

Cyanogen

Cyanogen (C_2N_2) is a low-boiling-point, poisonous liquid. It is the lowest member of a family whose next member is carbon subnitride (C_4N_4). They have flame temperatures when burned with oxygen or ozone that can approach the temperature of the surface of the sun. Some representative adiabatic conditions for these systems are collected in Appendix A-10 at the end of the book. These high temperatures result from the stability of the products rather than high enthalpies. Maximum temperatures are obtained by burning to $CO+N_2$ (not stoichoimetrically) to $N_2 + CO_2$. Temperature measurements on these flames by Thomas, Gaydon, and Brewer [1952] helped establish the correct dissociation energy for nitrogen. The system was explored in more detail by Conoway, Wilson, and Grosse [1953] and Kirshenbaum and Grosse [1956]. The pure flames do not appear particularly luminous, because there are no visible emitters among the products, but if a trace of an emitting species is added, they become oppressively bright. Of course the temperatures are reduced by the radiation loss.

The chemistry of cyanogen flames is unusual because H atoms are absent. Therefore, the radical generation must result from dissociation of cyanogen.

Silicon and the Silanes

Elemental silicon powder ignites when heated to a dull red heat in oxygen, but is only superficially attacked in air. The hydrides of silicon form a class with the general formula Si_nH_{2n+2}. They are called *silanes* after the lowest member of the series (SiH_4).

Silicon also forms combustible alkyl derivatives. Schalla, McDonald, and Gerstein [1955] studied the explosion limits of alkyl silanes. Gerstein, Wong, and Levine [1951] compared combustion characteristics of alkyl silanes with their hydrocarbon analogs and found alkyl silanes are significantly more reactive. Chung and Katz [1985] have studied the formation of SiO_2 particles in an opposed jet diffusion flame of hydrogen with added silane against oxygen. Koda and Fujiwara [1988] studied ignition and combustion in opposed jets of silane–nitrogen mixtures against air.

Silicon also forms silicon-oxygen ethers called *siloxanes* (R_3–Si–O–Si–R_3). They are stable nontoxic compounds that are used for a variety of purposes. Cross-linked siloxanes are called *silicones*. They form a family of liquids and rubberlike solids that find some medical applications. These compounds burn with luminous flames due to the formation of silica (SiO_2) and silicon monoxide (SiO).

Germanium, Tin, and Lead

Elemental germanium is more reactive than silicon; so it would be expected that it would also burn as a powder. Tin burns in air with a white flame, and finely divided lead powder is pyrophoric. The hydrides of germanium (GeH_4, Ge_nH_{2n+2} through Ge_6H_{14}) and tin (SnH_4) have been reported to be flammable. Lead forms an unstable hydride that presumably would be combustible.

Metallic Fuels

Many metals can be burned (see Fig. II-1). They will be discussed in order of valence and atomic number. Very few metals are sufficiently volatile to allow the formation of premixed flames. The high thermal conductivity of metals make them difficult to ignite in bulk, so most combustion studies are made on powders. Powdered metals are an industrial fire safety hazard, as Palmer [1973] discusses in his book on dusts. Bulk metals such as iron, magnesium, and aluminum will burn vigorously once ignited. The flames are blinding because solid products are formed. The most familiar example is the cutting of steel plates using oxygen. Grosse and Conoway [1967] give an excellent survey on the properties of metal dust flames with oxygen and discuss their industrial potential. Adiabatic flame temperatures and compositions for some representative metals with oxygen are given in Table A–6 of the appendix.

The hydrides of most metals are either involatile salts or poorly defined interstitial compounds (Cotton and Wilkinson [1988]). However, many of them can be burned as powders, as, for example, UH_3 (Palmer [1973]).

There are a number of carbonyls of the transition metals that are volatile and combustible (see Table XIII–3). They are very poisonous. They have been studied as inhibitors by Bonne, Jost, and Wagner [1962] and Lask and Wagner [1962] (see later material). Hastie's book [1975] provides an excellent summary of the flame chemistry of volatile metal compounds.

There are a host of metal alkyl compounds, many of which, such as triethyl alu-

Table XIII–3 Boiling Points of Some Volatile Metal Compounds[a]

Combustible		Inhibitors	
$V(CO)_6$	sublimes	$AlBr_3$	371
$Cr(CO)_6$	sublimes	$AsCl_3$	336
$Mo(CO)_6$	sublimes	$TiCl_4$	409
$W(CO)_6$	sublimes	OsF_6	319
$Mn_2(CO)_{10}$	sublimes	IrF_6	326
$Re_2(CO)_{10}$	sublimes	ReF_6	321
$Fe(CO)_6$	376 K	UF_6	329
$Co_2(CO)_8$	sublimes	CrO_2Cl_2	390
$Ni(CO)_4$	316 K	BCl_3	285
		$SnCl_4$	387

[a]This rather arbitrary list is chosen to illustrate the point that although most salts are very high boiling compounds, there are metallic compounds that boil around or below water. In combustion their behavior is usually inhibitory, but they can also act as combustible fuels.

minum, are relatively volatile. Studies of the aluminum alkyls in connection with supersonic combustion were reviewed by Olson and Setze [1958]. Much of this work is buried in government reports, but scattered studies were published, for example, the ignition study of Marsel and Kramer [1958]. The subject will not be pursued further.

The colors that mists of metal salt solutions add to flames are generally due to atomic lines since salts usually dissociate at flame temperatures. This has provided a tool for analytical chemistry since the time of Faraday. It has now developed into the atomic absorption technique as a routine analytical tool discussed by Hermann and Alkamade [1963]. The books of Gaydon [1979] and Mavrodineanu and Boiteux [1965] on flame spectroscopy provide excellent summaries. There is an extensive, though somewhat dated, bibliography by Mavrodineanu [1967].

The general topic of metal and metal salt vapors in flames was first extensively studied by Sugden's group at Cambridge. Parts of this work were mentioned in Chapter VIII in relation to estimating radical concentrations from metal and salt emission lines. Hastie's [1975] book on high temperature vapors also gives a good discussion of flame studies.

Monovalent Metals

The lower monovalent metals are called the *alkali metals.* They are soft low-melting-point silvery metals[1] with sufficient volatility combined with high enough burning

1. Metals as fuels. Many metals burn in air when reduced to micrometer-sized powder [Palmer (1973)], but a premixed flame requires that a fuel have sufficient volatility to make a combustible mixture. Most metals are involatile by this definition. We therefore will be concerned primarily with low-pressure flames of the relatively volatile alkali and alkaline earth metals. There are many metal organic compounds that are volatile and combustible. The carbonyls of the transition elements such as iron

velocity to form premixed flames. These are laboratory curiosities that require pressures below 1 torr. The alkali metals are reactive because they are radicals under conditions where premixed flames are possible. At higher pressures and lower temperatures they dimerize. They burn in air as diffusion flames. With the exception of lithium, they can take fire spontaneously in moist air. This occurs because they react with water to form hydrogen, liberating enough heat to ignite the hydrogen in air. This in turn heats the metal to its ignition point.

Polanyi [1932] first studied alkali metal combustion in his classic work on low-pressure flames with halogens. These studies were influential in the development of reaction kinetics and Eyring's theory of rate processes [1945]. Using low-pressure diffusion flames, Polanyi's group made a systematic study of the reactions of metals with halogenated compounds. They are extremely rapid; some of them show reaction cross sections higher than those associated with normal bimolecular collisions. This enhanced cross section is attributed to an intermolecular electron jump that occurs on near-miss collisions. It is called a "harpoon reaction" by Herschback in his crossed molecular beam studies of such reactions. Spectroscopic studies also provide information on the exit state of the reactants, and this has been put to good use in a number of studies (Polanyi [1972]). The details of state-to-state chemistry lie beyond the present scope. Lasers and crossed molecular beam techniques have superseded the study of these flame reactions; so the reader is referred to Bernstein's readable book [1982].

Alkali metal compounds are favorite trace additives for temperature determination for ionization studies. Their salts are also used as fire extinguishants. These questions are examined in a later section.

Divalent Metals

The lower divalent metals are called "alkaline earth" metals.

The alkaline earth metals are very electropositive elements that react rapidly with all oxidizers. The members of this family are beryllium (Be), magnesium (Mg), calcium (Ca), barium (Ba), and strontium (Sr). Second period divalent metals such as Zn and Cd are not considered volatile. They have lower vapor pressure than the alkali metals, but at high temperatures the reactions are so rapid that diffusion flames can be studied. For example, Palmer, Krugh, and Hsu [1975] studied the emission of flames of Ca, Ba, and Sr with NOCl and NOBr at low pressures (0.3–0.003 torr). This field was active during the early searches for visible chemical lasers, but there have been few recent combustion studies.

The lanthanide and actinide series in the periodic table are reactive metals that present a significant fire hazard, particularly in the form of dusts.

---

and nickel are volatile and burn. Flames are commonly used to dissociate compounds for the atomic absorption technique of analytical chemistry. These compounds are usually introduced as an aqueous mist and are peripheral to the combustion. Details can be found in Hastie's [1975] excellent book on high-temperature vapors.

On the practical side aluminum dust is a common enhancer used in solid rocket fuels. Aluminum increases rocket ranges because it has a high ratio of heat of combustion to molecular weight and its oxide is involatile so that energy is not lost by vaporization.

Other Oxidizers

Oxidizers other than molecular oxygen will be considered in three groups: (1) oxygen related, (2) oxides and oxyacids of nitrogen, and (3) oxides and oxyacids of the halogens. Other oxidizers include the halogens (Chapter XI) and the solids used in propellants and gunpowder. The many exotic high-energy oxidizers studied for potential use in rockets will not be considered here. This work is summarized in Lawless and Smith's [1975] book.

Oxygen-Related Oxidizers

Several compounds of oxygen are strong oxidizers in their own right. They are: (1) singlet delta molecular oxygen, (2) atomic oxygen, (3) ozone, and (4) hydrogen peroxide.

Singlet Delta Oxygen ($O_2\ {}^1\Delta$)

Normal molecular oxygen is in a triplet state having two unpaired electrons. However, there are other low-lying states. Their spectroscopy is discussed by Herzberg [1950]. The most easily excited is the singlet delta state O_2 (${}^1\Delta$), in which the electrons are paired. It is found in electric discharges and can be produced photolytically and chemically. Production and detection methods are reviewed in the book edited by Wasserman and Murray [1979]. Large amounts can be prepared using the reaction between concentrated, basic hydrogen peroxide and chlorine. It has a relatively short half-life; so its combustion properties have not been explored; however, it is used in the COIL (chemical oxygen iodine laser) system to produce an excited form of iodine that lases in the infrared. A survey of this field is given by Avizonis and Neumann [1992].

Oxygen Atoms

As might be expected, even low concentrations of atomic oxygen will support vigorous combustion. Studies of these systems have usually been made at low pressure since even at a few torr atom lifetime and concentration are limited by three-body and wall recombination. Atomic oxygen production is discussed in Chapter XII.

The spectroscopic studies of these flames showed many nonthermal distributions (Gaydon and Wolfhard [1952]; Ferguson and Broida [1956]). Oxygen atoms and other radicals were the object of intensive study as potential high-energy propellants, but this interest has waned. Minkoff [1960] edited a symposium on this work. Although direct combustion studies are relatively rare, the elementary reactions of atomic oxygen are well studied and have been reviewed by Herron and Huie [1973]. This is fortunate because reliable modeling of all oxygen flame systems requires these reactions.

Ozone

Oxygen has a second allotrope, ozone (O_3). Its reactions include: thermal decomposition ($O_3 + M^* \rightarrow O_2 + O + M$); reaction ($O_3$ + (H, O, OH) $\rightarrow O_2$ + (OH, O_2,

HO_2)]; and addition to olefinic and acetylinic bonds ($O_3 + R = R' \rightarrow$ products). Additionally they show all of the reactions characteristic of oxygen flames.

The ozone decomposition flame was discussed in Chapter XI. Pure ozone reacts explosively on contact with many organic compounds and supports combustion even more vigorously than molecular oxygen. There have, however, been few combustion studies using ozone. One low-pressure atomic flame reaction study was made of ozone with hydrogen atoms and nitrogen atoms by Garvin and Broida [1963]. With hydrogen atoms OH is produced in a relatively long-lived high vibrational states ($> v = 6$). Pure nitrogen atoms produce no visible radiation, but mixed nitrogen and hydrogen atoms reacted with ozone producing a red luminosity that they attributed to HNO.

Hydrogen Peroxide

H_2O_2 is a strong oxidizer that can support combustion of C/H/O compounds and other fuels. It is a pale blue viscous liquid that boils at 425 K and freezes at 274 K. It is available commercially in aqueous solutions with concentrations up to 90%. Pure material can be prepared by low-pressure fractional distillation. The liquid has been used as a monopropellant for torpedoes. Pitts and Finlayson [1975] and Lewis and Merrington [1958] used it to simulate rocket combustion. No premixed flame studies were found in the literature, probably because of the relatively low vapor pressure of H_2O_2. However, it should support low-pressure flames with most C/H/O fuels. The decomposition of hydrogen peroxide was discussed in Chapter XI.

Nitrogen Oxides and Acids

Nitrogen forms both fuels and oxidizers. The oxidizers are principally oxides and oxy acids, although some halogen compounds such as NF_3, N_2F_4, and NOCl also support combustion. There are a number of oxides of nitrogen. The lowest, nitrous oxide (N_2O), is a well-defined normal molecule. The next three in the series are radicals having unpaired electrons. They are: nitric oxide (NO), nitrogen dioxide (NO_2), and nitrogen trioxide (NO_3). In addition to these radicals all of their binary combinations have been detected (N_2O_2, N_2O_3, N_2O_4, and N_2O_5). These oxides form a number of acids with water, including hyponitrous acid ($H_2N_2O_2$), nitrous acid (HNO_2), and the common, familiar nitric acid (HNO_3). They are all strong oxidizers.

Several of these compounds are available commercially in high purity. Nitrous oxide (N_2O), often called laughing gas, is available in pressurized cylinders. Nitric oxide (NO) is also available in pressurized cylinders. Laboratory NO_2 is a low-boiling-point (295 K) equilibrium mixture of NO_2 and N_2O_4. It is available in steel cylinders. Concentrated nitric acid is a common laboratory chemical. The anhydrous nitric acid can be prepared and boils at 357 K and freezes at 232 K. The "pure" material is not simple since autoionization occurs ($2\,HNO_3 \rightarrow NO_2^+ + NO_3^- + H_2O$). The "red-fuming nitric acid" of commerce is a solution of NO_2 in the anhydrous acid.

The chemistry of nitrogen oxides and acids has been authoritatively reviewed by Jost and Russell [1944]. The H/O/N reaction system has been reviewed by Hanson and Salimian [1983]. Parker and Wolfhard [1956] explored the spectroscopy of flames of oxides of nitrogen with hydrogen, acetylene, and ethylene. Burning velocities of some of these flames are given in Figure XIII-12. The adiabatic combustion properties

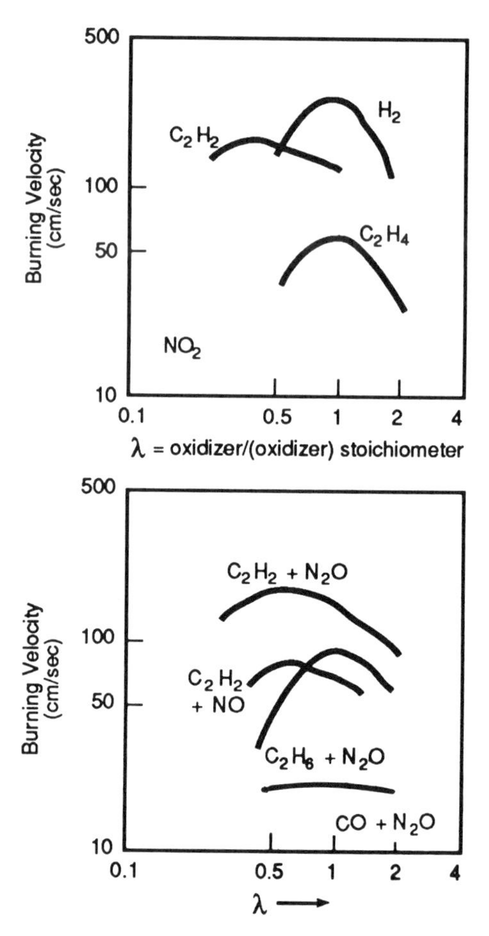

FIG. XIII-12 Burning velocity of several CHO fuels with various oxides of nitrogen. Burning velocity on a logarithmic scale is plotted against λ, the ratio of the oxidizer concentration to the stoichoimetric value. (Replotted from data by Parker and Wolfhard [1956]. Value for CO–N_2O taken from Simpson and Linnett [1957].)

of the flames of oxides of nitrogen with hydrogen are presented in Table A–7 in the appendix at the end of the book.

Nitrous Oxide

N_2O is the lowest oxide of nitrogen. It is the most widely studied oxide of nitrogen. Hydrogen (H_2) is the most widely studied fuel of the nitrous oxide flames. The burning velocity of a number of mixtures have been determined by Duval and Van Tiggelen [1967], Dixon-Lewis, Sutton, and Williams [1965a], and Parker and Wolfhard [1949] (Fig. XIII-12). Emission spectra were studied by Gaydon and Wolfhard [1949] and Wolfhard and Parker [1955]. Gray, McKinven, and Smith [1967] compared the burning velocities of hydrazine with oxygen, nitrous oxide, and nitric oxide (Fig. XIII-8).

The structure of several hydrogen–nitrous oxide flames have been measured beginning with the microprobe studies of Dixon-Lewis, Sutton, and Williams [1965a]. Radical concentrations were measured indirectly. They were interested in the hydrogen atom reaction with N_2O. Related studies of the reaction of N_2O in hydrocarbon flames were made by Fennimore and Jones [1962], and shock tube studies were made by Barton and Dove [1969]. Balakhnin, Vandooren, and Van Tiggelen [1977] and Vandooren, Balakhnin, Huber, and Van Tiggelen [1978] made complete structural measurements on several hydrogen flames using molecular beam sampling and mass spectrometry. They analyzed the data and derived species production rates. By interpreting the results in terms of elementary reactions, they were able to derive kinetic constants for N_2O decomposition and its reaction with oxygen atoms and hydrogen atoms. They believe that the driving radical producing reaction was the thermal decomposition. The flame showed two stages: a high-temperature stage where N_2O decomposed into O atoms that react to form molecular oxygen and an earlier nonluminous hydrogen–oxygen flame fed by the molecular oxygen diffusing back from the higher stage (see Fig. XIII-13). Hydrogen–nitrous oxide flames have been modeled by Coffee [1986] using a 27-step reaction mechanism. Results were compared with experiments of Dixon-Lewis et al. [1965a], Duval and Van Tiggelen [1977], Vanderhoff et al. [1986], and unpublished work of Cattolica. The outline of hydrogen–nitric oxide flame chemistry appears established. Several key reactions remain in doubt, however, such as the dissociation of N_2O and the third-body efficiencies.

Van Wontergeheim and Van Tiggelen [1956] measured the burning velocity of carbon monoxide–nitrous oxide mixtures. Simpson and Linnett [1957] studied mixtures with added water vapor, finding a modest positive dependence peaking at about 1.5%. The dropoff at higher concentrations presumably resulted from dilution effects exceeding the catalysis. The system does not appear to be as sensitive to the degree of $CO–O_2$ mixtures discussed in Chapter XII. Fuel-rich and stoichiometric carbon monoxide–nitrous oxide flames were studied by Kalaff and Alakmade [1972].

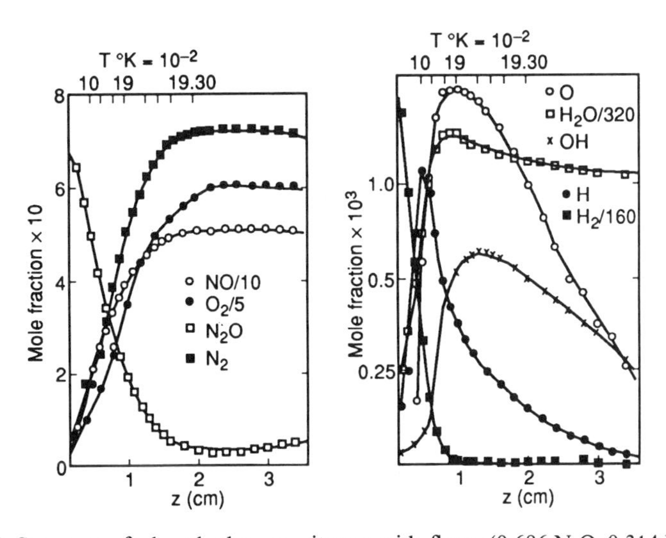

FIG. XIII-13 Structure of a lean hydrogen–nitrous oxide flame (0.686 N_2O, 0.314 H_2; P = 40 torr; v_0 = 280 mm/sec). (A) Reactants; (B) Products and radical. (Balakhnine, Vandooren, and Van Tiggelen [1977].)

The flames of methane (CH_4) with nitrous oxide have been studied. Flammability limits have been determined by Pannetier and Sicard [1955]. The structure of both lean and stoichiometric flames were determined by Harris [1983] using CARS spectroscopy. He was able to measure nitrogen, nitrous oxide and temperature profiles for the flames. Vanderhoff, Anderson, Kolter, and Bayer [1985] studied temperature profiles using laser fluorescence measurements on OH. Anderson, Decker, and Kotlar [1982] combined fluorescence and Raman spectroscopy to determine composition profiles for N_2, H_2O, and CH_4 (Raman) and NH, OH, CN, and NCO (fluorescence). Salmon, Lucht, Sweeney, and Laurendeau [1984] made LIF measurements of NH in these flames. Habeebullah, Alasfour, and Branch [1991] made a complete study using microprobe sampling for stable species combined with LIF for radicals. Vandooren, Branch, and Van Tiggelen [1991] compared molecular beam sampling structural studies of low-pressure stoichiometric methane flames with oxygen and N_2O.

Nitrous oxide flames with acetylene (C_2H_2) have been studied spectroscopically by Guillaume and Van Tiggelen [1984]. They determined temperature profiles of excited radicals. Darian and Vanpee [1987] studied infrared emission profiles of major species. By assuming these radiations were thermal, they were able to deduce concentration profiles for the major products. This is an excellent example of the application of high-resolution emission spectroscopy to provide quantitative measurements.

The ignition limits of the lower aliphatic hydrocarbons [methane (CH_4) through butane (C_4H_{10})] with nitrous oxide were studied by Pannetier and Sicard [1955]. Some burning velocities are reported. Spectroscopic studies were reported on heptane (C_7H_{16}–N_2O) flames by Gaydon and Wolfhard [1949].

Gaydon and Wolfhard [1949] reported spectra and the luminous structure of flames of methanol (CH_3OH), formaldehyde (H_2CO), formic acid (HCOOH), acetone (CH_3–CO–CH_3), and ethylene oxide (C_2H_4O).

The burning velocity of some ammonia (NH_3) and hydrazine (N_2H_4) flames with N_2O have been measured (Fig. XIII-8).

Nitric Oxide Flames

Parker and Wolfhard [1949] studied nitric oxide flames with fuels including hydrogen (H_2), carbon monoxide (CO), acetylene (C_2H_2), ethylene (C_2H_4), ethane (C_2H_6), ammonia (NH_3), and carbon disulfide (CS_2) (see Fig. XIII-12). Many studies of the formation of nitric oxide in hydrocarbon–oxygen flames have been made in connection with NO_x pollution.

Nitrogen Dioxide Flames

The burning velocity studies made with this oxide (Fig. XIII-12) were mentioned in the introduction. Albers and Homann [1958] studied low pressure, fuel-rich cyanogen–oxygen flames and found traces of polycyanogens (C_2N_2, 0.1 C_4N_2, 0.01 C_6N_2, and 0.005 C_8N_2). Thorne and Smith [1988] measured the structures of several cyanogen–NO_2 flames to provide a model for nitramine flames and propellants such as RDX (octahydro-1,2,5,7-trinitro-s-triazine) and HMX (octahydro-1,3,7-tetranitro-1,3,5,7 tetraazocine). Some of their results are shown in Figure XIII-14. The authors com-

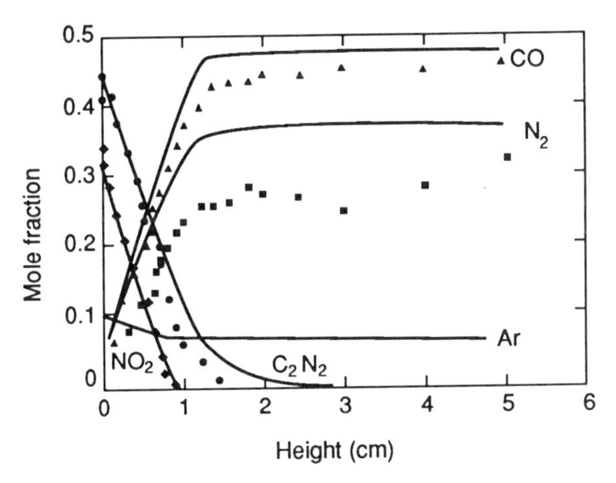

Fig. XIII-14 Profiles of a cyanogen nitric oxide flame. (Replotted from data of Smith and Thorne [1986]. See also Thorne and Melius [1991].)

ment with no apparent smile that this system was chosen for safety reasons (i.e., relative the HCN–NO_2 system).

Nitric Acid and Higher Oxides of Nitrogen

All of the higher oxides of nitrogen (N_2O_3, N_2O_5), as well as nitric acid (HNO_3), support flames. Fuming nitric acid, which is a solution of NO_2 in anhydrous nitric acid, has been used in liquid propellant rockets as the oxidizer for kerosene, dimethyl hydrazine [$N_2H_2(CH_3)_2$], and hydrazine (N_2H_4). These flames are hypergolic, and this tends to inhibit their study as premixed flames.

Halogen Oxides and Acids as Oxidizers

There a number of halogen oxides and oxyacids that could support combustion. However, only two of them, chlorine dioxide (ClO_2) and perchloric acid ($HClO_4$), have been systematically studied. As indicated in Chapter XI, both of these compounds support detonations, and the other members of this class are even less stable and more difficult to work with (Cotton and Wilkinson [1988]). The adiabatic combustion properties of ClO_2 and $HClO_4$ with hydrogen are given in Table A–14 in the appendix at the end of the book.

Chlorine Dioxide Flames

Chlorine dioxide (ClO_2) is a greenish-yellow gas that is very reactive and can explode. Its decomposition flame is discussed in Chapter XI. *The material is dangerous and should only be handled by experienced personnel using suitable safety precautions.* Laffitte, Combourieu, Hajal, Ben Caid, and Moreau [1967] and Combourieu, Moreau, Moreau, and Pearson [1970] studied flames of ClO_2 with methane and ammonia.

The acetylene flame (C_2H_2) has been extensively studied by Combourieu and Moreau [1975]. This forms a colorful flame system whose geometry changes with composition. Flammability limits, spectra, and burning velocities were determined over a range of compositions and pressures (Fig. XIII-15). The system is interesting because both components produce decomposition flames, and as a consequence all mixtures burn. They also measured temperature and concentration profiles of stable species using microprobes (Fig. XIII-16). Both stoichiometric and rich flames were studied. The authors discuss mechanisms and their relation to perchloric acid flames (see next section).

Perchloric Acid Flames

(Perchloric acid ($HClO_4$) forms a decomposition flame that is discussed in Chapter XI and also supports combustion. The concentrated acid (72%) is available as a laboratory

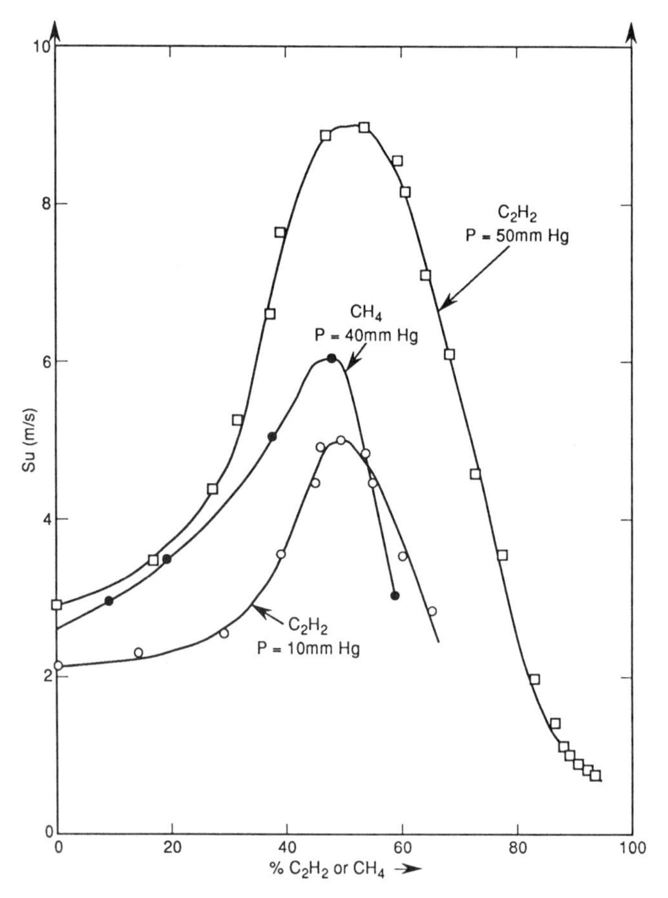

FIG. XIII-15 Effect of stoichoimetry and pressure on burning velocity of ClO_2 with C_2H_2 and CH_4. (Combourieu, Moreau, Hall, and Pearson [1969].)

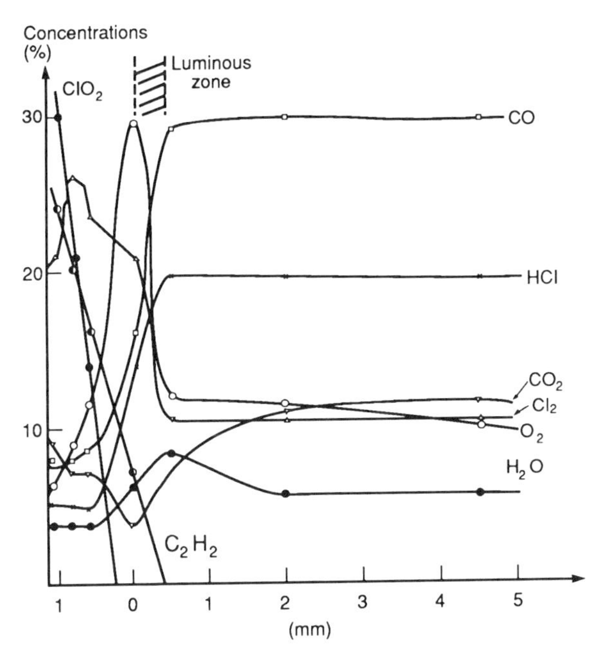

FIG. XIII-16 Composition profiles of stable species in a low pressure 0.7 ClO_2-0.3 C_2H_2 flame. Combourieu and Moreau [1975].

chemical. It ignites organic material often explosively. *This is a dangerous material to handle.* Hydrocarbon flames with $HClO_4$ are studied as models for the combustion of composite propellants that use ammonium perchlorate as the oxidizer. Pearson and his collaborators at the Rocket Propulsion Establishment, Wescott, have been particularly active. Their burning velocity studies on perchloric acid flames with H_2, CO, CH_4, C_2H_4, C_2H_6, and CH_3OH are presented in Figure XIII-17. They found a modest pressure dependence with the burning velocity increasing ten to twenty percent between atmospheric pressure and 40 torr with little dependence on stoichiometry. The flames generally show multiple luminous zones whose separation depends on stoichiometry.

Methane (CH_4) is the most extensively studied fuel. The Wescott group (Cummings and Hall [1965]; Heath and Pearson [1967]; Hall and Pearson [1969]; Pearson [1967a, 1968]) have measured the structure and analyzed the data in terms of species production rates. Korobeinichev, Orlov, and Shifon [1980] and Williams and Wilkins [1973] used a methane–perchloric acid flame to test their high-pressure molecular beam inlet sampling techniques. One of their studies involved looking at the structure of composites of ammonium perchlorate oxidizer and a rubber binder fuel. This is discussed briefly in the next section.

Pearson [1967b] studied the structure of several ethylene (C_2H_4) rich flames with perchloric acid. They report that excess ethylene acts primarily as a diluent, although in very rich flames some soot formation was observed.

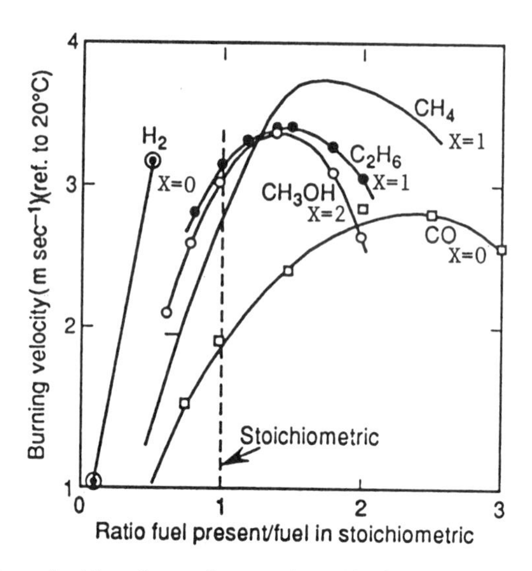

FIG. XIII-17 Burning velocities of some flames of perchloric acid with and without nitrogen dilution. X = the molar ratio of molecular nitrogen diluent to perchloric acid. Velocities are normalized to STP. Actual burner temperature was 210C. Cummings and Hall [1965].

Complex Fuels, Oxidizers, and Additives

The studies of mixed fuels, mixed oxidizers, and additives are so numerous that it will only be possible to present selected examples. Mixed reactants are common in practical combustion systems because the reactants rarely consist of pure components and purification costs money. The effects of mixed systems on combustion can stem from physical factors, thermal factors, or chemical factors. The chemical factors are usually the strongest, but all factors interact.

The motives for studying complex mixtures vary. Sometimes it is a drive to study "real life" situations. Often it has been the dream of finding some "magic ingredient" that would improve a fuel the way that tetraethyl lead improved gasoline or in the case of inhibition and extinction it may be the fireman's dream of a chemical bomb that would extinguish a fire the way an atomic bomb destroyed a city. There have been some remarkable practical additives, but caution should be used in extrapolating the results from small-scale experiments such as burning velocity and flame structure measurements to large-scale applications.

These systems are complex, and the generalizations given should only provide an initial guide. The only certain route to understanding is by direct experiments. As more reliable basic information becomes available, detailed modeling confirmed by experiments will become more useful and eventually become the route of choice.

As might be expected, the gross properties (burning velocity, ignition temperature, heat release, flame temperature, etc.) of mixed flame systems usually lie between those of the contributing systems, and averaging can be used. Anomalous situations where this is not the case are discussed in the following as additives. Averaging can be applied to mixed fuels, mixed oxidizers, or diluents. It requires a knowledge of two systems near the extremes. Averaging can be proportional to mole fraction or mass depending on the property. The mole fraction average is called *Le Chatelier's rule* when it is applied to limits of combustion.

$$Y_{ab} = \Sigma(X_a Y_a + X_b Y_b + \cdots)$$

This is reasonable where the properties are not too disparate. Where differences between properties of the two systems are large, an averaging that is proportional to inverse mole fraction may be more appropriate.

In these equations Y is the property in question; X is mole fraction; and a and b are the system or species indices.

$$Y_{ab} = [\Sigma(X_a/Y_a + X_b/Y_b + \cdots)]^{-1}$$

This is pure empiricism and therefore, should be tested experimentally. If applied intelligently it can yield useful practical relations. Where a theory or detailed modeling is available this should be used to determine appropriate averaging procedures.

Mixed Fuels

If two fuels vary widely in burning velocity, one may observe physical separation of the flame into two regimes or even the extreme where the rapid reactant goes to completion leaving the slower reactant intact. Such flames can be separated into two parts. This was demonstrated by Teclu [1891] and Smithels and Ingle [1892] on rich hydrocarbon flames in which they were able to separate the rapid fuel attack region from the slow carbon monoxide–hydrogen final oxidation. Their burner is discussed in Chapter III. This behavior was found by Berl and Dembrow [1952] in the diborane–propane–air system (Fig. XIII-18). The system was examined in detail by Breisacher, Dembrow, and Berl [1958].

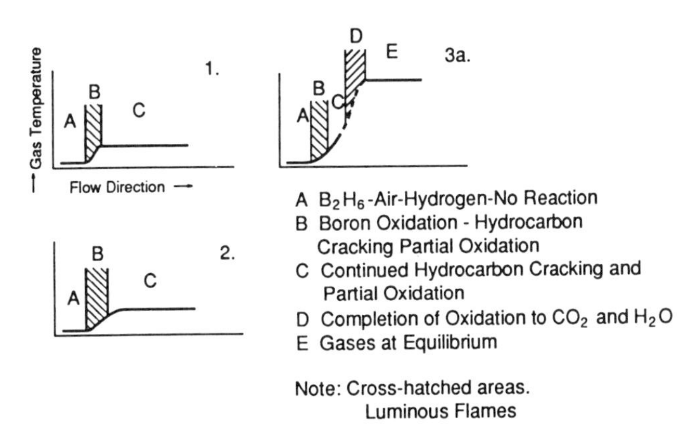

FIG. XIII-18 Separation of flame zones in several mixtures of diborane-propane with air. (A) low diborane concentration with propane passing through without ignition; (B) two separate diborane flames followed by delayed ignition of propane; (C) high diborane concentration resulting in immediated ignition of propane with combined flame zone. After Berl and Dembrow [1952] with modifications.

Mixed Oxidizers

In general terms the addition of a radical exchange oxidizer such as oxygen to a dissociation-controlled oxidizer such as chlorine will increase burning velocity and other combustion processes. Conversely, the addition of halogen compounds to oxygen flame systems tends to inhibit them. Within each group such as mixed halogens or mixed oxides the effects tend to be additive, with Le Chatelier's rule providing a reasonable first approximation. Chlorine–hydrogen flames are difficult to ignite when oxygen free for photoignition is also more difficult (see Chapter XI). This presumably is due to the oxygen opening up a lower activation energy path for radical production. Note that for $H + O_2 \Rightarrow OH + O$ Ea = 16.8 kcal/mole, while for chlorine dissociation Ea = 35 kcal/mole.

Additives

The addition of traces can have a profound effect on combustion. If combustion is increased, this is called *promotion;* if it is decreased, it is called *inhibition.*

Inhibition

Flame inhibition is an unusually strong reduction of burning velocity or other combustion property brought about by the addition of a trace material. Burning velocity reduction is associated with extinction, and these are relatively simple laboratory experiments that require only small amounts of materials. As a consequence, burning velocity reduction has become a common method to screen compounds for use as fire extinguishants. The field has been reviewed by several authors (Freidman [1961]; Fristrom [1967]; McHale [1969]; Creitz [1970]; Fristrom and Sawyer [1971]; Hastie [1973]; Williams [1981]). These articles provide extensive bibliographies. Several collections of symposia on fire suppression were published in the 1970s (Baratgov [1974]; Gann [1975a]), and the Fire Center of the U.S. National Bureau of Standards holds yearly symposia on fire topics. These can be obtained through the government printing office. Books on related topics are: on retardance by Lyons [1970b] and polymer flammability by the Products Research Committee [1980]. Hastie [1975] and Fenimore [1964] devote chapters in their books to the subject. In addition to the standard combustion literature, there are two specialized journals on fire topics: *Fire and Materials* and *Fire Technology.* Reviews and abstracts can be found in *Fire Research Abstract and Reviews* [1960–1974] and *Fire Technology Abstracts* [1972–1976; 1984–1987].

A number of experimental techniques are used to characterize inhibition. They include changes in:

1. Burning velocity;
2. extinction limits;
3. Blowoff limits and flame strength measurements; and
4. Minimum ignition energy and/or ignition delay.

Fristrom and Sawyer [1971] suggested a dimensionless index for air and oxygen flames that allowed some correlation of these measures.

$$\Phi = \left(\frac{\Delta v}{v}\right)\left(\frac{O_2}{I}\right) = \Sigma N_E \nu_{EI}$$

In this equation Φ is the inhibition index, the fractional change in burning velocity (or other inhibition parameter) normalized to the ratio of oxygen concentration to inhibitor concentration. N_E are the elemental inhibition indices. ν_{EI} is the number of atoms of element E in molecule I.

This is a normalized index that is the ratio of the fractional change of the property divided by the ratio of inhibitor concentration to the oxygen concentration. This assumes that the effects on combustion are related to the radical generation step $H + O_2 \Rightarrow OH + O$. The extension of these ideas to promotion and other flame chemistries is certainly possible, but requires a knowledge of the radical generation step in the system to be correlated.

Inhibition is commonly studied by measuring burning velocity. Studies were made by Burgoyne, Cullis, and Liebermann [1959] and Burdon, Burgoyne, and Weinberg [1955], followed by the study of Rosser, Wise, and Miller [1959], who pointed out that there was an additivity of inhibition effectiveness with halogen atom type and number.

This idea was elaborated to correlate the effects of adding halogen compounds to hydrocarbon air flames. Fristrom and Van Tiggelen [1979] proposed an additive atomic index that allows an estimation of the Fristrom–Sawyer inhibition index Φ from the elemental composition of an inhibiting molecule. This successful elaboration of the original observation of Rosser et al. suggests that radical chemistry is the dominant factor in inhibition by halogen compounds (see Fig. XIII-19).

Unfortunately the atomic indices show a systematic dependence on the flame temperature when various flame systems are compared. This dependence can be correlated with the activation energy required to break the hydrogen-halogen bond. A complete theory requires more detailed modeling.

Chlorine flames offer an interesting case. Burgoyne, Cullis, and Liebermann [1959] and Corbeels and Scheller [1965] studied the effect of additives on the burning velocity of hydrogen–chlorine flames. They both found examples of both promotion and inhibition (Fig. XIII-20). The additives were halogenated hydrocarbons, tin tetrachloride, and methane. All of the additives gave significant promotion of burning velocity when added in large enough concentrations. The mechanisms of these effects are obscure.

There a few metallic halides that are sufficiently volatile to allow their addition to flames (Table XIII–3). They act as inhibitors of oxygen flames. The few structural studies in the literature are aimed at inhibition. These compounds are important as flame retardant additives. This area is discussed in Lyon's excellent text on fire retardants [1970b] and Hastie's book on high-temperature species [1975].

Flame Structure Studies

The first composition profile measurement on an inhibited flame was made by Levy, Droege, Tighe, and Foster [1962]. The group at APL/JHU made several structural

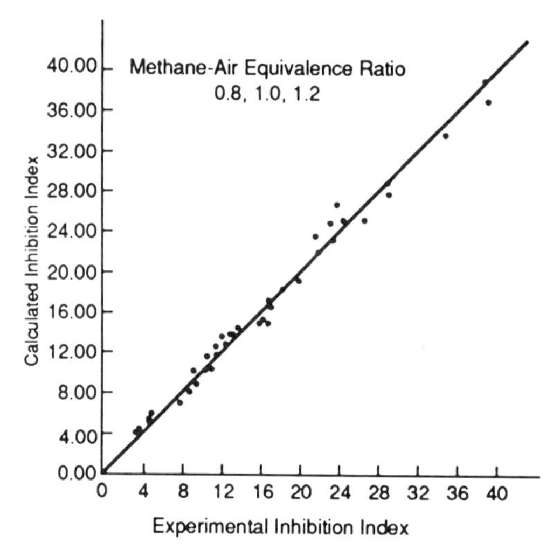

FIG. XIII-19 Calculated compared with experimental inhibition index showing the approximate atomic adaptivity of the inhibiting effect. The expeirmental index is defined as the fractional change in burning velocity normalized to the ratio of inhibitor concentration to oxygen concentration. The calculated index is defined as the sum over the number of atoms of a species in a molecule multiplied by its atomic index. Data is for 12 different inhibitors in three different flames. Fristrom and Van Tiggelen [1979].

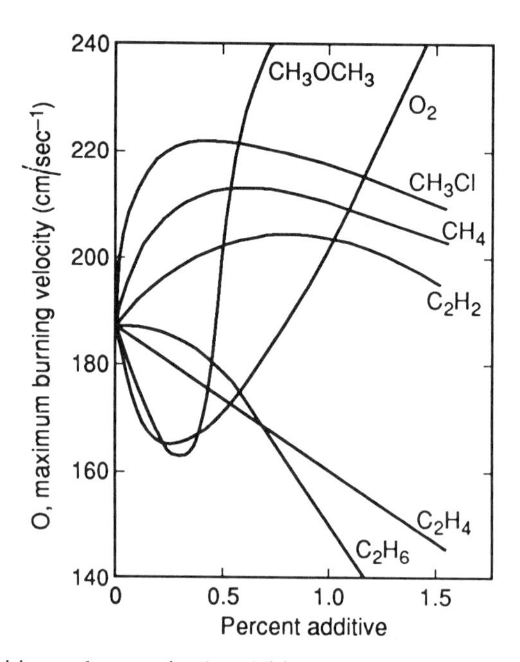

FIG. XIII-20 Inhibition and promotion by additives to a hydrogen–chlorine–nitrogen flame with thirty percent added nitrogen. (Burgoyne, Cullis, and Liebermann [1959].)

studies of inhibited flames (Wilson [1965]; Wilson [1967]; Wilson, O'Donovan, and Fristrom [1969]; Hunter, Grunfelder, and Fristrom [1975]) showing that inhibitors delayed the initiation of reaction so that reaction occurred in a narrower high-temperature region (Fig. XIII-21). Fenimore and Jones [1963c] and Hayes and Kaskan [1975] also studied the inhibition of methyl bromide on methane flames. Milne and Green [1965a, 1966, 1969, 1970] studied inhibition of methyl bromide and powders using flame structure measurements. Dixon-Lewis and Simpson [1976] made several experimental and modeling studies of hydrogen and hydrocarbon flames inhibited by halogenated compounds (Fig. XIII-22). They were able to establish the importance of the high-temperature stability of the halogen acid in the effectiveness of halogen inhibitors. The definitive structural studies on halogen inhibition of hydrocarbon flames were made by Biordi, Lazzara, and Papp [1973–1978] at the U.S. Bureau of Mines. In this classic series of papers, they studied a low-pressure flame inhibited by CF_3Br. Using flame structure analysis they were able to isolate a number of inhibition reactions and measure their rates. The radical concentration profiles were shifted toward higher temperatures, in agreement with earlier studies (Fig. XIII-23).

This work was successfully modeled by Westbrook [1980, 1981, 1983] at Lawrence Livermore. He employed their standard hydrocarbon model with the addition of eighty inhibition reactions.

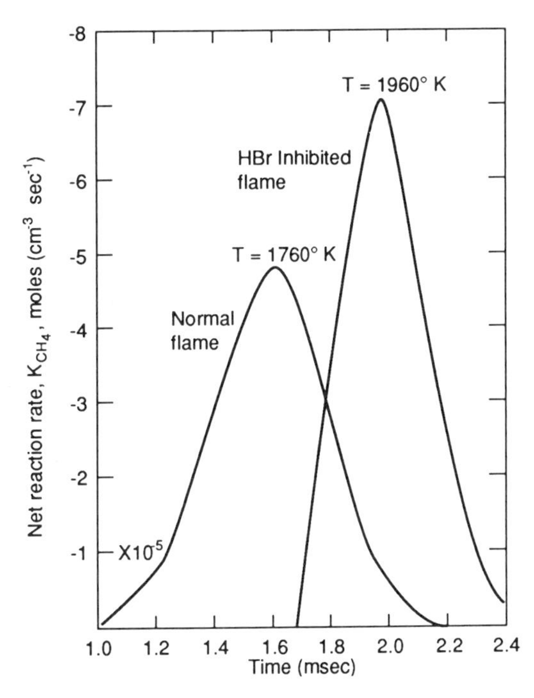

FIG. XIII-21 Shift of initial reaction to higher temperatures induced by addition of Br to an ethane–oxygen flame. Although burning velocity is depressed, the peak raction rate is actually increased by the shift to higher temperature. The reason is that burning velocity is proportional to the area under the curve, not the peak value. (Wilson, O'Donovan, and Fristrom [1969].)

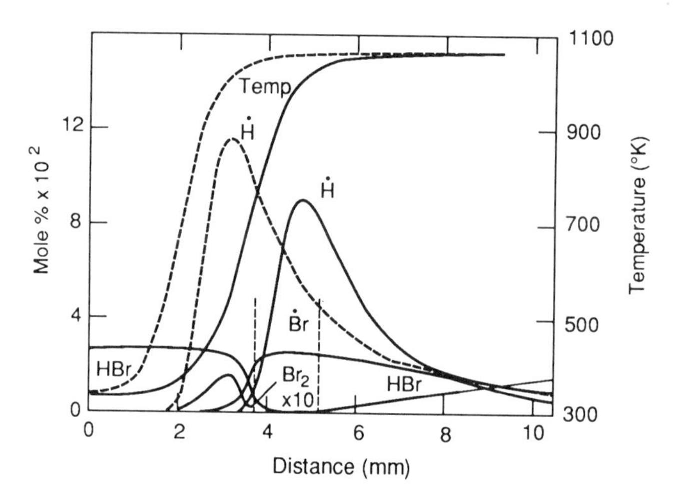

FIG. XIII-22 Effect of adding HBr to a hydrogen–oxygen flame showing the reduction of H atom concentration and the relation of the various bromine species. (Calculations by Dixon-Lewis and Simpson [1976].)

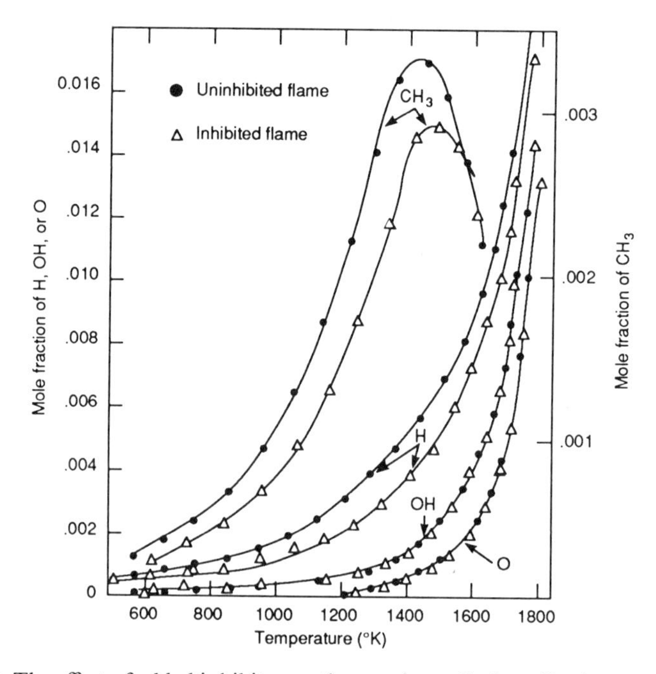

FIG. XIII-23 The effect of added inhibitor on the reactive radical profiles in a low-pressure methane–oxygen–argon flame with and without 0.35% added CF_3Br. (Biordi, Lazzara, and Papp [1975d].)

Brown and Fristrom [1978] applied zonal flame theory to the inhibition of the hydrogen–air system by hydrogen chlorine.

Pownall and Simmons [1971] measured the structure of a propane–oxygen flame inhibited by HBr.

Safieh, Vandooren, and Van Tiggelen [1982] studied the details of a carbon monoxide–hydrogen flame inhibited by CF_3Br. This complemented the Louvain study of a detonation in the same system (Libouton, Dorman, and Van Tiggelen [1975]). Palmer and Seery studied the structure of a carbon monoxide flame inhibited by molecular chlorine [1973].

The Fire Center at NBS studied the inhibition of flames by volatilized salts of antimony and phosphorous to simulate the gas phase reactions of flame retardance (Hastie [1973a]; McBee and Hastie [1975]; Lyons [1970b]). An example of one study is given in Figure XIII-24.

The structure of flames inhibited by alkali metal dusts have been studied by Milne and Green [1965a, 1966, 1969)] at Midwest Research and Knuth and Seeger [1982] at UCLA. These measurements were made using molecular beam inlet techniques. They discuss the problems in separating heterogeneous from homogeneous sampling.

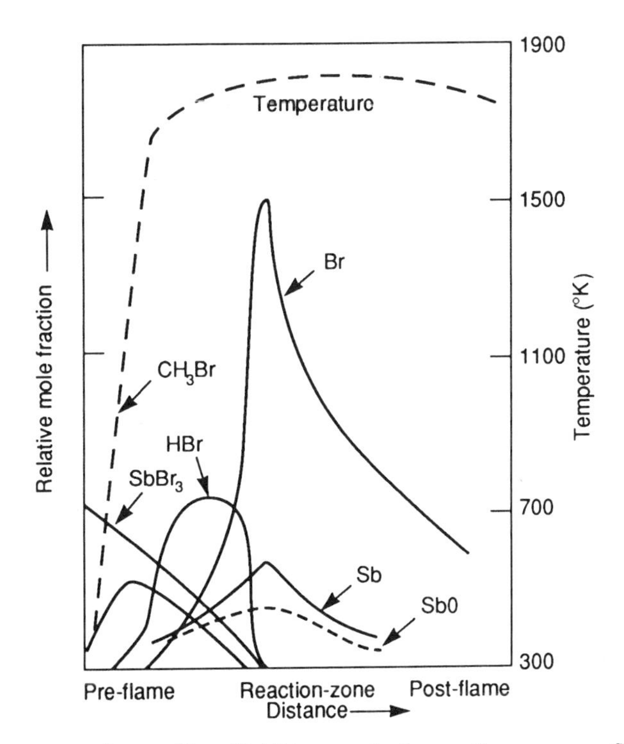

FIG. XIII-24 Concentration profiles of inhibitor species in a methane–oxygen flame inhbited by antimony tribromide. (Hastie [1973].)

Promotion

Combustion promotion is relatively neglected because most practical problems required moderating combustion. There have been tests of hypergolic additives to shorten ignition times in supersonic combustion situations.

The addition of hydrogen-containing compounds to nonhydrogenic fuels with oxygen has been discussed in the section on carbon monoxide flames in Chapter XII. Brokaw and Pease [1953] observed that similar effects occur with the CS_2 and C_2N_2. However, as observed by Milton and Keck [1984] and Yu, Law, and Wu [1986], the effect even occurs with hydrocarbons.

At Alma Alta in Kasakhstan Ksandropolo, Koles, Nikov and Odnoroq [1975] studied the structure of a number of flames inhibited by $C_2F_2Br_2$, diethyl amine, and promoted by cyclohexyl nitrate [1983]. The results were, as expected, complex. These chemistries are discussed in his books on flame chemistry (Ksandopolo [1980] and Ksandopolo and Dubinin [1987]). Unfortunately they are both in Russian. The second is under consideration for translation and publication in English.

Pollution

There are five major atmospheric pollutants generated by combustion: (1) NOx (nitrogen oxides and acids), (2) SOx (sulfur oxides and acids), (3) halogen acids and derivatives, (4) lead, and (5) carbonaceous solids such as soot, PNA (polynuclear aromatics), and PAH (polycyclic aromatic hydrocarbons). These have been the subject of a number of combustion studies, with the objective of minimizing their production in practical combustion devices. Each of these subjects has such an extensive literature that we can only comment on these areas, high lighting contributions by flame structure studies. There have been many colloquia and reviews on this subject, including those of Axworthy et al. [1975], Sawyer [1981], and Seeker [1991].

Nitrogen Oxides (NOx)

Nitric oxide is produced in small but significant amounts by combustion processes. When mixed with air NO_2 is formed, and this, in turn, reacts with hydrocarbons in the atmosphere forming peroxy acetyl nitrates (PAN), which are the eye-watering ingredients of the photochemical smog found in most urban areas. In addition, NO_2 reacts with water, forming nitric acid, which is a major component of the acid rain that is decimating northern hemisphere forests. Other nitrogen compounds, such as NH_3 and HCN, can also be formed and contribute to pollution. The only common benign form of nitrogen is the elemental molecule N_2.

There are three general sources of NO in combustion:

1. thermal reaction via the extended Zeldovitch [1946] mechanism: $O + N_2 \Rightarrow NO + N$; $N + O_2 \Rightarrow NO + O$; $N + OH \Rightarrow NO + H$;
2. "Prompt NO" was uncovered by Fenimore and Jones [1972, 1976] using flame structure measurements that showed NO formation before and in excess of that predicted by the Zeldovitch mechanism. Fenimore suggested that a likely mechanism would be the reaction of the CH radical with molecular nitrogen $CH + N_2 \Rightarrow HCN + N$, and that the HCN and N are converted into NO; and

3. Another path, suggested by Malte and Pratt [1975], is the formation of nitrous oxide in lean flames by a three-body recombination: $O + N_2 + M \Rightarrow N_2O + M^*$; $O + N_2O \Rightarrow NO + NO$; $H + N_2O \Rightarrow NO + NH$.

Combustion pollution has been reviewed by Sawyer [1981], Haynes [1977], Bowman and Seery [1972], Palmer and Seery [1973], the book on emissions edited by Cornelius and Agnew [1972], and Lange [1972]. All of these investigators have made significant individual contributions. Seery and his collaborators at United Technology have made a number of structural studies related to nitrogen paths. One study by Seery and Zabielski [1977] was of CO flames with added NH_3 to trace NO formation routes. Other representative studies include those by Hayhurst with McLean [1974], Hayhurst and Vince [1977]. Bachmaier, Eberius, and Just [1973] detected HCN in NO formation in hydrocarbon flames. Blawnens, Smets, and Peeters [1977] also proposed a mechanism for NO formation. The chemical kinetic aspects have been reviewed by Hanson and Salimian [1982].

One positive development in the field is the thermal de-NOx process invented by Richard Lyon of Exxon Research and Engineering. This consists of injecting NH_3 into exhaust gases containing NO and the RAPENOx process, which consists of injecting HNCO into exhaust gases. Both processes convert nitrogen oxides to molecular nitrogen. The Sandia (Livermore) combustion group suggests (Miller and Fisk [1987]) the mechanisms of Table XIII–4.

Sulfur Oxides (SOx)

Sulfur oxides have become significant polluters of the atmosphere, both directly and through enhancing the production of oxides of nitrogen. Their primary combustion source is sulfur in the coal burned in power plants and sulfur in petroleum and natural gas stocks. Pollution from coal burning areas plagued London from the time of Edward I (1272–1307), who issued the first air pollution control policy enforced by beheading (Sawyer [1981]). By 1950 the use of natural gas virtually eliminated the problem. In other areas, however, sulfur oxides have become a major contributor to acid rain through the formation of sulfuric acid. Sulfur oxides and acids are difficult

Table XIII–4 NO_x Reduction Processes

Thermal De-NOx
$H + NH_3 \Rightarrow NH_2 + H_2$
$NH_2 + NO \Rightarrow NNH + OH$
$NNH \Rightarrow N_2 + H$
RAPENOx
$HNCO + \{heat\} \Rightarrow NH + CO$
$NH + NO \Rightarrow H + N_2O$
$H + HNCO \Rightarrow NH_2 + CO$
$NH_2 + NO \Rightarrow N_2 + H_2O$
$NH_2 + NO \Rightarrow N_2H + OH$
$N_2H \Rightarrow N_2 + H$
$OH + CO \Rightarrow CO_2 + H$

to remove from the gas phase, and the best action is fuel pretreatment. In the case of coal this consists of mixing dolomitic limestone in the fuel bed and separating out the residual sulfates in the ash. For hydrocarbons the "doctor" desulfurization processes has been successful. The application of these methods must be increased if we are to protect the environment.

Sulfur flame chemistry was discussed earlier. The groups under Wendt [1979], at the University of Arizona, Owen I. Smith [1982, 1983] Knuth at UCLA, Malte at the University of Washington (Seattle) Kramlich, Malte, and Grosshandler [1981], Chen, Malte, and Thornton, [1985], and Levy [1983] at Battelle all have made significant contributions.

Sulfur combustion, pollution, and atmospheric chemistry has been reviewed by Cullis and Mulcahy [1972], Palmer and Seery [1973], Levy [1983], and Muller, Schofield, Steinberg, and Broida [1979].

Lead and Engine Knock

Lead has become a pollution problem because of the use of tetraethyl lead [$Pb(C_2H_5)_4$] and other lead alkyl compounds as antiknocks. Tetraethyl lead acts as a suppressant delaying premature ignition of the charge in internal combustion engines. The exact mechanism of operation is a matter of debate. Flame structure studies have been peripheral since the automotive industry considers this to be a development rather than a research problem. Agnew [1985] has reviewed this from the automotive industry standpoint. As a result of recent legislation, lead is being phased out in the United States. It is hoped that lead pollution from this source will become a page in history.

Cook and Simmons [1979] at Manchester have made several studies of propane flames with added tetraethyl lead and found that it has a significant effect on the heat release rate (Fig. XIII-25). They ascribed the effects to reduction in radical concentrations, particularly OH induced by the decomposition of the tetraethyl lead around 800 K, that leads to the formation of lead oxide particles which below 1100 K reduce radicals by heterogeneous recombination processes. Above this temperature they believe branching is delayed by H atom scavenging through $PbO + H \Rightarrow Pb + OH$.

Rocket Propulsion

Rocket propulsion is an important practical problem. It has been approached by a number of groups, and flame structure studies have made contributions. Most of these studies lie outside the focus of this book and cannot be included. However, as experimental and modeling techniques improve, these areas should continue to blossom. One example of the contribution of flame structure studies is that of Smith and Thorne [1986] on the structure of an NO_2–hydrazine flame (Fig. XIII-14). Another good example of such efforts is given by the flame structure studies of Korbeinechev's group at Novosibersk on perchloric acid flames and HMX flames. Recently, Korbeinichev et al. (1991) studied composite flames using an ingenious "sandwich" model for solid propellants. This is an example of a number of studies of complex systems using flame structure measurement techniques.

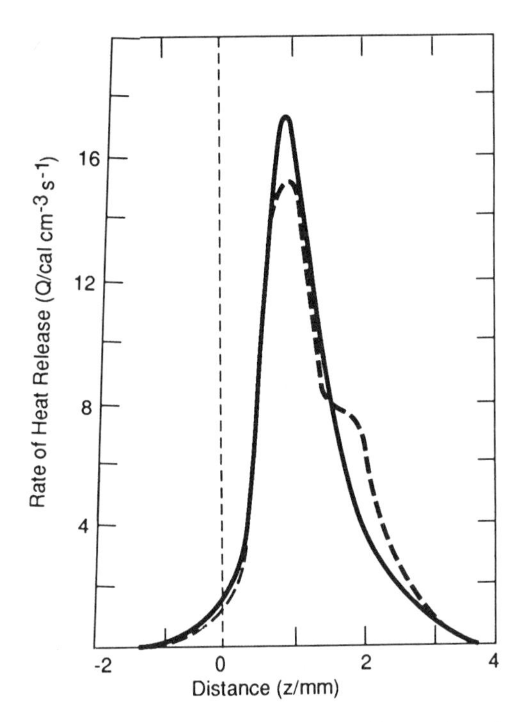

FIG. XIII-25 Effect of adding tetraethyl lead on the rate of heat release in a propane–air flame. (Cook and Simmons [1971].)

Incineration

The reduction of wastes by incineration is an important contribution of combustion science to society. Unfortunately, in recent years incineration has received a bad name because of the poor usage of overloaded, obsolescent facilities. During the past decade there has been a significant research effort supported primarily by the EPA and aimed at improving the situation. Seeker [1991], Senken [1987], and Exner [1982] have surveyed aspects of the effort with a book by Dillon [1981]. The journal *Hazardous Wastes* is devoted to this area.

Two examples of studies motivated by pollution problems are the study at UC Berkeley by Fisher, Koshland, Hall, Sawyer, and Lucas [1991] on the thermal decomposition of ethyl chloride, and that at IIT by Karra and Senkan [1987] on the structure of sooting mixtures of methyl chloride and methane flames (Chang and Senkin [1989]). They analyzed the structure to determine reaction paths for the destruction of chlorinated compounds.

One promising new technique is that of high pressure flames using supercritical water as a medium. These flames present a new challenge to the flame structure community.

APPENDIX A

TABLES OF PARAMETERS FOR SOME REPRESENTATIVE SYSTEMS

The following lists the tables included in this section:

Table A–1 Transport and Thermodynamic Parameters for Some Molecules

Parameters[a]	σ	ϵ/k	η	λ	D_{i,N_2}	α_{i,N_2}	H_i	G_i
Ar	3.45	116	223	4.2	0.20	0.05	0	0
HBr	3.35	112	185	23	0.12		−8.66	−12.8
Br_2	4.27	520	154	12	0.09		0	0
CO	3.71	88	165	60	0.22	0.00	−26.4	−32.0
CO_2	3.90	213	153	39	0.17	0.12	−94.0	−94.3
CH_4	3.80	144	112	80	0.23*	−0.07	−17.9	−12.2
HCl	3.30	360	144	35	0.14		−22.1	−22.8
CL_2	4.40	257	134	22	0.11		0	0
HF	3.15	330	111	51	0.13		−64.8	−65.3
F_2	3.65	112	233	66	0.20		0	0
H_2	2.92	38	85	434	0.78	−0.29	0	0
H_2O	2.64	809	90	40	0.24	−0.02	−57.8	−54.6
N_2	3.75	80	170	61	0.22	0.00	0	0
O_2	3.54	88	199	63	0.22	0.02	0	0
O_3	3.61	185			0.18		34.1	38.9
He	2.58	10	196	361	0.69	−0.33	0	0
NH_3	2.90	558	103	63	0.19			

Sources: *The Properties of Gases and Liquids,* by Reid, Prausnitz, and Poling [1977] (McGraw–Hill); *Molecular Theory of Gases and Liquids,* by Hirschfelder, Curtiss, and Bird [1964] (John Wiley); Fristrom and Monchick [1989].

[a] σ is the Lennard–Jones (LF) collision diameter in Angstroms from Hirschfelder, Curtiss, and Bird [1964]. The values for polar molecules HF, HCl, HBr, and H_2O are approximate values fitted neglecting polar contributions.

ϵ/k is the LF well depth in K. From Hirschfelder, Curtiss, and Bird [1964].

η is the viscosity in micropoise (10^{-8} g cm^{-1} s^{-1}) from the compilation of Yaws [1976].

λ is the thermal conductivity (10^{-5} cal cm^{-1} s^{-1} K) from experimental values collected by Fristrom and Westenberg [1965]. Values marked * were from the compilation by Yaws [1976].

D_{ij} is the binary diffusion coefficient (cm^2 s^{-1}) of the species with nitrogen under STP conditions taken from experimental values collected by Fristrom and Westenberg [1965]. Values marked * were calculated using the Svehela and McBride [1973] program.

α_{iN_2} is the binary thermal diffusion factor (dimensionless) for the species with nitrogen taken from the calculations of Fristrom and Monchick [1989].

H_i is the enthalpy of formation of the species (kcal/g mole) at STP taken from Stull and Prophet [1986].

G_i is the Gibbs energy of formation (kcal/g mole) at STP taken from Stull and Prophet [1986].

Values for LJ parameters of atoms and radicals can be estimated by assuming that the atoms are close to the nearest inert gas element, that OH and HO_2 are similar to water and methyl is similar to CH_4.

Table A–2a Conversion Factors for Second-Order Reactions

Converted Units and Given Units	cm^3/mol/sec	L/mol/sec	m^3/mol/sec	cm^3/molecule/sec	$(mm\ Hg)^{-1}\ s^{-1}$
cm^3/mol/sec	1	1000	1×10^{6}	6.023×10^{23}	62400 T
L/mol/sec	0.001	1	1000	6.023×10^{20}	62.40 T
m^3 mol/sec	1×10^{-6}	0.001	1	6.023×10^{17}	.0624 T
cm^3/molecule/sec	166×10^{-26}	166×10^{-23}	166×10^{-20}	1	1036×10^{-22} T
$(mm\ Hg)^{-1}\ s^{-1}$	$1603 \times 10^{-8}\ T^{-1}$	$0.01603\ T^{-1}$	$16.03\ T^{-1}$	$9653 \times 10^{15}\ T^{-1}$	1

Table A–2b Conversion Factors for Third-Order Reactions

Converted Units and Given Units	$cm^6\ mol^{-2}\ sec^{-1}$	$L^2\ mol^{-2}\ sec^{-1}$	$m^6\ mol^{-2}\ sec^{-1}$	$cm^6\ molecule^{-2}\ s^{-1}$	$(mm\ Hg)^{-2}\ sec^{-1}$
$cm^6\ mol^{-2}\ sec^{-1}$	1	1×10^{6}	1×10^{12}	3628×10^{44}	$3894 \times 10^{6}\ T^2$
$L^2\ mol^{-2}\ sec^{-1}$	1×10^{-6}	1	1×10^{-6}	3628×10^{38}	$3894\ T^2$
$m^6\ mol^{-2}\ sec^{-1}$	1×10^{-12}	1×10^{-6}	1	3628×10^{32}	$0.003894\ T^2$
$cm^{-6}\ molecule^2\ sec^{-1}$	276×10^{-50}	276×10^{-44}	276×10^{-38}	1	$107 \times 10^{-40}\ T^2$
$(mm\ Hg)^{-2}\ sec^{-1}$	$257 \times 10^{-12}\ T^{-2}$	$257 \times 10^{-6}\ T^{-2}$	$257\ T^{-2}$	$9318 \times 10^{34}\ T^{-2}$	1

Table A–3a Rate Coefficients for the Hydrogen–Halogen System*

Reaction	H_2–F_2	H_2–Cl_2	H_2–Br_2	H_2–I_2
$X_2 + M^* \rightarrow X + X + M$	0.21/17000	0.232/23600	2.35/21630	0.83/15300
$X + X + M \rightarrow X_2 + M^*$	10/0/0	2.23/−906	1.48/−856	2.36/−754
$H + X_2 \rightarrow HX + X$	0.88/1210	0.86/590	0.0023/1/220	4.3/−217
$X + HX \rightarrow H + X_2$	0.13/50700	1/23900	2.7/22350	0.5/18100
$X + H_2 \rightarrow H + HX$	0.15/0	0.15/2200	1.7/9650	1.7/16950
$H + HX \rightarrow H_2 + X$	0.27/15500	0.15/1750	700/−2/−200	0.45/330
$H + X + M \rightarrow HX + M^*$	0.08/−17700	0.25/−9950	0.76/−530	0.1/0
$HX + M^* \rightarrow H + X + M$	0.31/50000	0.45/41000	(1/44000)	(1/40000)
$H + H + M \rightarrow H_2 + M^*$	6500/−1/0	6500/−1/0	6500/−1/0	6500/−1/0
$H_2 + M^* \rightarrow H + H + M$	2.2/48500	2.2/48500	2.2/48500	2.2/48500

Table A–3b Rate Constants for Some CHO Reactions*

	Decomp.	H	O	OH	HO_2
H	—	a	c	d	
HO	24/50000	0.0005/275	0.12/−150	0.09/640	0.3/0
HO_2	13/23700	1.3/290	0.24/−60	0.22/−200	0.008/−180
H_2O	12/50000	0.97/10300	1.2/9500	—	—
O	—	c	f	—	0.24/0
O_2	3.4/58000	1.37/8016e	—	—	—
CH_2	—	0.56/−150	0.33/−150	0.18/0	0.2/0
CH_3	210/47000	200000(2)*	1/0	0.5/0	0.04/0
CH_4	0.09/46600	1.9/6110	1/5100	0.17/2500	0.04/10500
C_2H_2	E-5/31000	1/11600	0.22/1670	1.5/8500	100000/4000
C_2H_3	2.2/23400	0.19/−200	0.3/0	0.3/0	0.04/10500
C_2H_4	30/38000	1/6000	0.07/870	0.06/175	600000/4000
C_2H_5	0.07/19000	0.36/0	0.09/0	0.24/0	0.001/0
C_2H_6	26/36500	2/4850	0.05/2620	0.21/1470	0.01/8600
C_3H_6	2.3/37000	0.13/810	0.024/80	0.02/−250	0.04/8000
C_3H_8	42/38000	1.8/4350	1.8/3070	0.33/1150	0.06/8900
C_4H_{10}	0.29/30000	1/3950	3.5/3100	2.4/830	0.08/8800
HCO	1.4/4000	1.1/0	0.56/−130	0.4/0	0.3/0
OCH_2	1.2/8050	0.58/2280	1.2/2100	0.2/330	0.07/6200
CO	—	18(−0.7)*	1(0)*	a	3/12200
CO_2	0.09/65000	2.5/13300	0.27/26700	—	—

a–$CO+OH=>CO_2+H$ $\quad k=1.31\times10^{10}T^{1.81}\exp(590/T)$

THREE BODY REACTIONS

b–$H+H+M=>H_2+M^*$ $\quad k=7\times10^{15}T^{-0.875}$

c–$H+O+M=>OH+M^*$ $\quad k=1.6\times10^{16}T^{-1}$

d–$H+OH+M=>H_2O+M^*$ $\quad k=4.7\times10^{16}T^{-0.89}$

e–$H+O_2+M=>HO_2+M^*$ $\quad k=5.9\times10^{15}T^{-0.75}$

f–$O+O+M=>O_2+M^*$ $\quad k=2.2\times10^{15}T^{-0.89}$

g–$OH+OH=>H_2O_2+M^*$ $\quad k=5.6\times10^{16}T^{-2}$

*These are current rates from the NIST Reference Standard Data Program Data Chemical Kinetics Data Base 17 version 2.1 [1991] by Westley, Herron, Cvetanovic, Hampson, and Mallard. Rounded values given are for comparative purposes. *They are approximated and dated.* To save space and facilitate comparison most bimolecular reactions have been forced into the two-parameter fit $k = A\exp(-C/T)$, the exception being the OH + CO reaction and the three-body, where a three-parameter fit is used, $k = AT^B[\exp(-C/T)]$. This is also used for three-body recombinations. Recombinations are marked by *. Where there are competing channels the sum of all channels was chosen.

The format is: $A(B)/C$. The units of A are cm^{-3} (g)moles^{-1} sec^{-1} $\times\ 10^{-14}$, B is dimensionless, and C is in K. A conversion table to other units is given in Chapter X.

For serious work the most recent update should be consulted. This is available in compatible disk format from the National Institute of Standards and Technology (NIST) Standard Reference Data Program, Gaithersburg, MD 20899.

Table A–4 Physical and Combustion Properties for Some Representative Fuels with Air

Fuel	Formula	Specific Gravity Liquids	Boiling Point (K)	Stoichiometric: Quenching Distance (μm)	Stoichiometric: Min. Ignition Energy Joules × 10^5	Flammability Limit: Lean ER	Flammability Limit: Rich ER	Spontaneous Ignition Temperature (K)	Maximum Flame Velocity: ER (Equivalance Ratio)	Maximum Flame Velocity: V Max (m/sec)	Maximum Flame Velocity: Adiabatic Flame Temperature (K)
Acetaldehyde	CH_3,CHO	0.783	216	354	37.6						2288
Acetone	$(CH_3)_2CO$	0.792	330	591	115	0.59	2.33	834	1.31	0.49	2253
Acetylene	C_2H_2	0.621	225	118	3	0.39		578	1.33	1.63	2607
Allene (propadiene)	C_3H_4	0.658*	239						1.21	0.85	2544
Ammonia	NH_3	0.674*	240					965	1		2074
Benzene	C_6H_6	0.885	353	433	76	0.43	3.36	865	1.08	0.47	2363
n-Butane	C_4H_{10}	0.584	272	472		0.54	3.30	704	1.13	0.44	2248
1-Butene	C_4H_8	0.601	279			0.53	3.53	716	1.16	0.50	2305
1-Butyne	C_4H_6	0.650	281						1.20	0.67	2398
Carbon disulfide	CS_2	1.26	319	79	1.5	0.18	1.12	393	≈1.02	0.57	2257
Carbon monoxide	CO	0.803*	83			0.34	6.76	882	≈1.70	0.45	1388
Cyanogen	C_2N_3	0.866	252					1123	(1)		(2596)
Cyclohexane	C_6H_{12}	0.783	354	417	138	0.48	4.01	543	1.17	0.45	2250
Cyclopropane	C_3H_6	0.720	239	276	24	0.58	2.76	771	1.13	0.55	2370
n-Decane	$C_{10}H_{22}$	0.734	447			0.45	3.56	505	1.05	0.42	2286
Dimethyl ether	$(CH_3)_2O$	0.667*	249	354	45	0.50	3.30	623	1.19	0.53	2150
Dimethyl sulfide	$(CH_3)_2S$	0.846	311	472	76				(1)		(2251)
Di-*tert*-butyl peroxide	$C_6H_{18}O_2$			433	65						
Ethane	C_2H_6	0.548*	184	354	42	0.50	2.72	745	1.12	0.46	2244
Ethene	C_2H_4	0.547*	169	197	9.6	0.41	>6.1	763	1.15	0.79	2375
Ethyl acetate	$C_4H_9O_2$	0.901	350	669	142	0.61	2.36	759	≈1.00	0.37	
Ethyl alcohol	C_2H_5OH	0.789	351					665	1		2238
Ethylamine	$C_2H_5NH_2$	0.706	306	827	240						2270
Ethylene oxide	C_2H_4O	1.965	283	197	10.5			702	1.25	1.05	2177
n-Heptane	C_7H_{16}	0.688	371	591	115	0.53	4.50	520	1.22	.45	2214
n-Hexane	C_6H_{14}	0.664	341	551	95	0.51	4.00	534	1.17	.45	2238

continued

Table A–4 Physical and Combustion Properties for Some Representative Fuels with Air (continued)

				Stoichiometric		Flammability Limit			Maximum Flame Velocity		
Fuel	Formula	Specific Gravity Liquids	Boiling Point (K)	Quenching Distance (μm)	Min. Ignition Energy Joules × 10^5	Lean ER	Rich ER	Spontaneous Ignition Temperature (K)	ER (Equivalance Ratio)	V Max (m/sec)	Adiabatic Flame Temperature (K)
Hydrogen	H_2	0.0709*	20	98	2.0			844	≈1.70	3.06	2169
Hydrogen sulfide	H_2S	1.19	211	160	7.7			563	(1)		(2091)
Methane	CH_4	0.425*	107	390	33	0.46	1.64	905	1.06	.39	2236
Methyl alcohol	CH_3OH	0.793	337	280	21.5	0.48	4.08	743	≈1.01	.55	2222
Methyl formate	$C_2H_4O_2$	0.975	305	430	62				(1)		(2106)
n-Nonane	C_9H_{20}	0.722	424			0.47	4.34	507	(1)		(2286)
n-Octane	C_8H_{18}	0.707	399			0.51	4.25	513	(1)		(2276)
n-Pentane	C_5H_{12}	0.631	309	510	82	0.54	3.59	557	1.15	.44	2250
Propane	C_3H_8	0.508	231	310	30.5	0.51	2.83	777–	1.14	.45	2250
Propene	C_3H_6	0.522	225	310	28.2	0.48	2.72	831–	1.14	.50	2367
Propylene oxide (1,2-epoxy propane)	C_3H_6O	0.831	308	280	19	0.47			1.28	0.81	2275
1-Propyne	C_3H_4	0.706*	250	240					1.19	0.81	2544
Tetralin (tetrahydro naphtalene)	$C_{10}H_{12}$	0.971	480					696	1.01	0.38	2319
Toluene	$C_6H_5CH_3$	0.872	384			0.43	3.22	841–	1.05	0.40	2344
Turpentine (α-pinene)	$C_{10}H_{16}$	0.858	429					525			
Gasoline, 73-octane								299			
Gasoline, 100-octane								700–783	1.06	0.40	
Jet fuel JP-1[l]		0.81						522	1.07	0.39	
Jet fuel JP-3[l]		0.76									
Jet fuel JP-4[l]		0.78						534	1.07	0.40	
Jet fuel JP-5[l]		0.83						515			

Source: *Basic Considerations in the Combustion of Hydrocarbon Fuels with Air*, by Hibbard and Barnett, NACA Report 1300 [1959], with additions. Starred specific gravities are at the liquid boiling point taken from Reid, Prausnitz, and Poling [1987]. The adiabatic flame temperatures at the composition of maximum burning velocity were calculated by S. Favin of APL/JHU using a NASA program with JANAF thermodynamics.

Table A–5 Mass Transfer Driving Force "*B*" and Lower Oxygen Index "LOI" for Various Materials[a]

Liquids	*B*	LOI	Solids	*B*	LOI
Pentane	8.1	0.13	Polyethylene (PE) high den.	0.6	0.174
Decane	4.3	0.135	(PE) + 60% by wt. $Al_2O_3 \cdot 3H_2O$		0.205
Kerosene	3.9	0.13	Polypropylene (PP)	0.72	0.174
Diesel oil	3.9	0.13	(PP) Asbestos filled		0.3
Benzene	6.1	0.133	PMMA (Plexiglass)	1.55	0.174
Toluene	6.1	0.13	Polystyrene	0.45	0.181
Xylene	5.8	0.13	Polycarbonate (Lexan)		0.27
Methanol	2.7		P-Vinylfluoride (Tedlar)		0.226
Ethanol	3.3		Polyvinylchloride (PVC)	0.6	0.33
Acetone	5.1		Phenolic	0.09	
			Wood (maple)	0.15	

Sources: (for B) Spalding [1955] and (for LOI) Fenimore and Martin [1972].

[a]The numerator of Spalding's *B* index is approximately equal to the energy released by combustion per unit mass of *air* reacting, while the denominator is the energy required to vaporize unit mass of fuel. *r* is the stoichiometric mass fuel–air ratio. In the heat of combustion per unit mass of fuel vapor T_s is the surface temperature, T_s is ambient air temperature, *L* is the latent heat of vaporization of the fuel, T_i is the interior or bulk temperature of the liquid, C_c is the liquid specific heat.

The lower oxygen index (LOI) is defined as the minimum oxygen mole fraction that will support combustion in the LOI apparatus of Fenimore and Martin. (see Fig. II–). Values are taken from Fenimore, C., and Martin, F., "Burning of Polymers," in NBS Special Publication #357, L. Wall (Ed.), p. 159 (1972).

Table A–6 Adiabatic Flame Conditions for Some Representative Decomposition Flames[a]

N_2H_4	HN_3	NO	ClO_2
$T_F = 1339$ K	$T_F = 3369$ K	$T_F = 2665$ K	$T_F = 1765$ K
H, 2×10^{-6}	H, 0.165	NO, 0.035	Cl, 0.1313
H_2, 0.6666	H_2, 0.147	NO_2 4×10^{-5}	ClO, 0.0019
NH_3, 3.4×10^{-5}	N, 1×10^{-4}	N_2O, 2×10^{-6}	Cl_2, 0.245
N_2, 0.3333	NH, 2×10^{5}	O, 0.0211	O, 6.8×10^{-5}
	NH_2, 1×10^{-6}	O_2, 0.466	O_2, 0.622
	N_2, 0.688		

The source of the data of Tables A–7—A–14 are calculations by Stanley Favin of the Applied Physics Laboratory of the Johns Hopkins University. He used the NASA program by Svehela and McBride [1973]. The author would like to thank Mr. Favin for his many contributions to this book.

Table A–7 Adiabatic Combustion Conditions for Stoichiometric Hydrogen–Halogen Flames[a]

	$H_2 + F_2$	$H_2 + Cl_2$	$H_2 + Br_2$	$H_2 + I_2$
H	0.182	0.052	5×10^{-5}	$<10^{-8}$
H_2	0.0129	0.07	0.05	0.0824
X	0.208	0.07	0.08	7×10^{-6}
X_2	1×10^{-6}	4×10^{-4}	7×10^{-4}	0.0824
HX	0.0129	0.765	0.87	0.835
T_F	4006 K	2493 K	1544 K	545 K

[a]Pressure is 1 atm with initial temperature of chlorine and fluorine 298 K. For bromine and iodine the initial temperature was 400 K.

Table A–8 Adiabatic Combustion Properties of Some Sulfur-Containing Fuels with Oxygen[a]

Fuel	S	H_2S	CS_2	OCS
CO	—	—	0.18	0.2234
CO_2	—	—	0.066	0.1885
COS	—	—	1.7×10^{-5}	9.9×10^{-6}
H	—	0.053	—	—
H_2	—	0.065	—	—
HO	—	0.279	—	—
O	0.0786	0.04	0.122	0.0509
O_2	0.1092	0.082	0.138	0.1275
S	0.0215	0.067	0.194	0.00356
SO	0.245	0.003	0.175	0.0759
SO_2	0.544	0.300	0.298	0.331
SO_3	1.25×10^{-4}	SH = 0.002	6.9×10^{-5}	9.8×10^{-5}
S_2	0.00206	5×10^{-4}	9.6×10^{-4}	1.34×10^{-4}
S_2O	2.89×10^{-4}	—	1.2×10^{-4}	3.0×10^{-5}
T_F(K)	3234	3414	3348	3071

[a]Initial conditions are: initial temperature 298 K, pressure 1 atm. Sulfur is in the solid state, but the other fuels are gaseous. Concentrations are in mole fractions, temperatures are in K.

Table A–9 Adiabatic Combustion Properties of Some Nitrogen-Containing Fuels with Oxygen

Fuel	NH_3	N_2H_4	HCN
CO	—	—	0.269
CO_2	—	—	0.107
NO_2	$<10^{-6}$	$<10^{-6}$	6×10^{-6}
H	0.0281	0.059	0.057
HO_2	3×10^{-5}	5×10^{-5}	5×10^{-5}
H_2	0.087	0.087	0.0304
H_2O	0.559	0.402	0.093
N	2×10^{-6}	1×10^{-5}	3×10^{-7}
NO	0.0076	0.0125	0.00793
N_2	0.226	0.284	0.177
O	0.0107	0.0235	0.079
OH	0.054	0.074	0.0715
O_2	0.0278	0.338	0.0957
T_F(K)	2845	3037	3259

Table A–10 Cyanogen Flames with Oxygen and Ozone

System	$C_2N_2 + 2O_2$	$C_2N_2 + O_2$	$C_2N_2 + 2/3O_3$
C	10^{-9}	0.0045	0.012
CN	10^{-8}	0.00258	0.00421
CO	0.397	0.6514	0.631
CO_2	0.0867	3.8×10^{-5}	3.6×10^{-5}
C_2	$<10^{-9}$	1.8×10^{-5}	4.6×10^{-5}
N	$<10^{-9}$	0.0131	0.0291
NO	0.0302	3.1×10^{-4}	5.3×10^{-4}
N_2	0.2264	0.3213	0.309
O	0.154	0.0068	0.0158
O_2	0.1062	1.3×10^{-6}	3×10^{-6}
T(K)	3485	4855	5207

Table A–11 Adiabatic Flame Conditions for Some Volatile Hydrides with Oxygen

$B_2H_6 + 3\,O_2$ $T_F = 3350$ K	$P_{solid} + 5/4\,O_2$ $T_F = 3242$ K	$PH_3 + 11/4\,O_2$ $T_F = 3139$ K	$SiH_4 + 2\,O_2$ $T_F = 3043$ K
BO = 0.0545	O = 0.1084	H = 0.0609	H = 0.0435
BO_2 = 0.080	O_2 = 0.2000	$HO_2 = 8.4 \times 10^{-5}$	$HO_2 = 8.6 \times 10^{-5}$
B_2O_2 = 0.0004	$P = 2.8 \times 10^{-4}$	H_2 = 0.0665	H_2 = 0.0592
H = 0.148	PO = 0.161	H_2O = 0.304	H_2O = 0.3837
HBO = 0.0079	PO_2 = 0.531	O = 0.0572	O = 0.0438
HBO_2 = 0.137	$P_2 = 4.5 \times 10^{-6}$	OH = 0.105	OH = 0.0992
$HO_2 = 5 \times 10^{-5}$	—	O_2 = 0.104	O_2 = 0.1134
H_2 = 0.129	—	$P = 1.08 \times 10^{-4}$	SiO = 0.2393
H_2O = 0.200	—	$PH = 6 \times 10^{-6}$	SiO_2 = 0.0179
O = 0.067	—	PO = 0.0694	—
OH = 0.100	—	PO_2 = 0.2328	—
O_2 = 0.0409	—	—	—

Table A–12 Adiabatic Flame Conditions for Some Metals with Oxygen

$Li_{solid} + 1/4\,O_2$	$Sr_{solid} + 1/2\,O_2$	$Al_{solid} + 1/3\,O_2$	$Zr_{solid} + O_2$
Li = 0.0245	O = 0.1377	Al = 0.119	O = 0.2393
LiO = 00357	O_2 = 0.0335	AlO = 0.0217	O_2 = 0.01168
$Li_2 = 3.4 \times 10^{-4}$	Sr = 0.2048	$AlO_2 = 3 \times 10^{-5}$	$Zr = 8.5 \times 10^{-4}$
Li_2O = 0.493	SrO = 0.190	Al_2O_2 = 0.00772	ZrO = 0.261
$Li_2O_2 = 8.04 \times 10^{-4}$	SrO(L) = 0.434	O = 0.314	ZrO_2 = 0.3031
$O = 7.27 \times 10^{-3}$	—	O_2 = 0.0563	ZrO_2(L) = 0.1841
O_2 = 0.0449	—	Al_2O_3(L) = 0.2161	—
Li_2O(L) = 0.1655	—	—	—
T_F = 2711 K	T_F = 3807 K	T_F = 4005 K	T_F = 4278 K

Table A–13 Adiabatic Flame Conditions for Hydrogen with Some Nitrogen-Containing Oxidizers

	$H_2 + N_2O$	$H_2 + NO$	$H_2 + 1/2\ NO_2$	$H_2 + 2/5\ HNO_3$
H	0.0405	0.0805	0.0602	0.0343
HO_2	2.2×10^{-5}	4.1×10^{-5}	4.3×10^{-5}	3.3×10^{-5}
H_2	0.0828	0.1243	0.1238	0.0102
H_2O	0.3244	0.3523	0.492	0.6208
N	7×10^{-6}	1.5×10^{-5}	7×10^{-6}	$<10^{-7}$
NO	0.0122	0.0142	0.0102	0.00677
N_2	0.4476	0.2727	0.1669	0.1253
O	0.0156	0.033	0.0244	0.0134
OH	0.0525	0.0855	0.0833	0.0648
O_2	0.0244	0.0372	0.0388	0.0330
T_F	2965	3127	3028	2882

Table A–14 Adiabatic Combustion Conditions for Hydrogen with Some Chlorine Oxides and Acids

	ClO_2	$HClO_4$	$HClO_4 + 5\ H_2O$
Cl	0.0771	0.04	0.0172
ClO	1.2×10^{-4}	7×10^{-5}	3×10^{-5}
Cl_2	4.7×10^{-5}	2×10^{-5}	1×10^{-5}
H	0.0762	0.0537	0.0157
HCl	0.1972	0.1322	0.1139
HOCl	2.2×10^{-5}	1×10^{-5}	1×10^{-5}
HO_2	3.6×10^{-5}	4×10^{-5}	2×10^{-5}
H_2	0.1318	0.1222	0.0702
H_2O	0.3779	0.5195	0.7119
O	0.0277	0.0205	0.0056
OH	0.0801	0.0773	0.041
O_2	0.0316	0.035	0.0225
T_F (K)	3098	2992	2709

Table A–15 Structural Parameters for Two Ozone Decomposition Flames

	20% O_3	100% O_3
INLET STATE		
T_0 (K)	298	298
O_3	0.2	1
O_2	0.8	0
IGNITION STATE		
T_I (K)	680	885
O	7×10^{-5}	5×10^{-3}
O_2	0.989	0.335
O_3	0.09	0.66
PEAK RADICAL STATE		
T_R (K)	1070	2470
O	1.9×10^{-3}	0.065
O_2	0.998	0.935
FINAL ADIABADIC STATE		
T_F (K)	1110	2475
O	8×10^{-4}	0.060
O_2	0.999	0.94
v_0 (mm/sec)	370	4750
$(dT/dz)_I$ (K/mm)	290	2375

The source of the data of Tables A–15, A–19, and A–20 are calculations by Prof. Jurgen Warnatz of the University of Stuttgart, Germany. The author would like to thank Prof. Warnatz for permission to use them.

Table A–16 The Effects of Stoichiometry, Temperature, and Pressure on H_2–Br_2 Flames

	Lean	ER = 1	Rich	Temp.	Press.
v_0	9.0	22.5	31.3	51	57
zs	8E-3	7E-3	8E-3	8E-3	9E-2
INLET STATE					
P/T_0	1/323	1/323	1/323	1/400	0.1/300
H_2	0.4	0.505	0.65	0.65	0.65
Br_2	0.6	0.495	0.35	0.35	0.35
IGNITION STATE					
T_I	1052	1089	992	1007	916
H_2	0.081	0.13	0.335	0.32	0.36
Br_2	0.65	0.44	0.30	0.31	0.30
HBr	0.27	0.43	0.365	0.37	0.34
PEAK RADICAL STATE					
T_R	1318	1456	1357	1404	1296
H	4E-6	5E-5	8E-5	1E-4	6E-5
H_2	0.0076	0.057	0.31	0.31	0.31
Br	0.055	0.098	0.03	0.036	0.02
Br_2	0.18	4E-4	2E-5	3E-5	2E-5
HBr	0.76	0.85	0.66	0.656	0.67
FINAL ADIABADIC STATE					
T_F	1406	1544	1445	1491	1384
H	4E-6	5E-5	4E-5	7E-5	6E-5
H_2	8E-3	0.05	0.3	0.31	0.31
Br	0.055	0.08	0.011	0.016	0.02
Br_2	0.18	7E-4	4E-5	5E-5	2E-5
HBr	0.76	0.87	0.69	0.68	0.67

The source of these data is calculations by Fristrom, Favin, Linevsky, Vandooren, and Van Tiggelen [1992].

Table A–17 Structure Parameters for Hydrogen–Air Flames[a]

INLET STATE						
$[H_2]_O$	0.7	0.6	0.5	0.41	0.3	0.2
v_{0expt} (m/sec)	0.78	—	2.29	2.54	2.00	0.87
v_{0calc} (m/sec)	0.78	1.63	2.29	2.54	2.00	0.87
PEAK RADICAL STATE						
$[H]_{max} \times 100$	1.25	3.59	6.18	7.52	4.63	1.02
T(K)	1199	1343	1457	1538	1556	1392
$[O]_{max} \times 10^6$	6.56	45	174	458	1004	677
T(K)	1070	1172	1277	1384	1573	1478
$[OH]_{max} \times 10^5$	5.2	33.5	158	600	1526	793
T(K)	1040	1247	1602	1767	1984	1635
$[HO_2]_{max} \times 10^4$	3.4	5.16	5.88	6.09	6.5	6.11
T(K)	435	363	345	346	341	375
FINAL ADIABADIC STATE						
T_F (K)	1327	1640	1937	2185	2392	1832
H	$<10^{-6}$	0.00006	0.00521	0.00221	0.00021	$<10^{-6}$
H_2	0.613	0.472	0.324	0.185	0.193	0.00006
O	$<10^{-6}$	$<10^{-6}$	$<10^{-6}$	$<10^{-6}$	0.00048	0.00006
O_2	$<10^{-6}$	$<10^{-6}$	$<10^{-6}$	$<10^{-6}$	0.00334	0.0736
OH	$<10^{-6}$	$<10^{-6}$	0.00004	0.00047	0.00662	0.00040
H_2O	0.134	0.183	0.234	0.281	0.325	0.222

The source of the data of Tables A–17 and A–18 are calculations by Prof. G. Dixon-Lewis and co-workers at Leeds University, UK. Adiabatic values were added using calculations by S. Favin, APL/JHU. The author would like to thank Prof. Dixon-Lewis for permission to use the data.

[a]After Dixon-Lewis and Williams [1977].

Table A–18 Comparisons of Burning Velocity Calculations with Experiment[a]
INLET STATE

	60% Fuel/40% Air			40% Fuel/60% Air		
H_2/CO	0.435	0.117	0.00553	0.435	0.117	0.00553
v_{0expt}	1.51	0.68	0.21	2.17	0.71	0.21
v_{0calc}	1.37	0.73	0.195	2.03	0.69	0.196
PEAK RADICAL STATE						
$T(H_{max})$	1350	1475	1588	1582	1645	≈ 1780
H_{max}	0.030	0.0209	0.0003	>0.055	0.018	0.00155
O_{max}	0.0003	0.0014	0.00178	0.0061	0.0115	0.0085
OH_{max}	0.0003	0.00091	0.00019	0.0071	0.0039	0.00061
O_{max}	0.00058	0.0030	0.0027	0.0065	0.0115	0.0085
OH_{max}	0.00044	0.0010	0.0002	0.0084	0.0051	>0.00085
FINAL ADIABADIC STATE						
T_F	1676	1706	1725	2228	2310	2331
H_F	3.8×10^{-5}	2.8×10^{-5}	7×10^{-6}	0.0019	0.00073	0.000169
O_F	$<10^{-6}$	$<10^{-6}$	$<10^{-6}$	1×10^{-5}	9.4×10^{-5}	0.000136
OH_F	1.6×10^{-6}	1.7×10^{-6}	$<10^{-6}$	8.1×10^{-4}	9.9×10^{-4}	0.000271
H_{2F}	0.108	0.320	0.0015	0.080	0.0051	0.000223
H_2O_F	0.906	0.365	0.0021	0.237	0.420	0.00208
CO_F	0.364	0.440	0.470	0.0891	0.166	0.1715
CO_{2F}	0.0925	0.147	0.181	0.0495	0.243	0.2831
O_{2F}	$<10^{-6}$	$<10^{-6}$	$<10^{-6}$	1.6×10^{-5}	0.00032	0.00053

[a]Velocities are in meters per sec, temperatures are in K, and concentrations in mole fractions (or atm). Values are given for the peak concentrations of intermediates with the corresponding temperature. The initial temperature for all flames was 298 K and the pressure 1 atm. Experimental burning velocities were by Scholte and Vaags [1959].

Table A–19 STP Stoichiometric Atmospheric Air–Hydrocarbon Flames
INLET STATE

	$CH_4 = 0.0947$	$C_2H_6 = 0.0564$	$C_3H_8 = 0.0402$	$C_4H_{10} = 0.0312$
v_0(m/sec)	0.43	0.50	0.49	0.44
PEAK INTERMEDIATE STATE				
HO_2/T	$2 \times 10^{-4}/850$	$7 \times 10^{-4}/970$	$5 \times 10^{-4}/830$	$5 \times 10^{-4}/940$
H/T	0.0083/1910	0.010/1865	0.0115/1860	0.013/1870
OH/T	0.0083/1910	>0.009/>1900	>0.009/>1950	>0.01/>1900
O/T	0.0044/1910	0.0062/1865	0.0066/1860	0.0067/1870
H_2/T	0.017/1525	0.021/1450	0.016/1850	0.014/1460
CO/T	0.046/1730	0.054/1750	0.060/1740	0.059/1785
CH_4/T	—	0.0019/1350	0.0033/1210	0.0023/1295
CH_3/T	0.0036/1580	0.0025/1550	0.0030/1524	0.0064/1540
CH_2/T	$9 \times 10^{-6}/1825$	$3 \times 10^{-5}/1745$	$2 \times 10^{-5}/1750$	$2 \times 10^{-5}/1745$
CH_2O/T	0.0011/1625	0.00095/1550	0.0012/1580	0.0015/1605
CHO/T	$1.6 \times 10^{-5}/1664$	$2 \times 10^{-5}/1680$	$2 \times 10^{-5}/1670$	$2.3 \times 10^{-5}/1700$
C_2H_6/T	$1.8 \times 10^{-3}/1130$	—	0.0018/1180	0.0025/1215
C_2H_5/T	$1.1 \times 10^{-4}/1440$	$9 \times 10^{-4}/1385$	$2 \times 10^{-4}/1370$	$3.5 \times 10^{-4}/1365$
C_2H_4/T	0.0011/1450	0.005/1390	0.0033/1250	0.0038/1365
C_2H_3/T	$2.5 \times 10^{-4}/1450$	0.0012/1550	0.0013/1510	0.0017/1490
C_2H_2/T	$7 \times 10^{-4}/1450$	0.0021/1550	0.0013/1510	0.0017/1490
C_3H_8/T	$5 \times 10^{-4}/1310$	0.0024/1255	—	0.0013/1215
C_3H_7/T	$2 \times 10^{-7}/850$	$1 \times 10^{-6}/850$	$3 \times 10^{-5}/830$	$9 \times 10^{-7}/860$
C_3H_6/T	$3 \times 10^{-4}/1320$	0.0015/1455	0.0052/830	0.0056/1335
$\Sigma C_4/T$	$6 \times 10^{-6}/900$	0.0005/1200	$3 \times 10^{-5}/1170$	—
FINAL ADIABADIC STATE				
H/T_F	$3.9 \times 10^{-4}/2226$	$4.6 \times 10^{-4}/2260$	$4.7 \times 10^{-4}/2267$	$4.5 \times 10^{-4}/2266$
O/T_F	$2.2 \times 10^{-4}/2226$	$2.9 \times 10^{-4}/2260$	$3.1 \times 10^{-4}/2267$	$3.1 \times 10^{-4}/2266$
OH/T_F	0.00293/2226	0.00327/2260	0.00329/2267	0.00322/2266
H_2/T_F	0.0036/2226	0.00347/2260	0.00331/2267	0.00315/2266
H_2O/T_F	0.183/2226	0.1580/2260	0.1480/2267	0.1428/2266
O_2/T_F	0.0046/2226	0.00559/2260	0.00567/2267	0.00592/2266
CO/T_F	0.00895/2226	0.01163/2260	0.0125/2267	0.01272/2266
CO_2/T_F	0.0854/2226	0.09752/2260	0.1027/2267	0.1058/2266
NO/T_F	0.00224/2226	0.00187/2260	0.00233/2267	0.00234/2266

Table A–20 Some Atmospheric Stoichiometric Unsaturated Hydrocarbon–Air Flames[a]

INLET STATE	C_2H_2 = 0.077	C_2H_4 = 0.065	C_3H_6 = 0.044	C_4H_8 = 0.034
v_0 (m/sec)	0.80	0.57	0.56	0.55
PEAK INTERMEDIATE STATE				
HO_2/T	—	0.0013/925	0.0005/870	0.0006/860
H/T	—	0.013/1925	0.012/1900	0.012/1880
OH/T	—	0.01/1925	0.10/>1995	0.01/>1900
O/T	—	0.009/1925	0.0083/1900	0.0077/1880
H_2/T	—	0.015/1270	0.014/1550	0.012/1520
CO/T	—	0.069/1835	0.072/1800	0.069/1765
CH_4/T	—	0.002/1280	0.0032/1275	0.0026/1270
CH_3/T	—	0.003/1530	0.0032/1275	0.0026/1270
CH_2/T	—	8×10^{-5}/1800	1.6×10^{-5}/1800	10^{-5}/1780
CH_2O/T	—	0.0024/1000	0.0016/1580	0.0012/1610
CHO/T	—	4×10^{-5}/1530	3×10^{-5}/1740	10^{-5}/1660
C_2H_6/T	—	0.0016/1215	0.0025/1165	0.002/1115
C_2H_5/T	—	8×10^{-5}/1215	2×10^{-4}/1425	0.0002/1300
C_2H_4/T	—	—	0.0025/1420	0.0025/1425
C_2H_3/T	—	0.0013/1840	0.0010/1620	0.0015/1535
C_2H_2/T	—	0.005/1840	0.0010/1620	0.0015/1535
C_3H_8/T	—	0.0021/1240	0.0008/1280	0.0015/1175
C_3H_7/T	—	5×10^{-7}/975	2×10^{-5}/960	1×10^{-6}/850
C_3H_6/T	—	0.001/1405	—	0.0035/1065
$\Sigma C_4/T$	—	0.00035/1200	1.5×10^{-5}/1280	—
FINAL ADIABADIC STATE				
H/T_F	0.00186/2540	0.000859/2370	0.000777/2355	0.0006/2317
O/T_F	0.00224/2540	0.000694/2370	0.000622/2355	0.00047/2317
OH/T_F	0.00734/2540	0.00481/2370	0.00452/2355	0.00383/2317
H_2/T_F	0.00391/2540	0.00393/2370	0.00373/2355	0.00326/2317
H_2O/T_F	0.0698/2540	0.1219/2370	0.1224/2355	0.1236/2317
O_2/T_F	0.1644/2540	0.00896/2370	0.00851/2355	0.00745/2317
CO/T_F	0.4060/2540	0.2022/2370	0.1916/2355	0.1666/2317
CO_2/T_F	0.1163/2540	0.1087/2370	0.1099/2355	0.1127/2317
NO/T_F	0.00662/2540	0.00357/2370	0.00337/2355	0.00293/2317

[a]In this table velocities are in meters per second, temperatures are in K, and concentrations in mole fractions (or atm). Values are given for the peak concentrations of intermediates with the corresponding temperature. The values below T_F are the adiabatic equilibrium values. The initial temperature of all flames was 298 K and the pressure 1 atm. Burning velocities were calculated. Nitrogen is treated as a diluent. To estimate the importance of this approximation, the adiabatic concentration of NO at equilibrium is included. NO is the most important nitrogen compound.

APPENDIX B

REFERENCES ON BURNING VELOCITIES

Table B–1E References on Burning Velocities of Decomposition and Hydrogen–Halogen Flames

Flame (Figure, Table)	Reference
O_3 (Table XI–2)	Streng and Grosse [1957]
N_2H_4 (Fig. XI-1)	Gray, Lee, Leach, and Taylor [1957]; Adams and Cook [1960]
NH_3 (Fig. XI-3)	Laffitte, Hajal, and Combourieu [1965]
ClO_2 (Fig. XI-4)	Lafitte, Combourieu, Hajal, Ben Caid, and Moreau [1967]
$HClO_4$	Cummings and Pearson [1965]
C_2H_2, C_3H_4 (Table XI–5)	Cummings, Hall, and Straker [1962]
H_2–F_2 (Fig. XI-7)	Grosse and Kirshenbaum [1955]; Slootmaekers and Van Tiggelen [1958]; Vanpee, Cashin, Falabella, and Chintappili [1973]
H_2–Cl_2 (Fig. XI-7)	Bartholeme [1949B]; Rozlovski [1956]; Corbeels and Scheller [1965]
H_2–Br_2 (Fig. XI-7)	Cooley and Anderson [1952, 1955]; Cooley, Lasater, and Anderson [1952]; Garrison, Lasater, and Anderson [1949]; Huffstutler, Rode, and Anderson [1955]
Cl_2–F_2	Fletcher and Ambs [1968]; Ambs and Fletcher [1971]
c-C_4F_8–O_2	Fletcher and Kittelson [1968A, 1969]
C_2N_2–F_2	Vanpee, Vidaud, and Cashin [1974]

Table B–2 References on Burning Velocity of C/H/O Flames

Fuel	Figures	References
H_2	4,5,6	1, 3, 4, 8, 18, 20, 21, 24, 25, 28, 30, 31, 35, 39, 43, 44, 45, 52, 53, 62
CO	9,10	3, 11, 18, 32, 35, 52, 60, 63
OCH_2		15, 32, 46
CH_4	18	1, 2, 5, 7, 10, 12, 16, 19, 21, 22, 24, 26, 27, 29, 36, 40, 45, 47, 51, 55, 57, 58, 59, 61
C_2H_2	25	3, 8, 19, 23, 26, 27, 30, 31, 42, 50, 52, 56, 58, 59
C_2H_4	19	19, 26, 41, 59
C_2H_6	18	19, 26, 52, 59
C_3/C_4	18	19, 25, 26, 45, 52, 59
$>C_4$	34, 34	19, 25, 26, 30, 59
Alcohols	39	33, 44
Aromatics	39	33

[1] Agnew and Graiff [1961]
[2] Agrawal [1981]
[3] Andrews and Bradley [1972a,b]
[4] Andrews and Bradley [1973 a,b]
[5] Babkin and Kozachenko [1966]
[6] Badami and Egerton [1955]
[7] Barassin, A., Lisbet, R., Combourieu, J. and Laffitte [1967]
[8] Bartholeme [1939 a,b]
[9] Bone and Townend [1927]
[10] Bradley and Hundy [1971].
[11] Cherian, Rhodes, Simpson, and Dixon-Lewis [1981]
[12] Clingman and Pease [1955]
[13] Combourieu and Laffitte [1969].
[14] Cullsahw and Garside [1944]
[15] De Wilde and Van Tiggelen
[16] Diedrichsen and Wolfhard [1956]
[17] Dixon-Lewis [1970]
[18] Dixon-Lewis and Williams [1977]
[19] Dugger [1952,1959]
[20] Edse and Lawrence [1956]
[21] Edmunson and Heap [1988]
[22] Egerton and Lefebvre [1954].
[23] Friedman and Burke [1951]
[24] Garforth and Rallis [1973]
[25] Gerstein, Levine and Wong [1951a,b]
[26] Gibbs and Calcote [1959]
[27] Gilbert [1957]
[28] Gray and Smith [1980]
[29] Gulder [1982]
[30] Gunther and Janish [1972,1974]
[31] Gunther [1982]
[32] Hall, McCoubry and Wolfhard [1957]
[33]Hartman [1931]
[34] Hirano, Oda, Hirano and Akita [1982]
[35] Jahn [1934]
[36] Karpov and Sokolik [1961]
[37] Kaskan [1957,1961]
[38] Kuehl [1961]
[39] Kumar, Taman and Harrison [1983]
[40] Lindow [1968]
[41] Linnett and Hoare [1948]
[42] Linnett, Pickering, and Wheatley [1951]
[43] Liu and MacFarland [1983]
[44] Metghalchi and Keck [1980,1982]
[45] Milton and Keck [1984]
[46] Olenhove de Guertechin [1984]
[47] Passauer [1958]
[48] Price and Potter [1967]
[49] Raezer and Olson [1962]
[50] Rallie, Garforth and Steinz [1965]
[51] Reids, Mineur and McNaughton [1971]
[52] Schott and Vaags [1959]
[53] Senior [1961]
[54] Sharma, Agrawlet and Gupta [1981]
[55] Singer [1953]
[56] Smith [1937]
[57] Strauss, W. and Esde, R. [1959]
[58] Tsatsoranus [1978]
[59] Warnatz [1982] [1985]
[60] Wires, Watermeyer and Strehlow [1959]
[61] Yamaoka and Tsuji [1984]
[62] Yu, Law and Wu [1980]
[63] Yumlu [1967]

Table B–3 References on Burning Velocities of Non-C/H/O Fuels with Oxygen or Air

Fuel	Figure	Reference
H_2S	XIII-2	Vetter and Culick [1977]
B_2H_6, B_5H_9	XIII-4	Berl, Gayhart, Maier, Olsen, and Renich [1957]
NH_3	XIII-7	Murray and Hall [1951]
N_2H_4	XIII-8	Gray, Mackinven, and Smith [1967]
HCN	XIII-9	Cohen and Simpson [1957]
CS_2	XIII-10	Vetter and Culick [1977]

Table B–4 References on Burning Velocities of Nonoxygen Oxidizers

Oxidizer	Fuel	Figure	Reference
N_2O	H_2, C_2H_2, C_2H_4	XIII-10	Parker and Wolfhard [1949]
	CO	XIII-10	Simpson and Linnett [1957]
	NH_3, N_2H_4	XIII-8	Gray, Mackinven, and Smith [1957]
NO	CH_4, C_2H_4	XIII-10	Parker and Wolfhard [1949]
	NH_3, N_2H_4	XIII-8	Gray, Mackinven, and Smith [1967]
NO_2	H_2	XIII-10	Duval and Van Tiggelen [1967]
	CO	XIII-10	Simpson and Linnett [1957]
	C_2H_2, C_2H_4	XIII-10	Parker and Wolfhard [1949]
ClO_2	CH_4, C_2H_2	XIII-12	Combourieu, Moreau, Hall, and Pearson [1969]
$HClO_4$	H_2, CO, CH_4, C_2H_6, CH_3OH	XIII-13	Cummings and Hall [1965]

APPENDIX C

ABBREVIATIONS USED IN BIBLIOGRAPHY

References are listed using the abbreviations of *Chemical Abstracts* with the common exceptions listed here.

APZ	Annalen der Physik, Leipzeig
C&F	Combustion and Flame
CST	Combustion Science and Technology
CJC	Canadian Journal of Chemical Society
JACS	Journal of the American Chemical Society
JCS	Journal of the Chemical Society
JPC	Journal of Physical Chemistry
JCP	Journal of Chemical Physics
JFM	Journal of Fluid Mechanics
TFS	Transactions of the Faraday Society
PRS	Proceedings of the Royal Society (London) A
PPS	Proceedings of the Physical Society
ZfE	Zeitschrift für Elekrochemie
ZpC.	Zeitschrift für Physikalische Chemie
Buns.	Berichte der Bunsen Gesellschaft für Physikalische Chemie
1st Symp. to 24th Symp.	These refer to the International Symposia on Combustion sponsored biannually since 1954 by the Combustion Institute. The first two were sponsored by the American Chemical Society and published in *Industrial and Engineering Chemistry* and *Chemical Reviews,* respectively. They were reprinted in a single volume by the Combustion Institute in 1965. The third, fourth, and eighth symposia were published by Williams, Wilkins and Co., Baltimore; the fifth and sixth were published by Reinhold, New York; the ninth was published by Academic Press, New York; and the tenth and subsequent symposia were publsihed by the Combustion Institute, 5001 Baum Blvd., Pittsburgh, PA 15213. The series has been reprinted by the Combustion Institute. The full designation is: *Nth Symposium (International) on Combustion.*

BIBLIOGRAPHY

Books on Combustion Topics

Afgan, N., and Beer, J., (Eds.), *Heat Transfer in Flames,* John Wiley & Sons, New York (1974).

Barnard, J., and Bradley, J., *Flame and Combustion Phenomena,* 2nd ed., Methuen, London (1985).

Beams, J. W., *Physical Measurements in Gas Dynamics and Combustion,* Oxford Univ. Press, NY (1955).

Beer, J. M., and Chigier, N. A., *Combustion Aerodynamics,* Applied Science, London (1972).

Beer, J., and Thring M. W., (Eds), *Industrial Flame,* Vol. I, Edward Arnold, London, UK (1972).

Benson, R. S., and Whitehouse, N. D., *Internal Combustion Engines,* Vols. 1 & 2, Pergamon Press, New York (1979).

Benson, W. W., *Thermochemical Kinetics,* John Wiley & Sons, New York (1968).

Berl, W. G. (Ed.), *The Use of Models in Fire Research,* Publication 5786, National Academy of Sciences/National Research Council, Washington, D.C. (1961).

Bird, R. B., Stewart, W. E., and Lightfoot, E. N., *Transport Phenomena,* Wiley & Sons, New York (1962).

Blackshear, P., *Heat Transfer in Fires,* John Wiley & Sons, New York (1974).

Bone, W. A., and Townend, D. T. A., *Flame and Combustion in Gases,* Longmans Green & Co., New York (1927).

Bowman, C. T., and Birkeland, J. (Eds.), *Alternative Hydrocarbon Fuels: Combustion and Chemical Kinetics,* American Institute of Aeronautics and Astronautics, New York (1978).

Boyle, R., *The Skeptical Chymist,* Reprint, Dutton, New York (1967).

Bradley, J., *Shock Waves in Chemistry and Physics,* Methuen (1962).

Bradley, J., *Flame and Combustion Phenomena,* Methuen, London (1969).

Bragg, S. L., *Rocket Engines,* Geo. Newnes Ltd., London (1962).

Brame, J. S. S., and King, J. G., *Fuel,* Edward Arnold, London (1955).

Brown, S. C., *Benjamin Thompson, Count Rumford,* MIT Press, Cambridge, MA (1979).

Buckmaster, J. D., *The Mathematics of Combustion* (Vol. 2 of *Frontiers in Applied Mathematics*), Society for Industrial and Applied Mathematics, Philadelphia, PA (1985).

Buckmaster, J. D., and Ludford, G. S. S., *Theory of Laminar Flames,* Cambridge University Press, Cambridge, UK (1985).

Burstall, A. F., *A History of Mechanical Engineering,* MIT Press, Cambridge, MA (1965).

Charsley, E. L., and Warrington, S. B., *Thermal Analysis: Techniques and Applications,* CRC Press, Boca Raton, FL (1993).

Chedaille, J., and Braud, Y., "Measurements in Flames," in Vol. 1 of *Industrial Flames,* J. M. Beer and M. W. Thring (Eds.), Edward Arnold, London (1972).

Chigier, N. A. (Ed.), *Pollution Formation and Destruction in Flames,* Vol. 1, *Progress in Energy and Combustion Science,* Pergamon Press, London (1976).

Chigier, N. A., *Energy from Fossil Fuels and Geothermal Energy,* Vol. 2, *Progress in Energy and Combustion Science,* Pergamon Press, New York (1977).

Chigier, N.A. (Ed.), *Energy and Combustion Science,* Selected Papers from *Progress in Energy and Combustion Science,* Student ed. 1, Pergamon Press, London (1979).

Chigier, N. A., *Conservation and Combustion,* McGraw–Hill, New York (1980).

Chigier, N. (Ed.), *Combustion Measurements,* Hemisphere Publ. Co., Bristol, PA (1991).

Chomiak, J., *Combustion—A Study in Theory, Fact and Application,* Abacus Press, Gordon and Breach, NY (1990).

Compressed Gas Association, *Handbook of Compressed Gases,* 2nd ed., Van Nostrand Reinhold, New York (1981).

Cornelius, W., and Agnew, W. (Eds.), *Emissions from Continuous Combustion Systems,* Plenum Press, New York (1972).

Crosley, D. R. (Ed.), *Laser Probes for Combustion Chemistry,* American Chemical Society Symposium Series No. 134, American Chemical Society, Washington, D.C. (1980).

Cullis, C. F., and Hirschler, M. M., *The Combustion of Organic Polymers,* Clarendon Press, Oxford (1981).

Dalton, J., *New System of Chemistry,* Vol. 12, facsimile ed., William Dawson & Sons, London (1975).

Davy, H., *Collected Works,* Vol. VI (1840) pp. 3–130.01.

Demidov, P. G., *Combustion and Properties of Combustible Substances,* Wright Patterson AFB, AD 621738, Federal Clearing House for Scientific and Technical Information, Springfield, Va. (1956) (trans. from Russian).

DeSoete, G., and Feugier, A., *Aspects physiques et chimiques de la combustion,* Technip, Paris (1976).

Dillion, A. (Ed.), *Hazardous Waste Incineration Engineering,* Noyes Data Corp., Park Ridge, NY (1981).

Dolezal, R., *Large Boiler Furnaces,* Elsevier, New York (1971).

Dosanji, C. (Ed.), *Modern Optical Methods in Gas Dynamics Research,* Plenum Press, New York (1971).

Demtroder, W., *Analytical Spectroscopy Laser Spectroscopy,* Springer-Verlag (1983).

Drain, L. E., *The Laser Doppler Technique,* John Wiley and Sons, N. Y. (1980). 04-51.

Ducarme, J., Gerstein, M., and Lefebvre, A. H., *Progress in Combustion Science,* Vol. 1, Pergamon Press, London (1960).

Durao, D. F. G., Heitor, M. V., Whitelaw, J., and Witze, P. O., *Combustion Flow Diagnostics,* Kluwer Academic Publ., Norwell, MA (1991).

Eckert, E. R. G., and Drake, R. M., *Heat and Mass Transfer,* McGraw–Hill, New York (1959).

Edwards, J. B., *Combustion,* Ann Arbor Science, Ann Arbor, Mich. (1974).

Faraday, M., *Chemical History of A Candle,* Viking Press, New York (1960).

Faraday, M., *Experimental Researches in Chemistry and Physics,* Reprint, Cultureiet Civilization, Brussels (1969).

Fenimore, C. P., *Chemistry in Premixed Flames,* Pergamon Press, New York (1964).

Ferguson, C., *Internal Combustion Engines,* John Wiley & Co., New York (1986).

Ferri, A. (Ed.), *Fundamental Data Obtained From Shock Tube Experiments,* Pergamon Press, NY (1961).

Fickett, W., and Davis, W., *Detonation,* Univ. of California Press, Berkeley, CA (1976).

Field, M. A., Gill, D. W., Morgan, B. B., and Hawksley, F. G. W., *Combustion of Pulverized Coal,* Institute of Fuel, London (1967).

Forsythe, W. E., *Optical Pyrometry, Temperature—Its Measurement and Control In Science and Industry,* Reinhold Publ. Co., New York (1941), p. 1115.

Francis, W., *Fuels and Fuel Technology,* Vols. 1 & 2, Pergamon Press, London (1965).
Fristrom, R. M., and Westenberg, A. A., *Flame Structure,* McGraw–Hill, New York (1965).
Gann, R., *Halogenated Fire Suppressants,* ACS Series No. 6, American Chemical Society, Washington, D.C. (1975).
Gardner, W. C., *Chemistry of Combustion Reactions,* Springer-Verlag, New York (1983).
Gaydon, A. G., *Spectroscopy and Combustion Theory,* Chapman–Hall, London (1948).
Gaydon, A. G., *Spectroscopy of Flames,* Chapman and Hall, London, 2nd ed. (1974); 4th ed. (1979).
Gaydon, A. G., and Hurle, I., *The Shock Tube in High Temperature Chemical Physics,* Chapman–Hall, London (1963).
Gaydon, A. G., and Wolfhard, H. G., *Flames,* 4th ed., Chapman–Hall, London (1979).
Glassman, I., *Combustion,* Academic Press, New York (1977).
Glassman, I., and Sawyer, R. F., *Performance of Chemical Propellants,* AGARDOGRAPH No. 129, Pergamon Press, Oxford (1974).
Goldsmith, S. (Ed.), *Modern Developments in Fluid Dynamics,* Clarendon Press, Oxford (1938).
Goodger, E. M., *Combustion Calculations,* Macmillan, London (1977).
Goulard, R., *Combustion Measurements,* Academic Press, New York (1976).
Green, E. F., and Toennies, P., *Chemical Reactions in Shock Waves,* Arnold (1964).
Gruschka, H., and Wecken, F., *Gas Dynamic Theory of Detonation,* Gordon and Breach, NY (1971).
Gunther, R., *Verbrennung und Feuerungen,* Springer–Verlag, Berlin, West Germany (1974).
Harker, J. H., and Allen, D. A., *Fuel Science,* Oliver & Boyd, Edinburgh, Scotland (1972).
Hastie, J., *High Temperature Vapors,* Academic Press, New York (1975).
Hermann, R., and Alkamade, C., *Chemical Analysis by Flame Photometry,* Interscience, New York (1963).
Hibbard, R., and Barnett, C. M. (Ed.), *Basic Considerations in the Combustion of Hydrocarbon Fuels with Air,* NACA Report 1300, Supt. Documents, U.S. Govt. Printing Office, Washington, D.C. (1959).
Hilado, C., *Flammability of Cellulosic Materials,* Technomic Press, Westport, Conn. (1973).
Himus, G. W., *The Elements of Fuel Technology,* Leonard Hill Ltd., London (1958).
Hirschfelder, J. O., Curtiss, C. F., and Bird, R. B., *Molecular Theory of Gases and Liquids,* Wiley, New York (1954); 2nd Ed. (corrected, with notes), Wiley, New York, 562 (1964).
Howard, J., *Fluidized Beds,* Applied Science Publ., NY (1983).
Jost, W., *Explosion and Combustion Processes in Gases,* McGraw–Hill, New York (1946).
Jost, W. (Ed.), *Low Temperature Oxidation,* Gordon and Breach Science Publ., NY (1965).
Kanury, A. M., *Introduction to Combustion Phenomena,* Gordon and Breach, New York (1975).
Kays, W. M., *Convective Heat and Mass Transfer,* McGraw–Hill, New York (1966).
Kennedy, L. A. (Ed.), *Turbulent Combustion,* American Institute of Aeronautics and Astronautics, New York (1978).
Khitrin, L. N., *The Physics of Combustion and Explosion,* Israel Program for Scientific Translations, OTS-61-31205, Jerusalem (1962) (trans. from Russian).
Khitrin, L. N. (Ed.), *Combustion in Turbulent Flow,* Israel Program for Scientific Translations, Jerusalem (1963).
Kondratiev, V. N., *Rate Constants of Gas Phase Reactions,* trans. by L. J. Holtschlag, R. M. Fristrom (Ed.), COM-72-10014, National Standard Reference Data System, National Bureau of Standards, U.S. Department of Commerce, Washington, D.C. (1973).
Korobeinichev, O. P. (Ed.), *Flame Structure,* 2 vols., Proceedings of the Third International Seminar on Flame structure held at Alma-Atta Kasakhstan (1991).
Ksandopulo, G. E., *Flame Chemistry,* Chemistry, Moscow (1980) (in Russian).
Ksandopulo, G., and Dubin, V., *The Chemistry of Gas Phase Combustion,* "Chemistry," Moscow (1987).

Kuo, K. K., *Principles of Combustion,* Wiley, New York, (1986).

Kuvshinoff, B. W., Fristrom, R. M., Ordway, G. L., and Tuve, R. L. (Eds.), *Fire Sciences Dictionary,* John Wiley & Sons, New York (1977).

Kuznetsov, V. R., and Sabel'nikov, V. A., *Turbulence and Combustion*, Nauka, Moscow (1986).

Lafitte, P., *La propagation des flammes dans les melanges gazeuz: les deflagrations,* Hermann, Paris (1939).

Lapp, M., and Penney, C. M., *Laser Raman Gas Diagnostics,* Plenum Press, New York (1974).

Lavoisier, A., *Essays Physical and Chemical,* trans. by T. Henry, Reprint, F. Cass Ltd., London (1970).

Lawn, C. J., *Principles of Combustion Engineering for Boilers,* Academic Press, New York (1987).

Lawton, J., and Weinberg, F. J., *Electrical Properties of Flames,* Clarendon Press, Oxford, UK (1969).

Lefebvre, A. H. (Ed.), *Gas Turbine Combustor Design Problems,* Hemisphere, New York (1980).

Lefebvre, A., *Gas Turbine Combustion,* Hemisphere Publ. Co., Bristol, PA (1991).

Lefebvre, A., *Atomization and Sprays,* Hemisphere Publ. Co., Bristol, PA (1991).

Lewis, B., and von Elbe, G., *Combustion, Flames and Explosions of Gases,* Academic Press, New York (1938); 3rd ed. (1987).

Lichty, L. C., *Combustion Engine Processes,* McGraw–Hill, New York (1967).

Lyons, J., *The Chemistry and Uses of Fire Retardants,* Wiley Interscience, New York (1970).

Lyons, J., *Fire,* Milton Freeman Scientific American, New York (1986).

Lyons, J. W., Becker, W. E., Clayton, J. W., Fristrom, R. M., Glassman, I., Graham, D. L., Emmons, H. W., McDonald, D. W., and Nadeau, H. G., *Fire Research on Cellular Plastic,* Products, Research Committee, c/o J. Lyons, N.B.S., Washington, D.C. (1980).

Madavi, J. N., and Amann, C. (Eds.), *Combustion Modeling in Reciprocating Engines,* Plenum Press, New York (1980).

Madorgsky, S. L., *Thermal Degradation of Organic Polymers,* Interscience Publishers, New York (1964).

Markstein, G., *Non-Steady Flame Propagation (AGARDOGraph No. 75),* MacMillan, New York (1964).

Mavrodineanu, R., *Bibliography on Flame Spectroscopy, Analytical Applications,* 1800–1966, Miscellaneous Publication No. 281, National Bureau of Standards, Washington, D.C. (1967).

Mavrodineanu, R., and Boiteux, H., *Flame Spectroscopy,* Wiley, NY (1965).

Mettleton, M. A., *Gaseous Detonations: Their Nature, Effects and Control,* Chapman–Hall, NY (1987).

Ministry of Power, *The Efficient Use of Fuel,* Her Majesty's Stationary Office, London (1958).

Minkoff, G. J., and Tipper, C. F. H., *Chemistry of Combustion Reactions,* Butterworths, London (1962).

Mullins, B. P., *Spontaneous Ignition of Liquids,* Butterworths, London (1957).

Muragai, M. P., *Similarity Analysis in Fire Research,* Oxford (1976).

Murgai, M. P., *Natural Convection from Combustion Sources,* Oxford (1976).

Niessen, W. R., *Combustion and Incineration Processes,* Marcel Dekker, New York (1987).

Onofri, M., and Tesei, A. (Eds.), *Fluid Dynamics of Combustion Theory* Longman Scientific and Technical, Essex, UK (1991).

Oppenheim, A. K. (Ed.), *Combustion and Propulsion,* Fourth AGARD Colloquium, Pergamon Press, New York (1961).

Oran, E., and Boris, J., *Numerical Simulation of Reactive Flow,* Elsevier, NY (1987).

Palmer, H. B., and Beer, J. M. (Eds.), *Combustion Technology,* Academic Press, New York (1974).

Palmer, K. N., *Dust Explosions and Fire,* Chapman-Hall, London (1973).
Penner, S. S., *Introduction to the Study of Chemical Reactions in low Systems,* Butterworths, London (1955).
Penner, S. S., *Chemistry Problems in Jet Propulsion,* Pergamon Press, New York (1957).
Penner, S. S., *Chemical Rocket Propulsion and Combustion Research,* Gordon and Breach, New York (1962).
Peters, N., and Warnatz, J. (Eds.), *Numerical Methods in Laminar Flame Propagation,* Viewag, Wiesbaden, Ger. (1982).
Predvoditelev, A. S. (Ed.), *Gas Dynamics and Physics of Combustion,* Israel Program for Scientific Translations, Jerusalem (1962) (trans. from Russian).
Priestley, J., *Lectures on Combustion,* Reprint, Kraus, New York (1969).
Products Research Committee (see Lyons et al.).
Pungor, E. (trans. R. Chalmers), *Flame Photometry Theory,* D. Van Nostrand Ltd., London, UK (1967).
Robertson, A. C., *Fire Standards and Safety,* STP 614, American Society for Testing Material, Philadelphia (1977).
Rossler, D. E., *Transport Processes in Chemically Reacting Flow,* Butterworths, London (1986).
Rumford, Count, *Collected Works of Count Rumford,* Vols. 1 & 2, S. C. Brown (Ed.), Reprint, MIT Press, Cambridge, Mass. (1968).
Ryabov, I. V., Baratov, A. N., and Petrov, I. I. (Eds.), *Problems in Combustion and Extinguishment,* TT 71-580001, National Bureau of Standards and National Science Foundation, U.S. Government Printing Office, Washington, D.C. (1974) (trans. from Russian).
Schelkin, K. I., and Troshin, Y. K. (trans. B. Kuvshinoff and L. Holtschlag), *Gas Dynamics of Combustion,* Mono Book Co., Baltimore (1965).
Schlichting, H., *Boundary Layer Theory,* 6th ed., McGraw–Hill, New York (1968).
Schroeder, M. J., and Buck, C. C., *Fire Weather,* U.S. Department of Agriculture Handbook 360, Superintendent of Documents, Washington, D.C. (1978).
Scurlock, A., and Grover, J., *Selected Combustion Problems,* Butterworths, London (1954).
Semenov, N., *Some Problems of Chemical Kinetics and Reactivity,* Vols. 1 & 2, Pergamon Press, London (1959).
Shchetinkov, Ye. S., *The Physics of Combustion of Gases,* Izd. Nauka, Moscow, USSR (1965).
Siegla, D., and Smith, G. (Eds.), *Particulate Carbon Formation during Combustion,* Plenum Press, New York (1981).
Singer, J. (Ed.), *Combustion Fossil Power Systems,* 3rd ed., Combustion Engineering Inc., Windsor, Conn. (1981).
Skinner, D. G., *The Fluidized Combustion of Coal,* Institute of Fuel, London (1970).
Smith, I. E. (Ed.), *Combustion in Advanced Gas Turbine Systems,* Pergamon Press, London (1968).
Smith, M. L., and Stinson, K. W., *Fuels and Combustion,* McGraw–Hill, New York (1952).
Smoot, L. D., and Pratt, D., *Pulverized Coal Combustion and Gassification,* Plenum Press, New York (1979).
Sokolik, A. S., *Self-Ignition, Flame and Detonations in Gases,* Israel Program for Scientific Translations, NASA TTF-125OTS-63-11179, Jerusalem (1963) (trans. from Russian).
Spalding, D. B., *Some Fundamentals of Combustion,* Butterworths, London (1955).
Spalding, D. B., *Combustion and Mass Transfer,* Macmillan, New York (1979).
Stambuleanu, A., *Flame Combustion Processes in Industry,* 2nd ed., Abacus Press, London (1985).
Starkman, E. (Ed.), *Combustion Generated Air Pollution,* Plenum Press, New York (1971).
Strehlow, R. A., *Fundamentals of Combustion,* International Textbook Co., Scranton, Pa. (1968); 2nd ed (1983).

Stull, D., and Prophet, A., (Eds.), *Joint Army/Navy/Air Force Thermochemical Tables,* NSRDS NBS No. 37, National Standard Reference Data System, National Bureau of Standards, U.S. Government Printing Office (1969); 2nd ed. (1986).

Thring, M. W., *The Science of Flames and Furnaces,* Chapman–Hall, 2nd ed. (1962).

Timnat, Y. M., *Advanced Chemical and Rocket Propulsion,* Academic Press, NY (1955).

Tine, G., *Gas Sampling and Chemical Analysis in Combustion,* Pergamon Press, London (1961).

Toong, T-Y., *Combustion Dynamics,* McGraw–Hill, New York (1983).

Tryon, G. (Ed.), *Handbook of Fire Protection,* National Fire Protection Association, Boston, Mass. (1980).

Tuve, R., *Principles of Fire Protection Chemistry,* National Fire Protection Association, Boston, Mass. (1976).

Van Tiggelen, A., et al., *Oxidations et combustions,* Vols. 1 & 2, Editions Technip, Paris (1968) (in French).

Vargaftol, N. B., *Handbook of Thermal Conductivity of Liquids and Gases,* CRC Press, Boca Raton, FL (1993).

Voltaire, F. M., in *Collected Essays, La nature et la propagations du feu* (French), Academy of Science, Paris (1752).

von Karman, T., and Penner, S. S., *Selected Combustion Problems,* Butterworths, London (1954).

Vulis, L. A., *Thermal Regimes of Combustion,* McGraw-Hill, NY (1961).

Warnatz, J., and Maas, U., *Technische Verbrennung,* Springer-Verlag, Berlin (1992).

Weinberg, F. J., *Optics of Flames,* Butterworths, London (1963).

Weinberg, F. J. (Ed.), *Advanced Combustion Methods,* Academic Press, NY (1986).

Williams, F. A., *Combustion Theory,* Addison–Wesley, Reading, MA (1965); 2nd ed. Benjamin /Cummings Publ., Inc., Menlo Park, CA (1985).

Zeldovich, Ya. B., Barenblatt, G. I., Librovich, V. B., and Makhviladze, G. M., *The Mathematical Theory of Combustion and Explosions,* Plenum Press, NY (1985).

Zeldovitch, Ya., and Kompaneets, A., *Theory of Detonations,* Academic Press, NY (1960).

Zhou Lixing, *Theory and Numerical Modeling of Turublent Gas–Particulate Flows and Combustion,* CRC Press, Boca Raton, FL (1992).

Zinn, B. T. (Ed.), *Experimental Diagnostics in Gas Phase Combustion Systems,* American Institute of Aeronautics and Astronautics, NY (1977).

REFERENCES

AAAS, "Guide to Scientific Instruments" (see section on gases), American Association for the Advancement of Science, published annually as an addendum to *Science.* (1992).

Abramovich, G. N., *Theory of Turbulent Jets,* MIT Press, Cambridge, Mass. (1963).

Adams, G. K., and Cook, G. B., "The Effect of Pressure on the Mechanism and Speed of the Hydrazine Decomposition Flame," *C&F* 4, 9 (1960).

Adams, G. K., and Scrivener, J., "Flame Propagation in Methyl and Ethyl Nitrate," *5th Symp.,* 656 (1955).

Adams, G. K., and Stock, G. W., "The Combustion of Hydrazine," *4th Symp.,* 239 (1953) 11–49.

Adrian, R. J., "Multipoint Optical Measurements of Simultaneous Vectors in Unsteady Flow—A Review," *Int. J. Heat Fluid Flow* **7,** 127 (1986).

Adrian, R. J., "Particle-Imaging Techniques for Experimental Fluid Mechanics," *Ann. Rev. Fluid Mech.* **23,** 261 (1991).

Afgan, N., and Beer, J., Eds., *Heat Transfer in Flames,* John Wiley & Sons, New York (1974).

Agnew, W. G., "Contributions of Combustion Science to Engine Design," *20th Symp.,* 1, (1985) 13–361.

Agnew, W., and Agnew, J., "Composition profiles of the diethyl ether–air two state reaction stabilized on a flat flame burner," *10th Symp.,* 123 (1965).

Agnew, J. T., and Graiff, L. B., "The Pressure Dependence of Laminar Burning Velocity by Spherical Bomb Method," *C&F* **5,** 209 (1961).

Agrawal, R., "Determination of Burning Velocity of Methane–Air Mixtures in a Constant Volume Vessel," *C&F* **42,** 243–52 (1981).

Akrich, R., Vovelle, C., and Delbourgo, R., "Flame Profiles Combustion Mechanisms of Methanol–Air Flames under Reduced Pressure," *C&F* **32,** 171–79 (1978).

Al-Alami, M. Z., and Kiefer, J. H., "Shock Tube Study of Propane Pyrolysis. Rate of Initial Dissociation from 1400 to 2300 K," *JPC* **87,** 499 (1983).

Albers, E., and Homann, K. H., "Untersuchen an Dicyan–Sauerstoff Flammen," *ZpC* **58,** 220–22 (1958).

Alcay, J. A., and Knuth, E. L., "Molecular Beam Time of Flight Mass Spectrometry," *RSI* **40,** 438 (1969).

Alden, M., Blomqvist, J., Edner, H., and Lundberg, H. "Raman Spectroscopy in the Analysis of Fire Gases," *Fire and Materials* **7,** 32 (1983).

Aldred, J. W., and Williams, A., "The Structure of Burning Spheres of Liquid Methanol," *C&F* **13,** 559 (1969).

Allen, R., and Rice, P., "The Explosion of Azomethane," *JACS* **57,** 316 (1935).

Alpher, R., and White, D., "Optical Refractivity of High-Temperature Gases. I. Effects Resulting from Dissociation of Diatomic Gases," *PF* **2,** 153 (1959).

Ambs, L., and Fletcher, E., "Structure of the Cl_2–F_2 Flame," *13th Symp.,* 685 (1971).

Amin, H., "The Effect of Heterogeneous Removal of O Atoms on the Measurement of NO_2 in Sampling Probes," *CST* **15,** 31 (1977).

Anderson, J. B., "Mechanism of the Bimolecular (?) H_2–I_2 Reaction," *JCP* **61,** 3390 (1974).

Anderson, J. W., "A Process for the Electroforming of Nozzles," *RSI* **19,** 822 (1949).

Anderson, J. W., and Fein, R. S., "Normal Burning Velocities and Bunsen Flame Temperatures from Stroboscopically Illuminated Particle Tracks," *JCP* **17,** 1268 (1949).

Anderson, J. W., and Friedman, R., "Critical Orifice Flowmeters," *RSI* **20,** 61 (1949).

Anderson, W. R., Decker, L. J., and Kotlar, A. J., "Concentration Profiles of NH and OH in Stoichiometric CH_4–N_2O Flame by LEF and Absorption," *C&F* **48,** 179 (1982).

Andrews, D., and Gray, P., "Combustion of Ammonia Supported by Oxygen, Nitrous Oxide, or Nitric Oxide: Laminar Flame Propagation in Binary Mixtures," *C&F* **8,** 113 (1964).

Andrews, G., and Bradley, D., "The Burning Velocity of Methane–Air Mixtures," *C&F* **19,** 275 (1972a).

Andrews, G., and Bradley, D., "The Determination of Burning Velocity—A Review," *C&F* **18,** 133 (1972b).

Andrews, G., and Bradley, D., "Determination of Burning Velocity of H_2–Air Flames," *C&F* **20,** 77 (1973a).

Andrews, G., and Bradley, D., "Limits of Flammability and Natural Convection for Methane–Air Mixtures," *14th Symp.,* 1119 (1973b).

Andrews, G. E., Bradley, S., and Lwakabama, S., "Turbulent Flame Review," *C&F* **24,** 285 (1975).

Anon., *International Critical Tables* (20 volumes), McGraw–Hill Book Co. (1922) 11–96.

Anon., *Ozone Chemistry and Technology,* ACS Advances in Chemistry Series #21, American Chemical Soc., Washington, DC (1959).

Anon., *10th Symp.,* "Decennial Index" (1965).

Anon., Section on Soot; section on Internal Combustion Engines, *18th Symp.,* 1091 (1981).

Anon., 20th Symp., Cumulative Subject Index for Volumes I–XX; Sections on: Combustion in Practical Systems: Pollutants (1985).

Anon., *Handbook of Chemistry and Physics,* The Chemical Rubber Co., Cleveland, OH, Yearly editions, 12–286.

Antcliff, R. R., and Jarrett, O., "Comparison of CARS Combustion Temperatures with Standard Techniques," Paper 83-1482 at AIAA Thermophysics Conference, Montreal, Canada, June 1 (1983).

Antoine, A., "The Mechanism of Burning of Liquid Hydrazine," *8th Symp.,* 1057 (1962).

Arden, E. A., and Powling. J., "The Methyl Nitrite Decomposition Flame," *6th Symp.,* 177 (1957).

Aronowitz, D., Santoro, R., Dryer, F., and Glassman, I., "Kinetics of the Oxidation of Methanol Experimental Results, Global Modeling and Mechanistic Concepts," *17th Symp.,* 633 (1978).

Astholz, D., Durant, J., and Troe, J., "Thermal Decomposition of Toluene and Benzil Radicals in Shock Waves," *20th Symp.,* 885 (1985).

Attal-Trötout, B., Bouchardy, P., Magre, P., Pöalat, M., and Taran, J. P., "CARS in Combustion: Prospects and Problems," *Appl. Phys. B.* **50,** 445 (1990).

Ausloos, P., and Van Tiggelen, A., "Velocity of Propagation of Flames of Mixtures of Gases Containing Ammonia" (in French), *Bull. Soc. Chim. Belg.* **60,** 433 (1951).

Ausloos, P., and Van Tiggelen, A., "Emission Spectra of OH, NO, NH and NH_2 in Flame" (in French), *Bull. Soc. Chim. Belg.* **61,** 569 (1952).

Avery, W. H., "Space Heating Rates and High Temperature Kinetics," *5th Symp.,* 85 (1955) 02–18.

Avizonis, P. V., and Neumann, D. K., "The Chemical Oxygen–Iodine Laser (COIL)," scheduled for Publication in J. Def. Res. (1992).

Axelsson, E., Brezinski, K., Dryer, F. L., Pitz, W. J., and Westbrook, C. K., "Chemical Kinetic Modeling of the Oxidation of Large Alkane Fuels: N-Octane and Iso Octane," *21st Symp.,* 783 (1988).

Axelsson, E., and Rosengren, L., "Iso-Octane Combustion in a Flat Flame," *C&F* **65,** 229 (1986).

Axworthy, A. E., Schneider, G. R., and Dayan, V. H., "Chemical Reactions in the Conversion of Fuel Nitrogen to NOx," *US-EPA Symposium on Stationary Combustion,* Atlanta, GA, Sept. 24–26 (1975).

Azatyan, V. V., Gershenson, U. M., Sarkissyan, E. N., Sachyan, G. A., and Nalbandian, A. B., "Investigation of Low Pressure Flames of Sulfur Containing Compounds by the ESR Method," *12th Symp.,* 989–94 (1969).

Babkin, V., and Kozachenko, L., "Study of Burning Velocity in Methane–Air Mixtures at High Pressure," Fizika Goreniza i., Vzryuva, **3,** 77–856 (1966) (English translation).

Bachmaier, F., Eberius, K. H., and Just, Th., "The Formation of NO and the Detection of HCN in Atmospheric Premixed Hydrocarbon–Air Flames," *CST* **7** (2), 77–84 (1973).

Bacon, F., "Of the Advancement of Learning," Reprint by Dutton, New York (1915).

Badami, G. N., and Egerton, Sir A., "Burning Velocity of CO," *PRC* **A228,** 297 (1955).

Bain, J., Vandooren, J., and Van Tiggelen, P. J., "Experimental Study of an Ammonia-Oxygen Flame," *21st Symp.,* 953 (1988).

Balakhnin, V., Egorov, V., Van Tiggelen, P. J., Azatyan, V., Gershenzon, Y., and Kondratiev, V., "ESR Study of Reaction Involving O Atoms in Rarefied CO Flames with Added H_2," *Kinetika i Kataliz* **9,** 676 (1968). English: *Kinetics and Catalysis* **9,** 559 (1969).

Balakhnin, V., Vandooren, J., and Van Tiggelen, P. J., "Reaction Mechanism and Rate Constants in Lean Hydrogen–Nitrous Oxide Flames," *C&F* **28,** 165 (1977).

Baldwin, R. R., Bennett, J. P., and Walker, R., "Rate Constants for Elementary Steps in Hydrocarbon Oxidation," *16th Symp.,* 1041 (1977).

Baldwin, R. R., Jackson, D., Walker, R., and Webster, S., "The Use of Hydrogen–Oxygen Reaction in Evaluating Reaction Velocity Constants, *10th Symp.,* 423 (1965).

Baluch, D. L., Drysdale, D. D., Duxbury, J., and Grant, S. J., *Evaluated Kinetic Data for High Temperature Reactions 3,* "Homogeneous gas phase reactions of the O_3 system, the CO–H_2 system and of sulphur containing species," Butterworths, Boston (1976).

Baluch, D., Drysdale, D. D., and Horne, D., *Evaluated Kinetic Data for High Temperatures Reactions,* Vol. 2, Butterworths (1973).

Baluch, D. L., Duxbury, J., Grant, S., and Montague, D., "Evaluated Kinetic Data for High Temperature Reactions Halogen and Cyanide Species," *J. Phys. and Chem. Ref. Data* **10,** suppl. 1 (1981).

Bamford, C. H., and Tipper, C. F. H. (eds.), *Comprehensive Chemical Kinetics* **1,** "The Practice of Kinetics," Elsevier Pub. Co., New York (1969).

Bamford, C. H., and Tipper, C. F. H., *Comprehensive Chemical Kinetics* **17,** "Combustion Chemistry," Elsevier, Oxford (1977).

Baratgov, A. N., "Review of the Investigation on the Chemical Inhibition of Flames," *Problems in Combustion and Extinguishment* (Ryabov, I. J., Baratov, A., and Petrov, I., eds.) p. 29, English translation for the NBS and NSF, Washington, DC, by the Amerind Pub. Co. Ltd., New Delhi, India (1974).

Barnard, J., and Bradley, J., Flame and Combustion Phenomena, 2nd ed., Methuen, London (1985).

Barnett, H. C., and R. R. Hibbard, eds., *Basic Consideration in the Combustion of Hydrocarbon Fuels with Air,* National Advisory Committee for Aeronautics, Report No. 1300, U. S. Government Printing Office, Washington, D.C. (1959).

Bartels, M., Hoyermann, K-H., and Sievert, R., "Elementary Reactions of Ethylene: the Reaction of OH with Ethylene and the Reaction of C_2H_4OH Radicals with H Atoms," *19th Symp.,* 61 (1982).

Bartholome, E., "Flame Velocity in Stationary Burning Flames," *Naturwiss.* **36,** 218 (1949a).

Bartholome, E., "Zur Methodik der Messung von Flammengeschwindigkeiten," *ZfE* **53,** 191 (1949b).

Barton, S. C., and Dove, J. E., "Mass Spectrometric Studies of Chemical Reactions in Shock Waves: The Thermal Decomposition of Nitrous Oxide," *CJC* **47,** 521 (1969).

Basevich, U. Ya., Kogarko, S. M., and Posvyanskii, V. S., "Reaction Kinetics in Propagation of Ethylene–Oxygen Flame," *Fisika Gorenya i Vzryva* **13,** 192–200 (1977) (English translation).

Beams, J. W., *Physical Measurements in Gas Dynamics and Combustion,* Oxford Univ. Press, NY (1955).

Bechtel, J. H., and Blint, R. J., "Structure of a Flame-Wall Interface by Laser Raman Scattering," *Appl. Phys. Lett.* **37,** 576 (1980).

Bechtel, J. H., Blint, R. J., Dasch, C., and Weinberger, D., "Atmospheric Pressure Premixed Hydrocarbon Air Flames: Theory and Experiment (Methane and Propane)," *C&F* **42,** 197 (1981).

Bechtel, J. H., and Teets, R. E., "Hydroxyl and its Concentration Profile in Methane–Air Flames," *Appl. Opt* **18,** 4138 (1980).

Beck, N. J., Chen, S. K., Uyehara, O. A., Winans, J. G., and Myers, P. S., "Temperature Measurements from Iodine Absorption Spectrum," *5th Symp.,* 412, (1955).

Beer, J. M., and Chigier, N. A., *Combustion Aerodynamics,* Applied Science, London (1972).

Beer, J., and Thring M. W., (eds.) *Industrial Flames,* Vol. I, Edward Arnold, London, UK (1972).

Bell, K., and Tipper, C., "Slow Combustion of Methyl Alcohol," *TFS* **53,** 982 (1957).

Benson, S. W., *Thermochemical Kinetics,* John Wiley & Sons, New York (1968).

Benson, S. W., and Srinavasan, R., "Complexities in the Reaction System: $H_2 + I_2 = 2HI$," *JCP* **23,** 200 (1955).

Benson, R. S., and Whitehouse, N. D., *Internal Combustion Engines,* Vols. 1 & 2, Pergamon Press, New York (1979).

Berets, D., Green, E., and Kistiakowsky, G. B., "Detonations of Hydrogen–Oxygen Mixtures," *JACS* **72,** 1086 (1950).

Berl, W. G., Ed., *The Use of Models in Fire Research,* Publication 5786, National Academy of Sciences/National Research Council, Washington, D.C. (1961).

Berl, W. G., Breisacher, P., Dembrow, D., Falk, F., O'Donovan, J. T., Rice, J., and Sigillito, V. G., "Combustion Characteristics of Monopropylpentaborane Flames," *10th Symp.,* 87 (1966).

Berl, W. G., and Dembrow, D., "Diborane–Propane Flames," *Nature* **170,** 367 (1952).

Berl, W. G., Gayhart, E. L., Maier, E., Olsen, H. L., and Renich, W. T., "Determination of the Burning Velocity of Pentaborane–Oxygen–Nitrogen Mixtures," *C&F* **1,** 420 (1957).

Berl, W. G., and Wilson, W. E., "Formation of BN in Diborane Hydrazine Flames," *Nature* **191,** 380 (1961).

Bernstein, R. B., *Chemical Dynamics via Molecular Beam and Laser Techniques,* Oxford University Press, N.Y. (1982).

Bertrand, C., Dussart, B., and Van Tiggelen, P. J., "Use of Electric Fields to Measure Burning Velocity," *17th Symp.,* 967 (1979).

Bian, J., Vandooren, J., and Van Tiggelen, P. J., "Experimental Study of the Formation of

Nitrous and Nitric Oxides in Hydrogen, Oxygen, Argon Flame Seeded with Nitric Oxide or Ammonia," *23rd Symp.*, 953 (1991).

Biedler, W. T., and Hoelscher, H. E., "Studies in a New Type of Flat Flame Burner," *Jet Prop.* **27,** 1257 (1957).

Biordi, J., "Molecular Beam Mass Spectrometry Applied to Determining the Kinetics of Reactions in Flames: II. A Critique of Rate Coefficient Determinations," *C&F* **26,** 57 (1976).

Biordi, J., "Investigating Flame Inhibition with Molecular Beam Mass Spectrometry," *J. Aeronaut. and Astro.* **53,** 125–52 (1977).

Biordi, J., Lazzara, C., and Papp, J., "Studies of CF_3Br Inhibited Methane Flames," *14th Symp.*, 367 (1973).

Biordi, J., Lazarra, C., and Papp, J., "Molecular Beam Mass Spectrometry Applied to Determining the Kinetics of Reactions in Flames I. Empirical Characterization of Flame Perturbation by Molecular Beam Sampling Probes," *C&F* **23,** 73 (1974).

Biordi, J., Lazzara, C., and Papp, J., "Flame Structure—The Effect of CF_3Br on Radical Concentration Profiles in Methane Flames," *ACS Symposium Series # 1F,* R. G. Gann (Ed.), 256–94 (1975a).

Biordi, J., Lazzara, C., and Papp, J., "The Inhibition of Low Pressure Quenched Flames by CF_3Br," *C&F* **24,** 401 (1975b).

Biordi, J., Lazzara, C., and Papp, J., "Inhibition Using Molecular Beam Mass Spectrometry," U.S. Bur. Mines Rep. RI8029 (1975c).

Biordi, J., Lazzara, C., and Papp, J., "The Structure of Inhibited Flames II: Kinetics and Mechanisms," *15th Symp.*, 917 (1975d).

Biordi, J., Lazzara, C., and Papp, J., "Mass Spectrometric Observations of Difurocarbene and its Reactions in Inhibited Methane Flames," *JPC* **80,** 1042 (1976).

Biordi, J., Lazzara, C., and Papp, J., "Flame Structure Studies of CF_3Br Inhibited Methane Flames 3—The Effect of 1% CF_3Br on Rate Constants and Net Reaction Rates," *JPC* **81,** 1139 (1977a).

Biordi, J., Lazzara, C., and Papp, J., "An Examination of the Partial Equilibrium Hypothesis," *16th Symp.*, 1097 (1977b).

Biordi, J., Lazzara, C., and Papp, J., "Studies of CF_3BR Inhibited Methane Flames 4—Reactions of Inhibitor Related Species in Flames Continuing Initially 1.1% CF_3Br," *JPC* **82,** 125 (1978).

Bird, R. B., Stewart, W. E., and Lightfoot, E. N., *Transport Phenomena,* Wiley & Sons, New York (1962).

Bittner, J., and Howard, J., "Composition Profiles and Reaction Mechanisms in a Near Sooting Premixed Benzene/Oxygen/Argon Flame," *18th Symp.*, 1105 (1981).

Bittner, J., and Howard, J., "Mechanism of Hydrocarbon Decay in Fuel Rich Secondary Reaction Zones," *19th Symp.*, 211 (1982).

Blackshear, P., *Heat Transfer in Fires,* John Wiley & Sons, New York (1974).

Blauwens, J., Smets, B., and Peeters, J., "Mechanism of 'Prompt' NO formation in Hydrocarbon Flames," *16th Symp.*, 1055 (1977).

Bledjian, L., "Computation of Time-Dependent Laminar Flame Structure," *C&F* **20,** 5 (1973).

Blint, R. J., "Flammability Limits for Exhaust Gas Diluted Flames," *22nd Symp.*, 1547 (1989).

Blumenberg, B., Hoyermann, K-H, and Sievert, R., "Primary Products in the Reactions of O Atoms with Simple and Substituted Hydrocarbons," *16th Symp.*, 841 (1977).

Bockhorn, H., Fetting, F., and Mende, J. C., "The Laminar Flame Velocities of Propane/Ammonia Mixtures," *C&F* **18,** 471 (1972).

Bodenstein, M., "Hydrogen–Bromine Reaction," *ZPC* **29,** 295 (1899).

Bodenstein, M., and Lind, S. C., "The Rate of Formation of Hydrogen Bromide from its Elements," *ZPC* **57,** 168 (1907).

Bone, W. A., and Townend, D. T. A., *Flame and Combustion in Gases,* Longmans Green & Co., New York (1927).

Bonne, U., Grewer, T., and Wagner, H. Gg., "Messungen in der Reaktionszone von Wasserstoff–Sauerstoff und Methan–Sauerstoff Flammen," *ZpC.(Frankfort)* **24,** 93 (1960).

Bonne, U., Homann, K., and Wagner, H. Gg., "Carbon Formation of Rich Flames," *10th Symp.,* 503 (1965).

Bonne, U., Jost, W., and Wagner, H. Gg., "$Fe(CO)_6$ in Methane–Oxygen Flames," *Fire Res. Abs. and Rev.* **4,** 6 (1962).

Botha, J. P., and Spalding, D., "The Laminar Flame Speed of Propane Air Mixtures with Heat Extraction from the Flame," *PRC* **A225,** 71 (1954).

Boushehri, A., Bzowski, J., Kistin, J., and Mason, E. A., "Equilibrium and Transport Properties of Eleven Polyatomic Gases at Low Density," *J. Phys. and Chem. Ref. Data* **16,** 445 (1987).

Bowman, C. T., "An Experimental and Analytical Investigation of the High-Temperature Oxidation Mechanisms of Hydrocarbon Fuels," *CST* **2,** 161 (1970).

Bowman, C. T., "Non-Equilibrium Radical Concentrations in Shock Ignition Methane Oxidation," *15th Symp.,* 869 (1975a).

Bowman, C. T., "Kinetics of Pollutants in Combustion," *Progress in Energy and Combustion* **1,** 33 (1975b).

Bowman, C. T., "A Shock Tube Study of the High Temperature Oxidation of Methanol," *C&F* **25,** 343, (1975c).

Bowman, C. T., and Birkeland, J. (eds.), *Alternative Hydrocarbon Fuels: Combustion and Chemical Kinetics,* American Institute of Aeronautics and Astronautics, New York (1978).

Bowman, C. T., and Seery, D. J., "Investigation of NO Formation Kinetics: the CH_4–O_2–N_2 Reaction," *Emission from Continuous Combustion Systems* (W. Cornelius and W. Agnew, eds.), Plenum Press, London, 123 (1972).

Boyle, R., *The Skeptical Chymist,* Reprint, Dutton, New York (1967).

Boys, S. F., and Corner, J., "The Structure of the Reaction Zone in Flames," *PRS* **A197,** 90 (1949).

Bradley, D., and Hundy, G.,"Hot Wire Anemometers in Closed Vessel Explosions," *13th Symp.,* 575 (1971).

Bradley, D., and Hundy, G., "The Burning Velocities of Methane-Air Mixtures Using Hot Wire Annemometry in Closed Vessel Explosions," *13th Symp.,* 575 (1971).

Bradley, D., Liau, A. K. C., Missaghi, M., "Response of Compensated Thermocouples in Fluctuating Temperatures: Computer Simulation, Experimental Results and Mathematical Modeling," *CST* **64,** 119, (1989).

Bradley, J. N., *Shock Waves in Chemistry and Physics,* Methuen, London (1962).

Bradley, J. N., Jones, G. A., Skirrow, G., and Tipper, C., "Low Temperature Flames of Acetaldehyde and Propionaldehyde: A Mass Spectrometric Study," *C&F* **10,** 259 (1966).

Bradley, J. N., *Flame and Combustion Phenomena,* Methuen, London (1969).

Bradshaw, J. D., Omenetto, N., Zizak, G., Brower, J. N., and Winefordner, J. D., "Five Laser-Excited Fluorescence Methods for Measuring Spatial Flame Temperatures: Theoretical Basis," *Appl. Opt.* **19,** 2709 (1980).

Bragg, S. L., *Rocket Engines,* Geo. Newnes Ltd., London (1962).

Brame, J. S. S., and King, J. G., *Fuel,* Edward Arnold, London (1955).

Bray, K. N. C., "A Sudden Freezing Analysis for Non-Equilibrium Nozzle Flows," *ARSJ* **31,** 831 (1961).

Bregdon, B., and Kardirgan, M., "ESR Determination of H Atoms in Flames," *18th Symp.,* 408 (1981).

Breisacher, P., Dembrow, D., and Berl, W. G., "Flame Front Structures of Lean Diborane–Air and Diborane–Hydrocarbon–Air Mixtures," *7th Symp.,* 894 (1958).

Breton, J., "Detonations Studies," Théses Faculté de Sciences de Nancy, Ann. Off. Comb., **11,** 487 (1936) [quoted in Lewis and von Elbe [1987], p. 585].

Brezinsky, K., and Dryer, F. L., "A Flow Reactor Study of the Oxidation of Iso-Butylene and an Iso-Butylene/n-Octane Mixture," *CST* **45,** 225 (1986).

Brokaw, R. S., "Approximate Formulas for Viscosity and Thermal Conductivity of Gas Mixtures," NASA Tech Note D-2502, Lewis Research Center, Cleveland, OH (1964).

Brokaw, R. S., and Gerstein, M., "Correlations of Burning Velocity, Quenching Distances and Minimum Ignition Energies for Hydrocarbon–Oxygen–Nitrogen Systems," **6th Symp.,** 66 (1957).

Brokaw, R. S., and Pease, R. N., "The Effect of Water on the Burning Velocities of Cyanogen–Oxygen–Argon Mixtures," *JACS* **75,** 1454 (1953).

Brouwer, L., Muller-Markgraf, W., and Troe, J., "Identification of the Primary Reaction Products in the Thermal Decomposition of Aromatic Hydrocarbons," *20th Symp.,* 799 (1985).

Brown, N. J., Eberius, K. H., Fristrom, R. M., Hoyermann, K. H., and Wagner, H. Gg., "Low Pressure Hydrogen/Oxygen Flame Studies," *C&F* **33,** 151 (1978).

Brown, N. J., and Fristrom, R. M., "A Two Zone Model of Flame Propagation Applied to H_2 + Air Flames and HCl Inhibited Flames," *Fire and Materials* **7,** 117 (1978).

Brown, N. J., Fristrom, R. M., and Sawyer, R., "A Simple Premixed Flame Model Applied to H_2–Air," *C&F* **23,** 269 (1974).

Brown, S. C., *Benjamin Thompson, Count Rumford,* MIT Press, Cambridge, MA (1979).

Brown, W., Porter, R., Verlin, J., and Clark, A., "A Study of Acetylene–Oxygen Flames," *12th Symp.,* 1035 (1969).

Buckmaster, J. D., *The Mathematics of Combustion,* Vol. 2, *Frontiers in Applied Mathematics,* Society for Industrial and Applied Mathematics, Philadelphia, PA (1985).

Buckmaster, J. D., and Ludford, G. S. S., *Theory of Laminar Flames,* Cambridge University Press, Cambridge, UK (1985).

Bulewicz, E., James, C., and Sugden, T. M., "Photometric Investigations of Alkali Metals in H_2 Flame Gases. I. The Use of Resonance Radiation in the Measurement of Atomic Concentrations," *PRS* **A227,** 312 (1955)

Bulewicz, E. M., James, C. G., and Sugden, T. M., "Photometric Investigations of Alkali Metals in Hydrogen Flame Gases. II. Excess Concentrations of H Atoms in Burnt Gas Mixtures," *PRS* **A235,** 890 (1956).

Bulewicz, E. M., and Padley, P. J., "A Cyclotron Resonance Study of Ionization in Low Pressure Flames," *9th Symp.,* 638 (1963).

Bulewicz, E. M., and Sugden, T. M., "Recombination of Hydrogen Atoms and Hydroxyl Radicals in Hydrogen Flames," *TFS* **54,** 1855 (1958).

Bunsen, R., "Bunsen Burner," *Poggendorffs Ann.* **131,** 161 (1866).

Burcat, A., "Shock Tube Investigation of Ignition in $C_3H_8/O_2/Ar$ Mixtures," *13th Symp.,* 745 (1971).

Burcat, A., "Cracking of Propylene in a Shock Tube," *Fuel* **54,** 87 (1975).

Burcat, A., Scheller, K., and Lifshitz, A., "Shock Tube Investigation of Comparative Ignition Delay Times for C_1–C_5 Hydrocarbons," *C&F* **16,** 29 (1971).

Burdon, F. A., and Burgoyne, J. H., "Acetylene Decomposition Flame," *PRS* **A199,** 328 (1949).

Burdon, M. C., Burgoyne, J. H., and Weinberg, F. J., "Effect of Methyl Bromide on Combustion of Some Fuel Air Mixtures," *5th Symp.,* 647 (1955).

Burgoyne, J. H., Cullis, C. F., and Liebermann, M. J., "Influence of Additives on the Reactions in Hydrogen–Chlorine Flames," *12th Symp.,* 943 (1959).

Burke, S. P., and Schumann, T. E. W., "Diffusion Flames," *1st Symp.,* 2, (1928).

Burlbaw, E. J., and Armstrong, R. L., "Rotational Raman Interferometer Measurements of Flame Temperatures," *Appl. Opt.* **22,** 2860 (1983).

Burstall, A. F., *A History of Mechanical Engineering,* MIT Press, Cambridge, MA (1965).

Bystrova, V., and Librovich, V. B., "A Flat Flame of Poor Mixtures of CS_2–O_2 Above a Porous Burner," *Physics of Combustion and Shock Waves* (Eng. Trans.) **13,** 435–45 (1977).

Calcote, H., "Electrical Properties of Flames in Transverse Electric Fields," *3rd Symp.,* 245 (1949).

Calcote, H., "A Control and Vaporizing System for Small Liquid Flows," *Anal. Chem.* **22,** 1058 (1950).

Calcote, H., and King, I., "Study of Ionization in Flames by the Langmuir Probe," *5th Symp.,* 423 (1955).

Calcote, H., "Ion and Electron Profiles in Flames," *9th Symp.,* 622 (1963).

Callery Chemical Co., *Boranes and Related Compounds,* Callery Chemical Co., Pittsburgh, PA, 2nd ed. (1958).

Camac, M., "Vibrational Relaxation in Oxygen–Argon Mixtures," *JCP* **34,** 448 (1961).

Cambray, P., "Measuring Termocouple Time Constants—A New Method," *CST* **42,** 111 (1984).

Campbell, A. B., and Jennings, B. H., *Gas Dynamics,* McGraw–Hill, New York (1958).

Campbell, E. S., *Univ. of Wis. Rep. CM 849,* Madison, WI (1955).

Campbell, E. S., "Theoretical Study of the Hydrogen–Bromine Flame," *6th Symp.,* 213 (1957).

Campbell, E. S., "A Theoretical Analysis of Chemical and Physical Processes in an Ozone Flame," *J. Computat. Phys.* **27,** 410 (1978).

Campbell, E., and Fristrom, R. M., "Thermodynamics, Transport and Kinetics of the Hydrogen–Bromine System. A Survey of Properties for Flame Theory Calculations," *Chem. Rev.* **58,** 178 (1958).

Campbell, E. S., Hirschfelder, J. A., and Schalit, L. M., "Deviation from the Kinetic Steady-State Approximation in a Free Radical Flame," *7th Symp.,* 332 (1967).

Carhart, H., Williams, K., and Johnson, J., "The Vertical Tube Reactor: A Tool for the Study of Flame Processes," *7th Symp.,* 392 (1959).

Carito, L., and Sawyer, R., "Combustion Thermodynamics," in *Combustion Generated Air Pollution,* E. Starkman, Ed., Plenum Press, New York (1971).

Carlier, M., Corre, C., Minetti, R., Pauwels, J-F., Ribaucour, M., and Sochet, L-R., "Autoignition of Butane: A Burner and A Rapid Compression Machine Study," *23rd Symp.,* 1753 (1991).

Carlier, M., Pauwels, J. F., and Sochet, L. R., "Application of ESR Techniques to the Study of Gas Phase Oxidation and Combustion Phenomena," *Oxidation Communications* **6,** 141 (1984).

Carter, C. D., King, G. B., and Laurendeau, N. M., "Laser Induced Fluorescence Measurements of OH in Laminar $C_{22}/O_2/N_{22}$ Flames at High Pressure," *CST* **71,** 263 (1990).

Cashin, K., Chinteapalli, P., Vanpee, M., and Vidaud, P., "Emission Spectra for a F_2–NH_3 Flame," *C&F* **22,** 337 (1974).

Cashion, J. K., and Polanyi, J. C., "Infrared Chemiluminescence from the Gaseous Reaction Atomic H* plus Cl_2," *JCP* **29,** 455 (1958).

Cashion, J. K., and Polanyi, J. C., "Resolved Infrared Emission Spectrum of the Reaction Atomic H* plus Cl_2," *JCP* **30,** 1097 (1959).

Cathonnet, M. J., Boettner, J., and James, H., "Experimental Study and Numerical Modeling of High Temperature Oxidation of Propane and n-Butane," *18th Symp.,* 903 (1981).

Cattolica, R. J., Cavolowsky, J., and Mataga, T. G., "Laser-Fluorescence Measurements of Nitric Oxide in Low Pressure $H_2/O_2/NO$ Flames," *22nd Symp.,* 1165 (1989).

Cattolica, R. J., and Stephanson, D. A., "Two Dimensional Imaging of Flame Temperature using Laser-induced Fluorescence," p 714 in *Dynamics of Flames and Reactive Systems, Progress in Astronautics and Aeronautics Series,* **95,** AIAA, NY (1984).

Cattolica, R. J., Stepowski, D., and Cottereau, M., "Laser Fluorescence Measurements of C in a Low Pressure Flame," *J. Quant. Spectrosc. Rad. Transfer* **32,** 363 (1984).

Cattolica, R. J., and Vosen, S. R., "Two-dimensional Measurements of OH in a Constant Volume Combustion Chamber," *20th Symp.,* The Combustion Institute, Pittsburgh, 1273 (1984).

Cattolica, R. J., Yoon, S., and Knuth, E. L., "OH Concentrations in an Atmospheric Methane-Air Flame from Molecular Beam Mass Spectrometry and Laser Absorption Spectroscopy," *CST.* **28,** 225 (1982).

Caulfield, H. J., and Lu, Sun, *The Applications of Holography,* Wiley, New York (1970).

Chan, C., and Daily, J. W., "Measurement of Temperature in Flames Using Laser-Induced Spectroscopy of OH," *Appl. Opt.* **19,** 1963 (1980).

Chandran, S. B. S., Komerath, N. M., Grisson, W. M., Jagoda, J. I., and Strahle, W. C., "Time Resolved Thermometry by Simultaneous Thermocouple and Rayleigh Scattering Measurements in a Turbulent Flame," *CST.* **44,** 47 (1985).

Chang, R. K., and Long, M. B., "Optical Multichannel Detection," p. 179 in *Light Scattering in Solids,* Cardinam, N., and Guntherodt, G. (eds.) Springer-Verlag, NY (1982).

Chang, W. D., and Senkan, S. M., "Chemical Structure of Fuel-Rich, Premixed, Laminar Flames of Trichloroethylene," *22nd Symp.,* 1453 (1989).

Chapman, D. L., "Shock and Detonation Waves," *Phil. Mag.* **47,** (5) 90, (1899).

Chapman, S., and Cowling, T. G., *The Mathematical Theory of Non-Uniform Gases,* Cambridge Univ. Press (1939).

Chappell, G. A., and Shaw, H., "A Shock Tube Study of the Pyrolysis of Propylene. Kinetics of the Vinyl–Methyl Bond Rupture," *JPC* **72,** 4672 (1968).

Charsley, E. L., and Warrington, S. B., *Thermal Analysis: Techniques and Applications,* CRC Press, Boca Raton, FL (1993).

Chase, J., and Weinberg, F. J., "The Acetylene Decomposition Flame and the Deduction of Reaction Mechanism from 'Global' Flame Kinetics," *PRS* **A275,** 411 (1963).

Chen, A. T., Malte, P., and Thornton, M., "Sulfur–Nitrogen Interactions in Stirred Flames," *20th Symp.,* 769 (1985).

Chen, Hao-Lin, "Applications of Laser Absorption Spectroscopy," p. 261 in *Laser Spectroscopy and its Applications,* Radziemski, L. J., Solarz, R. W., and Paisner, J. A. (eds.) Marcel Dekker, Inc., NY (1987).

Chen, R., and Goulard, R., "Retrieval of Arbitrary Concentration and Temperature Fields by Multangular Scanning Techniques," *J. Quant. Spect.* **16,** 819 (1976).

Chendaille, J., and Braud, Y., *Measurements in Flames,* Crane, Russak, New York (1972).

Chendaille, J., and Braud, Y., "Measurements in Flames," in Vol. 1 of *Industrial Flames,* J. M. Beer and M. W. Thring (eds.) Edward Arnold, London, UK (1972).

Cheng, R. K., Popovich, M. M., Robben, F., and Weinberg, F. J., "Associating Particle Tracking with Laser Fringe Anemometry," *J. Phys. Chem.* **13,** 315 (1980).

Cheng, S., Zimmerman, M., and Miles, R., "Homogeneous LDV Using Sodium Seed," *App. Phys. Lett.* **43,** 143 (1983).

Cherian, M. A., Rhodes, P., Simpson, R. J., and Dixon-Lewis, G., "Kinetic Modeling of the Carbon Monoxide Flame," *18th Symp.,* 385 (1981a).

Cherian, M. A., Rhodes, P., Simpson, R. J., and Dixon-Lewis, G., "Structure, Chemical Mechanism and Properties of Premixed Flames in Mixtures of Carbon Monoxide, Nitrogen, and Oxygen with Hydrogen and Water Vapor," *Philos. Trans. Roy. Soc. London* **A303,** 181 (1981b).

Chiang, C., and Skinner, G. B., "Resonance Absorption Measurements of Atom Concentrations in Reacting Gas Mixtures. 5. Pyrolysis of C_3H_8 and C_3D_8 Behind Shock Waves," *18th Symp.,* 915 (1981).

Chiang, F. P., and Reid, G. T. (Eds.), *Optics and Lasers in Engineering* **9,** 161, Elseviere, NY (1988).
Chiger, N.A. (Ed.), Pollution Formation and Destruction in Flames, Vol. 1, *Progress in Energy and Combustion Science,* Pergamon Press, London (1976).
Chiger, N. A., *Energy from Fossil Fuels and Geothermal Energy,* Vol. 2, *Progress in Energy and Combustion Science,* Pergamon Press, New York (1977).
Chiger, N. A., (Ed.), *Energy and Combustion Science,* Selected Papers from *Progress in Energy and Combustion Science,* Student ed. 1, Pergamon Press, London (1979).
Chiger, N. A., *Conservation and Combustion,* McGraw–Hill, New York (1980).
Chiger, N. A. (Ed.), *Combustion Measurements,* Hemisphere Publ. Co., Bristol, PA (1991).
Chomiak, J., *Combustion—A Study in Theory, Fact and Application,* Abacus Press, Gordon and Breach, New York (1990).
Chou, M. S., Dean, A. M., and Stern, D., "Laser Absorption Measurements on OH, NH $+NH_2$ in NH_3/O_2 Flames: Determination of an Oscillator Strength for NH_2," *JCP* **76,** 5334 (1982).
Christiansen, J. A., "On the Reaction Between H_2 and Br_2," *Kgl. Dansk. Videnskab. Selsk., Math.-Fys. Medd.* **1,** 14 (1919).
Chung, S-L, and Katz, J. L., "The Counterflow Diffusion Flame burner: A New Tool for the Study of the Nucleation of Refractory Compounds," *C&F* **61,** 271 (1985).
Clark, R. W., *Benjamin Franklin,* Random House, NY (1983).
Clark, T., and Dove, J., "Shock Tube Studies of the H + C_2H_6 Reaction," *CJC* **51,** 2147 (1973).
Clingman, W. H., Jr., Brokaw, R. S., and Pease, R. N., "Burning Velocities of Methane with Nitrogen–Oxygen, Argon–Oxygen, and Helium–Oxygen Mixtures," *4th Symp.,* 310 (1953).
Clingman, W. H., and Pease, R., "Burning Velocity of Methane-Air Mixtures," *JACS* **78,** 1755 (1955).
Clyne, M., and Thrush, B., "Reaction of H Atoms with NO," *TFS* **57,** 1305 (1961).
Coffee, T., "Kinetic Mechanisms for Premixed Laminar, Steady State H_2–N_2O Flame," *C&F* **65,** 53 (1986).
Cohen, L. M., and Simpson, P., "Burning Velocities of HCN in Air and Oxygen," *C&F* **1,** 60 (1957).
Cole, J., Bittner, J., Longwell, J., and Howard, J., "Formation Mechanisms of Aromatic Compounds in Aliphatic Flames," *C&F* **56,** 51 (1984).
Colket, M., "Pyrolysis of Acetylene and Vinyl Acetylene in a Shock Tube," *21st Symp.,* 851 (1988).
Colket, M. B., Chiappetta, L., Guile, R. N., Zabielski, M. F., and Seery, D. J., "Internal Aerodynamics of Gas Sampling Probes," *C&F* **44,** 3 (1982).
Combourieu, J., and Laffitte, P., "Burning Velocity of Methane-Air Mixtures," *Bull. de la Soc. Chimique de France* **7,** 2521 (1969).
Combourieu, J., and Moreau, G., "A Study of ClO_2 Flames," *C&F* **24,** 381 (1975).
Combourieu, J., Moreau, R., Hall, A. R., and Pearson, G. S., "ClO_2 and $HClO_4$ Flames," *C&F* **13,** 596 (1969).
Combourieu, J., Moreau, G., Moreau, R., and Pearson, G., "Ammonium Perchlorate Combustion Analogue: Ammonia–Chlorine Dioxide Flames," *Amer. Inst. Aero. & Astro. J.* **8,** 594 (1970).
Compressed Gas Association, *Handbook of Compressed Gases,* 2nd ed., Van Nostrand Reinhold, New York (1981).
Comte-Bellot, G., "Hot Wire Annemometry," *Ann. Rev. of Fluid Mech.* **8,** 209 (1977).
Conway, J. B., Wilson, R. H., Jr., and Grosse, A. V., "Temperature of the Cyanogen–Oxygen Flame," *JACS* **75,** 499 (1953).

Cook, S. J., and Simmons, R. F., "The Effect of Hydrogen Bromide on the Structure of Propane–Oxygen Flames Diluted with Argon," *13th Symp.*, 585 (1971).

Cook, S. J., and Simmons, R. F., "The Effect of Tetraethyl Lead on the Structure of Propane–Oxygen–Argon Flames," *17th Symp.*, 891 (1979).

Cooke, D. F., Dodson, M. G., and Williams, A., "A Shock-Tube Study of the Ignition of Methanol and Ethanol with Oxygen," *C&F* **16,** 233 (1971).

Cooke, D. F., and Williams, A., "Shock Tube Studies of the Ignition and Combustion of Ethane and Slightly Rich Methane Mixtures with Oxygen," *13th Symp.*, 757 (1971).

Cooks, R. G., Glish, G. L., McLuckey, S. A., and Kaiser, R. E., "Ion Trap Mass Spectrometry," *Chem. Eng. News* **69** (12), March 25 (1991).

Cookson, R., Dunham, P., and Kilham, J., "Non-catalytic Coatings for Thermocouples," *C&F* **8,** 168 (1964).

Cool, T. A., Bernstein, J., Song, Xiao-Mei, and Goodwin, P. M., "Profiles of HCO and CH_3 in CH_4/O_2 and C_2H_4/O_2 Flames by Resonance Ionization," *22nd Symp.*, 1421 (1989).

Cooley, S. D., and Anderson, R. C., "Flame-Propagation Studies Using the Hydrogen–Bromine Reaction," *Ind. & Eng. Chem.* **44,** 1402 (1952).

Cooley, S. D., and Anderson, R. C., "Burning Velocities in Deuterium–Bromine and Hydrogen–Bromine Mixtures," *JACS* **77,** 235 (1955).

Cooley, S. D., Lasater, J. A., and Anderson, R. C., "Effect of Composition on Burning Velocities in Hydrogen–Bromine Mixtures," *JACS,* **74,** 739 (1952).

Corbeels, R., and Scheller, K., "Observations on the Kinetics of H_2–Cl_2 Flames," *10th Symp.*, 65 (1965).

Cornelius, W., and Agnew, W., (Eds.), *Emissions from Continuous Combustion Systems,* Plenum Press, New York (1972).

Corning Glass Co., "Pyrex^{tM} Glass Pipe Catalog" (1990).

Corre, C., Minetti, R., Pauwels, J. F., and Sochet, L. R., "About the Structure of a Two-Stage Butane Air Flame," in *Flame Structure,* Vol. I, O. P. Korobeinichev (Ed.), Novosibersk, NAUKA (1991), p. 74.

Cotton, F., and Wilkinson, G., *Advanced Inorganic Chemistry Interscience Publishers,* John Wiley (2nd ed., 1966; 4th ed., 1988).

Courtney-Pratt, J., "Advances in High Speed Photography," *High Speed Photography and Photonics Newsletter* **3,** Winter 1-6 (1983).

Coward, H. F., and Jones, G. W., "Burning Velocity of Methane–Oxygen Mixtures Diluted with Helium, Argon and Nitrogen," *JACS* **49,** 112 (1927).

Coward, H. F., and Jones, G. W., "Limits of Flammability of Gases and Vapors," *U. S. Bureau of Mines Bulletin #503,* Washington, DC (1954).

Cramerossa, F., and Dixon-Lewis, G., "Ozone Decomposition in Relation to the Problem of the Existence of Steady-State Flames," *C&F* **16,** 243–51 (1971).

Creighton, J. R., and Lund, C. M., "Modeling Study of Flame Structure in Low Pressure Laminar Pre-Mixed Methane Flames," *Proc. 10th Materials Research Symp.* (1980).

Creitz, E. C., "A Literature Survey of the Chemistry of Flame Inhibition," *J. of Res. NBS* **74A,** 521 (1970).

Creitz, E. C., "Flame Inhibition," *Fire Technology* **8,** 131 (1972).

Cremers, D. A., and Radziemski, L. J., "Laser Plasmas for Chemical Analysis," p. 351, in *Laser Spectroscopy and its Applications,* Radziemski, L. J., Solarz, R. W., and Paisner, J. A. (Eds.) Marcel Dekker, Inc., NY (1987).

Cribb, P., Dove, J., and Yamazaki, S., "A Shock Tube Study of Methanol Pyrolysis," *20th Symp.*, 779 (1985).

Crosley, D. R. (Ed.), *Laser Probes for Combustion Chemistry,* American Chemical Society Symposium Series No. 134, American Chemical Society, Washington, D.C. (1980).

Crosley, D. R., "Collisional Effects on Laser-Induced Fluorescence Flame Measurements," *Opt. Eng.* **23,** 475 (1984).
Cullis, C. F., and Hirschler, M. M., *The Combustion of Organic Polymers,* Clarendon Press, Oxford (1981).
Cullis, C. F., and Mulcahy, M. F. R., "The Kinetics of Combustion of Gaseous Sulphur Compounds," *C&F* **18,** 225 (1972).
Cullshaw, G. W., and Garside, J. E., "A Study of Burning Velocity (Pressure Dependence of Ethylene Air Mixtures)," *3rd Symp.,* 204 (1949).
Cummings, G. A. McD., and Hall, A. R., "Perchloric Acid Flames I: Methane and Other Fuels," *10th Symp.,* 1365 (1965).
Cummings, G. A. McD., Hall, A. R., and Straker, R., "Decomposition Flames of Acetylene and Methyl Acetylene," *8th Symp.,* 503 (1962).
Cummings, G. A. McD., and Pearson, G., "Perchloric Acid Decomposition Flame," *C&F* **8,** 199 (1965).
Curtiss, C. F., and Tonsager, W., "Atom–Diatomic Kinetic Theory Cross Sections," *JCP* **89,** 3795 (1985).
Daigreault, G. R., Morris, M. D., and Schneggenburger, R. G., "Quantitative Determination of Deuterium in Laser Fusion Targets by Inverse Raman Spectroscopy," *Appl. Spectrosc.* **37,** 443 (1983).
Dailey, J., "Laser Fluorescence," in *Laser Methods in Combustion,* Crosley, D. (Ed.), Am. Chem. Soc. (1982).
Dailey, J., and Chan, C., "Laser-Induced Fluorescence Measurement of Sodium in Flames," *C&F* **33,** 47 (1978).
Dainton, F. S., *Chain Reactions,* Methuen, London (1956).
D'Allessio, A., Di'Lorenzo, A., Beretta, F., and Ovenitozzi, C., "Soot in Fuel-Rich Methane–Oxygen Premixed Flames at Atmospheric Pressure," *14th Symp.,* 941 (1973).
Dalton, J., *New System of Chemistry,* Vol. 12, facsimile ed., William Dawson & Sons, London (1975).
Damkohler, G., "Combustion," *Der Chemieingenieur,* Vol. III, p. 1 Springer, Berlin (1940).
Damkohler, G., "Combustion," *ZfE* **46,** 601 (1940), English Translation, NACA Tech. Memo #1112 (1947).
Darian, S. T., and Vanpee, M. W., "A Spectroscopic Study of the Premixed Acetylene Nitrous Oxide Flame," *C&F* **70,** 65 (1987).
Davidson, D. F., Chang A., and Hanson, R. K., "Laser Photolysis Shock Tube for Combustion Kinetics Studies," *22nd Symp.,* 1877 (1989).
Davy, H., *Collected Works,* Vol. VI, (1840), pp. 3–130.
Day, M. J., Stamp, D. V., Dixon-Lewis, G., and Thompson, K., "Inhibition of Hydrogen–Air and Hydrogen–N_2O Flames by Halogen Compounds," *PRS* **A330,** 199 (1972).
Day, M. J., Stamp, D., Thompson, K., and Dixon-Lewis, G., "Inhibition of Hydrogen–Air and Hydrogen–Nitrous Oxide Flames by Halogen Compounds," *13th Symp.,* 705 (1971).
Day, M. J., Thompson, K., and Dixon-Lewis, G., "Reactions of HO_2 and OH," *14th Symp.,* 47 (1973).
Dean, A. M., Chou, M., and Stern, D., *The Chemistry of Combustion Processes, ACS Symposium #249,* Sloane, T. (Ed.), Washington, D.C., 71 (1984).
Dean, A., Hardy, J., and Lyon, R., "Kinetics and Mechanism of Ammonia Oxidation," *19th Symp.,* 97 (1982).
Deckers, J., and Van Tiggelen, A., "Ion Identification in Flames," *7th Symp.,* 254 (1959).
Demerdache, A., and Sugden, T. M., *The Mechanism of Corrosion by Fuel Impurities,* H. R. Johnson and J. D. Littler (Eds.), Butterworths, London, 12 (1963).
Demidov, P. G., *Combustion and Properties of Combustible Substances,* Wright Patterson AFB,

AD 621738, Federal Clearing House for Scientific and Technical Information, Springfield, Va. (1956) (trans. from Russian).
Demtroder, W., *Laser Spectroscopy,* Springer-Verlag, NY (1981).
Demtroder, W., *Analytical Laser Spectroscopy,* Springer-Verlag, New York (1983).
Dennison, D., and Wells, F., *JCP* **19,** 541 (1951).
Denton, M. B., and Malmstadt, H. V., "Tuneable Organic Dye Laser as an Excitation Source for Atomic Flame Florescence Spectroscopy," *Appl. Phys. Lett.* **18,** 485 (1971).
DeSoete, G., and Feugier, A., *Aspects physiques et chimiques de la combustion,* Technip, Paris (1976).
De Wilde, E., and Van Tiggelen, P. J., "Burning Velocities in Mixtures of Methyl Alcohol, Formaldehyde and Formic Acid with Oxygen," *Bull. Soc. Chim. Belges.* **77,** 67 (1968).
Diederichsen, J., and Wolfhard, H., "The Effect of Pressure on the Burning Velocity of Methane Flames," *TFS* **52,** 1102 (1956).
Dillion, A. (Ed.), *Hazardous Waste Incineration Engineering,* Noyes Data Corp., Park Ridge, N.Y. (1981).
Dils, R. R., and Tichenor, D. A., "A Fiberoptic Probe for Measuring High Frequency Temperature Fluctuations Gases," Sandia National Laboratories, Livermore, CA, *SAND83-8871* (February 1984).
Dixon-Lewis, G., "Temperature Distribution in Flame Reaction Zones," *4th Symp.,* 263 (1954).
Dixon-Lewis, G., "Flame Structure and Flame Reaction Kinetics I. Solution of Conservation Equations and Application to Rich Hydrogen–Oxygen Flames," *PRS* **A298,** 486 (1967).
Dixon-Lewis, G., "Flame Structure and Flame Reaction Kinetics II. Transport Phenomena in Multicomponent Systems," *PRS* **A307,** 111 (1968).
Dixon-Lewis, G., "Burning Velocities in Hydrogen–Air Mixtures," *C&F* **15,** 197 (1970).
Dixon-Lewis, G., "Flame Structure and Flame Reaction Kinetics. VII. Reactions of Traces of Heavy Watcr, Deuterium and Carbon Dioxide Added to Rich $H_2/O_2/N_2$ Flames," *PRS* **A330,** 219 (1972).
Dixon-Lewis, G., "Transport Phenomena, Chemical Mechanism and Laminar Flame Properties," *Flames as Reactions In Flow,* Sorgotto, I. (Ed.) p. 113, Ubstutyti du Unouabtu/ chemici dell 'Universita' Padova, Italy (1974).
Dixon-Lewis, G., "Kinetic Mechanism and Structure and Properties of Premixed Flames in Hydrogen–Oxygen–Nitrogen Mixtures," *Phil. Trans. Roy. Soc. London* **A292,** 45 (1979a).
Dixon-Lewis, G., "Mechanism of Inhibition of Hydrogen–Air Flames by Hydrogen Bromide and its Relevance to the General Problem of Flame Inhibition," *C&F* **36,** 1 (1979b).
Dixon-Lewis, G., "Aspects of the Kinetic Modeling of Methane Oxidation in Flames," *Coll. Int. Berthelot-Vieille Mallard Le Chatelier (Actes), 1st.,* 1, 284–89 (1981).
Dixon-Lewis, G., "Spherically Symmetric Flame Propagation in Hydrogen–Air Mixtures," *CST* **34,** 1 (1983).
Dixon-Lewis, G., "The Hydrogen-Oxygen-Nitrogen System," *Archivum Combustionis* **4,** 279 (1984).
Dixon-Lewis, G., "The Structure of Laminar Flames," *23rd Symp.,* 305 (1991).
Dixon-Lewis, G., David, T., Gaskell, P. H., Fukutani, S., Jinno, H., Miller, J., Kee, R. J., Smooke, D., Peeters, N., Effelsberg, E., Warnatz, J., and Behrendt, F., "Calculation of the Structure and Extinction Limit of a Methane–Air Counterflow Diffusion Flame in the Forward Stagnation Region of a Porous Cylinder," *20th Symp.,* 1893 (1985).
Dixon-Lewis, G., Goldsworthy, F. A., and Greenberg, J. B., "IX. Calculation of the Properties of Multiradical Premixed Flames," *PRS* **A346,** 261 (1976).
Dixon-Lewis, G., Greenberg, J. B., and Goldsworthy, F. A., "Reactions in the Recombination Region of Hydrogen and Lean Hydrocarbon Flames," *15th Symp.,* 717 (1975).

Dixon-Lewis, G., and Islam, S. M., "Modeling and Burning Velocity Measurement," *19th Symp.*, 283 (1982).

Dixon-Lewis, G., Isles, G. L., and Walmsley, R., "Flame Structure and Flame Reaction Kinetics: VIII. Structure, Properties, and Mechanism of a Rich $H_2/O_2/N_2$ Flame at Low Pressure," *PRS* **A331,** 571 (1973).

Dixon-Lewis, G., and Missaghi, M., "Structure and Extinction Limits of Counterflow Diffusion Flames of Hydrogen–Nitrogen Mixtures in Air," *22nd Symp.*, 1481 (1989).

Dixon-Lewis, G., and Shepherd, I. G., "Some Aspects of Ignition by Localized Sources and Spherical and Cylindrical Flames," *15th Symp.*, 1483 (1975).

Dixon-Lewis, G., and Simpson, R., "Flame Inhibition by Halogen Compounds," *16th Symp.*, 1111 (1976).

Dixon-Lewis, G., Sutton, M., and Williams, A., "Some Reactions of H Atoms and Simple Radicals at High Temperatures," *10th Symp.*, 495 (1965a).

Dixon-Lewis, G., Sutton, M. M., and Williams, A., "Reactions Contributing to the Establishment of the Water Gas Equilibrium when Carbon Dioxide is Added to a $H_2/O_2/N_2$ Flame," *TFS* **61,** 266 (1965b).

Dixon-Lewis, G., Sutton, M. M., and Williams, A., "Flame Structure and Reaction Kinetics: IV. Experimental Investigations of a Fuel Rich $H_2/O_2/N_2$ Flame Atmospheric Pressure," *PRS* **A317,** 227 (1970).

Dixon-Lewis, G., and Williams, A., "Observation of the Structure of the Slow Burning Hydrogen–Oxygen Flames," *9th Symp.*, 576 (1961).

Dixon-Lewis, G., and Williams, A., "The Combustion of CH_4 in Premixed Flames," *11th Symp.*, 951 (1967).

Dixon-Lewis, G., and Williams, A., "The Reactions of Hydrogen and Carbon-Monoxide with Oxygen," *Comprehensive Chemical Kinetics* **17,** 1 (1977).

Dolezal, R., *Large Boiler Furnaces,* Elsevier, New York (1971).

Dosanji, C. (Ed.), *Modern Optical Methods in Gas Dynamics Research,* Plenum Press, New York (1971).

Dove, J. E., and Warnatz, J., "Calculation of Burning Velocity and Flame Structure in Methanol Air Mixtures," *Buns.* **87,** 1040 (1983).

Dowdy, D. R., Smith, D. B., Taylor, S. C., and Williams, A., "The Use of Expanding Spherical Flames to Determine Burning Velocities and Stretch Effects on Hydrogen Air Mixtures," *23rd Symp.*, 325 (1991).

Drain, L. E., *The Laser Doppler Technique,* John Wiley and Sons, N.Y. (1980).

Drake, M. C., and Blint, R. J., "Structure of Opposed Flow Diffusion Flames with $CO/H_2/N_2$ Fuel" *CST* **61,** 187 (1988).

Drake, M. C., Lapp, M., Penney, C. M., Warshaw, S., and Gerhold, B. W., "Measurements of Temperature and Concentration Fluctuations in Turbulent Diffusion Flames Using Pulsed Raman Spectroscopy," *18th Symp.*, 1521 (1981).

Drake, M. C., Ratcliffe, J. W., Blint, B. J., Carter, C. D., and Laurendeau, N. M., "Measurements and Modeling of Flame Front Nitric Oxide Formation and Super Equilibrium Radical Concentrations in Laminar High-Pressure Premixed Flames," *23rd Symp.*, 387 (1991).

Dreier, T., Lange, B., Wolfrum, J., Zahn, M., Behrendt, F., and Warnatz, J., "Comparison of CARS Measurements and Calculations of the Structure of Laminar CH_4–Air Counterflow Diffusion Flames," *Buns.* **90,** 1010 (1986).

Druet, S. A., and Taran, J-P. E., "CARS Spectroscopy," *Prog. Quantum Electron.* **7,** 1 (1981).

Dryer, F. L., and Brezinsky, K., "Flow Reactor Octane Studies," *CST* **45,** 199 (1985).

Dryer, F. L., and Glassman, I., "The High Temperature Oxidation of CO and CH_4," *14th Symp.*, 987 (1973).

Dryer, F. L., and Glassman, I., "Combustion Chemistry of Chain Hydrocarbons," in C. T. Bowman and J. Birkeland (Eds.), *Alternative Hydrocarbon Fuels: Combustion and Chemical Kinetics. Progress in Astronautics and Aeronautics 62,* 255 (1979).

Ducarme, J., Gerstein, M., and Lefebvre, A. H., *Progress in Combustion Science,* Vol. 1, Pergamon Press, London (1960).

Dudderar, T. D., Meynart, R., and Simpkins, P. G., "Full Field Laser Metrology for Fluid Flow Measurement," in *Optics and Lasers in Engineering* **9,** 161–325, Chiang, F. P., and Reid, G. T. (Eds.), Elseviere, New York (1988).

Dudderar, T. D., and Simkins, P. G., "Summary of a Talk on Two-Dimensional Fluid Velocity Measurement by Laser Speckle Velocimetry," *AIAA 19th Fluid Dynamics, Plasma Dynamics and Lasers Conference, AIAA-87-1375,* American Institute of Aeronautics and Astronautics, New York (June 1987).

Duff, R., Knight, H., and Wright, H., "Detonation of Acetylene," *JCP* **2,** 1618 (1954).

Dugger, G. L., "Effect of Initial Temperature on the Flame Speed of Methane, Ethylene and Propane–Air Flames," *NACA Report,* 1061 (1952).

Durago, D. F. G., Heitor, M. V., Whitelaw, J. J. H., and Witze, P. O., *Combustion Flow Diagnostics,* Kluwer Academic Publ., Norwell, MA, NATO/ASI series (1991).

Durago, D. F. G., and Whitelaw, J. H., "Critical Review of Laser Annomometry," in *Experimental Diagnostics in Gaseous Combustion Systems,* Zinn, B. (Ed.), A.I.A.A., NY (1977).

Durant, W., *The Story of Civilization* (eight volumes), Simon & Schuster, New York, 130 (1944–1963), 01-010.

Durie, R. A., "Fluorine Diffusion Flames," *PPS* **A65,** 1225 (1952).

Durst, F., Melling, A., and Whitelaw, J. H., *Principles and Practice of Laser-Doppler Anemometry,* Academic Press (1976).

Dushman, S., *Scientific Foundation of Vacuum Techniques,* John Wiley, New York, 2nd Ed. (1962).

Duval, A., and Van Tiggelen, A., "Burning Velocities of Flames with Nitrous Oxide," *Bull. Acad. Roy. Belg.* **53,** 366 (1967).

Easley, G. L., *Coherent Raman Spectroscopy,* Pergamon Press, Oxford (1981).

Eberius, K. H., Hoyerman, K., and Wagner, H. G., "Experimental and Mathematical Study of a Hydrogen/Oxygen Flame," *13th Symp.,* 713 (1971).

Eberius, K. H., Hoyerman, K-H., and Wagner, H. G., "Structure of Lean Acetylene Oxygen Flames," *14th Symp.,* 147 (1973).

Echekki, T., and Mungal, M. G., "Flame Speed Measurements at the Tip of a Slot Burner. Effects of Flame Curvature and Hydrodynamic Stretch," *23rd. Symp.,* 455 (1991).

Eckart, C., "The Thermodynamics of Irreversible Processes," *Phys. Rev.* **58,** 267 (1940).

Eckbreth, A. C., "BOXCARS: Crossed-Beam Phase-Matched CARS Generation in Gases," *Appl. Phys. Lett.* **32,** 421 (1978).

Eckbreth, A. C., "Recent Advances in Laser Diagnostics for Temperature and Species Concentration in Combustion," *18th Symp.,* 1471 (1981).

Eckbreth, A. C., Bonczyk, P. A., and Verdieck, J. F., "Combustion Diagnostics by Laser Raman and Fluorescence Techniques," *Prog. Energy Comb. Sci.* **5,** 253 (1979).

Eckbreth, A. C., and Hall, R. J., "CARS Thermometry in a Sooting Flame," *C&F* **39,** 133 (1980).

Eckbreth, A. C., and Hall, R. J., "CARS Concentration Sensitivity With and Without Nonresonant Background Suppression," *CST* **22,** 175 (1981).

Eckbreth, A. C., Hall, R. J., Shirley, J. A., and Verdieck, J. F., "Spectroscopic Investigations of CARS for Combustion Applications," *Adv. Laser Spectrosc.* **1,** 101 (1982).

Eckert, E. R. G., and Drake, R. M. *Heat and Mass Transfer,* McGraw–Hill, New York (1959).

Edmundson, H., and Heap, M. P., "The Burning Velocity of Hydrogen–Air Flames," *C&F* **16,** 161 (1971).

Edwards, J. B., *Combustion,* Ann Arbor Science, Ann Arbor, Mich. (1974).

Egerton, A., and Lefebvre, A., "The Burning Velocity of Methane-Air Mixtures," *PRC* **222A**, 206 (1954).

Egolfopoulos, F. N., and Law, C. K., "An Experimental and Computational Study of the Burning Rates of Ultra Lean to Moderately Rich $H_2/O_2/N_2$ Laminar Flames with Pressure Variation," *23rd Symp.,* 333 (1991).

Egolfopoulos, F. N., Zhu, D. L., and Law, C. K., "Experimental and Numerical Determination of Laminar Flame Speeds: Mixtures of C_2 Hydrocarbons with Oxygen and Nitrogen," *23rd Symp.,* 471 (1991).

El Wakel, M. M., "Interferometry in Flames," in *Combustion Measurements,* R. Goulard (Ed.), Academic Press (1976).

Elder, M. L., and Winefordener, J. D., "Temperature Measurements in Flames: A Review," *Anal. At. Spectrosc.* **6,** 293 (1983).

Ellis, G. C., *Memoir of Sir Benjamin Thompson, Count Rumford,* Macmillan, London (1876).

Emmons, H., "The Basic Theory of Gas Dynamic Discontinuities Fundamentals of Gas Dynamics," Vol. 3, *High Speed Aerodynamics,* Princeton University Press, Princeton. p. 433 (1959), 01–037.

Emmons, H., "Fluid Mechanics and Combustion," *13th Symp.,* 1 (1971).

Emmons, H., "The Growth of Fire Science," *Fire Safety J.* **3,** 95 (1980–1981).

Engstrom, R. W., *RCA Photomultiplier Handbook,* RCA Corp., Lancaster, PA (1980).

Enskog, D., "Kinetic Theory of Processes in Gases" (in German), *Archiv för Mathematik, Astronomi, och Fysik* **16** (1922).

Ettre, L. S., *Introduction to Open Tubular Columns,* Perkin–Elmer Corp., (1979).

Evans, M., "Current Theoretical Concepts of Steady State Flame Propagation," *Chem. Rev.* **51,** 363 (1952).

Evans, S., and Simmons, R. F., "A Structural Study of a Propane–Oxygen Diffusion Flame Diluted with Argon," *22nd Symp.,* 1433 (1989).

Ewing, K., Hughes, T., and Carhart, H. W., "Evidence for Extinction by Thermal Mechanisms?," *Fire Tech.* **24,** 195 (1989).

Exner, J. H. (ed.), *Detoxification of Hazardous Wastes,* Ann Arbor Science (1982).

Eyring, H., Glasstone, S., and Laidler, K., *Theory of Rate Processes,* McGraw–Hill, New York (1945).

Falcone, P., Hanson, R., and Kruger, C., "Measurement of NO in Combustion Gases Using a Tuneable Diode Laser," Paper 79-54, Western States Combustion Institute Meeting (1979).

Faraday, M., *Chemical History of A Candle,* Viking Press, New York (1960).

Faraday, M., *Experimental Researches in Chemistry and Physics,* Reprint, Cultureiet Civilization, Brussels (1969).

Faris, G., and Byer, R., "Beam Deflection Optical Tomography of a Flame," *Opt. Lett.* **12,** 155 (1987).

Faris, G., and Byer, R., "Quantitative 3D Optical Tomography Imaging," *Science* **238,** 1700 (1988).

Farrow, R., Lucht, R., Flower, W., and Palmer, R., "CARS Measurement of Temperatures and Acetylene Spectra in Sooting Flames," *20th Symp.,* 1307 (1986).

Farrow, R., and Rahn, L., "Spatially Resolved IR Absorbtion Measurements: Application of an Optical Stark Effect," *Opt. Letters* **6,** 525 (1981).

Farrow, R. L., and Rakestraw, D. J., "Detection of Trace Molecular Species Using Degenerate Four-Wave Mixing," *Science* **257,** 1894 (1992).

Feiser, L., and Fieser, M., *Textbook of Organic Chemistry,* Heath and Co., Boston (1950) 11–77.

Fells, I., and Rutherford, A., "Effect of Additives on the Burning Velocity of CH_4–Air Flames," *C&F* **13,** 130 (1966).

Fenimore, C. P., *Chemistry in Premixed Flames,* Pergamon Press, New York (1964).

Fenimore, C. P., "Formation of NO in Premixed Hydrocarbon Flames," *13th Symp.,* 197, 373–80 (1971).

Fenimore, C. P., "Formation of NO from Fuel Nitrogen in Ethylene Flames," *C&F* **19,** 289 (1972).

Fenimore, C. P., "Reactions of Fuel Nitrogen in Rich Flame Gases," *C&F* **26,** 249 (1976).

Fenimore, C. P., and Jones, G. W., "Radical Recombination and Heat Evolution of H_2-O_2 Flames," *10th Symp.,* 489 (1956).

Fenimore, C. P., and Jones, G. W., "Determination of Oxygen Atoms in Lean Flames by Reaction with Nitrous Oxide," *JPC* **62,** 178 (1958a).

Fenimore, C. P., and Jones, G. W., "The Water-Catalyzed Oxidation of CO by Oxygen at High Temperature," *JPC* **61,** 651; "Reaction of H Atoms with CO_2 in Flame Gases," 1578 (1958b).

Fenimore, C. P., and Jones, G. W., "Formation of CO in Methane Flames by Reaction with Oxygen Atoms with Methyl Radical," *JPC* **65,** 1532 (1961a).

Fenimore, C. P., and Jones, G. W., "Rate of Reaction of Methane with H Atoms and OH Radicals in Flames," *JPC* **65,** 2200 (1961b).

Fenimore, C. P., and Jones, G. W., "Oxidation of Ammonia in Flames," *JPC* **65,** 298 (1961c).

Fenimore, C. P., and Jones, G. W., "Rate of the Reaction $O + N_2O = 2NO$," *8th Symp.,* 127 (1962).

Fenimore, C. P., and Jones, G. W., "The Destruction of C_2H_2 in Flames with Oxygen," *JCP* **39,** 1514 (1963a).

Fenimore, C. P., and Jones, G. W., "The Decomposition of Ethylene and Ethane in Premixed Hydrocarbon–Oxygen–Hydrogen Flames," *9th Symp.,* 597 (1963b).

Fenimore, C. P., and Jones, G. W., "Flame Inhibition by Methyl Bromide," *Comb. and Flame* **7,** 323 (1963c).

Fenimore, C. P., and Jones, G. W., "Rate of Destruction of Acetylene in Flame Gases," *JCP* **41,** 1887 (1964).

Fenimore, C. P., and Jones, G. W., "Sulfur in the Burnt Gas of H_2/O_2 Flame," *JPC* **69,** 3593 (1965).

Fenimore, C. P., and Martin, F. J., "The Burning of Polymers," in *Oxidation and Burning of Organic Materials,* National Bureau of Standards, Special Publication, No. 357, Washington, DC (1972).

Fenn, J., and Calcote, C. F., "Activation Energies in High Temperature Combustion," *4th Symp.,* The Combustion Institute, Pittsburgh, PA, p. 213 (1953).

Ferguson, C., *Internal Combustion Engines,* John Wiley & Co., New York (1986).

Ferguson, R., and Broida, H., "Atomic Flames: Spectra Temperatures and Products," *5th Symp.,* 754 (1956).

Ferri, A. (Ed.), *Fundamental Data Obtained From Shock Tube Experiments,* Pergamon Press, NY (1961).

Field, M. A., Gill, D. W., Morgan, B. B., and Hawksley, F. G. W., *Combustion of Pulverized Coal,* Institute of Fuel, London (1967).

Fisher, C., "A Study of Rich Ammonia/Oxygen/Nitrogen Flames," *C&F* **30,** 143 (1977).

Fisher, E. M., Koshland, C. P., Hall, M. J., Sawyer, R. F., and Lucas, D., "Experimental and Numerical Study of the Thermal Destruction of C_2H_5Cl," *23rd Symp.,* 895 (1991).

Fisher, S. S., and Knuth, E. L., "Properties of Low Density Free Jets Measured Using Molecular Beam Techniques," *Am. Inst. Aero. Astro. J.* **1,** 1174 (1969).

Fletcher, E., "Fluorocarbon Studies. VII. Competitive Reactions of Fluorocarbons Burning with Fluorine," *C&F* **51,** 193 (1983). Also see National Bureau of Standards Publication # 3357, 1553–58, June 1972.

Fletcher, E., and Ambs, L., "Fluorocarbon Combustion Studies—The Combustion of Perfluoroethane, Perfluoropropane and Perfluorocyclobutane with Chlorine Trifluoride," *C&F* **8,** 275 (1964).

Fletcher, E., and Ambs, L. L., "The Chlorine–Fluorine Flame," *C&F* **12,** 112 (1968).

Fletcher, E., and Kittelson, D., "Perflurocylobutane–Oxygen Spatial Velocity and Limits," *C&F* **12,** 1664 (1968a).

Fletcher, E., and Kittelson, D., "Fluorocarbon Studies. III. Detonation Velocities and Limits of Perfluorocyclobutane–Fluorine Mixtures," *C&F* **12,** 119 (1968b).

Fletcher, E., and Kittelson, D., "Burning Velocities of Perfluoro-Cyclobutane–O_2 Mixtures," *C&F* **13,** 434 (1969).

Foner, S., and Hudson, R., "Radicals in Flames," *JCP* **21,** 1374 (1953).

Forsythe, W. E., *Optical Pyrometry, Temperature—Its Measurement and Control In Science and Industry,* Reinhold Publ. Co., New York (1941), p. 1115.

Francis, W., *Fuels and Fuel Technology,* Vols. 1 & 2, Pergamon Press, London (1965).

Franck-Kamenetzki, D. A., *Diffusion and Heat Exchange in Chemical Kinetics,* Princeton University Press, Princeton (1955).

Frank, P., Bhaskaran, K., and Just, Th., "Acetylene Oxidation: The Reaction C_2H_2 + O at High Temperatures," *21st Symp.,* 885 (1988).

Franze, C., and Wagner, H. Gg., "The Burning Velocity of Carbon-Monoxide-Hydrogen Flames with Air," *Z., Elektrochem.* **60** 525 (1956).

Fraser, I. M., and Winefordner, J. D., "Laser-Excited Atomic Fluorescence Spectroscopy," *Anal. Chem.* **43,** 1693 (1971).

Frazier, G., "Hydrogen Bromine Flame Studies," *Ph.D. Thesis,* Dept. Chem. Eng., Johns Hopkins Univ., Baltimore, MD (1962).

Frazier, G., Fristrom, R. M., and Wehner, J. F., "Microstructure of a H_2–Br_2 Flame," *Am. Inst. Chem. Eng. J.* **9,** 689–93 (1963).

Frazier, G., and Wendt, J., "The Hydrogen–Bromine Flame," *J. Chem. Eng. Sci.* **24,** 95–111 (1969).

Friedman, R., "Measurement of the Temperature Profile in a Laminar Flame," *4th Symp.,* 259 (1953).

Friedman, R., "A Study of the Ethylene Decomposition Flame," *5th Symp.,* 596 (1955).

Friedman, R., "Survey of Chemical Inhibition," *Fire Res. Abs. and Rev.* **3,** 128 (1961).

Friedman, R., "The Role of Chemistry in Fire Problems," *Fire Res. Abs. and Rev.* **13,** 187 (1971).

Friedman, R., and Burke, E., "Measurement of a Temperature Distribution in a Low Pressure Flat Flame," *JCP* **22,** 824 (1954).

Friedman, R., and Cyphers, J., "Gas Sampling in a Low Pressure Propane–Air Flame," *JCP* **22,** 1875 (1955).

Friedman, R., and Macek, A., "Combustion Studies of Single Aluminum Particles," *9th Symp.,* 703 (1963).

Friedman, R., and Nugent, R., "Premixed Carbon Monoxide Combustion," *7th Symp.,* 311–16 (1959).

Fristrom, R. M., "The Ballistic Switch: A Means Toward High Intensity, High Speed Repetitive Illumination," *Phot. Eng.* **4,** 74 (1953).

Fristrom, R. M., "Applicability of One Dimensional Models to Three Dimensional Laminar Bunsen Flame Fronts," *JCP* **24,** 888–94 (1956).

Fristrom, R. M., "The Structure of Laminar Flames," *6th Symp.,* 96 (1957).

Fristrom, R. M., "Premixed Spherical Flames," *C&F* **2,** 102 (1958).

Fristrom, R. M., "Scavenger Probe Sampling—A Method for Studying Gas Phase Free Radicals," *Science* **140,** 297 (1963a).
Fristrom, R. M., "Radical Concentrations and Reactions in a Methane–Oxygen Flame," *9th Symp.,* 560 (1963b).
Fristrom, R. M., "Experimental Techniques for Studying Flame Structures," *JHU/APL BB300* (1963c).
Fristrom, R. M., "The Definition of Burning Velocity and a Geometric Interpretation of Flame Curvature," *PF* (1965).
Fristrom, R. M., "Flame Chemistry," *Survey of Progress in Chemistry,* **3,** 55, A. F. Scott (Ed.), Academic Press (1966).
Fristrom, R. M., "Combustion Suppression—Review and Comments," *Fire Res. Abs. and Rev.* **9,** 125 (1967).
Fristrom, R. M., "Chemical Factors in the Inhibition and Extinction of C/H/O Flames by Halogenated Compounds as Interpreted with a Zonal Model," Presented at Fall Meeting of Western States Section of UCLA, Los Angeles, CA (1980).
Fristrom, R. M., "Comments on Quenching in Microprobes," *C&F* **50,** 239 (1983).
Fristrom, R. M., "The Ozone Flame Revisited," APL, The Johns Hopkins Univ. RCP internal report prepared for submission to Combustion and Flame. [1994]
Fristrom, R. M., Avery, W. A., Prescott, R., and Mattuck, A., "Flame Zone Studies by the Particle-Track Technique. I. Apparatus and Techniques," *JCP* **22,** 106 (1954).
Fristrom, R. M., Favin, S., Linevsky, M. J., Vandooren, J., and Van Tiggelen, P. J., "A Study of Halogen Flames Using a Zonal Flame Model," to be published *Bull. Soc. Chim. Belg.* (1992).
Fristrom, R. M., Grunfelder, C., and Avery, W. H., "Microstructure of C_2 Hydrocarbon–Oxygen Flames," *7th Symp.,* 204 (1959). Also see Fristrom, R. M., and Grunfelder, C., "Temperature Profiles of Some C_2 Hydrocarbon/Oxygen Flames," Unpublished work, APL/JHU Johns Hopkins Rd., Laurel, MD 20207 (1968).
Fristrom, R. M., Grunfelder, C., and Favin, S., "Characteristic Profiles in a Low Pressure Laminar Lean Methane–Oxygen Flame," *JPC* **64,** 1386 (1960).
Fristrom, R. M., Grunfelder, C., and Favin, S., "Characteristic Profiles, Matter and Energy Conservation in a One-Twentieth Atmosphere Methane–Oxygen Flame," *JPC* **65,** 87 (1961).
Fristrom, R. M., Jones, A. R., Schwar, M. J. R., and Weinberg, F. J., "Particle Sizing by Interference Fringes and Signal Coherence in Doppler Velocimetry," *Faraday Symp. of Chem. Soc. #7,* 183 (1971).
Fristrom, R. M., and Linevsky, M. J., "Eddy Burner for Flame-Flow Studies," Poster Session, *20th Symp.,* University of Michigan (1985).
Fristrom, R. M., Linevsky, M. J., Hoshall, H., and Vandooren, J., "An Improved Shock Tube," submitted to *Rev. Sci. Inst.* (1993).
Fristrom, R. M., and McLean, W., "Scavenger Probe Studies Using Deuterium as a Scavenger," Unpublished Results Combustion Facility Sandia National Laboratory, Livermore, CA (1981).
Fristrom, R. M., and Monchick, L., "Two Simple Approximations for the Thermal Diffusion Factor," *C&F* **71,** 89 (1989).
Fristrom, R. M., Prescott, R., and Grunfelder, C., "Flame Zone Studies III—Techniques for Determining Composition Profiles of Flame Fronts," *C&F* **1,** 102 (1957).
Fristrom, R. M., Prescott, R., Neumann, R., and Avery, W. H., "Temperature Profiles in Propane–Air Flame Fronts," *4th Symp.,* 267 (1953).
Fristrom, R. M., and Sawyer, R. F., "Flame Inhibition Chemistry," Paper #12 in Proceedings of AGARD Conference, *Aircraft Fuels, Lubricants and Fire Safety,* available from NASA Langley Field, VA 23365, Report Distribution and Storage Unit (1971).

Fristrom, R. M., and Van Tiggelen, P. J., "An Interpretation of the Inhibition of C–H–O Flames by C–H–X Compound," *7th Symp.*, 773 (1979).

Fristrom, R. M., and Westenberg, A. A., "Microstructure and Material Transport in a Laminar Propane–Air Flame Front," *C&F* **1,** 217 (1957).

Fristrom, R. M., and Westenberg, A. A., "Conservation of Matter and Energy in the One-Tenth Atmosphere Methane Flame," *JPC* **64,** 1393 (1960).

Fristrom, R. M., and Westenberg, A. A., "Experimental Chemical Kinetics from Methane–Oxygen Flame Structure," *8th Symp.*, 438 (1963).

Fristrom, R. M., and Westenberg, A. A., *Flame Structure,* McGraw–Hill, New York (1965).

Fristrom, R. M., and Westenberg, A. A., "Molecular Transport Properties for Flame Studies," *Fire Res. Abs. and Rev.* **8,** 155 (1968).

Fristrom, R. M., Westenberg, A. A., and Avery, W. H., "The Study of the Mechanism of Propane–Air Reaction Based on the Analysis of Flame Front Profiles," *Revue de l'Institut Français du Petrole et Annales des Combustibles Liquides XI, No. 4,* 544 (April 1958).

Fujii, N., and Asaba, T., "Shock Tube Study of the Reaction of Rich Mixtures of Benzene and Oxygen," *14th Symp.*, 433 (1973).

Fuller, L., Parks, D., and Fletcher, E., "Flat Flames in Tubes for Measuring Burning Velocity," *C&F* **13,** 455 (1969).

Gann, R. G. (Ed.), *Halogenated Fire Suppressants,* American Chemical Society, Washington, DC (1975). Contents: (a) C. L. Ford, An Overview of Halon 1301 Systems, 1; (b) M. J. Miller, The Relevance of Fundamental Studies of Flame Inhibition to the Development of Standards for the Halogenated Extinguishing Agent Systems, 64; (c) N. J. Alvares, CF_3Br Suppression of Turbulent, Class-B Fuel Fires, 94; (d) J. W. Hastie and C. L. McBee, Mechanistic Studies of Halogenated Flame Retardants: The Antimony–Halogen System, 118; (e) K. Seshadri and F. A. Williams, Effect of CF_3Br on Counterflow Combustion of Liquid Fuel with Diluted Oxygen, 149; (f) R. F. Kubin, R. H. Knipe, and A. S. Gordon, Halomethane and Nitrogen Quenching of Hydrogen and Hydrocarbon Diffusion and Premixed Flames, 183; (g) R. W. Scheafer, N. J. Brown, and R. F. Sawyer, The Effect of Halogens on the Blowout Characteristics of an Opposed Jet Stabilized Flame, 208; (h) L. W. Hunter, C. Grunfelder, and R. M. Fristrom, Effects of CF_3Br on a H_2–O_2–Ar Diffusion Flame, 234; (i) J. C. Biordi, C. P. Lazzara, and J. F. Papp, The Effect of CF_3Br on Radical Concentration Profiles in Methane Flames, 256; (j) G. B. Skinner, Inhibition of the Hydrogen–Oxygen Reaction by CF_3Br and CF_2BrCF_2Br, 295; (k) R. G. Gann, Initial Reactions in Flame Inhibition by Halogenated Hydrocarbons, 318; (l) N. J. Brown, Halogen Kinetics Pertinent to Flame Inhibition: A Review, 341; (m) E. R. Larsen, Halogenated Fire Extinguishants: Flame Suppression by a Physical Mechanism?, 376; (n) E. T. McHale and A. Mandl, The Role of Ions and Electrons in Flame Inhibition by Halogenated Hydrocarbons: Two Views, 403; (o) S. Galant and J. P. Appleton, Theoretical Investigation of Inhibition Phenomena in Halogenated Flames, 406.

Gardiner, W. C., "Observations of Induction Times in the Acetylene–Oxygen Reaction," *JCP* **35,** 2252 (1961).

Gardiner, W. C., *Chemistry of Combustion Reactions,* Springer-Verlag, New York (1984).

Garforth, A., and Rallis, C., "Laminar Burning Velocity of Stoichiometric Methane–Air Pressure and Temperature Dependence," *C&F* **31,** 53 (1978).

Garrison, H. R., Lasater, J. A., and Anderson, R. C., "Studies on the Hydrogen–Bromine Flame," *3rd Symp.*, 155 (1949).

Garvin, D., and Broida, H. P., "Atomic Reactions Involving H, N with O_3," *9th Symp.*, 678 (1963).

Gay, R. L., Young, W. S., and Knuth, E., "Molecular Beam Sampling of H_2O, CO and NO in One-Atmosphere Methane–Air Flames," *C&F* **24,** 391 (1975).

Gaydon, A. G., *Spectroscopy and Combustion Theory,* Chapman–Hall, London (1948).

Gaydon, A. G., "The Use of Shock Tubes for Studying Combustion Processes," *11th Symp.*, 1 (1967).
Gaydon, A. G., *Spectroscopy of Flames,* Chapman and Hall, London, 2nd ed. (1974);
Gaydon, A. G., and Broida, H., "Luminous Reaction Between CO and O," *TFS* **49,** 1190 (1958).
Gaydon, A. G., and Hurle, I., *The Shock Tube in High Temperature Chemical Physics,* Chapman–Hall, London (1963).
Gaydon, A. G., and Wolfhard, H. G., "Spectroscopic Studies of Low Pressure Flames (N_2O with: H_2, CH_4, CH_3OH, CH_2O, HCOOH, C_2H_2, C_7H_8, C_2H_5OH, CH_3–CO–CH_3), *3rd Symp.*, 504 (1949).
Gaydon, A. G., and Wolfhard, H. G., "Spectra of Flames Supported by Free Atoms," *PRS* **A 213,** 366 (1952).
Gaydon, A. G., and Wolfhard, H. G., *Flames,* 4th ed., Chapman–Hall, London (1979).
Gehring, M., Hoyermann, K.-H., Schacke, H., and Wolfrum, J., "Direct Studies of Some Elementary Steps for the Formation and Destruction of Nitric Oxide in the H–N–O System," *14th Symp.*, 99 (1973).
Gerstein, M., "The Structure of Laminar Flames," *7th Symp.*, 35 (1953).
Gerstein, M., "A Study of Alkyl Silane Flames," *7th Symp.*, 903 (1959).
Gerstein, M., Levine, O., and Wong, E., "Flame Propagation: II. The Determination of Fundamental Burning Velocities Hydrocarbons by a Revised Tube Method," *JACS* **73,** 418 (1951a).
Gerstein, M., Levine, O., and Wong, E., "Fundamental Flame Velocities of Hydrocarbons," *Ind. and Eng. Chem.* **43,** 2770 (1951b).
Gerstein, M., McDonald, G. E., and Schalla, R. L., "Decomposition Flame Studies with Ethylene Oxide," *4th Symp.*, 375 (1953).
Gerstein, M., Wong, E. L., and Levine, O., "Flame Velocities of Four Alkyl Silanes," *NACA RM E51A08,* (1951).
Gibbs, G., and Calcote, H., "Effect of Molecular Structure on Burning Velocity," *J. Chem. Eng. Data* **4,** 226 (1959).
Gibbs, J. W., "On the Equilibrium of Heterogeneous Substances," *Trans. Conn. Acad. Sci.* **3,** 228 (1876).
Gibbs, J. W., *Collected Works of J. W. Gibbs,* Longmans Green & Co., Inc., NY (1928).
Gilbert, M., "The Influence of Pressure on Flame Speed," *6th Symp.*, 74 (1957).
Gilbert, M., and Altman, D., "Effect of Isotopic Substitution in Hydrogen-Bromine Flames," *JCP* **25,** 377 (1956).
Gilbert, M., and Altman, D., "Chemical Steady State in HBr Flames," *6th Symp.*, 2222 (1957).
Gilbert, M., and Lobdell, J. H., "Resistance Thermometer Measurements in a Low Pressure Flame," *4th Symp.*, 285 (1953).
Glassman, I., *Combustion,* Academic Press, New York (1977).
Glassman, I., Dryer, F. L., and Cohen, R., "Studies of Hydrocarbon Oxidation in a Flow Reactor," *2nd International Symposium Flames as Reactions in Flow,* I. Sorgotto (Ed.) Consiglio Nazionale Delle Riche, Rome, Italy (1976).
Glassman, I., Hansel, J., and Eklund, T., "Hydrodynamic Effects in the Flame Streading Ignitability and Burning of Liquid Fuels," *C&F* **13,** 99 (1969).
Glassman, I., and Sawyer, R. F., *Performance of Chemical Propellants,* AGARDOGRAPH No. 129, Pergamon Press, Oxford (1974).
Glasstone, S., Laidler, K., and Eyring, H., *The Theory of Rate Processes,* McGraw–Hill, New York (1941).
Gleick, J., *Chaos: Making a New Science,* Viking, New York (1987).
Glumac, N. G., and Goodwin, D. G., "Diamond Growth in a Novel Low Pressure Flame," *Appl. Phys. Lett.* **60,** 2695 (1992).

Goldsmith, J. E. M., "Two Step Saturated Fluorescence Detection of Atomic Hydrogen in Flames," *Opt. Lett.* **10,** 116 (1985).
Goldsmith, J. E. M., "Resonant Multiphoton Optogalvanic Detection of Atomic Hydrogen in Flames," *Opt. Lett.* **7,** 437 (1982).
Goldsmith, J. E. M., "Multiphoton Excited Fluorescence Measurements of Atomic Hydrogen in Low Pressure Flames," *22nd Symp.,* 1403 (1989).
Goldsmith, J. E. M., and Anderson, J., "Imaging of Atomic Hydrogen in Flames with Two Step Saturated Fluorescence Detection," *Appl. Opt.* **24,** (1985).
Goldsmith, J. E. M., Miller, J. A., Anderson, R. J. M., and Williams, L. R., "Multiphoton Excited Flourescence Measurements of Absolute Concentration Profiles of Atomic Hydrogen in Low-Pressure Flames," *23rd Symp.,* 1821 (1991).
Goldsmith, S. (Ed.), *Modern Developments in Fluid Dynamics,* Clarendon Press, Oxford (1938).
Goodger, E. M., *Combustion Calculations,* Macmillan, London (1977).
Gordon, A., and Ruven-Smith, S., "Sampling of a Candle Flame," *JCP* **22,** 769 (1956).
Goshgarian, B., and Solomon, W. C., "Nozzle Beam Mass Spectrometer System for Studying One Atmosphere Flames," *Report TR 72-30,* Air Force Rocket Propulsion Laboratory, Edwards, CA (1972).
Goulard, R., *Combustion Measurements,* Academic Press, New York (1976).
Gouy, G., "Measurement of Burning Velocity" (in French), *Ann. Chim. Phys.* **18,** 27 (1879).
Grabner, L., and Hastie, J. W., "Flame Boundary Layer Effects in Line-of-Sight Optical Measurements," *C&F* **44,** 15–25 (1982).
Graham, R., and Gutman, D., "Reaction O + CS_2 Temperature?," *JPC* **81,** 267 (1977).
Gray, B. F., "Carbon Monoxide Glowing Combustion," *Specialists Periodical Reports, Reaction Kinetics* **1,** 312 (1975).
Gray, P., "Chemistry and Combustion," *23rd Symp.,* 305 (1991).
Gray, P., and Kay, J. C., "Stability of the Decomposition Flame of Liquid Hydrazine," Research (Lond.) **8,** 3 (1955).
Gray, P., Lee, J. C., Leach, H. A., and Taylor, D., "The Propagation and Stability of the Decomposition Flame of Hydrazine," *6th Symp.,* 255 (1957).
Gray, P., McKinven, R., and Smith, D. B., "Combustion of Hydrogen and Hydrazine with Nitrous Oxide and Nitric Oxide: Flame Speeds and Flammability Limits of Ternary Mixtures at Sub-Atmospheric Pressures," *C&F* **11,** 217 (1967).
Gray, P., and Pratt, M. J., "Reduction of Nitric Oxide in Flames and the Decomposition Flame of Methyl Nitrite," *6th Symp.,* 183 (1957).
Gray, P., and Smith, D., "Isotopic Effects on the Flame Speeds of Hydrogen and Deuterium Flames," *Chem. Comm.,* 146 (1980).
Grayson, M. (Ed.), *Encyclopedia of Chemical Technology,* 22 Vols., Wiley (1978); 3rd ed. (1988).
Green, E. F., and Toennies, P., *Chemical Reactions in Shock Waves,* Arnold (1964).
Greenwood, N. N., and Earnshaw, A., *Chemistry of the Elements,* Pergamon Press (1984).
Groeger, W., and Fenn, J. B., "Microjet Burners for Molecular Beam and Combustion Studies," *RSI* **59,** 1971 (1988).
Gross, R., and Bott, J., *Handbook of Chemical Lasers,* John Wiley and Sons, NY (1976).
Grosse, A. V., and Conway, J. B., "Combustion of Metals in Oxygen," *Industrial and Engineering Chemistry* **50,** 663 (1958)
Grosse, A. V., and Kirshenbaum, A. D., "The Premixed Hydrogen–Fluorine Flame and its Burning Velocity," *JACS* **77,** 5012 (1955).
Gruschka, H., and Wecken, F., *Gas Dynamic Theory of Detonation,* Gordon and Breach, NY (1971).
Guillaume, P. J., and Van Tiggelen, P. J., "Spectroscopic Investigation of Acetylene–Nitrous Oxide Flames," *20th Symp.,* 751 (1984).

Gukasyan, P., Mantashyan, A., and Sayadyan, R., "Detection of High Concentration of Radicals in the Zone of the Cold Flame in the Reaction of the Oxidation of Propane," *Comb. Expl. and Shock Waves* (Eng. Trans.) **12,** 706 (1976).

Gulder, O., "Laminar Burning Velocities of Methanol, Ethanol and Isobutane–Air Mixtures," *19th Symp.,* 275–81 (1982); *C&F* **56,** 261 (1984); *CST* **33,** 179–92 (1983).

Gunther, R., *Verbrennung und Feuerungen,* Springer-Verlag, Berlin, West Germany (1974).

Gunther, R., "Turbulence Properties of Flames and Their Measurement," *Prog. Energy Comb. Sc.* **9,** 105 (1983).

Gunther, R., and Janisch, G., "Messwerte de Flammengeschwindigkeit von Gasen," *Chemi-Ing-Technik* **43,** 975 (1959).

Gustafson, E. K., McDaniel, J., and Byler, R. L., "The Use of Raman Techniques for Measuring Velocity," *IEEJ Quantum Electron. QE-17,* 2258 (1981).

Guthrie, A., *Vacuum Technology,* John Wiley, New York (1963).

Habeebullah, M., Alasfour, F., and Branch, M., "Structure and Kinetics of CH_4–N_2O Flames," *23rd Symp.,* 371 (1991).

Hajal, I., and Combourieu, J., "Detonation of Hydrazoic Acid," *Comp. Rend.* **253,** 2346 (1961).

Hajal, I., and Combourieu, J., "Detonation of Hydrazoic Acid," *Comp. Rend.* **255,** 509 (1962).

Hall, A., McCoubrey, J., and Wolfhard, H., "Some Properties of Formaldehyde Flames," *C&F* **1,** 53 (1952).

Hall, A. R., and Pearson, G., "$HC10_4$-IX Two Flame Structure with Hydrocarbons," *12th Symp.,* 1025 (1969).

Hall, A. R., and Wolfhard, H. G., "Hydrazine Decomposition Flames at Sub-Atmospheric Pressures," *TFS* **52,** 1520 (1956).

Hall, A. R., and Wolfhard, H. G., "Multiple Reaction Zones in Low Pressure Flames with Ethyl and Methyl Nitrate, Methyl Nitrite and Nitromethane," *6th Symp.,* 190 (1957).

Halpern, M., and Ruegg, A., "Sampling in Burners," *J. Res. Natl. Bur. Std.* **60,** 29 (1958).

Hamilton, C., and Schott, G. L., "Post Induction Kinetics in Shock Initiated H_2–O_2 Reaction," *11th Symp.,* 635 (1967).

Hanson, R. K., "Combustion Diagnostics: Planar Imaging Techniques," *21st Symp.,* 1677 (1986).

Hanson, R. K., and Salimian, S., "Report on N–O–H Chemistry," High Temperature Gas Dynamics Laboratory, Stanford University, CA (1983); ibid., Chapter 6, in *Combustion Chemistry,* W. Gardner, Jr. (Ed.), Springer Verlag, N.Y. (1982).

Hanson, R. K., Seitzman, J. M., and Paul, P. H., "Planar Laser-Fluorescence Imaging of Combustion Gases," *Appl. Phys. B.* **50,** 441 (1990.)

Haraguchi, H., Smith, B., Weeks, S., Johnson, D. J., and Winefordner, J. D., "Measurement of Small Volume Flame Temperatures by the Two-line Atomic Fluorescence Method," *Appl. Spec.* **31,** 156 (1977).

Harker, J. H., and Allen, D. A., *Fuel Science,* Oliver & Boyd, Edinburgh, Scotland (1972).

Harris, L., "CARS Spectra from Lean and Stoichiometric CH_4–N_20 Flames," *C&F* **53,** 103 (1983).

Harris, T. D., and Lytle, F. E., "Analytical Applications of Laser Absorption and Emission Spectroscopy," p. 369, in *Ultrasensitive Laser Spectroscopy,* Kliger, D. S. (Ed.) Academic Press, NY (1983).

Hartley, D. L., "Ramonography," in *Laser Raman Gas Diagnostics,* p. 311, M. Lapp and C. M. Penney (Eds.), Plenum Press (1974).

Hartman, E., "Normal Burning Velocity of Some Organic Molecules with Air," Thesis, Karlsruhe (Ger.) (1931).

Harvey, A. B. (Ed.), *Chemical Applications of Nonlinear Raman Spectroscopy,* Academic Press, NY (1981).

Hassa, C., Paul, P. H., and Hanson, R. K., "Laser Induced Fluorescence Modulation Techniques for Velocity Measurements in Gas Flows," *Exp. Fluids* **5,** 240 (1987).

Hastie, J. W., "Mass Spectrometric Analysis of 1 Atm. Flames; Apparatus and the CH_4–O_2 System," *C&F* **21,** 187 (1973a).

Hastie, J. W., "Mass Spectrometric Studies of Flame Inhibition: Analysis of Antimony Trihalides in Flames," *C&F* **21,** 49 (1973b).

Hastie, J. W., "Chemical Aspects of Flame Inhibition," NBS Special Publ. 411, Fire Safety Research, NBS, Gaithersburg, MD (1973c).

Hastie, J. W., "Molecular Basis of Flame Inhibition," *J. Res. NBS* **77A,** 733 (1973d).

Hastie, J. W., "Mass Spectrometer Evidence for HO_2 in Atmospheric Flames," *Chem. Phys. Letters* **26,** 338–42 (1974).

Hastie, J. W., *High Temperature Vapors,* Academic Press, New York (1975).

Hautman, D. J., Ph.D. Thesis, Mechanical and Aerospace Engineering Dept., Princeton University, NJ (1980).

Hayes, K. F., and Kaskan, W. E., "Inhibition by CF_3Br of CH_4–Air Flames Stabilized on a Porous Burner," *C&F* **24,** 405 (1975).

Hayhurst, A. N., and Kittelson, D. B., "Heat and Mass Transfer Considerations in Electrically Heated Thermocouple of Iridium and Iridium/Rhodium Alloy in an Atmospheric Flames," *C&F* **28,** 301 (1977).

Hayhurst, A. N., and Kittelson, D. B., "Probe Sampling," *C&F* **28,** 137 (1977).

Hayhurst, A. N., Kittelson, D. B., and Telford, N. R., "Distortions in Probe Sampling," *C&F* **28,** 123 (1973).

Hayhurst, A. N., and McLean, H., "Mechanism for Producing NO from N_2 in Flames," *Nature* **251,** 303 (1974).

Hayhurst, A. N., and Telford, N. R., "Modeling Probe Sampling," *Proc. Roy. Soc. Lond.* **A322,** 483 (1971).

Hayhurst, A. N., and Vince, I. M., "Production of 'Prompt' Nitric Oxide and Decomposition of Hydrocarbons in Flames," *Nature* **266,** 524 (1977).

Haynes, B. S., "Kinetics of Nitric Oxide Formation in Combustion," in *Progress in Astronautics and Aeronautics,* C. T. Bowman and J. Birkeland (Eds.), **62,** American Institute of Aeronautics and Astronautics (1975); Also, Thesis, University of New South Wales, Sidney, Australia (1973).

Haynes, B. S., "Reactions of Ammonia and Nitric Oxide in the Burnt Gases of Fuel-Rich Hydrocarbon–Air Flames," *C&F* **28,** 81 (1977a).

Haynes, B. S., "The Oxidation of Hydrogen Cyanide in Fuel-Rich Flames," *C&F* **28,** 113 (1977b).

Haynes, B. S., "Production of Nitrogen Compounds from Molecular Nitrogen in Fuel-Rich Hydrocarbon–Air Flames," *Fuel,* **56,** 199–203, April (1977c).

Haynes, B. S., Iverach, D., and Kirov, N. Y., "The Behavior of Nitrogen Species in Fuel-Rich Hydrocarbon–Air Flames," *15th Symp.,* 1103 (1975).

Haynes, B. S., and Wagner, H. Gg., "Soot Formation Review," *Prog. Energy and Comb.* **3,** 229 (1981).

Heath, G. A., "Perchloric Acid Flames," *IV Ministry of Aviation RPE Tech. Rept.,* 65/6 (August 1965).

Heath, G. A., and Pearson, G., "$HClO_4$-IV Chemical Structure of Methane Flames," *11th Symp.,* 967 (1967).

Hecker, Wm. C., "A Theoretical Study of Kinetics, Propagation, and Suppression of Methane–Air Flames," *Master of Science Thesis,* Chemical Engineering Department, Brigham Young University, Provo, UT (1975).

Heimerl, J., and Coffee, T., "Detailed Modeling of Flames-I Ozone," *C&F* **39,** 301 (1980).

Heiss, A., Dumas, G., and Ben-Aim, R., "Oscillations de relaxation de flames froid," *Second*

International Symposium on Flames as Reactions in Flow Systems, Padova, Italy, December 1975, published by the Consiglio Nazionale delle Recherche, Rome (1976).

Hellwig, K., and Anderson, R. C., "Calculations of Burning Velocities for Hydrogen–Bromine and Hydrogen–Chlorine Flames," *JACS* **77,** 232 (1955).

Henkel, M. J., Hummel, H., and Spaulding, W., "Theory of Flame Propagation III—Numerical Integrations," *3rd Symp.,* 135 (1949).

Hennessy, R., Robinson, C., and Smith, D. B., "A Comparative Study of Methane and Ethane Flame Chemistry by Experiment and Detailed Modeling," *21st Symp.,* 761 (1988).

Henrich, K. F. F., "Characterization of Particles," *NBS Special Publication,* 533, Washington (1980).

Hermann, R., and Alkamade, C., *Chemical Analysis by Flame Photometry,* Interscience, New York (1963).

Herriott, G., Eckert, R., and Albright, L., "Kinetics of Propane Pyrolysis," *Am. Inst. Chem. Eng. J.* **18,** 84 (1972).

Herron, J. T., and Huie, R. E., "Rate Constants for the Reactions of Atomic Oxygen with Organic Compounds in the Gas Phase," *J. Phys. Chem. Ref. Data* **2,** 467 (1973).

Herron, J. T., and Peterson, R. B., "Optical Determination of Hydrogen Atom Concentration behind a Gas Sampling Orifice," *23rd Symp.,* 1855 (1991).

Hertzberg, M., "Flame Stretch Extinction in Flow Gradients: Flammability Limits under Natural Convection," *20th Symp. (International) on Combustion,* 1967–1974 (1985).

Herzberg, G., *Infra-red and Raman Spectra,* Van Nostrand, New York (1945).

Herzberg, G., *Molecular Spectra and Molecular Structure, I. Diatomic Molecules,* Van Nostrand, New York (1950).

Herzberg, G., *Electronic Spectra of Polyatomic Molecules,* Van Nostrand, New York (1966).

Herzberg, G., *The Spectra and Structure of Simple Free Radicals,* Cornell University Press, Ithaca, NY (1971).

Herzfeld, K., "Relaxation of Partial Temperatures," *Temperature–Its Measurement and Control in Science and Industry 2,* Reinhold Publ. Co., New York (1955), p. 233, 2nd ed. (1990).

Hesselink, L., "Digital Image Processing in Flow Visualization," *Ann Rev. Fluid Mech.* **20,** 421 (1988).

Hibbard, R., and Barnett, C. M. (Ed.), *Basic Considerations in the Combustion of Hydrocarbon Fuels with Air,* NACA Report 1300, Supt. Documents, U.S. Govt. Printing Office, Washington, D.C. (1959).

Hibbard, R., and Pinkel, B., "Correlation of Maximum Fundamental Flame Velocity with Hydrocarbon Structure," *JACS* **73,** 1622 (1951).

Hicks, J. A., "The Low Pressure Decomposition Flame of Ethyl Nitrate," *8th Symp.,* 487 (1962).

Hilado, C., *Flammability of Cellulosic Materials,* Technomic Press, Westport, Conn. (1973).

Hill, P. G., and Hung, J., "Laminar Burning Velocities of Stoichiometric Mixtures of Methane with Propane and Ethane Additives," *CST* **60,** 7 (1988).

Himus, G. W., *The Elements of Fuel Technology,* Leonard Hill Ltd., London (1958).

Hinshelwood, C. N., *Kinetics of Chemical Change,* 123, Oxford Univ. Press, Oxford, U.K. (1941).

Hinshelwood, C. N., *The Nature of Physical Chemistry,* Oxford Univ. Press, Oxford (1948).

Hinshelwood, C. N., and Williamson, A. T., *The Reaction Between Hydrogen and Oxygen,* Oxford, London, UK (1935).

Hirano, M., Oda, K., Hirano, T., and Akita, K., "Burning Velocities of Methanol–Air–Water Gaseous Mixtures," *C&F* **48,** 191–210 (1982).

Hirschfelder, J. O., "Theory of Flames II: Ignition Temperature and Other Approximations," *Phys. Fluids* **2,** 565 (1959).

Hirschfelder, J. O., and Curtiss, C. F., "Propagation of Flames and Detonations," *Advances in Chemical Physics,* Interscience Publishers, Inc., New York, **III,** 69–70 (1961).

Hirschfelder, J. O., Curtiss, C. F., and Bird, R. B., *Molecular Theory of Gases and Liquids,* Wiley, New York (1954); 2nd Ed. (corrected, with notes), Wiley, New York, 562 (1964).

Hirschfelder, J. O., Curtiss, C. F., and Campbell, D. E., "The Theory of Flame Propagation, IV," *JCP* **57,** 403–14 (1953).

Hirschfelder, J. O., and McCone, A., "Theory of Flames Produced by Unimolecular Reactions I: Accurate Numerical Solutions," *Phys. Fluids* **2,** 551 (1959).

Hoelscher, H., and Beidler, W. T., "Studies in a New Type of Flat Flame Burner," *Jet Prop.* **27,** 1257 (1957).

Holve, D., and Sawyer, R., "Diffusion Controlled Combustion of Polymers," *15th Symp.,* 351 (1975).

Holzrichter, K., and Wagner, H. G., "On the Thermal Decomposition of Ammonia Behind Shock Waves," *18th Symp.,* 769 (1982).

Homann, K. H., "Formation of Large Molecules, Particulates and Ions in Premixed Hydrocarbon Flames," *20th Symp.,* 857–70 (1985).

Homann, K. H., Krome, G., and Wagner, H. G., "OCS Oxidation in a Low Pressure Flow Tube," *Buns.* **73,** 967 (1969).

Homann, K. H., and MacLean, D. I., "Influence of Hydrogen on the Combustion of Halogenated Hydrocarbons with Fluorine," *C&F* **14,** 409 (1970).

Homann, K. H., and MacLean, D. I., "Structure of the Dichloro-difluoro-methane Flame with Fluorine," *Buns.* **75,** 3645 (1971a).

Homann, K. H., and MacLean, D. I., "Concentration Profiles of the Flames of H_2, NH_3, C_2H_2, and C_2H_4 with F_2," *Buns.* **75,** 945 (1971b).

Homann, K. H., Mochizuki, M., and Wagner, H. G., "Uber den Reaktionsablauf in fetten Kohlenwasserstoff-Sauerstoff-Flammen (acetylene, ethylene, propane and benzene)," *ZpC* **NF 37,** 299 (1963).

Homann, K. H., and Poss, R., "Effect of Pressure on the Inhibition of Ethylene Flames," *C&F* **18,** 300 (1972).

Homer, J. B., and Kistiakowsky, G. B., "Oxidation and Pyrolysis of Ethylene in Shock Waves," *JCP* **47,** 5290 (1967).

Hottel, H. C., "Modeling Principles in Relation to Fire," *Fire Res. Abs. and Rev.* **1,** 2 (1960).

Hottel, H. C., "Combustion and Energy for the Future," *14th Symp.,* 1–25 (1973).

Hottel, H. C., and Hawthorne, W., "Diffusion in Laminar Flame Jets," *3rd Symp.,* Institute, Pittsburgh, PA, p. 254 (1949).

Hottel, H. C., Williams, G. C., Nerheim, N. M., and Schneider, G. R., "Kinetic Studies in Stirred Reactors: Combustion of Carbon Monoxide and Propane," *10th Symp.,* 111 (1965).

Hougen, O. A., and Watson, K. M., *Chemical Process Principles III,* Wiley, 977 (1947).

Howard, J. B., "Kinetic Measurements Using Flow Tubes," *JPC* **83,** 3 (1979).

Howard, J. B., "Carbon Addition and Oxidation Reactions in Heterogeneous Combustion and Soot Formation—A Review," *23rd Symp.,* The Combustion Institute, Pittsburgh PA, p. 1107 (1991).

Howard, J. B., *Fluidized Beds,* Applied Science Publ., NY (1983).

Howard, J. B., McKinnon, J. T., and Johnson, F. R., "Fullerenes in Flames," *Nature* **352** (1991).

Howard, J. B., Williams, G. C., and Fine, D., "Kinetics of CO Oxidation in Post-Flame Gases," *14th Symp.,* 975 (1973).

Hoyermann, K-H., Priv. Comm. (1977).

Hoyermann, K-H., "Primary Products of Elementary Reactions in Hydrocarbon Oxidation" (in German), *Habatilition Schrift,* Göttingen, Ger. (1979).

Hoyermann, K-H., and Sievert, R., "The Reactions of Alkyl Radicals with Oxygen Atoms: Identification of Primary Products at Low Pressure," *17th Symp.,* 517 (1979).

Hsium, S. S. Y., Line, H. D. S. Y., and Lin, M. C., "CO Formation in Early Stage High Temperature Benzene Oxidation under Fuel-Lean Conditions in Shock Tube; Kinetics of Thermal Decomposition of Benzene," *20th Symp.*, 623–30 (1985).

Huffman, D. R., "Solid C_{60}," *Physics Today,* 22 (1992).

Huffstutler, M. C., Rode, J. A., and Anderson, R. C., "Effects of Diluents on Burning Velocities in Hydrogen–Bromine Mixtures," *JACS* **77,** 809 (1955).

Huie, R. E., and Herron, J. T., "Reaction of Atomic Oxygen (O^3P) with Organic Compounds," *Progress in Reaction Kinetics* **8,** 1, Oxford, Pergamon Press (1975).

Hunter, L., Grunfelder, C., and Fristrom, R., "Effects of CF_3Br on a Hydrogen-Oxygen-Argon Diffusion Flame," *Halogenated Fire Suppressants* (ed. R. Gann), p 234, Am. Chem. Soc. Washington D. C. (1975).

Hussain, G., and Norrish, R. G. W., "Photoinitiation," *PRS* **A723,** 145 (1963).

Illes, V., "The Pyrolysis of Gaseous Hydrocarbons, III. Kinetics and Mechanism of the Thermal Decomposition of Propane," *Acta Chim. Acad. Sci. Hung.* **67,** 41 (1971).

Istratov, A. G., and Librovich, V. B., "Calculation of the Rate of Normal Flame Propagation in Hydrogen–Chlorine Mixtures," *Dokl. Akad. Nauk (SSSR)* **143,** 1380 (1962).

Iverach, D., Basden, K. S., and Kirov, N. Y., "Formation of Nitric Oxide in Fuel-Lean and Fuel-Rich Flames," *14th Symp.*, 767–75 (1973).

Jahn, G., *Der Zündvorgang in Gasgemischen,* Thesis, Oldenbourg, Berlin (1934).

Japar, S. M., and Szkarlat, A. T., "Measurement of Diesel Exhaust Particulate Using Photoacoustic Spectroscopy," *CST* **24,** 215 (1981).

Jenkins, D. R., Spalding, D. B., and Yumlu, V. S., "The Combustion of Hydrogen and Oxygen in a Steady Flow Adiabatic Stirred Reactor," *14th Symp.*, 779 (1973).

Jennings, W., *Gas Chromatography with Glass Capillary Columns,* Academic Press, 2nd ed. (1992).

Jeong, K-M., and Kaufman, F., "Kinetics of Hydroxyl Radical with Methane and Nine Chlorine and Fluorine Substituted Methanes. 1 Experimental Results, Comparisons and Applications," *JCP* **85,** 1808 (1982).

Johnson, G. M., Smith, M. Y., and Mulcahy, M. R. F., "The Presence of NO_2 in Premixed Flames," *17th Symp.*, 647 (1979).

Johnston, H. S., "Gas Phase Reaction Kinetics of Neutral Oxygen Species," *NSRDS-NBS-20* (September 1968).

Joklik, R., Daily, J., and Pitz, W. J., "Measurement of CH Concentrations in an Acetylene/Oxygen Flame and Comparisons to Modeling," *21st Symp.*, 895 (1988).

Jones, G. W., "Decomposition Flames of Acetylene," *U.S. Bureau of Mines Investigation No. 3755,* (1944).

Jones, W. M., and Davidson, N., "Ozone Decomposition in Shock Tube," *JACS* **84** (1962).

Jost, W., *Explosion and Combustion Processes in Gases,* McGraw–Hill, New York (1946).

Jost, W., "Reactions of Adiabatically Compressed Hydrocarbon Air Mixtures," *3rd Symp.*, The Combustion Institute, Pittsburgh, PA, 424 (1949).

Jost, W. (Ed.), *Low Temperature Oxidation,* Gordon and Breach Science Publ., NY (1965).

Jost, W., Schacke, H., and Wagner, H. G., "The Water Gas Equilibrium," *Zeit. Physik. Chem. Frankfurt* **45,** 47 (1965).

Jouget, E., *Mecanique des Explosifs,* O. Doin, Paris (1917).

Kaiser, E. I., Rothschild, W., and Lavoie, G., "Effect of Fuel–Air Ratio and Temperature in the Structure of Laminar Propane–Air Flames," *CST* **33,** 123–34 (1983).

Kaiser, R., *Gas Chromatography* (two vols.), Butterworths (1963).

Kalaff, P., and Alkemade, C., "Characteristics of Premixed Laminar CO–N_{20} Flames," *C&F* **19,** 257 (1972).

Kallend, A. S., Purnell, J. H., and Shurlock, B. C., "The Pyrolysis of the Propylene," *PRS* **A300,** 120 (1967).

Kantrowitz, A., and Trimpi, R., "A Sharp Focusing Schlieren System," *J. Roy. Aeron. Soc.* **17,** 311 (1950).

Kanury, A. M., *Introduction to Combustion Phenomena,* Gordon and Breach, New York (1975).

Karlovitz, B., "A Turbulent Flame Theory Derived from Experiments," in *Selected Combustion Problems (AGARDOGRAPH),* Butterworths, London, 248 (1954), Hawthorne, W., Fabri, J., and Karlovitz, B. (Eds.).

Karlovitz, B., Dennison, D., Knapschafe, D., and Wells, F., "Studies on Turbulent Flow," *4th Symp.,* The Combustion Institute, Pittsburgh, PA, 613 (1953).

Karpov, V. and Sokolik, A. *Proc. Acad. Sci. USSR Phys. Chem Sec.* **138,** 457 (1961).

Karra, S. B., and Senkan, S. M., "Chemical Structures of Sooting $CH_3Cl/CH_4/O_2/Ar$ and $CH_4/O_2/Ar$ Flames," *CST* **54,** 333 (1987).

Kasem, M., Quin, M., and Senkan, S. M., "Effects of Microprobing on the Chemical Structure of Fuel-Rich $1,2C_2H_2Cl_2/CH_4/O_2/Ar$ Flames," *CST* **67,** 147 (1989).

Kaskan, W. E., "The Dependence of Flame Temperature on Mass Burning Velocity," *6th Symp.,* Reinhold Publ. Co., New York (1957), p. 134.

Kaskan, W. E., "Hydroxyl Concentrations in Rich Hydrogen–Air Flames Held on Porous Burners," *C&F* **2,** 29 (1958).

Kaskan, W. E., "Excess Radical Concentrations and the Disappearance of Carbon Monoxide in Flame Gases from Some Lean Flames," *C&F* **3,** 49 (1959).

Kaskan, W. E., and Hughes, D., "Mechanism of Decay of Ammonia in Flame Gases from an NH_3/O_2 Flame," *C&F* **20,** 381 (1973).

Kaskan, W., and Reuther, J., "Limiting Equivalence Ratios, Dissociation and Self Inhibition in Premixed Quenched Fuel-Rich Hydrocarbon Flames," *16th Symp.,* The Combustion Institute, Pittsburgh, PA (1977).

Kaskan, W. E., and Schott, G. L., "Requirements Imposed by Stoichiometry in Dissociation–Recombination Reactions," *C&F* **6,** 73 (1962).

Kassel, L. S., *Kinetics of Homogeneous Gas Reactions,* ACS Monograph American Chemical Soc., (1932).

Kaufman, F., "Chemical Kinetics and Combustion: Intricate Paths with Simple Steps," *19th Symp.,* (1982).

Kaufmann, F., *"Rates of Elementary Reactions" Science,* **230,** 393 (1985).

Kawakami, T., Okajima, S., and Iinuma, K., "Measurment of Slow Burning Velocity by Zero-Gravity Method," *22nd Symp.,* The Combustion Institute, Pittsburgh PA, p. 1609 (1989).

Kays, W. M., *Convective Heat and Mass Transfer,* McGraw–Hill, New York (1966).

Kee, R. J., Dixon-Lewis, G., Warnatz, J., Coltrin, M., and Miller, J., "A FORTRAN Computer Code Package for the Evaluation of Gas Phase Multicomponent Transport Properties," *SAND RR-8246,* Sandia National Laboratories, Livermore, CA (1986).

Kee, R. J., Miller, J., Evans, G. H., and Dixon-Lewis, G., "A Computational Model of the Structure and Extinction of Strained, Opposed Flow Premixed Methane–Air Flames," *22nd Symp.,* The Combustion Institute, Pittsburgh, PA, p. 1479 (1989).

Kee, R. J., Miller, J. A., and Jefferson, T. H., "CHEMKIN, A General Purpose Problem-Independent, Transportable Fortran Chemical Kinetics Program Package," Sandia National Laboratories Report, SAND 808003 (1980). See also Kee, R. J., Warnatz, J., and Miller, J. A., "A Fortran Computer Program Package for the Evaluation of Gas-Phase Viscosities, Conductivities and Diffusion Coefficients," Sandia National Laboratories Report, *SAND 83 8209,* Livermore, CA (1983).

Kee, R. J., Rupley, F. M., and Miller, J. A., "The CHEMKIN Thermodynamic Data Base," *SAND87-8215, UC4,* Sandia National Laboratory, Livermore, CA (1987).

Kennard, E. H., *The Kinetic Theory of Gases,* McGraw–Hill, New York (1938).

Kennedy, L. A. (Ed.), *Turbulent Combustion,* American Institute of Aeronautics and Astronautics, New York (1978).

Kennel, C., Gottgens, J., and Peters, N., "The Basic Structure of Lean Propane Flames," *23rd Symp.,* The Combustion Institute, Pittsburgh, PA (1991).

Kent, J. H., "A Non-Catalytic Coating for Platinum–Rhodium Thermocouple," *C&F* **14,** 279 (1970).

Kern, R., Singh, H., Esslinger, M., and Weinkler, P., "Product Profiles of Toluene, Benzene, Butadiene and Acetylene," *19th Symp.,* 1351 (1983).

Kern, R., Wu, C. K., Skinner, G., Rao, V., Kiefer, J., Towers, J., and Mizerka, L., "Collaborative Shock Tube Studies of Benzene Pyrolysis," *20th Symp.,* 789 (1985).

Khitrin, L. N., *The Physics of Combustion and Explosion,* Israel Program for Scientific Translations, OTS-61-31205, Jerusalem (1962) (trans. from Russian).

Khitrin, L. N. (Ed.), *Combustion in Turbulent Flow,* Israel Program for Scientific Translations, Jerusalem (1963).

Kichakiff, G., Hanson, R., and Howe, R., "Simultaneous Multiple Point Measurements of OH in Combustion Gases Using Planar Laser Induced Fluorescence," *20th Symp.,* 1265 (1984).

Kiefer, J. H., Al-Alami, M. Z., and Budach, K. A., "A Shock Tube, Laser Schlieren Study of Propene Pyrolysis at High Temperature," *JCP* **86,** 808 (1982).

Kiefer, J. H., and Lutz, R. W., "Recombination of Oxygen Atoms at High Temperatures as Measured by Shock-Tube Densitometry," *JCP* **42,** 1709 (1965).

Kiefer, J. H., and Lutz, R. W., "The Effect of Oxygen Atoms on the Vibrational Relaxation of Oxygen (Ozone Decomposition in Shock Tubes)," *11th Symp.,* 307 (1967).

Kimball-Linne, M. A., Kychakoff, G., and Hanson, R. K., "Fiberoptic Absorption/Fluorescence Combustion Diagnostics," *CST* **21,** 307 (1986).

Kirshenbaum, A. D., and Grosse, A. V., "The Combustion of Carbon Subnitride C_4N_2 and a Chemical Method for the Production of Continuous Temperatures in the Range 5000–6000 K," *JACS* **78,** 2020 (1956).

Kistiakowsky, G. B., and Kydd, P.H., "A Study of the Reaction Zone (in Gaseous Detonations) by Gas Density Measurements," *JCP* **25,** 824 (1958).

Kitagawa, T., "Emission Spectrum of the Flame of Chlorine Burned in Hydrogen," *Rev. Phys. Chem. (Japan)* **8,** 71 (1934).

Kitagawa, T., "Chlorine Flames," *Rev. Phys. Chem. Japan* **12,** 135–47 (1938).

Klainer, S. M. (Ed.), Special issue devoted to Fluorescence Spectroscopy, *Opt. Eng.* **20,** 507 (1983).

Klaukens, H., and Wolfhard, H. G., "Temperature Profile of a Low Pressure Acetylene–Oxygen Flame," *PRS* **A193,** 512 (1948).

Klein, G., "A Contribution to Flame Theory," *PRS* **A240,** 389 (1957).

Knewstubb, P., and Sugden, T. M., "Mass Spectrometry of Ions Present in Hydrocarbon Flames," *7th Symp.,* 247 (1959).

Knight, H., and Venable, D., "Apparatus for Precision Flash Radiography of Shock and Detonation Waves in Gases," *Rev. Sci. Inst.* **25,** Vol. 29, p. 92 (1958).

Knuth, E. L., "Relaxation Processes in Flow Systems," *Proc. 1972 Heat Transfer and Fluid Mechanics Institute,* Landis and Hordemann (Eds.), Stanford University Press, Stanford, CA, p. 89 (1972).

Knuth, E. L., "Molecular Beam Inlet Sampling," *Engine Emissions: Pollutant Formation and Measurement,* Springer, J. and Patterson, F. (Eds.), Plenum Press, NY, p. 319 (1973).

Knuth, E. L., Ni, W-F, and Seeger, C., "Molecular Beam Sampling Study of Extinguishment of Methane Air Flames by Dry Chemicals," *CST* **28,** 247 (1982).

Koda, S., and Fujiwara, O., "Silane Combustion in an Opposed Jet Diffusion Flame," *21st Symp.,* 1861 (1988).

Kohse-Hoinghaus, K., ''Quantitative Laser-Induced Fluorescence: Some Recent Developments in Combustion Diagnostics,'' *Appl. Phys. B* **50,** 455 (1990).

Kohse-Hoinghaus, K., Jeffries, J. B., Copeland, R. A., Smith, G. P., and Crosley, D. R., ''The Quantitative LIF Determination of OH Concentrations in Low Pressure Flames,'' *22nd Symp.,* The Combustion Institute, Pittsburgh, PA, p. 1857 (1989).

Koike, T., and Gardiner, W. C., Jr., "Thermal Decomposition of Propane," *JCP* **82,** 2005 (1980).

Kokochashvili, V. I., "The Combustion of Hydrogen–Bromine Mixtures," *Zhur. Fiz. Khim.* **25,** 444–52 (1951).

Kolesnikov, B. Y., and Ksandopolo, G. I., "Concentration Profile of Hydrogen Atoms in the Low Temperature Zone of a Propane–Air Flame Front," *Doklady Akademii Nauk SSSR* **216** (5) (1974) (English translation).

Kondratiev, V. N., *Rate Constants of Gas Phase Reactions,* trans. by L. J. Holtschlag, R. M. Fristrom (Ed.), COM-72-10014, National Standard Reference Data System, National Bureau of Standards, U.S. Department of Commerce, Washington, D.C. (1973).

Korobeinichev, O. P., "Flame Structure," *Combustion Explosion and Shock Waves* **24,** 565 (1988).

Korobeinichev, O. P. (Ed.), *Flame Structure* (two vols.) (one section on solid state flames) Novosibirsk, Nauka, Siberian Branch, Proceedings of the Third International Seminar on Flame Structure held at Alma-Atta Kasakhstan (1991b).

Korobeinichev, O. P., Orlov, V. N., and Shifon, N. Y., "Mass Spectrometric Study of Flames of Perchloric Acid in Lean and Rich Mixtures with Methane," *Physics of Combustion and Shock Waves,* (English Trans. of Phys. Gor. i Vzr.) 1835, 563 (1980).

Korobeinichev, O. P., Orlov, V. N., and Shifon, N. Ya., "Spectroscopic Study of the Chemical Structures of Flames of Perchloric Acid with Lean and Rich Mixtures of Methane," *Fizika Goreniya i Vrzryva* **18,** 77–83 (1982) (translated into English in *Combustion, Explosions and Shock Waves*).

Korobeinichev, O. P., Paletsky A. A., Makhov, G. A., Eropov, A. V., Kuibida, L. V., and Chernov, A. A., "Methods of Probing Mass-Spectrometry and Laser Spectroscopy and their Application for Study of Gaseous and Condensed Systems Flame Structure," Paper #8 at *Joint meeting of the Soviet and Italian Sections of the Combustion Institute,* Tacchi Editore Pisa, Italy (1990)

Korobeinichev, O. P., Tereschenco, A. G., Shvartsberg, V. N., Chernov, A. A., Makhov, G. A., and Zabolotny, A. E., "A Study of Flame Structure of Sandwich Systems Based on Ammonium Perchlorate, HMX and Polybutadiene Binder," p. 262, in *Flame Structure,* O. Korbeinechev (Ed.), NAUKA, Siberian Branch, Novosibirsk, Russia (1991).

Kovasznay, L., *Turbulence Measurements in High Speed Aerodynamics and Jet Propulsion,* Vol. IX, Sec. F, R. Landenberg (Ed.), Princeton University Press, N.J. (1954).

Kovasznay, L., "Turbulent Combustion," *Jet Prop.* **26,** 485 (1956).

Kovasznay, L., "The Turbulent Boundary Layer," *Ann. Rev. Phys. Fluids* **2,** 100 (1970).

Kowalik, R. M., and Kruger, C. H., "Laser Fluorescence Temperature Measurements," *C&F* **34,** 135 (1979).

Kozlof, G. I., "On the High Temperature Oxidation of Methane," *7th Symp.,* 142 (1959).

Kramlich, J., Heap, M., Seeker, R., and Samulson, G. S., "Flame Mode Destruction of Hazardous Waste Compounds," *20th Symp.,* The Combustion Institute, Pittsburgh, PA, p. 1991 (1985).

Kramlich, J., and Malte, P. C., "Modeling and Measurement of Sample Probe Effects on Pollutant Gases Drawn from Flame Zones," *CST* **18,** 91 (1978).

Kramlich, J., Malte, P. C., and Grosshandler, W. L., "Reaction of Fuel Sulfur in Hydrocarbon Combustion," *18th Symp.,* 151 (1981).

Kroto, H., "Space, Stars, CBO and Soot," *Science* **242,** 1139 (1988).

Krumwiede, K. A., Norton, D. A., Johnson, G. W., Thompson, R. E., Breenk, B. P., and Quan, V., "A Probing Study of NO Formation in the Flame Zone of a 175 mW Gas Fired Utility Boiler," *8th Annual Meeting of the Air Pollution Control Association,* Boston, MA, June 15–20 (1975).

Ksandopulo, G. E., "Energy Aspects of the Oxidation of Propane Near a Flame Front," *Combustion Explosion and Shock Waves* **11,** 350–54 (1975).

Ksandopulo, G. E., *Flame Chemistry,* Chemistry, Moscow (1980) (in Russian).

Ksandopulo, G. E., and Dubinin, V., *The Chemistry of Gas Phase Combustion,* "Chemistry," Moscow (1987).

Ksandopulo, G. E., et al., "III—Oxidation of Propane in the Presence of Diethyl Amine," *Combustion Explosion and Shock Waves* **11,** 113–17 (1975).

Ksandopulo, G. E., et al. "VIII—Effects of Cyclohexyl Nitrate on the Structure of Hexane Flames," *Combustion Explosion and Shock Waves* **19,** 13–16 (1983).

Ksandopulo, G. E., and Kolesnikov, B. Y., in *Collection of Articles on Chemistry* (in Russian), No. 3, Izd. KGU, Alma Ata, 636 (1973).

Ksandopulo, G. E., Kolesnikov, B. Y., and Dubinin, V., "Low Temperature Zone of the Front of Hydrocarbon Flames: The Oxidation of Hexane," *Fisika Goreniza i Vzryva* **13,** 641–44 (1977) (English translation).

Ksandopulo, G. E., Kolesnikov, B. Y., and Odnoroq, D. S., "Low Temperature Zone in Hydrocarbon Flames—Oxidation of Propane Near the Front of a Flame," *Comb. Expl. and Shock Waves* (English translation) **10,** 757–61 (1974).

Ksandopulo, G. E., Kolesnikow, Z. B., and Odnoroq, D. S., "Low Temperature Zone of the Front of Hydrocarbon Flames II: Oxidation of Propane in the Presence of C_2F4Br_2," *Combustion Explosion and Shock Waves* **11,** 114–17 (1975).

Ksandopulo, G. E., Sagindykof, A., Kudaibergenov, S., and Mansurov, Z., "Profiles of Atomic Hydrogen and Peroxide Radicals at the Front of Propane-Air Flames," *Combustion Explosion and Shock Waves* **11,** 714–19 (1975).

Kuehl, D. K., "Laminar Burning Velocities of Propane–Air Mixtures," *8th Symp.,* 510 (1961).

Kumanagi, S., and Isoda, H., "New Aspects of Droplet Combustion," *7th Symp.,* 523 (1959).

Kumar, R., Tamm, H., and Harrison, W., "Combustion of Hydrogen–Steam–Air Mixtures Near Lower Flammability Limits," *CST* **33,** 167 (1983).

Kuo, K. K., *Principles of Combustion,* Wiley, New York (1986).

Kutznetov, V. R., and Sabelnikov, V. A., *Turbulence and Combustion,* Nauka, Moscow (1986).

Kuvshinoff, B. W., Fristrom, R. M., Ordway, G. L., and Tuve, R. L. (Eds.), *Fire Sciences Dictionary,* John Wiley & Sons, New York (1977).

Kychakoff, G., Hanson, R. K., and Howe, R. D., "Simultaneous Multiple Point Measurements of OH in Combustion Gases Using Planar LIF," *20th Symp.,* 1265 (1984).

Kychakoff, G., Howe, R. D., Hanson, R. K., Drake, M. C., Pitz, R. W., Lapp, M., and Penney, C. M., "Visualization of Turbulent Flame Fronts with Planar Laser Induced Fluorescence," *Science* **224,** 382 (1984).

Kychakoff, G., Howe, R. D., Hanson, R. K., and McDaniel, J. C., "Quantitative Visualization of Combustion Species in a Plane," *Appl. Opt.* **21,** 3225 (1982).

Laffitte, P., *La propagation des flammes dans les melanges gazeuz: les deflagrations,* Hermann, Paris (1939).

Laffitte, P., Combourieu, J., Hajal, I., Ben Caid, M., and Moreau, R., "Characteristics of Chlorine Dioxide Decomposition Flames at Reduced Pressure," *11th Symp.,* 941 (1967).

Laffitte, P., Hajal, I., and Combourieu, J., "The Decomposition Flame of Hydrazoic Acid," *10th Symp.,* 79 (1965).

Lahaye, J., and Prado, G. (Eds.), *Soot in Combustion Systems and Its Toxic Properties,* Plenum (1983).

Laidler, K. J., Sagert, N. H., and Wojiechowski, B. W., ''Kinetics and Mechanisms of the Thermal Decomposition of Propane. I. The Uninhibited Reaction,'' *PRS* **A270,** 242 (1962).

Laidler, K. J., and Wojiechowski, B. W., "Kinetics of the Thermal Decomposition of Propylene and of Propylene-Inhibited Hydrocarbon Decompositions," *PRS* **A259,** 257 (1960).

Lakshmish, K. N., Paul, P. J., Rajan, N. K. S., Goyal, G., and Mukunda, H. S., "Behavior of Methane–Oxygen–Nitrogen Mixtures Near Flammability Limits," *21st Symp.,* The Combustion Institute, Pittsburgh, PA, p. 1573 (1989).

Lange, H. B., "No Formation in Premixed Combustion," *American Institution of Chemical Engineers Symposium,* Series 126, **68,** 17–25 (1972).

Langmuir, I., "Atomic Hydrogen Torch," *Ind. Eng.Chem.* **25,** 404 (1927).

Lapp, M., ''Flame Temperatures from Vibrational Raman Scattering,'' p. 107 in *Laser Raman gas Diagnostics,* Lapp, M., and Penney, C. M. (Eds), Plenum Press, NY (1974).

Lapp, M., Drake, M. C., Penney, C. M., and Pitz, R. W., ''Flame Images from Pointwise Raman Data,'' *J. Quant. Spectros. Radiant. Transfer* **40,** 363 (1988).

Lapp, M., and Penney, C. M., *Laser Raman Gas Diagnostics,* Plenum Press, New York (1974).

Lapp, M., and Penney, C. M., ''Raman Measurements in Flames,'' p 204 (ch. 6) in *Advances in Infrared and Raman Spectroscopy,* Clark, R. J. H. and Hester, R. E. (Eds.), Hayden and Son ltd., London (1977).

Lapp, M., and Rich, J. A., ''Electrical conductivities of Seeded Flame Plasmas in Strong electric Fields,'' *Phys. Fluids* **6,** 806 (1963.)

Larsen, E., "Halogenated Fire Extinguishants: Flame Extinction by a Physical Mechanism," *Halogenated Fire Suppressants,* R. G. Gann (Ed.), 376 (1975).

Lask, G., and Wagner, H. G., "Influence of Additives on the Velocity of Flames," *8th Symp.,* 432 (1962).

Laurendeau, N. M., and Goldsmith, J. E. M., "Comparison of Hydroxyl Concentration Profiles Using Five Laser Induced Fluorescence Methods in a Lean Subatmospheric $H_2/O_2/N_2$ Flame," *CST* **63,** 139 (1989).

Lauterborn, W., and Vogel, A., "Modern Optical Techniques in Fluid Mechanics," *Ann. Rev. Fluid Mech.* **16,** 223 (1984).

Lavoisier, A., *Essays Physical and Chemical,* trans. by T. Henry, Reprint, F. Cass Ltd., London (1970).

Law, C. K., and Egolfopoulos, F. N., "A Kinetic Criterion of Flammability Limits," *23rd Symp.,* The Combustion Institute, Pittsburgh, PA, 413 (1991).

Law, C. K., Ishizuka, S., and Cho, P., "On the Opening of Premixed Bunsen Flame Tips," *CST* **28,** 89 (1982).

Lawless, E. W., and Smith, I. C., "Inorganic High Energy Oxidizers," Marcel Dekker, Inc., New York (1975).

Lawn, C. J., *Principles of Combustion Engineering for Boilers,* Academic Press, New York (1987).

Lawton, J., and Weinberg, F. J., *Electrical Properties of Flames,* Clarendon Press, Oxford, UK (1969).

Layokun, S. K., "Oxidative Pyrolysis of Propane," *Ind. Eng. Chem. Process Des. Dev.* **18,** 241 (1979).

Layokun, S. K., and Slater, D. H., "Mechanism and Kinetics of Propane Pyrolysis," *Ind. Eng. Chem. Process Des. Dev.* **18,** 232 (1979).

Lazzara, C. P., Biordi, J. C., and Papp, J. F., "Flame Inhibition," *C&F* **21,** 371 (1973).

Lefebvre, A. (Ed.), *Gas Turbine Combustor Design Problems,* Hemisphere, New York (1980).

Lefebvre, A., *Gas Turbine Combustion,* Hemisphere Publ. Co., Bristol, PA (1991a).

Lefebvre, A., *Atomization and Sprays,* Hemisphere Publ. Co., Bristol, PA (1991b).

Letokhov, V. S., Shank, C. V., Shen, Y. R., and Walther, H., *Laser Science and Technology (An International Handbook),* Gordon and Breach Science Publishers, Philadelphia, PA (1992).

Levy, A., "Problems in SO_2, NOx and Soot Control in Combustion," *19th Symp.,* 1223 (1983).

Levy, A., Droege, J. W., Tighe, J., and Foster, J. F., "Inhibition of Lean Methane Flames," *C&F* **6,** 524–33 (1962).

Levy, A., and Merriman, E., "The Microstructure of Hydrogen Sulfide Flames," *C&F* **9,** 229 (1965).

Levy, A., and Weinberg, F., "The Propagation of Flat Flames," *7th Symp.,* 296 (1958).

Levy, A., and Weinberg, F., "Examination of Reaction Rate Laws in Ethylene–Air Flames," *C&F* **3,** 229 (1959).

Levy, J., Taylor, B., Longwell, J., and Sarofim, A., "C_1 and C_2 Chemistry in Rich Mixture, Ethylene, Air Flames," *19th Symp.,* 167 (1983).

Lewis, B., "Remarks," *AGARD Selected Topics on Combustion Problems,* p. 177, Butterworths, London, UK (1954).

Lewis, B., and von Elbe, G., "Stability and Structure of Burner Flames," *JCP* **11,** 75 (1943).

Lewis, B., and von Elbe, G., *Combustion, Flames and Explosions of Gases,* Academic Press, New York (1938); 3rd ed. (1987).

Lewis, G. N., and Randall, M., revised by Pitzer, K. S., and Brewer, L., *Thermodynamics,* McGraw–Hill, New York (1961).

Lewis, J. D., and Merrington, A., "The Combustion of N-Heptane Spray in the Decomposition Products of Concentrated Hydrogen Peroxide," *7th Symp.,* 953 (1958).

Lewis, J. W. L., and Selman, J. D., "Laser Induced Schlieren Effect in Sodium–Nitrogen Mixture," Paper 83-1467 in *AIAA 18th Thermophysics Conference,* Montreal, Canada, June (1983).

Lias, S. G., "NIST/EPA/NIH Mass Spectral Database 1992 Version," available as magnetic tape in ASCI or binary versions, National Institutes of Standards and Technology, Gaithersburg, MD 20899.

Libby, P. A., and Williams, F. A. (Eds.), *Turbulent Reacting Flows,* Springer-Verlag, Berlin (1980).

Libouton, J. C., Dormal, M., and Van Tiggelen, P. J., "The Role of Chemical Kinetics in the Structure of Detonation Waves," *15th Symp.,* 73 (1975).

Lichty, L. C., *Combustion Engine Processes,* McGraw–Hill, New York (1967).

Lienhard, J., "Notes on the Origins and Evolution of the Subject of Heat Transfer," *Mech. Eng.* **105,** 20 (1983).

Lifshitz, A., and Frenklach, M., "Mechanism of the High Temperature Decomposition of Propane," *JCP* **79,** 686 (1975).

Lifshitz, A., Scheller, K., and Burcat, A., "Decomposition of Propane Behind Reflected Shocks in a Single Pulse Shock Tube," *Proc. Seventh Int. Shock Tube Symp.,* 690 (1973).

Lindow, R., "Burning Velocities" (in German), *Brennstoff Wärme Kraft* **20,** 8 (1968).

Linevsky, M. J., and Carabetta, R., "CW Laser Power from Carbon Bisulfide Flames," *App. Phys. Lett.* **22,** 228 (1973).

Linevsky, M. J., and Carabetta, R. A., "Chemical Laser Systems," p. 43, *Interim Technical Report 74SD4259,* Space Sciences Laboratory, General Electric Co., PO Box 8555, Philadelphia, PA 19101 (Sept. 1974).

Linevsky, M. J., Fristrom, R. M., and Smith, J. R., "The Simulation of Some Turbulent Combustion Processes Using an Acoustic Burner," Poster Session, *20th Symp.,* University of Michigan, Ann Arbor, Mich. (1985).

Linnett, J., "Methods of Measuring Burning Velocities," *4th Symp.,* The Williams & Wilkins Co., Baltimore, MD, 20 (1953).

Linnett, J. W., and Hoare, M., "Burning Velocities in Ethylene–Air–Nitrogen Mixtures," *3rd Symp.,* 195 (1948).

Linnett, J. W., Pickering, H., and Wheatley, P., *Trans. Farad. Soc.* **47,** 974 (1951).

Linnett, J. W., Reuben, B. G., and Wheatley, T. F., "A Photoelectric Study of the Low Pressure Explosion Limit of the Wet Carbon-Monoxide Oxygen Reaction," *C&F* **12,** 325 (1968).

Lixing, Zhou, *Theory and Numerical Modeling of Turbulent Gas-Particle Flows and Combustion,* CRC Press, Boca Raton, FL (1992).

Longwell, J., "Synthetic Fuels and Combustion," *16th Symp.,* 1 (1977).

Longwell, J., "The Formation of Polycyclic Aromatic Hydrocarbons by Combustion" (Followed by other papers on Soot and PAH), *19th Symp.,* 1339 (1983).

Longwell, J., and Weiss, M., "High Temperature Reaction Rates in Hydrocarbons," *Ind. Eng. Chem.* **47,** 1634 (1955).

Lossing, F. P., Ingold, K. U., and Tickner, A. W., "The Ethylene Oxide Decomposition Flame," *Disc. Faraday Soc.* **14,** 34 (1953).

Lovachev, L. A., and Kaganova, Z. I., "Calculation of the Characteristics of Bromine–Hydrogen Flames," *Dokl. Akad. Nauk SSSR* **188,** 1087 (1969).

Lucas, D., Brown, N. J., and Newton, A. S., "The Measurement of Ammonia in Lean Combustion Exhaust Gases," *CST* **52,** 139 (1987).

Lucas, D., Petterson, R., Hurlbut, F. C., and Oppenheim, A. K., "Effects of Transient Combustion Phenomena on Molecular Beam Sampling," *JCP* **88,** 4584 (1984).

Lucht, R. P., "Applications of Laser-Induced Fluorescence Spectroscopy for Combustion and Plasma Diagnostics," p. 623, in *Laser Spectroscopy and its Applications,* Radziemski, L. J., Solarz, R. W., and Paisner, J. A. (Eds.), Marcel Dekker, Inc., NY (1987).

Lucht, R. P., Farrow, R. L., and Palmer, R. E., "Acetylene Measurements in Flames by Coherent Anti Stokes Raman Scattering," *CST* **45,** 261 (1986).

Lucht, R. P., Laurendeau, N. M., and Sweeney, D. W., "Temperature Measurement by Two-Line Saturated OH Fluorescence in Flames," *Appl Opt.* **21,** 3729 (1982).

Lucht, R. P., Salmon, J. T., King, G. B., Sweeney, D. W., and Laurendeau, N. M., "Two-Photon-Excited Fluorescence Measurement of Hydrogen Atoms in Flames," *Opt. Lett.* **8,** 365 (1983).

Lucht, R. P., Sweeney, D. W., and Laurendeau, N. M., "Laser Saturated Florescence Measurements of OH Concentration in Flames," *C&F* **50,** 189 (1983).

Lucht, R. P., Sweeney, D. W., and Laurendeau, N. M., "Laser-Saturated Fluorescence Measurements of OH in Atmospheric Pressure $CH_4/O_2/N_2$ Flames under Sooting and Non-Sooting conditions," *CS&T* **42,** 259 (1985).

Lucht, R. P, Sweeney, D. W., Laurendeau, N. M., Drake, M. C., Lapp, M., and Pitz, R. W., "Single Pulse, Laser-Saturated Fluorescence Measurements of OH in Turbulent Flames," *Opt. Lett.* **9,** 90 (1984).

Lui, D., and MacFarlane, R., "Burning Velocities of Hydrogen–Air–Water Flames," *C&F* **49,** 59 (1983).

Lyon, R. K., and Benn, R., "Kinetics of the NO–NH_3–O_2 Reaction," *17th Symp.,* 601 (1979).

Lyons, J. W., "Mechanism of Fire Retardation with Phosphorous: Some Speculation," *J. Fire and Flammability* **1,** 302 (1970a).

Lyons, J. W., *The Chemistry and Uses of Fire Retardants,* Wiley Interscience, New York (1970b).

Lyons, J. W., "Fire," *Scientific American,* Milton Freeman, New York (1986).

Lyons, J. W., Becker, W. E., Clayton, J. W., Fristrom, R. M., Glassman, I., Graham, D. L., Emmons, H. W., McDonald, D. W., and Nadeau, H. G., *Fire Research on Cellular Plastic,* Products, Research Committee, c/o J. Lyons, N.B.S., Washington, D.C. (1980).

Ma, A. S. C, Spalding, P. B, and Sun, R. T. M., "Application of the 'Eskimo' Model," *19th Symp.,* 393 (1982).

Maas, U., Rafel, B., Wolfrum, J., and Warnatz, J., "Ozone Ignition and Combustion," *Progr. Astronaut. Aeronaut (Part II),* 335 (1986a).

Maas, U., Rafel, B., Wolfrum, J., and Warnatz, J., "Observation and Ozone Ignition and Combustion," *J. Appl. Phys.* **B37,** 189 (1986b).

Maas, U., Rafel, B., Wolfrum, J., and Warnatz, J., "Observation and Simulation of Laser Induced Ignition Processing O_2–O_3 and H_2–O_2 Mixtures," *21st Symp.,* 1869 (1988).

Maas, U., and Warnatz, J., "Ignition Processes in Carbon-Monoxide–Hydrogen–Oxygen Mixtures," *22nd Symp.,* The Combustion Institute, Pittsburgh, PA, 1695 (1988).

Macek, A., "Combustion of Boron Particles: Experiment and Theory," *14th Symp.,* 1401 (1973).

Maclatchy, C. S., Clements, R. M., and Smy, P. R., "Experimental Investigation of the Effect of Microwave Radiation on a Propane–Air Flame," *C&F* **45,** 161–69 (1982).

Maclean, D. I., and Tregay, G. W., "Spectroscopic Studies of Some Low Pressure Hydrogen Fluorine Flames," *14th Symp.,* The Combustion Institute, **157,** 2 (1973).

Maclean, D. I., and Wagner, H. Gg., "Structure of Reaction Zones of Ammonia–Oxygen and Hydrazine Decomposition Flames," *11th Symp.,* 871–78 (1967).

Madavi, J. N., and Amann, C. (Eds.), *Combustion Modeling in Reciprocating Engines,* Plenum Press, New York (1980).

Madorgsky, S. L., *Thermal Degradation of Organic Polymers,* Interscience Publishers, New York (1964).

Maeda, M. (Ed.), *Laser Dyes,* Academic Press (1984).

Mahnen, G., "Contribution a l'etude du mechanisme des deflagration flames ethylene–oxygen," *These de Doctorate,* University Catholique de Louvain, Belgium, 115 pp., 21 figs. (1973).

Maitland, A., and Dunn, M. H., *Laser Physics,* North Holland Pub. Co., Amsterdam–London (1969).

Mallard, E., and Le Chatelier, L., "Combustion des mélanges gaseux explosifs," *Ann. Mines* **4,** 274 (1883).

Malte, P., and Kramlich, J. C., "Further Observations of the Effects of Sample Probes on Pollutant Gases Drawn from Flame Zones," *CST* **22,** 263 (1980).

Malte, P., and Pratt, D., "The Measurement of Oxygen and Nitric Oxide in a Jet Stirred Combustor," *15th Symp.,* The Combustion Institute, Pittsburgh, PA, p. 1991 (1975).

Manton, J., von Elbe, G., and Lewis, B., "Burning Velocity Measurements in a Bomb," *4th Symp.,* 358 (1953).

Margolis, S. B., "Time-Dependent Solution of a Premixed Laminar Flame," *J. Computat. Phys.* **27,** 410–27 (1978).

Markstein, G., "Composition Traverses in Curved Laminar Flame Fronts," *7th Symp.,* 289 (1958).

Markstein, G., *Non-Steady Flame Propagation (AGARDOGraph No. 75),* Macmillan, New York (1964).

Marreno, T. R., and Mason, E. A., "Gaseous Diffusion Coefficients," *J. Phys. Chem. Data* **16,** 445 (1972).

Marsel, J., and Kramer, L., "Spontaneous Ignition Properties of Metal Alkyls," *7th Symp.,* 906 (1958).

Mason, J. M., and Theby, E. A., "SiO_2 Coated Thermocouples," *CST* **36,** 205 (1984).

Matula, R. A., Orloff, D. I., and Agnew, J. T., "Burning Velocities of Fluorocarbon–Oxygen Mixtures," *C&F* **14,** 97 (1970).

Mavrodineanu, R., *Bibliography on Flame Spectroscopy, Analytical Applications,* 1800–1966, Miscellaneous Publication No. 281, National Bureau of Standards, Washington, D.C. (1967).

Mavrodineanu, R., and H. Boiteux, *Flame Spectroscopy,* Wiley, NY (1965).

Mayer, S. W., Schieler, L., and Johnston, H. S., "Computation of High Temperature Rate Constants for Bimolecular Reaction of Combustion Products," *11th Symp.,* 837 (1967).

McBee, C. L., and Hastie, J. W., "Mechanistic Studies of Phosphorous Containing Retardants," *Proceedings 1975 International Symposium on Flammability Fire Retardants,* 253 (1975).

McHale, E. T., "Survey of Vapor Phase Agents for Combustion Suppression," *Fire Res. Abs. and Rev.* **11,** 90 (1969).

McKenna Products, Inc., P. O. Box 331, Pittsburgh, CA 94565.

Meier, U., Kohse-Hoinghaus, J., and Just, Th., "Discussion of Two-Photon Laser-Excited Fluorescence as a Method for Quantitative Detection of Oxygen Atoms in Flames," *22nd Symp.,* The Combustion Institute, Pittsburgh, PA, p. 1887 (1989).

Melius, C., and Binkley, J. S., "Thermochemistry of the Decomposition of Nitramines in the Gas Phase," *21st Symp.,* 1953 (1988).

Merriman, E., and Levy, A., "SO_3 Chemistry—H_2S and OCS Flames," *13th Symp.,* 427–36 (1971).

Mertzkirch, W. F., *Flow Visualization,* Academic Press, New York (1974).

Mertzkirch, W. F., *Flow Visualization II,* Hemisphere, New York (1982).

Merzhanov, A., "Combustion Processes and Synthesis of Refractory Materials," *23rd Symp.,* The Combustion Institute, Pittsburgh, PA, p. 1723 (1991).

Metghalchi, M., and Keck, J., "Burning Velocity of Propane–Air Mixtures at High Temperature and Pressure," *C&F* **38,** 143 (1980).

Metghalchi, M., and Keck, J., "Laminar Burning Velocity of Iso-Octane, Indolene and Methanol–Air Mixtures at High Temperature and Pressure," *C&F* **48,** 191 (1982).

Mettleton, M. A., *Gaseous Detonations: Their Nature, Effects and Control,* Chapman–Hall, NY (1987).

Michaud, P., Delfau, J., and Barassin, A., "Positive Ion Chemistry of Sooting Acetylene Flames," *18th Symp.,* 3 (1981).

Michelson, W., "Measurement of Flame Propagation Rates" (in German), *Ann. Physik.* **37,** 1 (1899).

Miles, R. B., Connors, J. J., Markovitz, E., Howard, P. J., and Roth, G. J., "Instantaneous Profiles and Turbulence Statistics of Supersonic Free Shear Layers by Raman Excitation Plus Laser-Induced Electronic Fluorescence Relief Velocity Tagging of Oxygen," *Exp. Fluids* **8,** 17 (1989).

Miller, B., and Hanson, R. K., "Homogeneous LDV Using Iodine Seeding," *Appl. Phys. Lett.* **43,** 143 (1983).

Miller, J., and Fisk, G., "Combustion Chemistry," *Chem. and Eng. News* **22,** Aug. 31 (1987).

Miller, J., Mitchell, R., Smooke, M., and Kee, R., "Toward a Comprehensive Chemical Kinetic Mechanism for the Oxidation of Acetylene: Comparison of Model Predictions with Results from Flame and Shock Tube Experiments," *19th Symp.,* 167 (1982).

Miller, J., Smooke, M., Green, R. M., and Kee, R. J., "Kinetic Modeling of the Oxidation of Ammonia in Flames," *CST* **34,** 149 (1983).

Miller, J. H., and Taylor, P. M., "Methyl Radical Concentrations and Production Rates in a Methane/Air Diffusion Flame," *CST* **51,** 139 (1987).

Miller, J. A., Volponi, J. V., Durant, J. L., Goldsmith, J. E. M., Fisk, G. A., and Kee, R. J., "The Structure and Reaction Mechanisms of Rich, Non-Sooting Acetylene, Oxygen, Argon Flames," *23rd Symp.,* The Combustion Institute, Pittsburgh, PA, p. 187 (1991).

Miller, W. J., Evans, M., and Skinner, D. G., "Flame Inhibition," *C&F* **7,** 163 (1963).

Miller, W. J., and Palmer, H. P., "Spectra of Alkali Metal–Organic Halide Flames," *9th Symp.,* The Combustion Institute, Pittsburgh, PA, p. 90 (1963).

Millikan, R., Roller, D., and Watson, E., *Mechanics, Molecular Physics and Sound,* Ginn & Co., New York (1937).

Milne, T., and Green, F. T., "Mass Spectrometric Sampling of 1 Atm. Flames," *10th Symp.*, 153 (1965).
Milne, T. A., and Green, F. T., "Mass Spectrometric Studies of Reactions in Flames II—Quantitative Sampling of Free Radicals from One-Atmosphere Flames," *JCP* **44,** 2444 (1966).
Milne, T. A., and Green, F., "Molecular Beams in High Temperature Chemistry," in *Advances in High Temperature Chemistry,* Academic Press, NY, 107–58 (1969).
Milne, T. A., Greene, C. L., and Benson, D. K., "The Use of the Counterflow Diffusion Flame in Studies of Inhibition Effectiveness of Gaseous and Powdered Agents," *C&F* **15,** 255 (1970).
Milton, B., and Keck, J., "Laminar Burning Velocities in Stoichiometric Hydrogen and Hydrogen–Hydrocarbon Gas Mixtures," *C&F* **58,** 13 (1984).
Minetti, R., Corre, C., Pauwels, J-F., Devolder, P., and Sochet, L-R., "On the Reactivity of Hydroperoxy Radicals and Hydrogen Peroxide in a Two Stage Butane–Air Flame," *C&F* **85,** 263 (1991).
Ministry of Power, *The Efficient Use of Fuel,* Her Majesty's Stationary Office, London (1958).
Minkoff, G. J., *Frozen Free Radicals,* Interscience Publishers, N.Y. (1960).
Minkoff, G. J., and Tipper, C. F. H., *Chemistry of Combustion Reactions,* Butterworths, London (1962).
Mitani, T., and Williams, F. A., "Studies of Cellular Flames of Hydrogen–Oxygen–Nitrogen Mixtures," *C&F* **39,** 169–90 (1980).
Mitani, T., and Williams, F. A., "A Model for the Deflagration of Nitramines," *21st Symp.,* 965 (1988).
Mitchell, R., "Numerical Solution of a Confined Diffusion Flame," *SAND 83-8624,* Sandia Livermore Laboratories, Livermore, CA (1983).
Mitchell, R. E., Sarofin, A. F., and Yu, R., "Nitric Oxide and Hydrogen Cyanide Formation in Methane/Air Diffusion Flames," *CST* **21,** 157 (1980).
Monchick, L., and Mason, E. A., "Transport Properties of Polar Gases I," *JCP* **35,** 1671 (1961a).
Monchick, L., and Mason, E. A., "Transport Properties of Polar Gases II," *JCP* **36,** 2746 (1961b).
Monger, J. M., Baumgartner, H. J., Hood, G. C., and Sanborn, C. E., "Explosive Limits of Hydrogen Peroxyde Vapor," *J. Chem. Eng. Data* **9,** 119 (1964a).
Monger, J. M., Baumgartner, H. J., Hood, G. C., and Sanborn, C. E., "Detonations in Hydrogen Peroxyde Vapor," *J. Chem. Eng. Data* **9,** 107 (1964b).
Moore, D. W., "A Pneumatic Probe Method for Measuring High Temperature Gases," *Aero. Eng. Rev.* **7,** No. 5 (1958).
Morgan, G. H., and Kane, W. R., "Some Effects of Inert Diluents on Flame Speeds and Temperatures," *4th Symp.,* p. 313, The Combustion Institute, Pittsburgh, PA (1953).
Morley, C., "The Formation and Destruction of Hydrogen Cyanide from Atmospheric and Fuel Nitrogen in Rich Atmospheric Pressure Flames," *C&F* **27,** 189 (1976).
Morley, C., "Photolytic Perturbation Method to Investigate the Kinetics of Hydrocarbon Oxidation near 800K," *22nd Symp.,* The Combustion Institute, Pittsburgh, PA p. 911 (1989).
Morrison, R., and Boyd, R. K., *Organic Chemistry,* Allyn and Bacon Inc., Boston, MA, 3rd ed. (1973).
Morrison, R., and Scheller, K., "Flame Inhibition," *C&F* **18,** 3 (1972).
Morse, J. S., Cundy, V. A., and Lester, T. W., "Chemical, Temperature and Net Reaction Rate Profiles of Laminar Carbon Tetrachloride/Methane/Air Flames," *CST* **66,** 59 (1989).
Mulcahy, M. F. R., "Experimental Methods for the Direct Investigation of Elementary Reactions of Atoms and Radicals in *Gas Kinetics,* Chapter 4, p. 122, Thomas Nelson and Sons, London (1982).

Mullaney, G. J., "Temperature Determination in Flames by X-Ray Absorption Using a Radioactive Source," *Rev. Sci. Inst.* **29,** 87 (1958).

Muller, C. H. III, Schofield, K., Steinberg, M., and Broida, H. P., "Sulphur Chemistry in Flames," *17th Symp.,* 867–77 (1979).

Müller-Dethlefs, K., and Schlader, A. F., "The Effect of Steam on Flame Temperature, Burning Velocity and Carbon Formation in Hydrocarbon Flames," *C&F* **27,** 205 (1976).

Müller-Dethlefs, K., and Weinberg, F. J., "Burning Velocity Measurements Based on Laser Rayleigh Scattering," *17th Symp.,* p. 985 (1979).

Mullins, B. P., *Spontaneous Ignition of Liquids,* Butterworths, London (1957).

Muragai, M. P., *Similarity Analysis in Fire Research,* Oxford (1976a).

Muragai, M. P., *Natural Convection from Combustion Sources,* Oxford (1976b).

Murray, J. R., "Lasers for Spectroscopy", p. 91, in *Laser Spectroscopy and its Applications,* Radziemski, L. J., Solarz, R. W., and Paisner, J. A. (Eds.), Marcel Dekker, Inc., NY (1987).

Murray, R. C., and Hall, A. R., "Flame Speeds in Hydrazine Vapors and in Mixtures of Hydrazine and Ammonia with Oxygen," *Trans. Farad. Soc.* **47,** 43 (1951).

Myerson, A. L., "The Reduction of Nitric Oxide in Simulated Combustion Effluents by Hydrocarbon–Oxygen Mixtures," *15th Symp.,* 1085–92 (1975).

Navier, C. L. M. H., "Memoire sur les lois du movement des fluides," *Mem. Acad. Sci. Inst. Fr.* **6,** 389 (1822).

Nenninger, J. E., Kridiotis, A., Chomiak, J., Longwell, J. P., and Sarofim, A. F., "Characterization of a Torroidal Well Stirred Reactor," *20th Symp.,* The Combustion Institute, Pittsburgh, PA, 473 (1985).

Nieser, P. R., Wurster, M., and Haas, U., "Application of LAMMA in Aerosol Research," *Fresenius Z. Anal. Chem.* **308,** 260 (1981).

Niessen, W. R., *Combustion and Incineration Processes,* Marcel Dekker, New York (1987).

Nomaguchi, T., and Koda, S., "Spark Ignition of Methane and Methanol in Ozonized Air," *21st Symp.,* The Combustion Institute, Pittsburgh PA, p. 1677 (1989).

Norrish, G., "The Study of Combustion Reactions by Photochemical Methods," *10th Symp.,* The Combustion Institute, Pittsburgh, PA, p. 1 (1965).

NSRD, *DIPPR* (Design Institute for Physical Property Data), "Database of 39 properties of 1000 compounds," NSRD/NIST (1989).

Odnoroq, D. S., Kolesnikov, B. Ya., Kamyshaev, D. K., and Ksandopolo, G. E., "Structure of the Flame Front of Propane Inhibited by Diethylamine and Dibromotetrafluorenthane," *Mater. Soveshch. Mekh. Inqibirovaniza Tsephnykj Gazov. Reakts.,* 101 (1971).

Ohmann, H., *Ber. Deutsch. Chem. Gesellsch.* **53,** 1427–29 (1920).

Oldenhove de Guertechin, L., "Formaldehyde Combustion," Ph.D. Thesis, Catholic University of Louvain (1982).

Oldenhove de Guertechin, L., Vandooren, J., and Van Tiggelen, P. J., "Stabilisation et Analyse par Spectrometrie de Masse des Flammes Unidimensionelles CH_2O/O_2 a Basse Pression," *J. de Chim. Phys.(Belg.)* **80,** 583–94 (1983a).

Oldenhove de Guertechin, L., Vandooren, J., and Van Tiggelen, P. J., "Radical Reactions in Formaldehyde Flames," *Bull. Soc. Chim. Belg.* **92,** 663–64 (1983b).

Oldenhove de Guertechin, L., Vandooren, J., and Van Tiggelen, P. J., "Kinetics in a Lean Formaldehyde Flame," *C&F* **64,** 127 (1986).

Olsen, D., and Calcote, H., "Ions in Fuel-Rich and Sooting Acetylene and Benzene Flames," *18th Symp.,* 453 (1981).

Olsen, H. L., and Gayhart, E. L., "Measurements by the Flame Kernel Technique," *APL/JHU CM 855,* Laurel, MD (Nov. 1955).

Olsen, H. L., Gayhart, E. L., and Edmonson, R. B., "Propagation of Spark Ignited Flames," *4th Symp.,* William Wilkins and Co., Baltimore, p. 144 (1953).

Olson, D., Tanzawa, T., and Gardiner, W. C., "Thermal Decomposition of Ethane," *Int. J. Chem. Kin.* **11,** 23 (1979).

Olson, I. B. M., and Smooke, M. O., "Computational Modeling of a Premixed Formaldehyde Flame," *JCP* **93,** 3107 (1989).

Olson, J., and Anderson, L., "Time Dependent Solution of Premixed Laminar Flames with a Known Temperature Profile," *J. Comp. Phys.* **59,** 369 (1985).

Olson, W. T., and Setze, P. C., "Some Combustion Problems of High Energy Fuels," *7th Symp.,* 883 (1958).

Omega Corp., "Catalog," Omega Corporation, NY, Omega Engineering Corp., One Omega Drive, PO Box 4047, Stamford, Conn. 06907-0047.

Omenetto, N., Browner, R., and Winefordner, J., " 'Color Temperature' Atomic Fluorescence Method for Flame Temperature Measurement," *Anal. Chem.* **44,** 1683 (1972).

Oppenheim, A. K. (Ed.), *Combustion and Propulsion,* Fourth AGARD Colloquium, Pergamon Press, New York (1961).

Oran, E., and Boris, J., "Detailed Modeling of Combustion Systems," *Progress in Energy and Combustion Science* **7,** 1–72 (1981).

Oran, E., and Boris, J., "Detailed Modeling of Combustion Systems," *18th Symp.,* 1641 (1982).

Orr, C., "Combustion of Hydrocarbons Behind a Shock Wave," *9th Symp.,* 1034 (1963).

Pacsko, G., Lefdal, P., and Peters, N., "Reduced Reaction Schemes for Methane, Methanol and Propane Flames," *21st Symp.,* 739 (1988).

Padley, P. J., and Sugden, T. M., "Chemiluminescence and Radical Recombination in Hydrogen Flames," *7th Symp.,* The Combustion Institute, Pittsburgh, PA, p. 235 (1959).

Palmer, H. B., and Beer, J. M. (Eds.), *Combustion Technology,* Academic Press, New York (1974).

Palmer, H. B., Krugh, W. D., and Hsu, C-J., "Chemiluminescent Spectra and Light Yields from Several Low Pressure Diffusion Flames of Alkaline Earth Metal Vapors," *15th Symp.,* 951 (1975).

Palmer, H. B., and Seery, D. J., "CO–Air Flames Inhibition by Chlorine," *C&F* **4,** 213 (1963).

Palmer, H. B., and Seery, D. J., "Chemistry of Pollutant Formation in Flames," *Ann. Rev. Phys. Chem.* **24,** 235 (1973).

Palmer, K. N., *Dust Explosions and Fire,* Chapman–Hall, London (1973).

Pandya, T. P., and Weinberg, F. J., "The Study of the Structure of Laminar Diffusion Flames by Optical Methods," *9th Symp.,* Academic Press, New York, p. 587 (1963).

Pannetier, G., and Guedeney, F., "The Stationary Decomposition of Hydrazine Hydrate," *Compt. Rend.* **238,** 898 (1954).

Pannetier, G., and Sicard, A., "Regions of Flammability of Mixtures of Hydrocarbons with Nitrous Oxide," Action of Nitrous Oxide as Oxidant," *5th Symp.,* 620 (1955).

Papp, J., Lazzara, C., and Biordi, J., "Kinetics of Reactions in Flames. II. A Critique of Rate Coefficient Determinations," *C&F* **23,** 73 (1974).

Papp, J., Lazzara, C., and Biordi, J., "Computational Methods for Analyzing Flame Microstructure Data," *Report of Investigation #8109,* U.S. Bureau of Mines, Department of Interior, Pittsburgh, PA (1975).

Parker, W., and Wolfhard, H. G., "Some Characteristics of Flames Supported by NO and NO_2," *4th Symp.,* The Combustion Institute, Pittsburgh, PA, p. 420 (1949).

Parker, W., and Wolfhard, H. G., "Properties of Diborane Flames," *Fuel* **35,** 323 (1956).

Parks, D., and Fletcher, E., "Fluorocarbon Combustion Studies V—Trifluorochloromethane and Trifluorobromoethane–Fluorine Flames," *C&F* **13,** 487 (1969).

Partington, J. R., *History of Chemistry,* (four volumes), Macmillan, London (1961a).

Partington, J. R., *A Short History of Chemistry,* Macmillan, London (1961b).

Partington, J. R., *A Textbook of Inorganic Chemistry,* Macmillan Book Co., NY, 5th ed. (1939).

Passauer, H., "The Burning Velocity of Methane-Air Mixtures" (in German), *Gasund Wasserfadch* **73** 31 (1958).

Pauling, L., *The Nature of the Chemical Bond,* Cornell Univ. Press, Ithaca, NY (1939).

Pauling, L., *General Chemistry,* Dover Reprints, NY (1988).

Pauwels, J. F., Carlier, M., Devolder, P., and Sochet, L-R., "An ESR Study of Labile Species in Methanol Air Flames Doped with Hydrogen Sulfide," *JCP* **90,** 4377 (1986).

Pauwels, J. F., Carlier, M., Devolder, P., and Sochet, L-R., "Experimental and Numerical Analysis of a Low Pressure Stoichiometric Methanol Air Flame," *CST* **64,** 97 (1989).

Pauwels, J. F., Carlier, M., Devolder, P., and Sochet, L-R., "Influence of Equivalence Ratio on the Structure of Low-Pressure Premixed Methanol Air Flames," *C&F* **82,** 163 (1990).

Pauwels, J. F., Carlier, M., and Sochet, L-R., "Analysis by Gas-Phase ESR of H, O, OH and Halogen Atoms in Flames," *JCP* **86,** 4330 (1982).

Pauwels, J. F., Carlier, M., and Sochet, L-R., "Etude Experimentale et Modelisation de Flammes de Premelange," *Bull. Soc. Chim. Belg.* **99,** 503 (1990a).

Pauwels, J. F., Carlier M., and Sochet, L-R., "Modelisation de flammes de premelange unidimensionnelles Analyse de sensibilite et reduction de mechanismes," *JCP* **87,** 583 (1990b).

Peacock, F., and Weinberg, F. J., "Methods for the Study of Hydrogen–Bromine Kinetics," *8th Symp.,* The Combustion Institute, 458–72 (1961).

Pearson, G. S., "Perchloric Acid Flames IV—Methane Rich Flames," *C&F* **11,** 89 (1967a).

Pearson, G. S., "Perchloric Acid Flames V: Ethylene Rich Flames," *C&F* **11,** 97 (1967b).

Pearson, G. S., "Perchloric Acid Flames VIII—Methane Rich Flames with Oxygen," *C&F* **12,** 54 (1968).

Pease, R. N., *Equilibrium and Kinetics of Gas Reactions,* 102, Princeton Univ. Press (1942).

Peeters, J., and Mahnen, G., "The Structure of Ethylene–Oxygen Flames: Reaction Mechanism and Rate Constants," *Combustion Institute European Symposium, Sheffield,* Academic Press, NY, 53 (1973a).

Peeters, J., and Mahnen, G., "Reaction Mechanisms and Rate Constants of Elementary Steps in Methane–Oxygen Flames," *14th Symp.,* The Combustion Institute, Pittsburgh, PA, 133 (1973b).

Penner, S. S., *Introduction to the Study of Chemical Reactions in Low Systems,* Butterworths, London (1955).

Penner, S. S., *Chemistry Problems in Jet Propulsion,* Pergamon Press, New York (1957).

Penner, S. S., *Quantitative Molecular Spectroscopy and Gas Emissivities,* Chap. 16, Addison–Wesley, Reading, Mass. (1959).

Penner, S. S., *Chemical Rocket Propulsion and Combustion Research,* Gordon and Breach, New York (1962).

Penner, S. S., Wang, C. P., and Bahadori, N. Y., "Laser Diagnostics Applied to Combustion Systems," *20th Symp.,* 149 (1984).

Peters, N., "Flamlet Model in Turbulent Combustion," *21st Symp.,* p. 1231, (1988).

Peters, N., and Warnatz, J. (Eds.), *Numerical Methods in Laminar Flame Propogation,* Viewag, Wiesbaden, Ger. (1982).

Peters, N. and Williams, F., "The Asymptotic Structure of Stoichiometric CH_4–Air Flames," *C&F* **68,** 185 (1987).

Petersen, R., and Emmons, H., "Stability of Laminar Flames," *Phys. of Fluids* **4,** 456 (1961); see also Quinn, J. C., "Laminar Flame Front Thickness," *Report #5, Comb. Aero. Lab.,* Harvard, Cambridge, MA (1953).

Philbean, D., "The Descent of Hominoid and Hominoids," *Sci. Am.* **250,** 84 (1984).

Phillips, V. D., Brotherton, T. D., and Anderson, R. C., "Physical Characteristics and Stability of Some Hydrogen–Bromine Flames," *4th Symp.,* The Computer Institute, 701 (1953).

Phinney, R., "Mathematical Nature of the Freezing Point in an Expanding Flow," *Report RM-172,* Martin Marietta Co., Baltimore, MD (1964).

Pitts, J. N., and Finlayson, B., "Mechanisms of Photochemical Air Pollution," *Angew. Chem. Internat. Ed.* **14,** 1 (1975).

Pitz, W. J., Westbrook, C. K., Proscia, W., and Dryer, F. L., "Comprehensive Reaction Mechanisms for the Oxidation of n-Butane," *20th Symp.,* 831 (1985).

Polanyi, J., "Review on nonequilibrium radiation in reacting systems," *Acc. Chem. Res.* **5,** 161 (1972).

Polanyi, J., Lossing, F. P., et al., "The Decomposition of Ethylene Oxide," *Disc. Faraday Soc.* **14,** 115 (1953).

Polanyi, M., *Atomic Reactions,* Williams and Norgate, London (1932).

Poling, E. L., and Simmons, H., "Explosive Reaction of Diborane in Dry and Water-Saturated Air," *Ind. Eng. Chemistry* **35,** 1698 (1958).

Pollack, D. D., "Thermocouples in High Temperature Reactive Atmospheres," *CST* **42,** 111 (1984).

Poncelet, J., Deckers, J. and Van Tiggelen, A. "Comparative Study of Ionization in Acetylene-Oxygen and Acetylene-N_2O Flames," *7th Symp.,* 256 [1959].

Porter, R., Clark, A., Kaskan, W., and Browne, W., "A Study of Hydrocarbon Flames," *11th Symp.,* 907–17 (1957).

Potter, A., and Butler, J., "A Novel Combustion Measurement Based on the Extinction of Diffusion Flames," *ARS J.* **29,** 54 (1959).

Potter, A., and Butler, J., "Apparent Flame Strengths," *8th Symp.,* The Combustion Institute, Pittsburgh, PA 1027–34 (1963).

Powell, H. N., and Browne, W. G., "The Use of Coiled Capillaries in a Convenient Laboratory Flowmeter," *Rev. Sci. Inst.* **28,** 138 (1957).

Powling, J., "The Flat Flame Burner," See Sec. 2-1 of *Experimental Methods In Combustion Research,* J. Surugue (Ed.), Pergamon Press, New York (1961).

Pownall, C., and Simmons, R. F., "The Effect of Hydrogen Bromide on the Structure of Propane–Oxygen Flames Diluted with Argon," *13th Symp.,* 585–92 (1971).

Pratt, W. K., *Digital Image Processing,* Wiley-Interscience, NY (1978).

Predvoditelev, A. S. (Ed.), *Gas Dynamics and Physics of Combustion,* Israel Program for Scientific Translations, Jerusalem (trans. from Russian) (1962).

Prescott, R., Hudson, R., Foner, S., and Avery, W. H., "Composition Profiles in Premixed Laminar Flames," *JCP* **22,** 106 (1954).

Price, F. P., "Explosion Limits of Diborane–Oxygen Mixtures," *JACS* **72,** 5361 (1951).

Priestley, J., *Lectures on Combustion,* Reprint, Kraus, New York (1969).

Products Research Committee, *Fire Research on Cellular Plastics,* see J. Lyons, Chair., c/o NBS, Washington, D.C. (1980).

Prudnikov, A., "Flame Turbulence," *7th Symp.,* Combustion Institute, Pittsburgh, PA, 575 (1959).

Pungor, E. (trans R. Chalmers), *Flame Photometry Theory,* D. Van Nostrand Ltd., London, UK (1967).

Purcell, J., *From Hand Ax to Laser: Man's Growing Mastery of Energy,* Vanguard Press (1983).

Rabek, J. F., *Experimental Methods in Photochemistry and Photophysics* (2 vols.), John Wiley and Sons, New York (1982).

Radziemski, L. J., Solarz, R. W., and Paisner, J. A. (Eds.), *Laser Spectroscopy and its Applications,* Marcel Dekker, Inc., NY (1987).

Raezer, S. D., and Olsen, H. L., "Intermittent Thermometer, A New Method for Measuring Extreme Temperatures," *Report CM 985,* Applied Physics Laboratory/The Johns Hopkins University, Laurel, MD, AD-252682, Armed Services Information Agency, Arlington Hall Station, VA (1960).

Raezer, S. D., and Olsen, H. L., "Measurement of Laminar Flame Speeds of Ethylene–Air and Propane–Air Mixtures by the Double Kernel Method," *C&F* **6,** 227 (1962).

Rahn, L. A., and Farrow, R. L., "Measurements of Nitrogen Q-Branch Foreign Gas-Broadening Coefficients Relevant to Flames," *22nd Symp.,* Combustion Institute, Pittsburgh, PA, p. 1851 (1989).

Rahn, L. A., Farrow, R. L., Koszykowski, M. L., and Mattern, P. L., "Observation of an Optical Stark Effect on Vibrational and Rotational Transitions," *Phys. Rev. Lett.* **4,** 620 (1980).

Rahn, L. A., Zych, L. J., and Mattern, P. L., "Background Free CARS Studies of Carbon Monoxide in a Flame, *Opt. Commun.* **30,** 249 (1979).

Rallis, C., Arforth, A., and Steinz, J., "Laminar Burning Velocity of Acetylene–Air Mixtures by the Constant Volume Method: Dependence on Composition, Pressure and Temperature," *C&F* **9,** 345 (1965).

Ranade, M. B., Werle, D. K., and Wasan, D. T., "Dilution Microprobe," *J.Colloid Sci.* **56,** 42 (1976).

Rau, G., and Barzziv, E., "Deflection Mapping of Flames Using the Moire Effect," *Appl. Opt.* **23,** 2886 (1984).

Ravichandran, M., and Gouldin, F. C., "Retrieval of Asymmetric Temperature and Concentration Profiles from a Limited Number of Absorption Measurements," *CST* **60,** 231 (1988).

Rawlings, W. T., and Gardiner, W. C., "Rate Constant for $CO + O_2 = CO_2 + O$ from 1500 to 2500 K; A Reevaluation of Induction Times in the Shock-Initiated Combustion of Hydrogen–Oxygen–Carbon Monoxide–Argon Mixtures," *JCP* **78,** 497 (1974).

Razdau, M. K., "Two Component Velocity and Turbulence Measurements in a CO/Air Diffusion Flame," *CST* **46,** 45 (1986).

Rea, E. C., Jr., and Hanson, R. K., "Fully Resolved Absorption Fluorescence Lineshape Measurements of OH Using a Rapid Scanning Ring Dye Laser," *WSSCS 1983 Fall Meeting (1983),* UCLA, Los Angeles, CA, Paper 83 (1983).

Reed, H. G., and Herzfeld, C. M., "Theory of Flame Propagation in Solid Nitrogen at Low Temperature," *JCP* **32,** 1 (1960).

Rees, Y., and Williams, G. H., "Reactions of Organic Free Radicals with Nitrogen Oxides," *Advances in Free Radical Chemistry* **3,** 199–230 (1968).

Reid, R. C., Prausnitz, J. M., and Poling, B. E., *The Properties of Gases and Liquids,* McGraw–Hill, 3rd ed. (1977), 4th ed. (1987).

Reid, S., Mineur, J. and McNaughton, J., ''Burning Velocity of Methane-Air Mixtures'' *J. Inst. Fuel,* **44** 149 (1971).

Rensberger, K. J., Copeland, R. A., Wise, M. L., and Crosley, D. R., "NH and CH Laser-Induced Fluorescence in Low Pressure Flames: Quantum Yields from Time-Resolved Measurements," *22nd Symp.,* The Combustion Institute, Pittsburgh, PA, p. 1867 (1989).

Reuss, D. L., "Temperature Measurements in a Radially Symmetric Flame Using Holographic Interferometry," *C&F* **49,** 207 (1983).

Rice, J., and Sickman, R., "Studies on Azomethane Decomposition," *JCP* **4,** 245 (1936).

Ritter, E. R., and Bozzelli, J. W., "Reactions of Chlorinated Benzenes in H_2 and H_2O Mixtures: Thermodynamic Implications on Pathways to Dioxin," *CST* **74,** 117 (1990).

Robben, F., "Comparison of Density and Temperature Measurements Using Raman Scattering and Rayleigh Scattering," *Combustion Measurements in Jet Propulsion Systems* (Ed. R. Goulard), p. 179, Academic Press (1976).

Roberts, C., and Vanderslice, T., *Ultrahigh Vacuum and Its Applications,* Prentice–Hall, Inc., Englewood Cliffs, NJ (1966).

Robertson, A. C., *Fire Standards and Safety,* STP 614, American Society for Testing Material, Philadelphia, PA (1977).

Robinson, A., "Physicists Try to Find Order in Chaos," *Science* **218,** 554 (1982).

Rogg, B., and Wichman, I. S., "Approach to Asymptotic Analysis of the Ozone-Decomposition Flame," *C&F* **62**, 271–93 (1985).

Rogg, B., and Williams, F. A., "Structures of Wet CO Flames with Full and Reduced Kinetic Mechanisms," *22nd Symp.*, The Combustion Institute, Pittsburgh PA, p. 1441 (1989).

Rose, A., and Gupta, R., "Application of the Photothermal Deflection Technique to Flow-Velocity Measurements in a Flame," *Optics Lett.* **10**, 532 (1985).

Rosen, G., "An Action Principle for Laminar Flames," *7th Symp.*, The Combustion Institute, Pittsburgh, PA, 339 (1958).

Rosser, W. G., Inami, J., and Wise, H., "Flame Inhibition," *C&F* **7**, 107 (1963).

Rosser, W. G., Wise, H., and Miller, J., "Mechanism of Combustion Inhibition by Compounds Containing Halogen," *7th Symp.*, 175 (1959).

Rossler, D. E., *Transport Processes in Chemically Reacting Flow*, Butterworths, London (1986).

Roth, W., and Bauer, W., "The Combustion of Diborane–Oxygen Mixtures in the Second Explosion Limit," *5th Symp.*, 710 (1955).

Royal Society of Chemistry, *The Eight Peak Index of Mass Spectra*, Vols. 1–3, 3rd ed., Royal Society of Chemistry, University of Nottingham, England (1983).

Rozlovskiy, A. I., "Kinetics of the Dark Reaction of Hydrogen–Chlorine Mixtures—III: The Normal Burning of Hydrogen–Chlorine Mixtures," *Zh. Fiz. Khim.* **30**, 2489 (1956a).

Rozlovskiy, A. I., "Kinetics of the Dark Reaction of Hydrogen–Chlorine Mixtures," *Zh. Fiz. Khim.* **30**, 2713 (1956b).

Rumford, Count, *Collected Works of Count Rumford*, Vols. 1 & 2, S. C. Brown (Ed.), Reprint, MIT Press, Cambridge, Mass. (1968).

Russell, B., *History of Western Philosophy*, Simon & Schuster, New York (1967).

Ryabov, I. V., Baratov, A. N., and Petrov, I. I. (Eds.), *Problems in Combustion and Extinguishment*, TT 71-580001, National Bureau of Standards and National Science Foundation, U.S. Government Printing Office, Washington, D.C. (1974) (trans. from Russian).

Sachsteder, K. B., "The Implications of Experimentally Controlled Gravitational Accelerations for Combustion Science," *23rd Symp.*, The Combustion Institute, Pittsburgh PA, p. 1589 (1991).

Sadequi, M., and Branch, M., "A Continuous Flow, Gaseous Formaldehyde Generation System for Combustion Studies," *C&F* **17**, 325 (1988).

Safieh, H., Vandooren, J., and Van Tiggelen, P. J., "Experimental Study of Inhibition Induced by CF_3Br in a CO–H_2–O_2–Ar Flame," *19th Symp.*, 117 (1982).

Sagulin, A. B., "Explosion Temperatures of Gaseous Mixtures at Different Pressures," *Z. Physik. Chem.* **1b**, 275 (1928).

Salmon, J., Lucht, R., Sweeney, D., and Laurendeau, N., "Laser Saturated Fluorescence Measurements of NH in a Premixed Sub-atmospheric CH_4–N_2O–Ar Flame," *20th Symp.*, The Combustion Institute, Pittsburgh, PA, p. 1187 (1985).

Santoro, R. J., and Glassman, I., "Short Communication—A Review of Oxidation of Aromatic Compounds," *CST* **19**, 161 (1979).

Santoro, R., Semerjain, H., Emerson, P., and Goulard, R., "Optical Tomography for Flow Field Diagnosis," *Int. J. Heat and Mass Trans.* **24**, 1139 (1981).

Sarton, O., *A History of Science* O.U.D., Baltimore, MD (1953).

Sato, A., Hashiba, K., Hasatani, J., Sugiyama, S., and Kimura, J., "A Correctional Calculation Method for Thermocouple Measurements of Temperature in Flames," *C&F* **24**, 35 (1975).

Sawyer, R. F., "The Formation and Destruction of Pollutants in Combustion Processes: Clearing the Air on the Role of Combustion Research," *18th Symp.*, 1 (1981).

Sawyer, R. F., and Glassman, I., "Gas Phase Reactions of Hydrazine with Nitrogen Dioxide, Nitric Oxide and Oxygen," *11th Symp.*, The Combustion Institute, Pittsburgh, PA, p. 861 (1967).

Scarre, C. (Ed.), *Past Worlds: the Times Atlas of Archeology,* Hammond, Inc., Maplewood, NJ (1988).
Schalla, R. L., "Spontaneous Ignition Limits of Pentaborane," *Nat. Adv. Comm. Aero Res. Mem., E55BO3* (1957).
Schalla, R. L., McDonald, G. E., and Gerstein, M., "Combustion Studies of Alkyl Silanes," *5th Symp.,* 705 (1955).
Schefer, R., "Catalyzed Recombination of H_2/Air in the Boundary Layer of a Flat Plate. II. Numerical Calculation," *C&F* **45,** 171 (1982).
Schefer, R., Johnston, S., Kelly, J., Namazian, N., and Long, M., "Two Dimensional Raman Imaging of Methane in Flames," *Combustion Research Laboratory, Annual Report (1985),* Sandia National Laboratory (Livermore, CA), Sec. 4. Also see Combined Eastern and Western States Sections Combustion Institute Meeting, paper 2-1B, Ion Institute, Pittsburgh, PA (1983).
Schelkin, K. I., and Troshin, Y. K. (trans. B. Kuvshinoff and L. Holtschlag), *Gas Dynamics of Combustion,* Mono Book Co., Baltimore (1965).
Schenck, P. K., Travis, J., Turk, G., and O'Haver, T., "LEI (Laser Enhanced Ionization) Method for Velocity Measurements in Flames," *Appl. Spectrosc.* **36,** 168 (1982).
Schlichting, H., *Boundary Layer Theory,* 6th ed., McGraw–Hill, New York (1968).
Schoenung, S., *"Tuneable Diode Laser Absorption Probe Techniques for Species Measurements in Combustion Gases,"* Ph.D. Thesis, Mech. Eng., Stanford Univ., CA (1981).
Schoenung, S., and Hanson, R., "CO and Temperature Measurements in a Flat Flame by Laser Absorption Spectroscopy and Probe Techniques," *C&F* **24,** 227 (1981).
Schoenung, S., and Hanson, R., "Temporally and Spatially Resolved Measurements of Fuel Mole Fraction in a Turbulent CO Diffusion Flame," *19th Symp.,* p. 449 (1982).
Schofield, K., and Steinberg, M., "Review of Quantitative Atomic and Molecular Fluorescence in the Study of Detailed Combustion Processes," *Opt. Eng.* **20,** 501 (1981).
Scholte, T., and Vaags, P., "Burning Velocity of Hydrogen–Air Mixtures and Mixtures of Some Hydrocarbons with Air," *C&F* **3,** 495 (1959).
Schott, G. L., "Kinetic Studies of Hydroxyl Radicals in Shock Waves. III. The OH Concentration Maximum in the Hydrogen Reaction," *JCP* **32,** 710 (1960).
Schroeder, M. J., and Buck, C. C., *Fire Weather,* U.S. Department of Agriculture Handbook 360, Superintendent of Documents, Washington, D.C. (1978).
Schug, K., Santoro, R., Dryer, F. L., and Glassman, I., "Ethylene Oxidation," Western States Section Meeting, The Combustion Institute, Tuscon, AZ (1978).
Schumacker, B. W., and Gadamer, B., "Electron Beam Fluorescence Probe for Measuring the Local Gas Density in a Wide Field of Observation," *JCP* **36,** 659 (1958).
Schumb, W. C., Satterfield, C. N., and Wentworth, R., *Hydrogen Peroxide,* ACS Monograph No. 128, Reinhold, New York (1955).
Schwar, M. J. R., and Weinberg, F. J., "Laser Techniques in Combustion," *C&F* **13,** 335 (1969a).
Schwar, M. J. R., and Weinberg, F. J., "The Measurement of Velocity by Applying Schlieren-Interferometry to Doppler Shifted Laser Light," *PRS* **A311,** 469 (1969b).
Schwartzwald, R., Monkhouse, P., and Wolfrum, J., "Fluorescence Studies of OH and CN Radicals in Atmospheric Pressure Flames Using Picosecond Excitation," *22nd Symp.,* The Combustion Institute, Pittsburgh, PA, p. 1413 (1989).
Scott, A., "The Invention of the Balloon and the Birth of Chemistry," *Sci. Am.* **250,** p. 126 (1984).
Scurlock, A., and Grover, J., *Selected Combustion Problems,* Butterworths, London 215 (1954).
Seeker, W. R., "Waste Combustion," *23rd Symp.,* The Combustion Institute, Pittsburgh, PA, p. 867 (1991).

Seery, D. J., and Zabielski, M., "Mass Spectrometer Sampling of a CO–NH_3–O_2 Flame," *C&F* **28,** 93 (1977).

Sell, J., "Velocity Measurements," *Ap. Opt* **23,** 1586 (1984).

Semenov, N., *Chemical Kinetics and Chain Reactions,* Oxford, UK (1935).

Semenov, N., *Some Problems of Chemical Kinetics and Reactivity,* Vols. 1 & 2, Pergamon Press, London (1959).

Senior, D. A., "Burning Velocities of Hydrogen–Air and Hydrogen–Oxygen Mixtures," *C&F* **5,** 7 (1961).

Senkian, S. M., "On the Combustion of Chlorinated Hydrocarbons II—Chemical Kinetic Modeling of the Intermediate Zone of the Two-Stage Trichloroethylene–Oxygen–Nitrogen Flames," *CST* **38,** 197 (1984).

Senkian, S. M., "The Thermal Destruction of Halogenated Compounds," *Chem. Eng. Prog.* **12,** 58 (1987)

Sensor, D. W., Cundy, V. A., and Morse, J., "Chemical Species and Temperature Profiles of Laminar Dichloromethane–Methane–Air Flames Variation of Chlorine/Hydrogen Loading," *CST* **51,** 209 (1987).

Settles, G. S., "Modern Developments in Flow Visualization," *AIAAA J.* **23,** 1313 (1985).

Shakespeare, W., *Hamlet Prince of Denmark,* Act I, Scene 5 (1603).

Shandross, R. A., Longwell, J. P., and Howard, J. B., "Non Catalytic Thermocouple Coating for Low Pressure Flames," *C&F* **85,** 282 (1991).

Shapiro, A. H., *The Dynamics and Thermodynamics of Compressible Fluid Flow* (two volumes), Vol. 1, p. 47, Ronald Press, N.Y. (1953).

Sharma, P. K., Knuth, E. L., and Young, W. S., "Species Enrichment Due to Mach Number Focusing in a Molecular Beam-Mass Spectrometer Sampling System," *JCP* **64,** 4345 (1965).

Shchetinkov, Ye. S., *The Physics of Combustion of Gases,* Izd. Nauka, Moscow, USSR (1965).

Sherman, F. S., "Self Similar Development of Inviscid Hypersonic Free Jet Flow," *Report 6-90-63-61,* Lockheed Missiles and Space Co., Sunnyvale, CA (1963).

Shirodkar, A., "The Measurement of Temperature in a Coal Gas Flame by Alpha Particle Method," *Phil. Mag.* **15,** p. 426 (1933).

Sick, V., Arnold, A., Dieffel, E., Dreioer, T., Ketterle, W., Lange, B., Wolfrum, J., Thiele, K. U., Behrendt, F., and Warnatz, J., "Two Dimensional Laser Diagnostics and Modeling of Counterflow Diffusion Flames," *23rd Symp.,* The Combustion Institute, Pittsburgh, PA, p. 495 (1991).

Sidgwick, V., *The Chemical Elements and Their Compounds, Two Volumes,* Oxford (1950).

Siegla, D., and Smith, G. (Eds.), *Particulate Carbon Formation During Combustion,* Plenum Press, New York (1981).

Simmons, R. F., and Wolfhard, H. G., "The Influence of Methyl Bromide on Flames," *Trans. Farad. Soc.* **51,** 1211 (1955).

Simpson, C. F., and Linnett, J. W., "Burning Velocities of Mixtures of Carbon Monoxide–Nitrous Oxide, Nitrogen and Water," *6th Symp.,* 149 (1957).

Singer, J. M., "Burning Velocity Measurements on Slot Burners; Comparison with Cylindrical Burner Determinations," *4th Symp.,* 352 (1953).

Singer, J. (Ed.), *Combustion Fossil Power Systems,* 3rd ed., Combustion Engineering Inc., Windsor, Conn. (1981).

Singh, T., and Sawyer, R. F., "CO Reactions in the Afterflame Region of Ethylene/Oxygen and Ethane/Oxygen Flames," *13th Symp.,* 403 (1971).

Skinner, D. G., *The Fluidized Combustion of Coal,* Institute of Fuel, London (1970).

Skirrow, G., and Wolfhard, H., "Studies of ClF_3 Flames," *PRS* **232,** 78 (1955).

Slone, T. M., and Ratcliffe, J. W., "Time Resolved Mass Spectrometry of a Propogating Methane–Oxygen–Argon Flame," *CST* **33** (1983).

Slootmaekers, A., and Van Tiggelen, A., "Fluorine and Chlorine Flames," *Bull. Soc. Chim. Belg.* **67,** 135 (1958).

Smalley, R. E., "Buckminsterfullerene Bibliography," Chemistry Dept., Rice Univ., Houston, TX (1991)

Smeeton-Leah, A., and Carpenter, N., "The Estimation of Atomic Oxygen in Open Flames and The Measurement of Temperature," *4th Symp.,* 274 (1953).

Smith, D., and Agnew, J., "The Effect of Pressure on the Laminar Burning Velocity of Methan-Oxygen-Nitrogen Mixtures," *6th Symp.,* 83 (1951).

Smith, F. A., "Problems of Stationary Flames," *Chem. Rev.* **31,** 389 (1937).

Smith, H. M., *Principles of Holography,* Wiley, NY (1975).

Smith, I. E. (Ed.), *Combustion in Advanced Gas Turbine Systems,* Pergamon Press, London (1968).

Smith, J. R., "Rayleigh Temperature Profiles in a Hydrogen Diffusion Flame," *Proc. SPIE* **158,** Laser Spectroscopy, p. 84 (1978).

Smith, M. L., and Stinson, K. W., *Fuels and Combustion,* McGraw–Hill, New York (1952).

Smith, O. I., Priv. Comm., Chem. Eng. Dept., UCLA, (1988).

Smith, O. I., and Chandler, D. W., "An Experimental Study of Probe Distortions to the Structure of One Dimensional Flames," *C&F* **63,** 19 (1986).

Smith, O. I., and Thorne, L., "The Structure of Cyanogen–Nitrogen Dioxide Premixed Flames," Paper WSS/CI 86-34, presented at the Fall 1986 Meeting of the Western States Section/ The Combustion Institute, Tucson, AZ (1986).

Smith, O. I., Wang, S-N, Tseregounis, S., and Westbrook, C., "Sulfur Catalyzed Recombination of Atomic Oxygen in an $CO/O_2/Ar$ Flame," *CST* **30,** 241 (1983).

Smith, R. D., "Benzene Pyrolysis," *C&F* **35,** 179 (1979a).

Smith, R. D., "A Direct Mass Spectrometric Study of the Mechanism of Toluene Pyrolysis at High Temperatures," *JCP* **83,** 1553 (1979b).

Smith, S. R., and Gordon, A. S., "Precombustion Reactions in Hydrocarbon Diffusion Flames: The Paraffin Candle Flame," *JCP* **22,** 1159 (1954).

Smith, R. D., and Johnson, A. L., "Mass Spectrometric Study of the High Temperature Chemistry of Benzene," *C&F* **51,** 1 (1979).

Smith, S. R., and Gordon, A., "Studies of the Diffusion Flames of Some Simple Alcohols," *JPC* **60,** 1059 (1956).

Smith, S.R., and Gordon, A. R., "The Methane Diffusion Flame," *JPC* **60,** 759 (1956).

Smith, S. R., and Gordon, A., "The Diffusion Flames of the Butanols," *JPC* **61,** 553 (1958).

Smithels, A., and Ingle, H., "Flame Separation Burner," *Trans. Chem. Soc.* **61,** 204 (1892).

Smooke, M. D., "Solution of Burner-Stabilized Premixed Laminar Flames by Boundary Valve Methods," *JCP* **48,** 72 (1982).

Smooke, M. D., Crump, J., Seshadri, K., and Giovangigli, V., "Comparison between Experimental Measurements and Numerical Calculations of the Structure of Counterflow, Diluted Methane Air Premixed Flames," *23rd Symp.,* The Combustion Institute, Pittsburgh, PA, p. 463 (1991).

Smooke, M.D., Seshadri, K., and Puri, I. K., "The Structure and Extinction of Partially Premixed Flames Burning Methane in Air," *22nd Symp.,* p. 1555 (1989).

Smoot, L. D., Hecker, W., and Williams, G., "Prediction of Propagating Methane–Air Flames," *C&F* **26,** 323–42 (1976).

Smoot, L. D., and Pratt, D., *Pulverized Coal Combustion and Gassification,* Plenum Press, New York (1979).

Sokolik, A. S., *Self-Ignition, Flame and Detonations in Gases,* Israel Program for Scientific Translations, NASA TTF-125OTS-63-11179, Jerusalem (1963) (trans. from Russian).

Sommerfeld, M., "Turbulent Flames," *Jet Prop.* **26,** 486 (1956).

Sorenson, S., Myers, P., and Uyehara, O., "Ethane Kinetics in Spark Ignition Engine-Exhaust Gases," *13th Symp.*, 451 (1971).

Spalding, D. B., *Some Fundamentals of Combustion,* Butterworths, London (1955).

Spalding, D. B., "Theory of Flame Phenomena with a Chain Reaction," *Phil. Trans. Roy. Soc. London* **A249,** 1 (1956).

Spalding, D. B., *Combustion and Mass Transfer,* Macmillan, New York (1979).

Spalding, D. B., and Stephenson, P. L., "Laminar Flame Propagation in Hydrogen + Bromine Mixtures," *PRS* **A324,** 315 (1971).

Stambuleanu, A., *Flame Combustion Processes in Industry,* 2nd ed., Abacus Press, London (1985).

Starkman, E. S., Patterson, D. J., and Taylor, C. F. (Eds.), "Digital Calculations of Engine Cycles," *Progress in Technology Series,* Vol. 7, Society of Automotive Engineers, Inc. (1961).

Starkman, E. S., and Samuelsen, G., "Flame Propagation Rates in Ammonia–Air Combustion at High Pressure," *11th Symp.*, 1037 (1967).

Stehling, F., Frazee, J. D., and Anderson, R. C., "Carbon Formation from Acetylene," *6th Symp.*, 247–54 (1957).

Stein, S. E., "NIST Mass Spectral Database of Common Compounds," available on diskettes for PCs; Will run on Mackintosh with PC emulators (1992).

Steinberger, R., "The Nature of Burning of Nitrate Esters," *5th Symp.*, 205 (1955).

Steinberger, R., and Schaaf, V. P., "Decomposition Flame of Ethylene Glycol Dinitrite," *JPC* **62,** 280 (1958).

Stephenson, D., "Non Intrusive Profiles of Flames," *19th Symp.*, p. 993 (1978).

Stephensen, P. I., and Taylor, R. B., "Laminar Flame Propagation in the Hydrogen–Oxygen Nitrogen Mixtures," *C&F* **20,** 231 (1973).

Stepowski, D., "Autocalibration of OH Laser Induced Fluorescence Signals by Local Absorption Measurements in a Laminar, Methane Air Diffusion Flame," *23rd Symp.*, p. 1829 (1991).

Stepowski, D., and Cottereau, M. J., "Direct Measurement of OH Local Concentration in a Flame from the Fluorescence by a Single Laser Pulse," *Appl. Opt.* **18,** 354 (1979).

Stepowski, D., Puechberty, D., and Cottereau, M. J., "The Use of Laser-Induced Fluorescence of OH to Study the Perturbation of a Flame by a Probe," *18th Symp.*, 1567 (1981).

Steward, P. H., Rothem, T., and Golden, D. M., "Emulation of Rate Constants for Combustion Modeling," *22nd Symp.*, Pittsburgh, PA, p. 943 (1989).

Stover, C. M., "Method for Butt Welding Small Thermocouples," *Rev. Sci. Inst.* **3,** 605 (1960).

Strauss, W. A., and Edse, R., "Burning Velocity Measurements by the Constant-Pressure Bomb Method," *7th Symp.*, Butterworths, London, 377 (1959).

Strehlow, R. A., *Fundamentals of Combustion,* International Textbook Co., Scranton, Pa. (1968); 2nd ed. (1983).

Streng, A. G., and Grosse, A. V., "The Ozone to Oxygen Flame," *6th Symp.*, 265–73 (1957).

Strong, H. M., Bundy, F. P., and Larson, D. A., "Temperature Measurements on Complex Flames by Sodium Line Reversal and Sodium D-Line Intensity Contour Studies," *3rd Symp.*, p. 641 (1949).

Strong, J., *Procedures in Experimental Physics,* Prentice–Hall (1944).

Stull, D., "Chemical Thermodynamics and Fire Problems," *Fire Res. Abs. and Rev.* **13,** 161 (1971).

Stull, D., and Prophet, A. (Eds.), *Joint Army/Navy/Air Force Thermochemical Tables,* NSRDS NBS No. 37, National Standard Reference Data System, National Bureau of Standards, U.S. Government Printing Office (1969); 2nd ed. (1986).

Suetonius, G. S. (trans. R. Graves), *The Twelve Caesars,* Penguin Books, NY (1979).

Sugden, T. M., Bulewicz, E., and Demerdache, A., *Chemical Reaction in Lower and Upper Atmosphere,* Wiley, New York, 89 (1962).

Suits, C. G., "High Temperature Measurements in Gaseous Arcs," p. 720, in *Temperature: Its Measurement and Control in Industry,* Vol. 1, Reinhold Publ. Co., NY (1941).

Sullivan, J. H., "Rates of Reaction of Hydrogen with Iodine," *JCP* **30,** 1292 (1959).

Sullivan, J. H., "Rates of Reaction of Hydrogen with Iodine. II," *JCP* **36,** 1925 (1962).

Sullivan, J. H., "Mechanism of the 'Bimolecular' Hydrogen–Iodine Reaction," *JCP* **46,** 73 (1967).

Sullivan, J. H., "Rates of Forward and Reverse Rate Coefficients and the Equilibrium Constant," *JCP* **51,** 2288 (1969).

Svehela, R., and McBride, B., "Fortran IV Program Package for Thermodynamic and Transport Calculations in Complex Chemical Systems," *NASA Tech. Note TN S7185,* NASA, Washington, D.C., Jan. (1973).

Takahashi, F., Dryer, F. L., and Williams, F. A., "Combustion Behavior of Free Boron Slurry Droplets," *21st Symp.* (1988).

Takeyama, T., and Miyama, H., "A Shock Tube Study of the Ammonia–Oxygen Reaction," *11th Symp.,* 845 (1967).

Tanzawa, T., and Gardiner, W. C., Jr., "Thermal Decomposition of Acetylene," *17th Symp.,* 563 (1979).

Teclu, N., "Flame Separation Burner," *J. Prakt. Chem.* **44,** 246 (1891).

Tennekes, H., and Limley, J., *A First Course in Turbulence,* MIT Press, Cambridge, MA, 87 (1972).

Thomas, A., "A Theory of Turbulent Combustion," *C&F* **65,** 291–312 (1986).

Thomas, N., Gaydon, A. G., and Brewer, L., "Cyanogen Flames and the Dissociation Energy of N_2," *JCP* **20,** 369–74 (1952).

Thorne, J. L., Branch, M. C., Chandler, D. W., Kee, R. J., and Miller, J., "Nitric Oxide Hydrocarbon Interactions in Low Pressure Flames," *21st Symp.,* 965 (1988); see also Thorne, L. R., and Smith, O. I., "The Structure of Cyanogen–Nitrogen Dioxide Premixed Flames," presented at Western State; also Sandia Combustion Research Annual Report, 1985, 5–7.

Thorne, L. R., and Melius, C. F., "The Structure of Hydrogen Cyanide Nitrogen Dioxide Premixed Flames," *23rd Symp.,* p. 397 (1991).

Thorpe, C., *Bibliography of Ozone Technology,* Armour Research Foundation, Chicago (1955).

Thring, M. W., *The Science of Flames and Furnaces,* Chapman Hall, 2nd ed. (1962).

Tichenor, D. A., Mitchell, R. E., Hencken, K. R., and Niksa, S., "Simultaneous In Situ Measurement of the Size, Temperature and Velocity of Particles in a Combustion Environment," *20th Symp.,* 1213 (1984).

Timnat, Y. M., *Advanced Chemical and Rocket Propulsion,* Academic Press, NY (1955).

Tine, G., *Gas Sampling and Chemical Analysis in Combustion,* Pergamon Press, London (1961).

Tizard, H., and Pye, D., "Adiabatic Compression," *Phil. Mag.* **44,** 79 (1922).

Toong, T-Y., *Combustion Dynamics,* McGraw–Hill, New York (1983).

Toepler, A., "On the Schlieren Method," *Pogg. Ann.* **127,** 566; **128,** 126 (1866); **134,** 194 (1868).

Troe, J., "Toward a Quantitative Understanding of Elementary Combustion Reactions," *22nd Symp.,* p. 843 (1989).

Trolinger, J. D., "Flow Visualization Holography," *Optical Engineering* **14,** 470, (1975).

Trolinger, J. D., "Holographic Interferometry as a Diagnostic Tool for Reactive Flows," *CST* **13,** 229 (1976).

Townes, C., and Schallow, A., *Microwave Spectroscopy,* McGraw-Hill, NY (1955)

Tryon, G. (Ed.), *Handbook of Fire Protection,* National Fire Protection Association, Boston, Mass. (1980).

Tsang, W., and Hampson, R., "Chemical Kinetic Data Base for Methane Combustion," *NBSIR 84-2913,* National Institute of Science and Technology, Gaithersburg, MD (1985); *J. Phys. and Chem. Ref. Data* (1985). General kinetic data source for PCs is available as NIST Standard Reference Data Base 17 Chemical Kinetics, Westley, F., Herron, J., Cvetanovic, R., Hampson, R., and Mallard, W., Version 3 (1991).

Tsatsaronis, G., "Prediction of Propagating Laminar Flames in Methane–Oxygen," *C&F* **33,** 217 (1978). See also Doctoral Thesis, Univ. Achen, Germany.

Tseregounis, S., and Smith, O. I., "An Experimental Investigation of Fuel Sulfur–Fuel Nitrogen Interactions in Low-Pressure Flames," *12th Symp.,* 761 (1983).

Tsuji, H., "Counterflow Cylindrical Diffusion Flame," *11th Symp.,* p. 979 (1967).

Tsuji, H., "Structure and Extinction of Near Limit Flames," *19th Symp.,* p. 1533 (1982).

Tuve, R., *Principles of Fire Protection Chemistry,* National Fire Protection Association, Boston, Mass. (1976).

Ulrich, G., "Flame Propagation of Fine Particles," *Chem. & Eng. News* **62,** 22 (1984).

Valentini, J., "Laser Raman Techniques," p. 597, in *Laser Spectroscopy and its Applications,* Radziemski, L. J., Solarz, R. W., and Paisner, J. A. (Eds.), Marcel Dekker, Inc., NY (1987).

Valentini, J., Moore, D. S., and Bomser, D. S., "Collision-Free CARS Spectroscopy of Molecular Photofragments," *Chem. Phys. Lett.* **83,** 217 (1981).

Valerias, J., Gupta, A. K., and Senkan, S. M., "Laminar Burning Velocities of Chlorinated Hydrocarbon–Methane–Air Flames," *CST* **36,** 123 (1990).

Van Tiggelen, A., "Kinetics and Inhibition of the Inflammation of Methane," *Mem. Acad. Roy. Belg. (Cl. Sc.)* **27,** 1 (1952).

Van Tiggelen, A., et al., *Oxidations et combustions,* Vols. 1 & 2, Editions Technip, Paris (1968) (in French).

Van Tiggelen, A., and Vandooren, J., Priv. comm. on Data Reduction (1980).

Van Wonterghem, J., Slootmaekers, P. J., and Van Tiggelen, A., "Flame Propagation in Mixtures of Less Usual Fuels with Oxygen," *Bull. Soc. Chim. Belg.* **65,** 899 (1956).

Van Wonterghem, J., and Van Tiggelen, A., "Burning Velocity of CO–N_2O," *Bull. Soc. Chim. Belges.* **64,** 780 (1955).

Vanderhoff, J., Anderson, W. H., Kotlar, A. J., and Beyer, R., "Ar + Laser Excited Fluorescence of C_2 and CN Produced in Flames," *C&F* **49,** 197 (1982).

Vanderhoff, J., Anderson, W. H., Kotlar, A. J., and Beyer, R., "Raman and Fluorescence Spectroscopy in a Methane–Nitrous Oxide Flame," *20th Symp.,* 1299 (1985).

Vanderhoff, J., Bunte, S., Kotlar, A. J., and Beyer, R., "Temperature and Concentration Profiles in Hydrogen–Nitrous Oxide Flames," *C&F* **65,** 45 (1986).

Vandooren, J., Balakhnin, V., Huber, K. and Van Tiggelen, P. J., "Determination of the Absolute Concentration Profiles of Stabile and Labile Species in the Front of a Hydrogen/Nitrous Oxide Flame," *Kinetika i Kataliz,* **19,** 1377-83 (1978).

Vandooren, J., Balakhnin, V. and Van Tiggelen, P. J., "Mass Spectrometric Investigation of the Structure of Methanol Flames," *Archivum Combustionis,* **1,** 229 (1981).

Vandooren, J., and Bian, J., "Validation of the Hydrogen, Oxygen Reaction Mechanisms by Comparison with Experimental Structure of a Rich Hydrogen, Oxygen Flame," *23rd Symp.,* The Combustion Institute, Pittsburgh PA, p. 341, (1991).

Vandooren, J., Branch, M. C., and Van Tiggelen, P. J., "Comparisons of the Structure of Stoichiometric Methane Flames with Oxygen and N_2O as the Oxidizer," Priv. Comm. 1991. To be submitted to C&F.

Vandooren, J., Fristrom, R. M., and Van Tiggelen, P. J., "Experimental Study of the Kinetics of a Hydrogen–Chlorine–Argon Flame," submitted to *Bull. Soc. Chim. Belg. 101* 825 (1992).

Vandooren, J., Nelson da Crux, F., and Van Tiggelen, P. J. "The Inhibiting Effect of CF_3H on the Structure of a Stoichiometric H_2/CO/O_2/Ar Flame," *22nd Symp.*, p 1587 (1989).

Vandooren, J., Peeters, J., and Van Tiggelen, P. J., "Rate Constants of the Elementary Reaction of CO with OH," *15th Symp.*, 745 (1975).

Vandooren, J., and Van Tiggelen, P. J., "Reaction Mechanisms of Combustion in Low Pressure Acetylene-Oxygen Flames," *16th Symp.*, 1113-44 (1977).

Vandooren, J., and Van Tiggelen, P. J., "Experimental Investigation of Methanol Oxidation in Flames: Mechanisms and Rate Constants of Elementary Steps," *18th Symp.*, 473–83 (1981).

Vandooren, J., and Vanpee, M., "Infra Red Emission Spectroscopy in H_2/F_2 and CH_4/F_2 Premixed Flames," *Bull. Soc.Chim. Belg.* **97,** 797 (1988).

Vanpee, M., "Limits of Inflammation of Mixtures of Methane and Oxygen," *Bull. Soc. Chim. Belg.* **62,** 458 (1953).

Vanpee, M., Cashin, D., Falabella, B., and Chintappili, P., "Deflagration on the Combustion of H_2–F_2 Mixtures," *C&F* **20,** 443 (1973).

Vanpee, M., Cashin, K., and Mainiero, R., "Emission Spectra of an F_2–CH_4 Flame," *C&F* **33,** 1778 (1978).

Vanpee, M., Clark, A. H., and Wolfhard, H. G., "Flame Characteristics of Diborane–Hydrazine System," *9th Symp.*, 27 (1963); also "Characteristics of Diborane Flames," *Project Squid Tech. Rept. RMD-P,* (1963).

Vanpee, M., and Quang, L., "A Study of Premixed Hydrocarbon–Fluorine Flames," *17th Symp.*, 681 (1979).

Vanpee, M., Vidaud, P., and Cashin, K., "The Emission Spectra and Burning Velocity of the Premixed Cyanogen–Fluorine Flame," *C&F* **23,** 227 (1974).

vant Hoff, J. H. *Etudes du dynamique chimque,* Thesis, Paris (1884).

Vargaftol, N. B., *Handbook of Thermal Conductivity of Liquids and Gases,* CRC Press, Boca Raton, FL (1993).

Vaughn, S., Lester, T., and Merklin, J., *Proc. 13th International Symposium on Shock Tubes and Waves,* "References on Pyrolysis of Benzene," published by State University of New York Press, Albany, NY (1982).

Verdieck, J. F., and Bonczyk, P. A., "Laser-Induced Saturated Fluorescence Investigations of CH, CN, and NO in Flames," *Eighteenth Symposium (International) on Combustion,* The Combustion Institute, Pittsburgh, PA, p. 1559 (1981).

Verkat, C., Brezinsky, K., and Glassman, I., "High Temperature Oxidation of Aromatic Hydrocarbons," *19th Symp.*, 143–52 (1983).

Verwimp, J., and Van Tiggelen, A., "Reactions of Mixtures of Ammonia and Oxygen" (in French), *Bull. Soc. Chim. Belg.* **62,** 205 (1953).

Vest, C. M., *Holographic Interferometry,* J. Wiley and Sons, New York (1979).

Vest, C. M., "Tomography," in *Flow Visualization and Laser Velocity for Wind Tunnels,* Proceedings of NASA Workshop, Hunter, W., and Goughner, J. (Eds.), Hampton, VA, NASA CP 2243 (1982).

Vetter, A., and Culick, F. E. C., "Flame Speed of a Low Pressure CS_2/O_2 Flame," *C&F* **30,** 107–9 (1977).

Vielle, P., "The Shock Tube" (French), *Compt. rendu* **149,** 1228 (1889).

Vinckier, J., and Van Tiggelen, A., "Structure and Burning Velocity of Turbulent PreMixed Flames," *C&F* **12,** 561 (1968).

Volponi, J., McLean, W., Fristrom, R. M., and Munir, Z., "Determination of H, OH and O by Deuterium Scavenging in Low Pressure Acetylene–Oxygen–Argon Flames," *C&F* **65,** 243 (1986).

Voltaire, F. M., in *Collected Essays, La nature et la propagations du feu* (French), Academy of Science, Paris (1752).

Von Karman, T., and Penner, S. S., *Selected Combustion Problems,* Butterworths, London (1954).
Von Karman, T., "Status of the Theory of Flame Propagation," *6th Symp.,* 1 (1957).
Vulis, L. A., *Thermal Regimes of Combustion,* McGraw-Hill, NY (1961).
Wagner, H. Gg., "Soot Formation in Combustion," *17th Symp.,* 3 (1979).
Wagner, H. Gg., Bonne, U., and Grewer, Th., "Messungen in der Reaktionszone von Wasserstoff–Sauerstoff und Methan–Sauerstoff Flammen," *ZpC (Frankfort)* **26,** 93 (1960).
Wagner, P., and Dugger, G., "Structural Influences on Burning Velocity," *JACS* **77,** 227 (1955).
Waldman, L., "Uber die Kraft einer inhommogenen Gases auf Klein suspendierte Kugeln," *A. Fur Naturfors.,* Band **14A,** 589 (1959).
Walker, I., "Spontaneous Combustion," *Fire Res. Abs. and Rev.* **9,** 5 (1967).
Walker, R. E., The Johns Hopkins University Applied Physics Laboratory, unpublished results (1959).
Walker, R. E., de Haas, N., and Westenberg, A. A., "Measurements of Multicomponent Diffusion Coefficients for the CO_2–He–N_2 System Using the Point Source Technique," *JCP* **37,** 891 (1962).
Walker, R.E., and Westenberg, A. A., "Molecular Diffusion Coefficient Studies at High Temperature IV: CH_4-O_2, CO_2-O_2,H_2O-O_2, and H_2-O_2 Systems," *JCP* **32,** 436 (1960).
Walker, R. E., and Westenberg, A. A., "Absolute Low Speed Anemometer," *Rev. Sci. Inst.* **27,** 844 (1956).
Walker, R. E., and Westenberg, A. A., "Molecular Diffusion Studies at High Temperature I—The Point Source Method," *JCP* **29,** 1139 (1958a).
Walker, R. E., and Westenberg, A. A., "Molecular Diffusion Studies in Gases at High Temperature II (He–N_2 and CO_2–N_2)," *JCP* **29,** 1147 (1958b).
Walker, R. W., "Reactions of HO_2 Radicals in Combustion Chemistry," *22nd Symp.,* p. 883 (1989).
Wall, L. (Ed.), *The Mechanisms of Pyrolysis, Oxidation and Burning of Organic Materials,* NBS Special Publication #357, NBS, Washington, D.C. (1972).
Wang, T., Matual, R., and Farmer, R., "Combustion Kinetics of Soot Formation from Toluene," *18th Symp.,* 1149 (1981).
Ward, F., and Weinberg, F., "Photographic Recording of Slow Ions from Flames," *8th Symp.,* p. 217 (1962).
Warnatz, J., "Berechnung der Flammengeschwindigkeit und der Structur von Laminaren Flachen Flammen," *Habilitationschrift,* Darmstadt (1977).
Warnatz, J., "Calculation of the Structure of Laminar Flat Flames I: Flame Velocity of Freely Propagating Ozone Decomposition Flames," *Ber. Bunsenges. Phys. Chem.* **82,** 193 (1978).
Warnatz, J., "The Structure of Freely Propagating and Burner-Stabilized Flames in the H_2–CO–O_2 System," *Ber. Bunsenges. Phys. Chem.* **83,** 950 (1979).
Warnatz, J., "Auswertung Reaktionskinetischer Messungen mit Rechnen. Bunsenkolloquium: Neuere Methoden zur Untersuchung von Radikal-Reaktionen in der Gasphase," Göttingen (1980).
Warnatz, J., "Concentration, Pressure, and Temperature Dependence of the Flame Velocity in Hydrogen–Oxygen–Nitrogen Mixtures," *CST* **26,** 203–13 (1981a).
Warnatz, J., "The Structure of Laminar Alkane and Alkene and Acetylene Flames," *18th Symp.,* 369 (1981b).
Warnatz, J., "The Mechanism of High Temperature Combustion of Propane and Butane," *CST* **34,** 177–200 (1983a).
Warnatz, J., "Survey of Rate Coefficients in the C/H/O System," Chapter 5 in *Combustion Chemistry,* W. Gardiner (Ed.), Springer–Verlag, New York (1983b).
Warnatz, J., "The Chemistry of Combustion of Alkanes up to Octane," *20th Symp.,* 845 (1985).

Warnatz, J., Bockhorn, H., Moser, A., and Wenz, H., "Experimental Investigations and Computational Simulation of Acetylene Oxygen Flames from Near Stoichiometric to Sooting Conditions," *19th Symp.,* 197 (1982).

Warnatz, J., and Maas, U., *Technische Verbrennung,* Springer-Verlag, Berlin (1992).

Wasserman, H., and Murray, R., *Singlet Oxygen,* Academic Press (1979).

Webster, P., and Walsh, A. D., "Effect of Sulfur Dioxide on the Second Pressure Limit of Explosion of Hydrogen–Oxygen Mixtures," *10th Symp.,* 463 (1964).

Wehry, E. L. (Ed.), *Modern Fluorescence Spectroscopy,* (two vols.), Plenum Press, NY (1976).

Weinberg, F. J., "Measurement of Field Induced Ion Flows in Flames," *8th Symp.,* p. 204 (1962).

Weinberg, F. J., *Optics of Flames,* Butterworths, London (1963).

Weinberg, F. J., "The First Half Million Years of Combustion Research and Today's Burning Problems," *15th Symp.,* The Combustion Institute, Pittsburgh, PA (1975a).

Weinberg, F. J., "Optical Methods in Combustion Research," in *Flow Visualization II,* 3-4, Hemisphere, New York (1982).

Weinberg, F. J. (Ed.), *Advanced Combustion Methods,* Academic Press, New York (1986).

Weinberg, F. J., and Hardesty, D., "Burners Producing Large Excess Enthalpies," *CST* **8,** 201 (1974).

Weinberg, F. J., and Wong, W-Y, "A Laser Optical Method for the Direct Measurement of Normal Propagation Velocities of Phase Objects," *PRS* **A345,** 379 (1975).

Weinberg, F. J., and Wong, W-Y, "Optical Studies in Fire Research," *16th Symp.,* p. 799 (1977).

Wendt, J. O., Morocomb, J. T., and Corley, T. L., "Influence of Fuel Sulfur on Fuel Nitrogen Oxidation," *17th Symp.,* 671 (1979).

West, G. A., Barrett, J. J., Siebert, D. R., and Reddy, K. V., "Photoacoustic Spectroscopy," *Rev. Sci. Inst.* **54,** 797 (1983).

Westbrook, C. K., "An Analytical Study of the Shock-Tube Ignition of Mixtures of Methane and Ethane," *CST* **20,** 5 (1979).

Westbrook, C. K., "Inhibition of Laminar Methane–Air and Methanol–Air Flames by Hydrogen Bromide," *CST* **23,** 191 (1980).

Westbrook, C. K., "Flame Inhibition by Methyl Bromide," *UCRL 86456 Preprint,* Lawrence Livermore Laboratory, CA (1981).

Westbrook, C. K., "Inhibition of Hydrocarbon Oxidation in Laminar Flames and Detonations by Halogenated Compounds," *19th Symp.,* 127–43 (1983a).

Westbrook, C. K., "Numerical Modeling of Flame Inhibition by CF_3Br," *CST* **34,** 201–35 (1983b).

Westbrook, C. K., Adamczyk, A. A., and Lavoie, G. A., "A Numerical Study of Laminar Flame Wall Quenching," *C&F* **40,** 81 (1981).

Westbrook, C. K., Beason, D., and Alvares, N. J., "An Experimental and Theoretical Study of Flame Inhibition by Bromine Containing Compounds," *Proc. 1st Int. Specialists Meeting,* 13-298.

Westbrook, C. K., and Chase, L., "Chemical Kinetic and Thermochemical Data for Combustion," *UCID-1783,* Lawrence Livermore National Laboratory, Livermore, CA (Feb. 1982).

Westbrook, C. K., and Dryer, F. L., "Modeling Flame Properties of Methanol," *Proceedings: Alcohol Fuels Technology—Third International Symposium,* Asilomar, CA (1979a).

Westbrook, C. K., and Dryer, F. L., "A Comprehensive Mechanism for Methanol Oxidation," *CST* **20,** 125–40 (1979b).

Westbrook, C. K., and Dryer, F. L., "Prediction of Laminar Flame Properties of Methanol–Air Mixtures," *C&F* **37,** 171–82 (1980).

Westbrook, C. K., and Dryer, F. L., "Chemical Kinetics and Modeling of the Combustion Processes," *18th Symp.,* The Combustion Institute, Pittsburgh, PA, p. 749 (1981).

Westbrook, C. K., Dryer, F. L., and Schug, K., "A Comprehensive Mechanism for the Pyrolysis and Oxidation of Ethylene," *19th Symp.*, 296–303 (1983).

Westbrook, C. K., and Pitz, W. J., "Effects of Propane on Ignition of Methane–Ethane–Air Mixtures," *CST* **33,** 315 (1983).

Westbrook, C. K., and Pitz, W. J., "Chemical Kinetic Reaction Mechanism for Oxidation and Pyrolysis of Propane and Propyne," *CST* **37,** 117–52 (1984).

Westbrook, C. K., Warnatz, J., and Pitz, W. J., "Detailed Chemical Kinetic Reaction Mechanism of Iso-Octane, and *n*-Heptane over an Extended Temperature Range and Its Application to Engine Knock," *22nd Symp.*, The Combustion Institute, Pittsburgh, PA, p. 893 (1989).

Westenberg, A. A., and deHaas, N., "Gas Thermal Conductivity Studies at High Temperature. Line Source Technique Applied to N_2, CO_2, and N_2-CO_2 Mixtures," *Phys Fluids* **5,**296, (1962).

Westenberg, A. A., and Fristrom, R. M., "Methane–Oxygen Flame Structure II. Conservation of Matter and Energy in the One-Tenth Atmosphere Flame," *JCP* **64,** 1393 (1960).

Westenberg, A. A., and Fristrom, R. M., "IV-Chemical Kinetic Considerations," *JPC* **64,** 591 (1961).

Westenberg, A. A., and Fristrom, R. M., "H and O Atom Profiles in Flames using ESR," *10th Symp.*, 473 (1965).

Westenberg, A. A., Raezer, S., and Fristrom, R. M., "Interpretation of the Sample Taken by a Probe in a Laminar Concentration Gradient," *C&F* **1,** 467 (1957).

Westenberg, A. A., and Rice, J., Flame Generated Turbulence," *C&F* **3,** 459 (1959).

Westenberg, A. A., and Walker, R. E., "Transport Properties in Gases," in *Thermodynamic and Transport Properties of Gases, Liquids and Solids,* McGraw–Hill (1959).

Westley, J. T., Herron, J. T., Cvetanovic, R. J., Hampson, R. F., and Mallard, W. G., *NIST Chemical Kinetics Database,* version 2.0, Standard Reference Data Program, NIST, Gaithersburg, MD 20899 (1990). Version 4.0 (1992).

Westmoreland, P. R., Howard, J. B., and Longwell, J. P., "Tests of Published Mechanisms by Comparison with Measured Laminar Flame Structure in Fuel-Rich Acetylene Combustion," *21st Symp.*, p. 773 (1988).

Whatley, A., and Pease, R. N., "Observations on Thermal Explosions of Diborane–Oxygen Mixtures," *JACS* **76,** 1997 (1954).

White, J. N., and Gardiner, W. C., Jr., "An Evaluation of Methane Combustion Mechanisms. 2. Comparison of Model Predictions with Experimental Data from Shock-Initiated Combustion of C_2, H_2, C_2H_4 and C_2H_6," *JPC* **83,** 562 (1979).

Wibberly, I. J., and Phong-Anant, D., "A Simple Laboratory Feeder for Fine Particles" *CST* **49,** 93, (1986).

Wilde, K. A., "Boundary-Value Solutions of the One Dimensional Laminar Flame Propagation Equations," *C&F* **18,** 43–52 (1972).

Wilk, R., Pitz, W., Westbrook, C., Addagarlia, S., Miller, D., Czransky, N., and Green, R., "The Combustion of *n*-Butane and iso-Butane in an IC Engine: Comparison of Experiment and Modeling," *23rd Symp.*, p. 1047 (1991).

Wilkins, C. L., and Gross, M. L., "Fourier Transform Mass Spectrometry for Analysis," *Anal. Chem.* **53,** 1661A (1981).

Williams, F., *Combustion Theory,* Addison–Wesley, Reading, MA (1965); 2nd ed., Benjamin–Cummings Publ. Inc., Menlo Park, CA (1985).

Williams, F., "Mechanisms of Fire Spread," *16th Symp.*, 1281 (1976).

Williams, F., "A Review of Flame Extinction," *Fire Safety J.* **3,** 163–75 (1981).

Williams, G. J., and Wilkins, R. G., "Investigation of the Structure of Perchloric Acid Flames I—Molecular Beam Sampling Apparatus," *C&F* **21,** 325 (1973).

Wilson, E. B., jr., Decius, J. C., and P.C. Cross, *Molecular Vibrations,* McGraw-Hill, NY (1955).

Wilson, W. E., "Structure, Kinetics and Mechanism of an Methane–Oxygen Flame Inhibited by Methyl Bromide," *10th Symp.*, 47–54 (1965).

Wilson, W. E., O'Donovan, J. T., and Fristrom, R. M., "Flame Inhibition by Halogen Compounds," *12th Symp.*, 929 (1969).

Wires, R., Watermeier, L., and Strehlow, R., "Effect of Added H_2O on Dry CO–O_2 Flames," *JPC* **63,** 989–91 (1959).

Wohl, K., "Quenching Flashback and Blowoff," *4th Symp.*, p. 69 (1953).

Wohl, K., and Welty, F., "Spectrophotometric Traverses through Flame Fronts," *5th Symp.*, 746 (1955).

Wolfhard, H. G., and Parker, W. G., "A New Technique for Studying Diffusion Flames Spectroscopically at Normal Pressure," *PRS* **A162,** 722 (1949).

Wolfhard, H. G., and Parker, W. G., "A New Technique for Spectrascopic Examination of Flames at Normal Pressure," *Proc. Phys Soc. (London)* **65A,** 2 (1952).

Wolfhard, H. G., and Parker, W. G., "Spectra and Combustion of Flames Supported by Oxides of Nitrogen," *5th Symp.*, 718 (1955).

Wolfhard, H. G., and Parker, W. G., "The Influence of Methyl Bromide on Flames," *AGARD Selected Combustion Problems,* Butterworths, London, 328 (1956).

Wolfrum, J., "Chemical Kinetics in Combustion Systems: Specific Effects of Energy, Collision and Transport Processes," *20th Symp.*, 59 (1986).

Wood, R., *Physical Optics,* Macmillan (1934).

Wray, F. L., "Shock-Tube Study of the Recombination of O Atoms at High Temperatures," *JCP* **38,** 1518 (1963).

Yamaoka, I., and Tsuji, H., "Determination of Burning Velocity Using Counter Flow Flames," *20th Symp.*, 1883 (1984).

Yamaoka, I., and Tsuji, H., "Extinction of Near-Stoichiometric Flames Diluted with Nitrogen in a Stagnation Flow," *22nd Symp.*, p. 1565, (1989).

Yang, W. (Ed.), *Flow Visualization III,* Hemisphere, New York (1985).

Yastrebov, V. V., "The Physical Chemistry of Concentrated Ozone. VIII. Thermal Propagation of Flame in Gaseous Ozone Mixtures," *Russian J. of Phys. Chem.* **34,** 21–23 (1960).

Yaws, C. L., "Correlation Constants for Chemical Compounds," *Chem. Eng.*, Nov. issue, p. 153 (1976).

Yi, A., and Knuth, E. L., "Probe Induced Concentration Distortions in Molecular Beam Mass Spectrometry," *C&F* **63,** 369 (1986).

Yoon, S., and Knuth, E. L., "Species Enrichment Due to Mach Focusing. II. Diatomic Major Species," *Rarefied Gas Dynamics,* R. Campargue (Ed.), Commissariat a l'Energie Atomique, Paris, France, 639 (1979).

Yost, D. M., and Russell, H., Jr., *Systematic Inorganic Chemistry of the Fifth-and-Sixth Group Nonmetallic Elements,* Prentice–Hall, New York (1944).

Young, W. S., "Correlation of Chemical Freezing in an Expanding Jet," *AIAA J.* **13,** 1478 (1975).

Young, W. S., Rodgers, W. E., Cullian, C. A., and Knuth, E. L., "Supersonic Molecular Beams with Cycling Pressure Sources," *AIAA J.* **9,** 323 (1971).

Young, W. S., Rodgers, W., and Knuth, E. L., "A Dual-Disk Chopper for Time-of-Flight Measurements," *Rev. Sci. Inst.* **41,** 380 (1970).

Young, W. S., Wang, Y. G., Rodgers, W. E., and Knuth, E. L., *Technology Utilization Ideas for the 70's and Beyond,* 26, Science and Technology, American Astronautical Society, Tarzana, CA, p. 281, (1970).

Yu, F. T. S., *Optical Information Processing,* Wiley Interscience, New York (1982).

Yu, G., Law, C. K., and Wu, C. K., "Laminar Flame Speeds of Hydrocarbon + Air Mixtures with Hydrogen Addition," *C&F* **63,** 339 (1986a).

Yumlu, V. S., "Prediction of Burning Velocities of Carbon Monoxide–Hydrogen–Air Flames," *C&F* **11,** 190 (1967).
Zabieleski, M. I., Dodge, L. M., Colket, M., and Seery, D., "Optical and Probe Measurements of NO: A Comparative Study," *20th Symp.,* The Combustion Institute, Pittsburgh, PA, p. 1591 (1981).
Zacariah, M., and Smith, O. I., "Studies in Sulfur Chemistry in $H_2/O_2/SO_2$ Flames," *C&F* **69,** 125 (1987).
Zeegers, P. J., and Alkemade, C. Th. J., "Radical Recombinations in Acetylene–Air Flames," *C&F* **9,** 247 (1965).
Zeldovich, Ya. B., "The Oxidation of Nitrogen in Combustion Explosions," *Acta Physicochimica U.R.S.S.,* Vol. XXI, 577–628 (1946).
Zeldovich, Ya. B., *Combustion Theory,* National Advisory Committee for Aeronautics, Technical Report F-TS-1226-LA, Air Material Command, Washington, D.C. (1949) (trans. from Russian).
Zeldovich, Ya. B., "Chain Reactions in Hot Flames; An Approximate Theory of Burning Velocity," *Kinetika i Kataliz* **2,** 305 (1961).
Zeldovich, Ya. B., Barenblatt, G. I., Librovich, V. B., and Makhviladze, G. M., *The Mathematical Theory of Combustion and Explosions,* Plenum Press, NY (1985).
Zeldovitch, Ya. B., and Kompaneets, A., *Theory of Detonations,* Academic Press, NY (1960).
Zhou Lixing, *Theory and Numerical Modeling of Turbulent Gas–Particulate Flows and Combustion,* CRC Press, Boca Raton, FL (1992).
Zinn, B. T. (Ed.), *Experimental Diagnostics in Gas Phase Combustion Systems,* American Institute of Aeronautics and Astronautics, NY (1977).
Zizak, G., Omenetto, N., and Winefordner, J. D., "Laser-Excited Atomic Fluorescence Techniques for Temperature Measurements in Flames: A Summary," *Opt. Eng.* **23,** 749 (1984).

AUTHOR INDEX

SUBJECT INDEX